OP-AMPS AND LINEAR INTEGRATED CIRCUIT TECHNOLOGY

RAMAKANT A. GAYAKWAD

Missouri Institute of Technology
Kansas City, Missouri

PRENTICE-HALL, INC., Englewood Cliffs, New Jersey 07632

Library of Congress Cataloging in Publication Data

Gayakwad, Ramakant A.
Op-amps and linear integrated circuit technology.

Bibliography: p.
Includes index.
1. Operational amplifiers. 2. Integrated circuits. I. Title.
TK7871.58.6G38 1982 621.381′735 82-11227
ISBN 0-13-637355-0

Editorial/Production supervision
and interior design by *Mary Carnis*
Manufacturing buyer: *Gordon Osbourne*
Cover design by *Edsal Enterprises*

Printed in the United States of America

10 9 8 7 6 5 4 3 2 1

ISBN 0-13-637355-0

Prentice-Hall International, Inc., *London*
Prentice-Hall of Australia Pty. Limited, *Sydney*
Prentice-Hall Canada Inc., *Toronto*
Prentice-Hall of India Private Limited, *New Delhi*
Prentice-Hall of Japan, Inc., *Tokyo*
Prentice-Hall of Southeast Asia Pte. Ltd., *Singapore*
Whitehall Books Limited, *Wellington, New Zealand*

To my father and mother

Contents

Preface

This book is designed primarily for students taking courses in op-amps and linear integrated circuits. It may also help designers and hobbyists interested in knowing more about linear ICs so that these ICs may be used effectively in microcomputer interfacing. Linear ICs such as op-amps, converters, comparators, and detectors have become a fundamental part of today's electronic systems. Therefore, basic principles of operation, electrical specifications, and limitations of these linear ICs are essential to know. I believe this book is the right blend of theory and practice and presents the theory in a practical way, which is most needed in industry today.

This book requires a knowledge of amplifier analysis, algebra, and some trignometry. It presents an in-depth discussion of the internal construction and operating principles of the op-amp and other important linear ICs such as the 555 timer, 380 audio-power amplifier, 565 phase-locked loop, and 9400 V/F and F/V converter. Illustrations and timing diagrams used throughout the book will give the student a concrete idea of the operating principles and internal construction of the ICs. Each chapter contains extensive examples that emphasize the main topics. In addition, at the end of each chapter is a summary of major points, key topic questions, and many practical problems. Laboratory experiments at the end of most chapters demonstrate the use and operation of the ICs discussed.

This book is full of a variety of practical circuits not discussed elsewhere. For example, it unveils the mystery behind versatile devices such as op-amps and timers by analyzing their equivalent circuits using timing diagrams wherever applicable. This approach is especially helpful in understanding the electrical characteristics, limitations, and practical uses of ICs and eliminating the "black magic" surrounding op-amps and other ICs. In addition, all the circuits discussed, including those in experiments, have been built and tested to prove the theories behind them. Unlike other books in the field, this book has a separate chapter devoted to IC systems, in which a group of discrete ICs is used to form simple but practical systems. Such exercises are valuable and much needed in industry. These IC system projects will not only expose the student to practical aspects of system design but also help the student to gain confidence in building and troubleshooting the various ICs.

To help the reader understand the basic operational amplifier principles and analysis techniques, the book starts with a review of discrete differential amplifiers. In this chapter the basic amplifier principles are used in the analysis and design of direct-coupled amplifiers—differential and cascode. Chapter 1 also provides analysis and design of constant current sources and current mirrors. Chapter 2 presents a complete analysis of equivalent circuits of typical op-amps and their characteristics. Interpretation of 741 op-amp data sheets and a list of important parameters for ac and dc applications are the content of Chapter 3. Chapter 4 covers the effect of negative feedback with different feedback configurations. Chapter 5 provides the practical considerations for an op-amp and looks at different techniques to improve the accuracy and performance of op-amp circuits. Frequency response and frequency-dependent parameters of the op-amp are discussed in Chapter 6.

General op-amp applications are covered in Chapter 7, while Chapter 8 presents filter and oscillator applications using op-amps. Additional op-amp applications such as comparators and converters are discussed in Chapter 9. In addition, special functions such as frequency-to-voltage and voltage-to-frequency conversion, digital-to-analog and analog-to-digital conversion, and comparators using specially designed ICs are also included in Chapter 9. Chapter 10 deals with the analysis and applications of some of the most commonly used special-purpose ICs. These ICs include the FLT-UT universal filter, 555 timer, 565 phase-locked loop, 380 audio power amplifier, and the 7800 and 7900 voltage regulator series. Finally, Chapter 11 provides a basic knowledge of system construction and design using various ICs discussed previously. These practical systems demonstrate the operating principles and the validity of the ICs used. In addition, Chapter 11 provides some systems in block diagram form. These projects should prove practical, interesting, and challenging.

The linear ICs presented in this book represent the best overall choices for each type. I strongly believe that the practical approach taken here is more beneficial than pure theory for the majority of students. In short, the entire book is practically oriented: it proves the theories and reinforces the concepts involved in a practical way. It should be enjoyable to both students and teachers.

I am grateful to my colleague Maris Roze for his careful review of the manuscript. Being the author of an English book, he was able to provide many helpful suggestions.

I am also grateful to Edward Wilson, Vice President of Accreditation Bell and Howell Schools, for writing a forward. Many thanks to academic deans Duane Gowing and Thomas Colvin at Missouri Institute of Technology for their continued help and encouragement. I also wish to thank all the academic deans at the other Bell and Howell schools for their assistance in having the manuscript reviewed. Of course I must thank my colleagues at M.I.T. as well as the faculty at the other Bell and Howell schools for reviewing the manuscript, and giving me technical advice.

But most of all, I want to thank my wife, Pratibha, for her patience and understanding while I spent many, many hours in the preparation of this book.

Ramakant A. Gayakwad
Missouri Institute of Technology
Kansas City, Missouri

Foreword

Electronics engineering technology has undergone profound change over the past few years, and we can expect that such change will accelerate over the next decade. Advances in integrated circuit technology, especially those which led to the development of the microprocessor, have had significant impact on both industry and education.

Despite the power of the microprocessor, its capabilities are severely limited unless it can communicate with the predominantly analog "real world." This requirement for an interface between the analog and digital domains has motivated the semiconductor industry to continue development of linear integrated circuits of ever-increasing complexity. As more and more functions are integrated onto a single chip, the emphasis on traditional discrete device analysis must shift toward analysis of the more complex integrated devices. It would be a grave error, however, to neglect discrete devices altogether as they not only provide the basis for understanding the linear integrated circuit, but are still used extensively in specialized applications, especially those related to interfacing.

This text is an important contribution to electronics engineering technology education because it maintains a reasonable balance between discrete and integrated circuitry. Beyond its balanced approach, however, the text excels in its presentation of practical, detailed examples and carefully designed and tested laboratory experiments.

This text can serve as the basis for a strong fundamental linear electronics sequence in an electronics engineering technology program; in addition, it can provide students in related fields as well as people in industry with an excellent source of relevant material to assist in individual professional development programs.

Edward A. Wilson
Vice President, Accreditation
Bell & Howell Education Group

1

Differential and Cascode Amplifiers

1-1 INTRODUCTION

The first part of this chapter is devoted to an in-depth analysis of differential amplifier types using *r*-parameters. The cascaded differential amplifiers are presented in a stage-by-stage analysis. Detailed numerical examples are included in this analysis. The results of all types of differential amplifiers are summarized at the end of first part as a quick and convenient cross-reference. The latter part of the chapter presents the cascode amplifier analysis.

Since the differential amplifier is the basic building block of operational amplifiers, the discussion of differential amplifiers in this chapter sets the groundwork for analysis and design procedures for the operational amplifiers. The analysis of differential amplifiers not only clarifies the operation of the operational amplifier but also makes the characteristics of the operational amplifier easy to understand. In addition, the analysis procedure which is developed in this chapter provides a means of checking operational amplifier parameters such as voltage gain, input and output resistances which are specified on the manufacturers' data sheets.

1-2 DIFFERENTIAL AMPLIFIER

Let us reconsider the emitter-biased circuit. Figure 1-1 shows two identical emitter-biased circuits in that transistor Q_1 has the same characteristics as transistor Q_2, $R_{E1} = R_{E2}$, $R_{C1} = R_{C2}$, and the magnitude of $+V_{CC}$ is equal to the magnitude of $-V_{EE}$. Remember that the supply voltages $+V_{CC}$ and $-V_{EE}$ are measured with

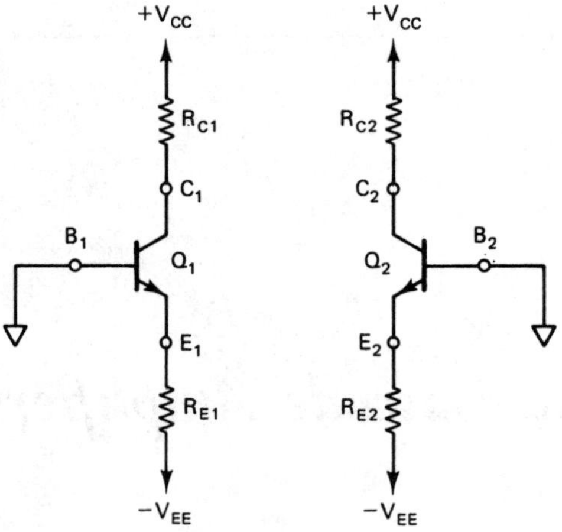

Figure 1-1 Two identical emitter-biased circuits.

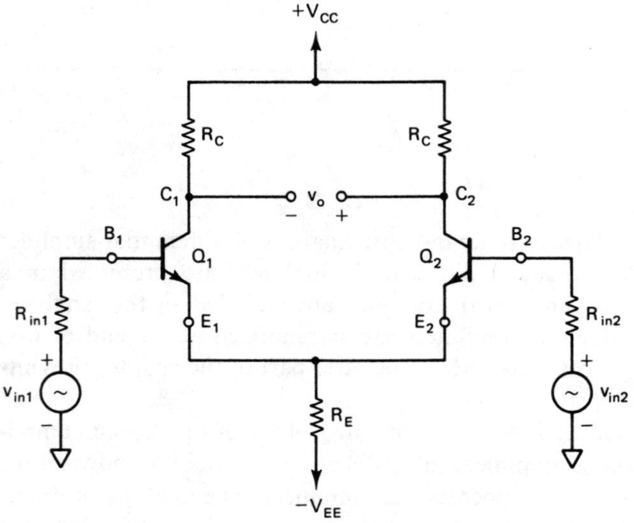

Figure 1-2 Dual-input, balanced-output differential amplifier.

respect to ground. To obtain a single circuit such as the one in Figure 1-2, we should reconnect these two circuits as follows:

1. Reconnect $+V_{CC}$ supply voltages of the two circuits since the voltages are of the same polarity and amplitude. Similarly, reconnect $-V_{EE}$ supply voltages.
2. Reconnect the emitter E_1 of transistor Q_1 to the emitter E_2 of transistor Q_2. (This reconnection places R_{E1} in parallel with R_{E2}.)
3. Show the input signal v_{in1} applied to the base B_1 of transistor Q_1 and v_{in2} applied to the base B_2 of transistor Q_2.
4. Label the voltage between the collectors C_1 and C_2 as v_o. (The v_o is the output voltage.)

The circuit of Figure 1-2 depicts these changes and is referred to as a *differential amplifier*. Note that in the circuit of Figure 1-2, $R_{C1} = R_{C2}$; hence their subscripts are omitted for simplicity. In addition, *the parallel combination of R_{E1} and R_{E2} is denoted by R_E.*

The differential amplifier of Figure 1-2, as its name implies, amplifies the difference between two input signals v_{in1} and v_{in2}. (The differential amplifier is also referred to as a *difference amplifier*.) The circuit show in Figure 1-2 is a *basic* differential amplifier circuit from which we can derive the remaining three differential amplifier circuit configurations.

1-3 DIFFERENTIAL AMPLIFIER CIRCUIT CONFIGURATIONS

The four differential amplifier configurations are the following:

1. Dual-input, balanced-output differential amplifier
2. Dual-input, unbalanced-output differential amplifier
3. Single-input, balanced-output differential amplifier
4. Single-input, unbalanced-output differential amplifier

The configurations listed above are defined by the number of input signals used and the way an output voltage is measured. If we use two input signals, the configuration is said to be *dual input*; otherwise, it is a *single-input* configuration. On the other hand, if the output voltage is measured between two collectors, it is referred to as a *balanced output*, because both collectors are at the same dc potential with respect to ground. However, if the output is measured at one of the collectors with respect to ground, the configuration is called an *unbalanced output*. Before proceeding any further with the circuit analysis, some general observations on the differential amplifier are in order.

Two matched semiconductors of the same type (BJTs or FETs) are required for the differential amplifier. All the components in the two emitter-biased circuits, which constitute a differential amplifier, must be matched in all respects for proper operation. Furthermore, the magnitude of the supply voltages $+V_{CC}$ and $-V_{EE}$ must be equal. Fortunately, a variety of matched transistor pairs is readily available for differential amplifier applications. A multistage amplifier with a desired voltage gain can be formed using a direct connection between successive stages of differential amplifiers. The beauty of the direct connection between stages is that it removes the lower cutoff frequency imposed by the coupling capacitors. Therefore, the differential amplifier is capable of amplifying dc as well as ac input signals. In instrumentation systems differential amplifiers are widely used to compare two input signals.

In the analysis to follow we use *r*-parameters instead of *h*-parameters for a number of reasons:

1. Because of the nature of *r*-parameters, ac analysis of differential amplifiers with *r*-parameters is simpler, more straightforward, and less cumbersome.
2. Unlike with *h*-parameters, there is no need to manipulate the *r*-parameters at different operating levels (different I_E and V_{CE} values) except for the r_e value.

3. The performance equations obtained are easy to remember since they are not as complex or as lengthy as h-parameter equations would be.
4. Above all, the results obtained using r-parameters compare favorably with the actual results.

1-4 DUAL-INPUT, BALANCED-OUTPUT DIFFERENTIAL AMPLIFIER

The circuit shown in Figure 1-2 is a dual-input, balanced-output differential amplifier. Notice that the two input signals (dual input), v_{in1} and v_{in2}, are applied to the bases B_1 and B_2 of transistors Q_1 and Q_2. The output v_o is measured between the two collectors, C_1 and C_2, which are at the same dc potential. Because of the equal dc potential at the two collectors with respect to ground, the output is referred to as a balanced output.

1-4.1 DC Analysis

To determine the operating point values (I_{CQ} and V_{CEQ}) for the differential amplifier of Figure 1-2, we need to obtain a dc equivalent circuit. The dc equivalent circuit can be obtained simply by reducing the input signals v_{in1} and v_{in2} to zero. The dc equivalent circuit thus obtained is shown in Figure 1-3. Note that the internal resistances of the input signals are denoted by R_{in} because $R_{in1} = R_{in2}$. Since both emitter-biased sections of the differential amplifier are symmetrical (matched in all respects), we need to determine the operating point for only one section. We shall determine the I_{CQ} and V_{CEQ} values for transistor Q_1 only. These I_{CQ} and V_{CEQ} values can then be used for transistor Q_2 also.

Applying Kirchhoff's voltage law to the base–emitter loop of the transistor Q_1 (see Figure 1-3):

$$-R_{in}I_B - V_{BE} - R_E(2I_E) + V_{EE} = 0 \tag{1-1}$$

but

$$I_B = \frac{I_E}{\beta_{dc}} \qquad \text{since } I_C \cong I_E$$

Thus the emitter current through transistor Q_1 is determined directly from Equation (1-1) as follows:

$$I_E = \frac{V_{EE} - V_{BE}}{2R_E + R_{in}/\beta_{dc}} \tag{1-2}$$

where $V_{BE} = 0.6$ V for silicon transistors
$V_{BE} = 0.2$ V for germanium transistors.

Generally, $R_{in}/\beta_{dc} \ll 2R_E$. Therefore, Equation (1-2) can be rewritten as

$$I_E = \frac{V_{EE} - V_{BE}}{2R_E} \tag{1-3}$$

From Equation (1-3) we see that the value of R_E sets up the emitter current in transistors Q_1 and Q_2 for a given value of V_{EE}. In other words, by selecting a proper value of R_E, we can obtain a desired value of emitter current for a known value of $-V_{EE}$.

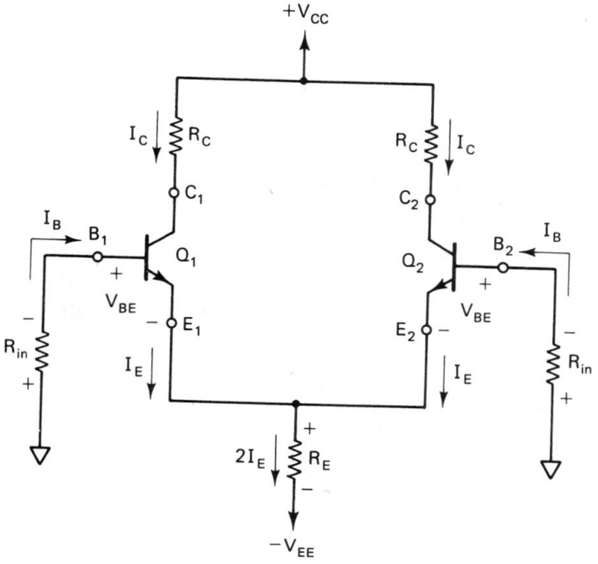

Figure 1-3 Dc equivalent circuit of the dual-input, balanced-output differential amplifier.

Notice that the emitter current in transistors Q_1 and Q_2 is independent of collector resistance R_C.

Next we shall determine the collector-to-emitter voltage V_{CEQ}. The voltage at the emitter of transistor Q_1 is approximately equal to $-V_{BE}$ if we assume the voltage drop across R_{in} to be negligibly small. Knowing the value of emitter current I_E ($= I_C$), we can obtain the voltage at the collector V_C, as follows:

$$V_C = V_{CC} - R_C I_C$$

Thus the collector-to-emitter voltage V_{CE} is

$$\begin{aligned} V_{CE} &= V_C - V_E \\ &= (V_{CC} - R_C I_C) - (-V_{BE}) \\ &= V_{CC} + V_{BE} - R_C I_C \end{aligned} \tag{1-4}$$

Hence, for both transistors we can determine the operating point values (I_{CQ} and V_{CEQ}) by using Equations (1-2) and (1-4), respectively. Note that $I_E = I_{CQ}$ and $V_{CE} = V_{CEQ}$.

Remember that the dc analysis equations (1-2) and (1-4) are applicable for all four differential amplifier configurations as long as we use the same biasing arrangement (as discussed above) for each of them. That is, knowing the dc analysis procedure for one configuration (dual input, balanced output), we do not have to repeat the dc analysis procedure for the remaining three configurations.

1-4.2 AC Analysis

To perform ac analysis to derive the expression for the voltage gain A_d and the input resistance R_i of the differential amplifier shown in Figure 1-2:

1. Set the dc voltages $+V_{CC}$ and $-V_{EE}$ at zero.
2. Substitute the small-signal T-equivalent models for the transistors.

Figure 1-4(a) shows the resulting ac equivalent circuit of the dual-input, balanced-output differential amplifier.

1-4.2(a) Voltage gain. Before we proceed in deriving the expression for the voltage gain A_d, the following should be noted about the circuit in Figure 1-4(a):

1. $I_{E1} = I_{E2}$; therefore, $r_{e1} = r_{e2}$. For this reason, the ac emitter resistance of transistor Q_1 and Q_2 is simply denoted by r_e.
2. The voltage across each collector resistor is shown out of phase by 180° with respect to the input voltages v_{in1} and v_{in2}. This polarity assignment is in accordance with the common-emitter configuration. As we have seen, the differential

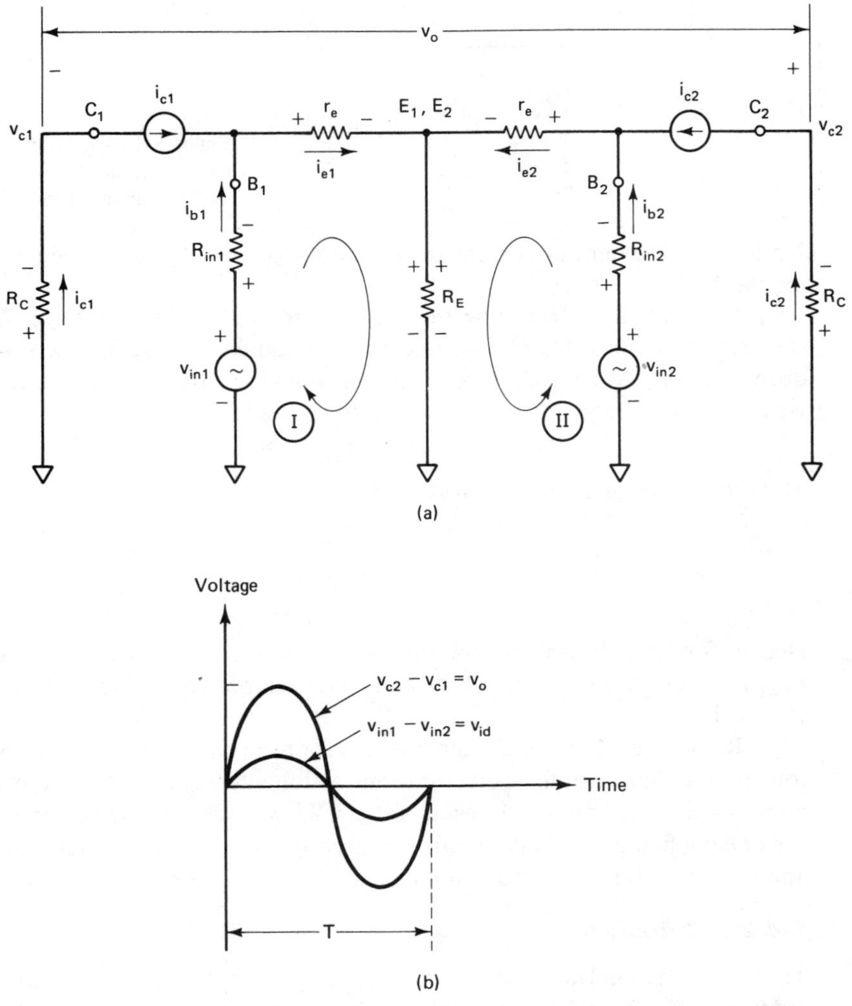

Figure 1-4 Dual-input, balanced-output differential amplifier. (a) Ac equivalent circuit. (b) Input and output waveforms.

amplifier is basically constructed in a common-emitter configuration using two identical common-emitter circuits.

3. Note the assigned polarity of the output voltage v_o. This polarity simply indicates that the voltage at collector C_2 is assumed to be more positive with respect to that at collector C_1, even though both of them are negative with respect to ground.

Writing Kirchhoff's voltage equations for loops I and II in Figure 1-4(a) gives us

$$v_{\text{in}1} - R_{\text{in}1} i_{b1} - r_e i_{e1} - R_E(i_{e1} + i_{e2}) = 0 \tag{1-5}$$

$$v_{\text{in}2} - R_{\text{in}2} i_{b2} - r_e i_{e2} - R_E(i_{e1} + i_{e2}) = 0 \tag{1-6}$$

Substituting current relations $i_{b1} = i_{e1}/\beta_{\text{ac}}$ and $i_{b2} = i_{e2}/\beta_{\text{ac}}$ yields

$$v_{\text{in}1} - \frac{R_{\text{in}1}}{\beta_{\text{ac}}} i_{e1} - r_e i_{e1} - R_E(i_{e1} + i_{e2}) = 0$$

$$v_{\text{in}2} - \frac{R_{\text{in}2}}{\beta_{\text{ac}}} i_{e2} - r_e i_{e2} - R_E(i_{e1} + i_{e2}) = 0$$

Generally, $R_{\text{in}1}/\beta_{\text{ac}}$ and $R_{\text{in}2}/\beta_{\text{ac}}$ values are very small; therefore, we shall neglect them here for simplicity and rearrange equations above as follows:

$$(r_e + R_E) i_{e1} + (R_E) i_{e2} = v_{\text{in}1} \tag{1-7}$$

$$(R_E) i_{e1} + (r_e + R_E) i_{e2} = v_{\text{in}2} \tag{1-8}$$

Equations (1-7) and (1-8) can be solved simultaneously for i_{e1} and i_{e2} by using Cramer's rule:

$$i_{e1} = \frac{\begin{vmatrix} v_{\text{in}1} & R_E \\ v_{\text{in}2} & r_e + R_E \end{vmatrix}}{\begin{vmatrix} r_e + R_E & R_E \\ R_E & r_e + R_E \end{vmatrix}}$$

$$= \frac{(r_e + R_E) v_{\text{in}1} - v_{\text{in}2} R_E}{(r_e + R_E)^2 - (R_E)^2} \tag{1-9a}$$

Similarly,

$$i_{e2} = \frac{\begin{vmatrix} r_e + R_E & v_{\text{in}1} \\ R_E & v_{\text{in}2} \end{vmatrix}}{\begin{vmatrix} r_e + R_E & R_E \\ R_E & r_e + R_E \end{vmatrix}}$$

$$= \frac{(r_e + R_E) v_{\text{in}2} - (R_E) v_{\text{in}1}}{(r_e + R_E)^2 - (R_E)^2} \tag{1.9b}$$

The output voltage is

$$\begin{aligned} v_o &= v_{c2} - v_{c1} \\ &= -R_C i_{c2} - (-R_C i_{c1}) \\ &= R_C i_{c1} - R_C i_{c2} \\ &= R_C(i_{e1} - i_{e2}) \qquad \text{since } i_c \cong i_e \end{aligned} \tag{1-10}$$

Substituting the current relations i_{e1} and i_{e2} in Equation (1-10), we get

$$v_o = R_C\left[\frac{(r_e + R_E)v_{\text{in}1} - (R_E)v_{\text{in}2}}{(r_e + R_E)^2 - (R_E)^2} - \frac{(r_e + R_E)v_{\text{in}2} - (R_E)v_{\text{in}1}}{(r_e + R_E)^2 - (R_E)^2}\right]$$

$$= R_C\frac{(r_e + R_E)(v_{\text{in}1} - v_{\text{in}2}) + (R_E)(v_{\text{in}1} - v_{\text{in}2})}{(r_e + R_E)^2 - (R_E)^2}$$

$$= R_C\frac{(r_e + 2R_E)(v_{\text{in}1} - v_{\text{in}2})}{r_e^2 + 2r_eR_E + R_E^2 - R_E^2}$$

$$= R_C\frac{(r_e + 2R_E)(v_{\text{in}1} - v_{\text{in}2})}{r_e(r_e + 2R_E)}$$

$$= \frac{R_C}{r_e}(v_{\text{in}1} - v_{\text{in}2}) \tag{1-11}$$

Thus a differential amplifier amplifies the *difference* between two input signals, as expected. Figure 1-4(b) shows the input and output waveforms of the dual-input, balanced-output differential amplifier. By defining $v_{id} = v_{\text{in}1} - v_{\text{in}2}$ as the difference in input voltages, we can write the voltage-gain equation of the dual-input, balanced-output differential amplifier as follows:

$$A_d = \frac{v_o}{v_{id}} = \frac{R_C}{r_e} \tag{1-12}$$

Notice that the voltage-gain equation of the differential amplifier is independent of R_E since R_E did not appear in the gain equation (1-12). Another point of interest is that this equation is identical to the voltage-gain equation of the common-emitter amplifier. The latter is obvious because the differential amplifier employs a common-emitter configuration.

1-4.2(b) Differential input resistance. Differential input resistance is defined as the equivalent resistance that would be measured at either input terminal with the other terminal grounded. This means that the input resistance R_{i1} seen from the input signal source $v_{\text{in}1}$ is determined with the signal source $v_{\text{in}2}$ set at zero. Similarly, the input signal source $v_{\text{in}1}$ is set at zero to determine the input resistance R_{i2} seen from the input signal source $v_{\text{in}2}$. Usually, the source resistances $R_{\text{in}1}$ and $R_{\text{in}2}$ are very small and hence will be ignored in the derivation of input resistances R_{i1} and R_{i2}. In equation form,

$$R_{i1} = \left.\frac{v_{\text{in}1}}{i_{b1}}\right|_{v_{\text{in}2}=0}$$

$$= \left.\frac{v_{\text{in}1}}{i_{e1}/\beta_{\text{ac}}}\right|_{v_{\text{in}2}=0}$$

Substituting the value of i_{e1} from Equation (1-9a), we get

$$R_{i1} = \frac{\beta_{\text{ac}}v_{\text{in}1}}{\dfrac{(r_e + R_E)v_{\text{in}1} - (R_E)(0)}{(r_e + R_E)^2 - (R_E)^2}}$$

$$= \frac{\beta_{\text{ac}}(r_e^2 + 2r_eR_E)}{r_e + R_E}$$

$$= \frac{\beta_{\text{ac}}r_e(r_e + 2R_E)}{r_e + R_E} \tag{1-13}$$

Generally, $R_E \gg r_e$, which implies that $(r_e + 2R_E) \cong 2R_E$ and $r_e + R_E \cong R_E$. Therefore, Equation (1-13) can be rewritten as

$$R_{i1} = \frac{\beta_{ac} r_e (2R_E)}{(R_E)}$$

$$= 2\beta_{ac} r_e \tag{1-14}$$

Similarly, the input resistance R_{i2} seen from the input signal source v_{in2} is defined as

$$R_{i2} = \left. \frac{v_{in2}}{i_{b2}} \right|_{v_{in1}=0}$$

$$= \left. \frac{v_{in2}}{i_{e2}/\beta_{ac}} \right|_{v_{in1}=0}$$

Substituting the value of i_{e2} from Equation (1-9b), we obtain

$$R_{i2} = \frac{\beta_{ac} v_{in2}}{\dfrac{(r_e + R_E)v_{in2} - (R_E)(0)}{(r_e + R_E)^2 - (R_E)^2}}$$

$$= \frac{\beta_{ac} r_e (r_e + 2R_E)}{r_e + R_E} \tag{1-15}$$

However, $r_e + 2R_E \cong 2R_E$, and $r_e + R_E \cong R_E$ if $R_E \gg r_e$. Therefore, Equation (1-15) can be rewritten as

$$R_{i2} = \frac{\beta_{ac} r_e (2R_E)}{R_E}$$

$$= 2\beta_{ac} r_e \tag{1-16}$$

1-4.2(c) Output resistance. Output resistance is defined as the equivalent resistance that would be measured at either output terminal with respect to ground. Therefore, the output resistance R_{o1} measured between collector C_1 and ground is equal to that of the collector resistor R_C (see Figure 1-4). Similarly, the output resistance R_{o2} measured at collector C_2 with respect to ground is equal to that of the collector resistor R_C. Thus

$$R_{o1} = R_{o2} = R_C \tag{1-17}$$

The current gain of the differential amplifier is undefined; therefore, the current-gain equation will not be derived for any of the four differential amplifier configurations. Furthermore, like the common-emitter amplifier, the differential amplifier is a small-signal amplifier; therefore, it is generally used as a voltage amplifier and not as a current or power amplifier.

1-4.3 Inverting and Noninverting Inputs

In the differential amplifier circuit of Figure 1-2 the input voltage v_{in1} is called the *noninverting input* because a positive voltage v_{in1} acting alone produces a positive output voltage. This can be seen from voltage-gain Equation (1-11). Similarly, the positive voltage v_{in2} acting alone produces a negative output voltage; hence v_{in2} is called the *inverting input* [see Equation (1-11)]. Consequently, the base terminal B_1

to which v_{in1} is applied is referred to as the *noninverting input terminal*, and the base terminal B_2 is called the *inverting input terminal.*

1-4.4 Common-Mode Rejection Ratio

An important characteristic of the dual-input, balanced-output differential amplifier is its ability to suppress undesired disturbances that might be amplified along with the desired signal. When the matched pair of transistors are used in the differential amplifier, the unwanted signals, such as 60-Hz noise or hum pickup, would appear as common to both input bases, and therefore the net output would theoretically be zero. The practical effectiveness of rejecting the common signal, however, depends on the degree of matching between the two common-emitter stages forming the differential amplifier. In other words, more closely equal are the currents in the input transistors Q_1 and Q_2, the better is the common-mode signal rejection (see Figure 1-5). When the same voltage is applied to both input terminals of a differential amplifier, the differential amplifier is said to operate in the common-mode configuration.

The ability of a differential amplifier to reject a common-mode signal is expressed by its common-mode rejection ratio (CMRR). It is the ratio of differential gain A_d to the common-mode gain A_{cm}. In equation form,

$$\text{CMRR} = \frac{A_d}{A_{cm}} \tag{1-18}$$

The common-mode voltage gain A_{cm} can be determined as follows. Apply a known voltage v_{cm} to both input terminals of the differential amplifier as shown in Figure 1-5, and measure the resultant output voltage v_{ocm}. Then, using Equation (1-19), calculate A_{cm}.

$$A_{cm} = \frac{v_{ocm}}{v_{cm}} \tag{1-19}$$

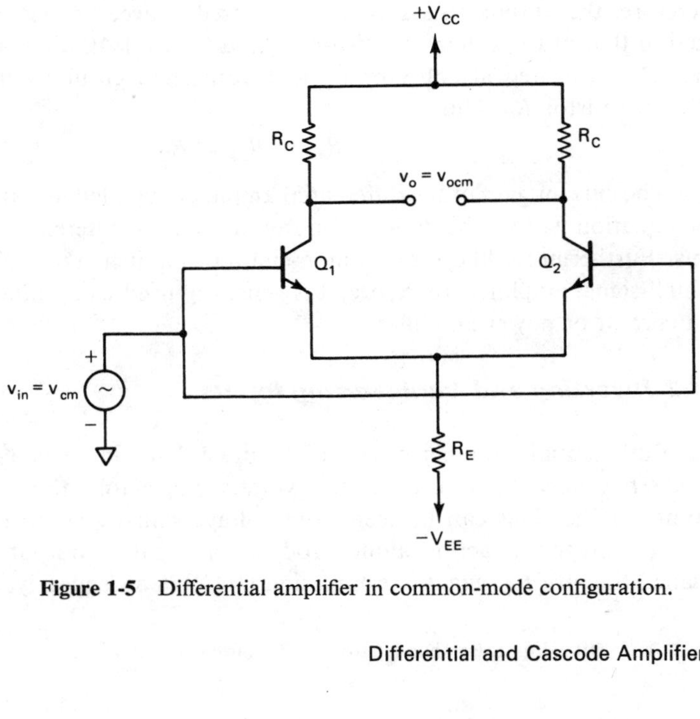

Figure 1-5 Differential amplifier in common-mode configuration.

Ideally, we expect A_{cm} to be zero, that is, $v_{ocm} = 0$ V. In other words, we would like to have CMRR of a differential amplifier ideally equal to infinity. Thus, in practice, it is advantageous to use differential amplifier with higher CMRR since this amplifier is better able to reject common-mode signals.

1-5 DUAL-INPUT, UNBALANCED-OUTPUT DIFFERENTIAL AMPLIFIER

In this configuration two input signals are used; however, the output is measured at only one of the two collectors with respect to ground. The output is referred to as an *unbalanced output* because the collector at which the output voltage is measured is at some finite dc potential with respect to ground. In other words, there is some dc voltage at the output terminal without any input signal applied. Let us assume that the output is measured at the collector of transistor Q_2 with respect to ground (see Figure 1-6).

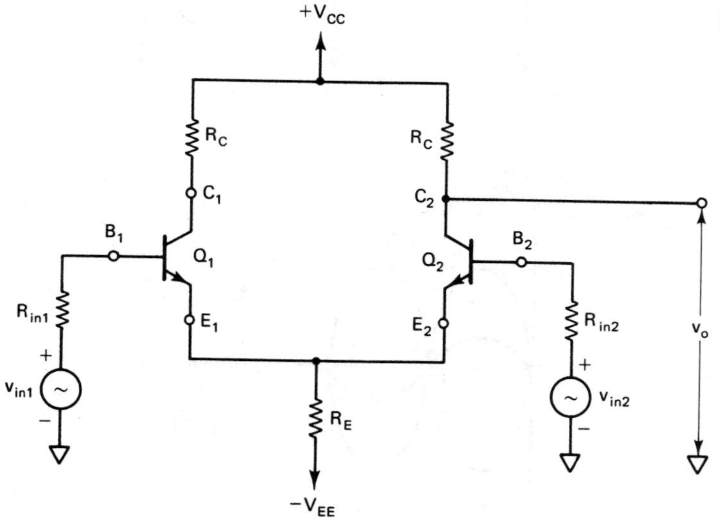

Figure 1-6 Dual-input, unbalanced-output differential amplifier.

1-5.1 DC Analysis

The dc analysis procedure for the dual-input, unbalanced-output differential amplifier is identical to that of the dual-input, balanced-output differential amplifier because both configurations use the same biasing arrangement. Therefore, the emitter current and emitter-to-collector voltage for the dual-input, unbalanced-output differential amplifier is determined from Equations (1-2) and (1-4), respectively. These equations are repeated here for convenience:

$$I_E = I_{CQ} = \frac{V_{EE} - V_{BE}}{2R_E + R_{in}/\beta_{dc}}$$

$$V_{CE} = V_{CEQ} = V_{CC} + V_{BE} - R_C I_{CQ}$$

1-5.2 AC Analysis

Figure 1-7(a) shows the ac equivalent circuit of the dual-input, unbalanced-output differential amplifier with small-signal T-equivalent models substituted for the transistors.

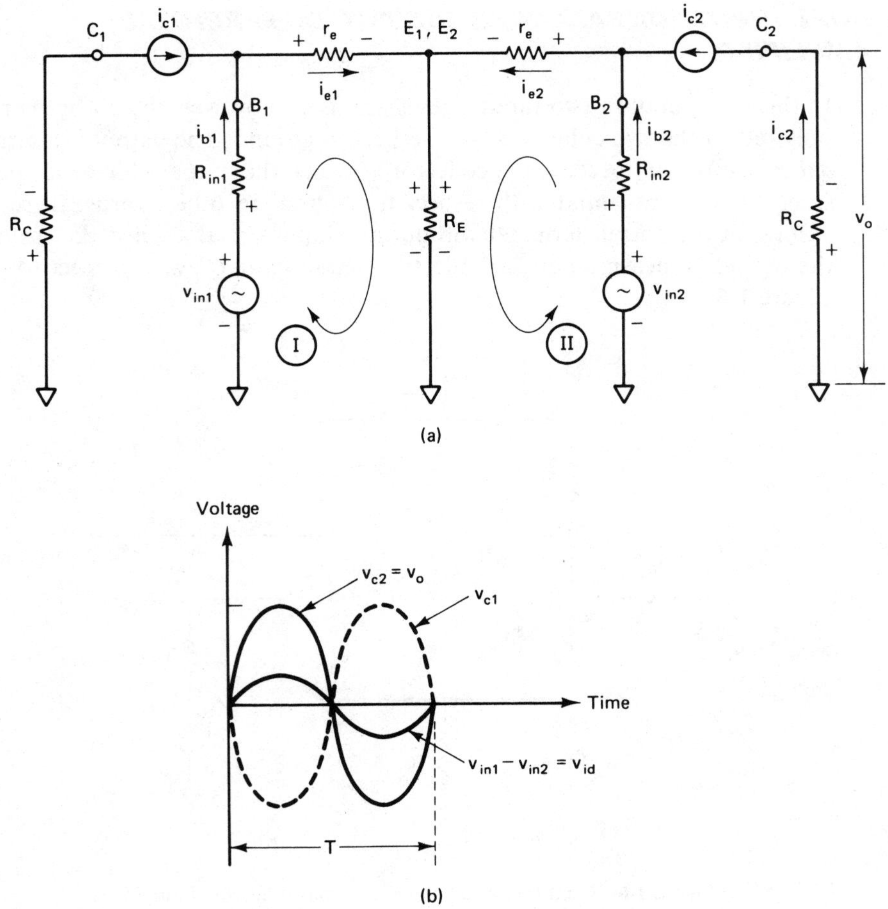

Figure 1-7 Dual-input, unbalanced-output differential amplifier. (a) Ac equivalent circuit. (b) Input and output waveforms, assuming that the dc output voltage is zero.

1-5.2(a) Voltage gain. Writing Kirchhoff's voltage equations for loops I and II in Figure 1-7(a) gives us

$$v_{\text{in}1} - R_{\text{in}1}i_{b1} - r_e i_{e1} - R_E(i_{e1} + i_{e2}) = 0$$

$$v_{\text{in}2} - R_{\text{in}2}i_{b2} - r_e i_{e2} - R_E(i_{e1} + i_{e2}) = 0$$

Since these equations are the same as Equations (1-5) and (1-6), the expressions for i_{e1} and i_{e2} will be the same as in Equations (1-9a) and (1-9b), respectively. That is,

$$i_{e1} = \frac{(r_e + R_E)v_{\text{in}1} - (R_E)v_{\text{in}2}}{(r_e + R_E)^2 - (R_E)^2}$$

$$i_{e2} = \frac{(r_e + R_E)v_{\text{in}2} - (R_E)v_{\text{in}1}}{(r_2 + R_E)^2 - (R_E)^2}$$

The output voltage is

$$v_o = v_{c2} = -R_C i_{c2} = -R_C i_{e2}, \qquad \text{since } i_c \cong i_e$$

Substituting the value of i_{e2} yields

$$v_o = -R_C \frac{(r_e + R_E)v_{\text{in}2} - (R_E)v_{\text{in}1}}{(r_e + R_E)^2 - (R_E)^2}$$

$$= R_C \frac{(R_E)v_{\text{in}1} - (r_e + R_E)v_{\text{in}2}}{r_e(r_e + 2R_E)}$$

Generally, $R_E \gg r_e$; hence $r_e + R_E \cong R_E$ and $r_e + 2R_E \cong 2R_E$. Therefore,

$$v_o = R_C \frac{(R_E)v_{\text{in}1} - (R_E)v_{\text{in}2}}{2r_e R_E}$$

$$= R_C \frac{R_E(v_{\text{in}1} - v_{\text{in}2})}{2r_e R_E}$$

$$= \frac{R_C}{2r_e}(v_{\text{in}1} - v_{\text{in}2}) \tag{1-20}$$

$$A_d = \frac{v_o}{v_{id}} = \frac{R_C}{2r_e} \tag{1-21}$$

Thus the voltage gain of the dual-input, unbalanced-output differential amplifier is half the gain of the dual-input, balanced-output differential amplifier [see Equations (1-12) and (1-21)]. The input and output waveforms of dual-input, unbalanced-output differential amplifier are shown in Figure 1-6(b). Notice also that in the dual-input, unbalanced-output differential amplifier the dc voltage at the output terminal is the error voltage in the desired output signal. Therefore, to reduce the undesired dc voltage to zero, this configuration is generally followed by a circuit called a *level translator* circuit. (Refer to Section 1-13 for more information.)

1-5.2(b) Differential input resistance. The only difference between the circuits in Figures 1-2 and 1-6 is in the way output voltage is measured. However, the input resistance seen from either input source does not depend on the way the output voltage is measured. Therefore, the input resistance R_i seen from either input source of the circuit of Figure 1-6 should be the same as that of the circuit in Figure 1-2 and is given by Equation (1-14). That is,

$$R_{i1} = R_{i2} = 2\beta_{\text{ac}} r_e$$

1-5.2(c) Output resistance. The output resistance R_o measured at collector C_2 with respect to ground is equal to the collector resistor R_C (see Figure 1-7). Thus

$$R_o = R_C$$

1-6 SINGLE-INPUT, BALANCED-OUTPUT DIFFERENTIAL AMPLIFIER

Figure 1-8 shows the single-input, balanced-output differential amplifier. In this figure the (single) input is applied to the base of transistor Q_1, and the output is measured between the two collectors which are at the same dc potential. Therefore, the output is said to be a *balanced output*.

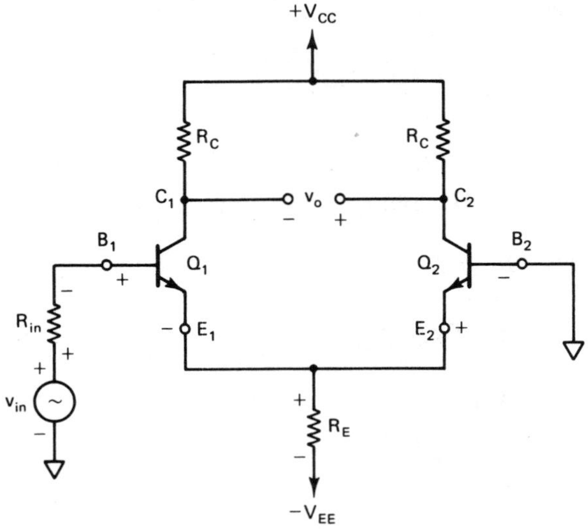

Figure 1-8 Single-input, balanced-output differential amplifier.

1-6.1 DC Analysis

The dc analysis procedure and bias equations for the single-input, balanced-output differential amplifier are identical to those of the two previous configurations because the dc equivalent circuit for all three configurations is the same. Thus the bias equations [Equations (1-2) and (1-4)] are

$$I_E = I_{CQ} = \frac{V_{EE} - V_{BE}}{2R_E + R_{\text{in}}/\beta_{\text{dc}}}$$

$$V_{CE} = V_{CEQ} = V_{CC} + V_{BE} - R_C I_{CQ}$$

1-6.2 AC Analysis

The ac equivalent circuit of the single-input, balanced output differential amplifier with a small-signal T-equivalent models substituted for the transistors is shown in Figure 1-9(a).

1-6.2(a) Voltage gain. Before we derive the expression for the voltage gain A_d, the following should be noted about the circuit in Figure 1-9(a).

1. During the positive half-cycle of the input signal (or when v_{in} is the positive voltage), the base–emitter voltage of the transistor Q_1 is positive and that of transistor Q_2 is negative, as shown in Figure 1-8. This means that the collector

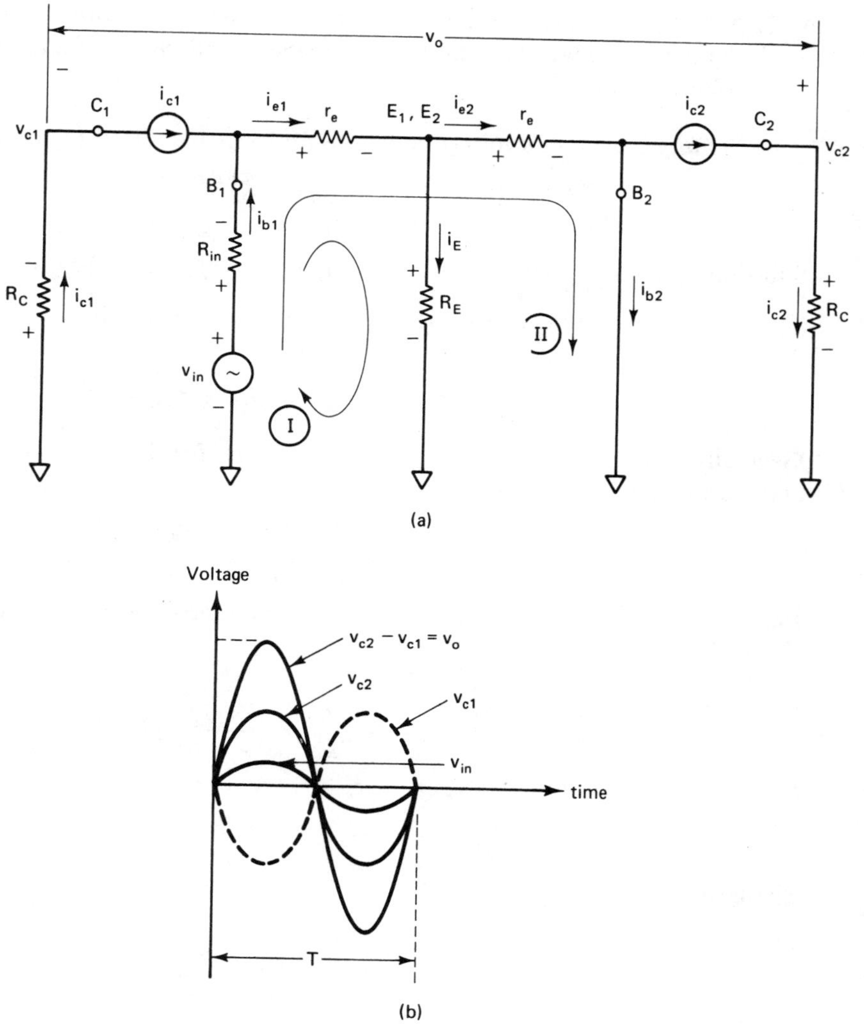

Figure 1-9 Single-input, balanced-output differential amplifier. (a) Ac equivalent circuit. (b) Input and output waveforms.

current in transistor Q_1 increases and that in transistor Q_2 decreases from the operating point value I_{CQ}. This change in collector currents during the positive half-cycle of the input signal is indicated in Figure 1-9(a), in which the currents of both sources i_{c1} and i_{c2} are shown to be in the same direction. In fact, during the negative half-cycle of the input signal (or when v_{in} is the negative voltage), the opposite action takes place; that is, the collector current of transistor Q_1 decreases and that in transistor Q_2 increases.

2. The polarity of voltage across each collector resistor is consistent with the direction of current sources i_{c1} and i_{c2}. That is, the voltage across the collector resistor of transistor Q_2 is positive, and that across the collector resistor of transistor Q_1 is negative with respect to ground [see Figure 1-9(a)].

3. Note the assigned polarity of the output voltage v_o. According to this polarity, the output voltage is equal to the voltage at collector C_2 minus the voltage at collector C_1.

Writing Kirchhoff's voltage equations for loops I and II of Figure 1-9(a) gives us

$$v_{in} - R_{in}i_{b1} - r_e i_{e1} - R_E i_E = 0 \tag{1-22}$$

$$v_{in} - R_{in}i_{b1} - r_e i_{e1} - r_e i_{e2} = 0 \tag{1-23}$$

Substituting current relations $i_E = (i_{e1} - i_{e2})$, $i_{b1} \cong i_{e1}/\beta_{ac}$, and $i_{b2} \cong i_{e2}/\beta_{ac}$, we get

$$v_{in} - \frac{R_{in}}{\beta_{ac}}i_{e1} - r_e i_{e1} - (R_E)(i_{e1} - i_{e2}) = 0$$

$$v_{in} - \frac{R_{in}}{\beta_{ac}}i_{e1} - r_e i_{e1} - r_e i_{e2} = 0$$

Generally, the R_{in}/β_{ac} value is very small; therefore, for simplicity we shall neglect it and rearrange the equations above as follows:

$$(r_e + R_E)i_{e1} - (R_E)i_{e2} = v_{in} \tag{1-24}$$

$$(r_e)i_{e1} + (r_e)i_{e2} = v_{in} \tag{1-25}$$

Equations (1-24) and (1-25) can be solved simultaneously for i_{e1} and i_{e2} by using Cramer's rule:

$$i_{e1} = \frac{\begin{vmatrix} v_{in} & -R_E \\ v_{in} & r_e \end{vmatrix}}{\begin{vmatrix} r_e + R_E & -R_E \\ r_e & r_e \end{vmatrix}} = \frac{r_e v_{in} + R_E v_{in}}{(r_e + R_E)r_e + R_E r_e}$$

$$= \frac{(r_e + R_E)v_{in}}{r_e(r_e + 2R_E)} \tag{1-26a}$$

Similarly,

$$i_{e2} = \frac{\begin{vmatrix} r_e + R_E & v_{in} \\ r_e & v_{in} \end{vmatrix}}{\begin{vmatrix} r_e + R_E & -R_E \\ r_e & r_e \end{vmatrix}} = \frac{(r_e + R_E)v_{in} - (r_e)v_{in}}{(r_e + R_E)r_e + R_E r_e}$$

$$= \frac{(R_E)v_{in}}{r_e(r_e + 2R_E)} \tag{1-26b}$$

The output voltage is

$$\begin{aligned} v_o &= v_{c2} - v_{c1} \\ &= R_C i_{c2} - (-R_C i_{c1}) \\ &= R_C(i_{c2} + i_{c1}) \\ &= R_C(i_{e2} + i_{e1}) \qquad \text{since } i_e = i_c \end{aligned} \tag{1-27}$$

Substituting the current relations i_{e1} and i_{e2} in Equation (1-27), we get

$$v_o = R_C\left[\frac{(R_E)v_{in}}{r_e(r_e + 2R_E)} + \frac{(r_e + R_E)v_{in}}{r_e(r_e + 2R_E)}\right]$$

$$= R_C \frac{(r_e + 2R_E)v_{in}}{r_e(r_e + 2R_E)}$$

$$= \frac{R_C}{r_e} v_{in} \tag{1-28}$$

$$A_d = \frac{v_o}{v_{in}} = \frac{R_C}{r_e} \tag{1-29}$$

Notice that the voltage gain of the single-input, balanced-output differential amplifier is equal to that of the dual-input, balanced-output differential amplifier [see Equations (1-12) and (1-29)]. The input and output waveforms of the single-input, balanced-output differential amplifier are shown in Figure 1-9(b).

1-6.2(b) Differential input resistance. The input resistance R_i seen from the input signal source is determined as follows:

$$R_i = \frac{v_{in}}{i_{b1}} = \frac{v_{in}}{i_{e1}/\beta_{ac}} = \frac{\beta_{ac}v_{in}}{i_{e1}}$$

Substituting the value of i_{e1} from Equation (1-26a), we get

$$R_i = \frac{\beta_{ac}v_{in}}{\frac{(r_e + R_E)v_{in}}{r_e(r_e + 2R_E)}} = \frac{\beta_{ac}r_e(r_e + 2R_E)}{r_e + R_E}$$

$$= 2\beta_{ac}r_e \qquad \text{if } R_E \gg r_e \tag{1-30}$$

1-6.2(c) Output resistance. Recall that the output resistance R_o is the equivalent resistance that would be measured at either output terminal with respect to ground. Therefore, the output resistance R_{o1} measured at the collector C_1 and the output resistance R_{o2} measured at the collector C_2 is given by Equation (1-17):

$$R_{o1} = R_{o2} = R_C$$

1-7 SINGLE-INPUT, UNBALANCED-OUTPUT DIFFERENTIAL AMPLIFIER

This differential amplifier configuration is not commonly used, for several reasons:

1. Although identical to the common-emitter amplifier, it not only requires more components but also yields less voltage gain than the former one.
2. There exists a dc output voltage without any input signal applied.

Nevertheless, we shall include the analysis and derivations of the performance equations in order to make the differential amplifier discussion complete. In this configuration an input signal can be applied to the base of either transistor and the output can be measured at either collector. In Figure 1-10 the input is applied to the base of transistor Q_1, and the output is measured at the collector of transistor Q_2.

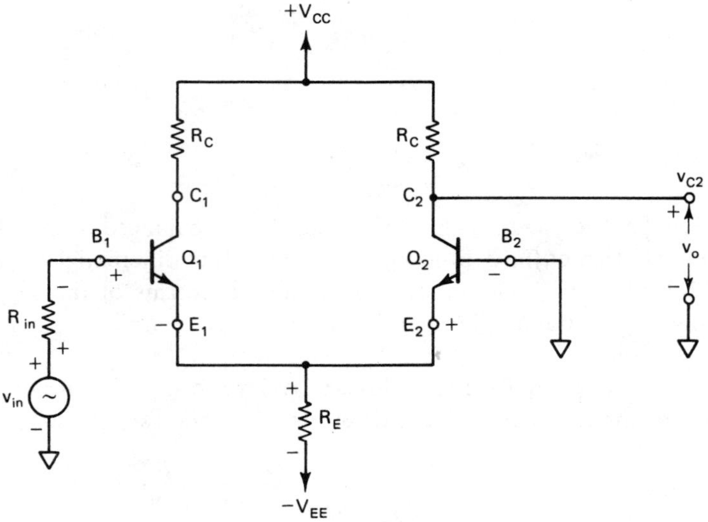

Figure 1-10 Single-input, unbalanced-output differential amplifier.

1-7.1 DC Analysis

The dc equivalent circuit for this configuration is identical to that of the dual-input, balanced-output configuration shown in Figure 1-3. Therefore, the I_{CQ} and V_{CEQ} equations obtained from Figure 1-3 are applicable to this configuration also. These equations are repeated here for convenience:

$$I_E = I_{CQ} = \frac{V_{EE} - V_{BE}}{2R_E + R_{\text{in}}/\beta_{\text{dc}}}$$

$$V_{CE} = V_{CEQ} = V_{CC} + V_{BE} - R_C I_{CQ}$$

1-7.2 AC Analysis

The ac equivalent circuit of single-input, unbalanced-output differential amplifier is shown in Figure 1-11(a).

1-7.2(a) Voltage gain. The equivalent circuit shown in Figure 1-11(a) is identical to that shown in Figure 1-9(a), except in the way in which the output voltage is measured. Therefore, the current relations (1-26a) and (1-26b) derived for the circuit of Figure 1-9(a) are also valid for the circuit shown in Figure 1-11(a). The output voltage is

$$v_o = v_{c2} = R_C i_{c2} = R_C i_{e2} \qquad \text{since } i_c \cong i_e$$

Substituting the i_{e2} value from Equation (1-26b) yields

$$v_o = R_C \frac{(R_E)v_{\text{in}}}{r_e(r_e + 2R_E)}$$

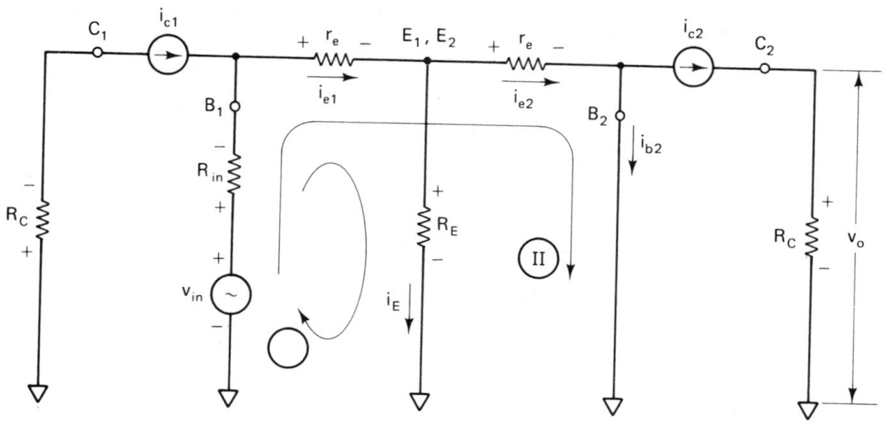

(a)

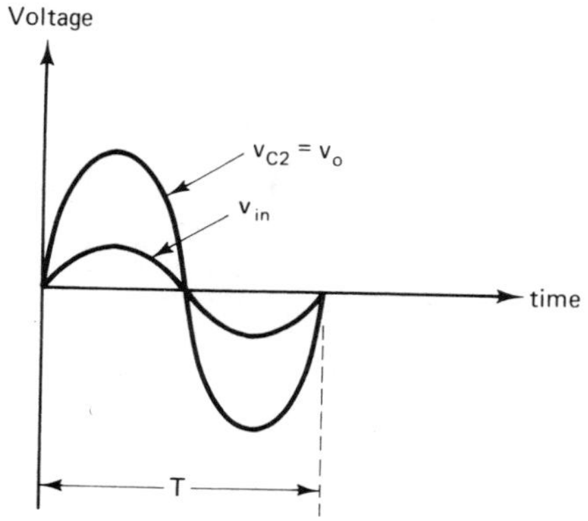

(b)

Figure 1-11 Single-input, unbalanced-output differential amplifier. (a) Ac equivalent circuit. (b) Input and output waveforms, assuming that the dc output voltage is zero.

$$= R_C \frac{R_E v_{in}}{r_e(2R_E)} \qquad \text{if } R_E \gg r_e$$

$$= \frac{R_C}{2r_e} v_{in} \tag{1-31}$$

$$A_d = \frac{v_o}{v_{in}} = \frac{R_C}{2r_e} \tag{1-32}$$

The input and output waveforms of a single-input, unbalanced-output differential amplifier are shown in Figure 1-11(b).

1-7.2(b) Differential input resistance. The input resistance R_i of this configuration is identical to that of the previous one. That is,

TABLE 1-1 PROPERTIES OF THE DIFFERENTIAL AMPLIFIER CIRCUIT CONFIGURATIONS

Configuration	Circuit	Voltage gain	Input resistance	Output resistance
1. Dual input, balanced output	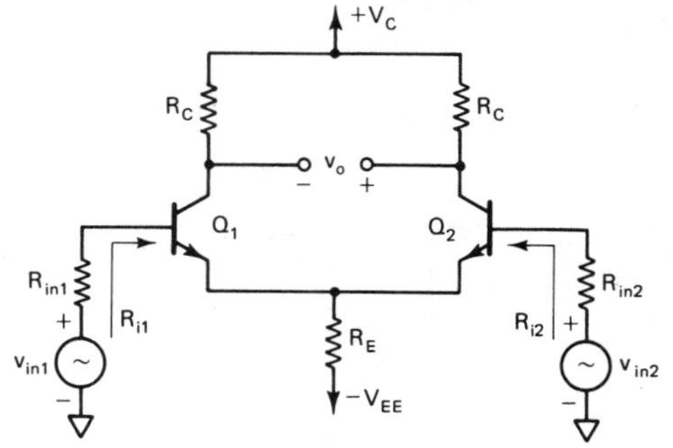	$A_d = \frac{R_C}{r_e}$	$R_{i1} = 2\beta_{ac} r_e$ $R_{i2} = 2\beta_{ac} r_e$	$R_{o1} = R_C$ $R_{o2} = R_C$
2. Dual input, unbalanced output	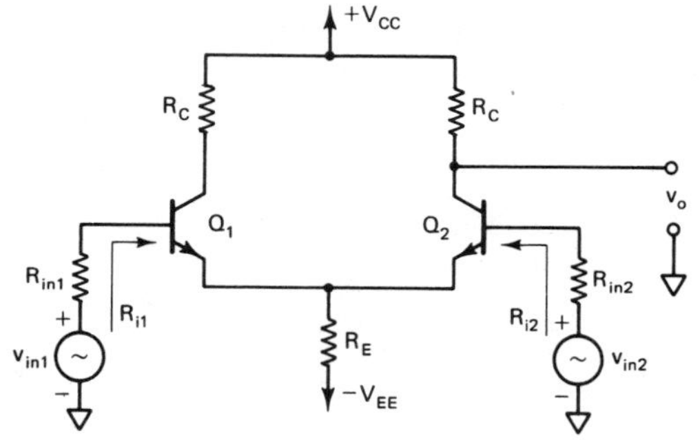	$A_d = \frac{R_C}{2r_e}$	$R_{i1} = 2\beta_{ac} r_e$ $R_{i2} = 2\beta_{ac} r_e$	$R_o = R_C$
3. Single input, balanced output	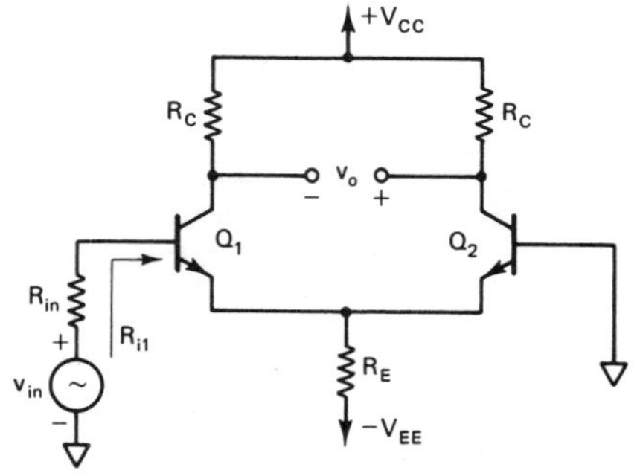	$A_d = \frac{R_C}{r_e}$	$R_i = 2\beta_{ac} r_e$	$R_{o1} = R_C$ $R_{o2} = R_C$
4. Single input, unbalanced output	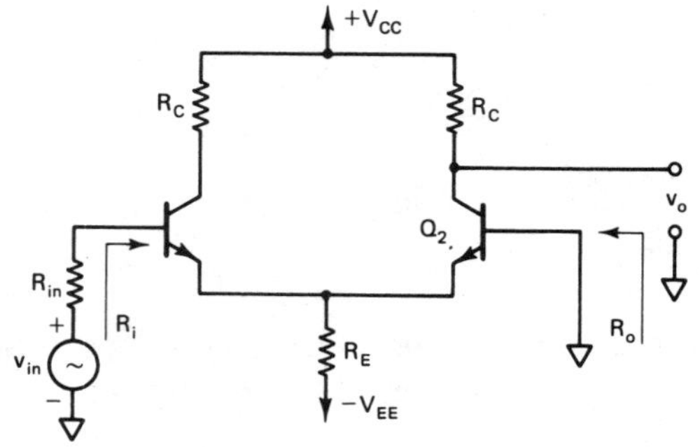	$A_d = \frac{R_C}{2r_e}$	$R_i = 2\beta_{ac} r_e$	$R_o = R_C$

$$R_i = 2\beta_{ac} r_e \qquad \text{if } R_E \gg r_e$$

1-7.2(c) Output resistance. The output resistance R_o measured at collector C_2 with respect to ground is equal to the collector resistor R_C (see Figure 1-10). Thus

$$R_o = R_C$$

Properties of all the four types of differential amplifiers discussed above are summarized in Table 1-1.

1-8 FET DIFFERENTIAL AMPLIFIERS

In the differential amplifier configurations just discussed we have used BJTs. But if we require very high input resistance, we can use FETs instead. Fortunately, the voltage-gain equations derived for these configurations using BJTs can also be used for configurations using FETs, except for the following replacements:

$$R_C \longrightarrow R_D$$

$$r_e \longrightarrow \frac{1}{g_m}$$

For instance, the voltage gain of the JFET dual-input, balanced-output differential amplifier obtained from Equation (1-12) is

$$A_d = \frac{v_o}{v_{id}} = \frac{R_D}{1/g_m} = g_m R_D \tag{1-33}$$

1-9 DIFFERENTIAL AMPLIFIER WITH SWAMPING RESISTORS

By using external resistors R'_E in series with each emitter, the dependence of voltage gain of the differential amplifiers on variations in r_e can be reduced. Using R'_E also increases the linearity range of the differential amplifiers. Figure 1-12 shows the dual-input, balanced-output differential amplifier with emitter resistors R'_E. Generally, the value of R'_E is large enough to swamp the effect of r_e. For this reason R'_E is sometimes referred to as a *swamping resistor*.

The emitter current in each transistor can be determined by writing Kirchhoff's voltage equation around the base–emitter loop of transistor Q_1 with $v_{in1} = v_{in2} = 0$ V:

$$-R_{in} I_B - V_{BE} - R'_E I_E - R_E(2I_E) + V_{EE} = 0$$

(Note that we have omitted the subscripts because of symmetry between two halves of the differential amplifier.)

Substituting $I_B \cong I_E/\beta_{dc}$ and simplifying, we get

$$I_E = \frac{V_{EE} - V_{BE}}{2R_E + R'_E + R_{in}/\beta_{dc}} \tag{1-34}$$

Remember that the emitter current Equation (1-34) is also valid for the three remaining differential amplifier configurations with emitter resistor R'_E. The voltage between collector and emitter [Equation (1-4)] remains unchanged and will be repeated here

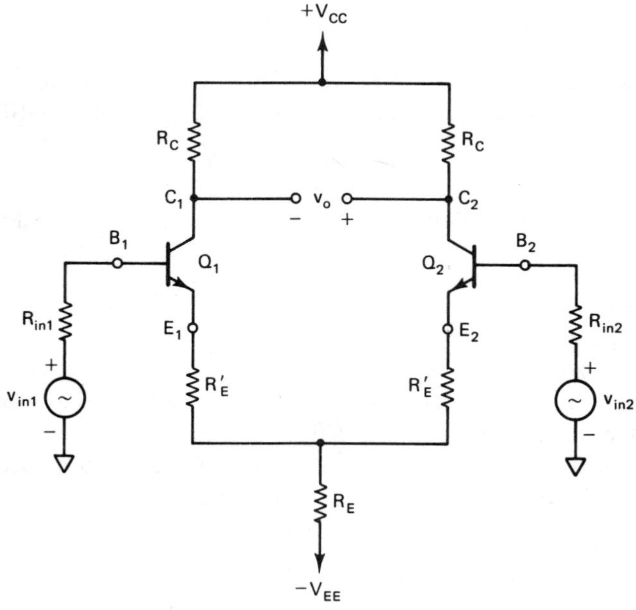

Figure 1-12 Dual-input, balanced-output differential amplifier with emitter resistors R'_E.

for convenience:

$$V_{CE} = V_{CC} + V_{BE} - R_C I_C$$

When external resistors R'_E are added to each emitter lead, the new voltage gain and new input resistance can be obtained for any configuration simply by replacing r_e by $(r_e + R'_E)$. For instance, the voltage gain of the dual-input, balanced-output differential amplifier of Figure 1-12 becomes

$$A_d = \frac{v_o}{v_{id}} = \frac{R_C}{r_e + R'_E} \tag{1-35}$$

and the new input resistance becomes

$$R_{i1} = R_{i2} = 2\beta_{ac}(r_e + R'_E) \tag{1-36}$$

Keep in mind that the output resistance with or without R'_E is the same [see Equation (1-17)]. That is,

$$R_{o1} = R_{o2} = R_C$$

Thus the use of R'_E reduces the voltage gain but increases the input resistance, as can be seen from Equations (1-35) and (1-36), respectively.

Example 1-1

The following specifications are given for the dual-input, balanced-output differential amplifier of Figure 1-2: $R_C = 2.2\ \text{k}\Omega$, $R_E = 4.7\ \text{k}\Omega$, $R_{\text{in}1} = R_{\text{in}2} = 50\ \Omega$, $+V_{CC} = 10$ V, $-V_{EE} = -10$ V, and the transistor is the CA3086 with $\beta_{dc} = \beta_{ac} = 100$ and $V_{BE} = 0.715$ V typical.

(a) Determine the I_{CQ} and V_{CEQ} values.
(b) Determine the voltage gain.
(c) Determine the input and output resistances.

Solution

(a) To determine I_{CQ} and V_{CEQ} values, we have to substitute the known values in Equations (1-2) and (1-4) as follows:

$$I_{CQ} = I_E = \frac{V_{EE} - V_{BE}}{2R_E + R_{\text{in}}/\beta_{\text{dc}}} = \frac{10 - 0.715}{9.4\text{ k}\Omega + 50/100} = 0.988\text{ mA}$$

$$V_{CEQ} = V_{CC} + V_{BE} - R_C I_{CQ}$$
$$= 10 + 0.715 - (2.2\text{ k}\Omega)(0.988\text{ mA})$$
$$= 8.54\text{ V}$$

(b) The ac emitter resistance

$$r_e = \frac{25\text{ mV}}{I_E\text{ mA}} = \frac{25\text{ mV}}{0.988\text{ mA}} = 25.3\ \Omega$$

Therefore, substituting the known values in voltage-gain equation (1-12), we obtain

$$A_d = \frac{v_o}{v_{id}} = \frac{R_C}{r_e} = \frac{2.2\text{ k}\Omega}{25.3} = 86.96$$

(c) The input resistance seen from each input source is given by Equation (1-14):

$$R_{i1} = R_{i2} = 2\beta_{\text{ac}} r_e = (2)(100)(25.3) = 5.06\text{ k}\Omega$$

The output resistance seen looking back into the circuit from each of the two output terminals is given by Equation (1-17):

$$R_{o1} = R_{o2} = 2.2\text{ k}\Omega$$

Example 1-2

Repeat Example 1-1 for the (a) dual-input, unbalanced-output and (b) single-input, balanced-output differential amplifiers.

Solution

(a) Since the component values remain unchanged and the biasing arrangement is same, the I_{CQ} and V_{CEQ} values as well as input and output resistance values for the dual-input, unbalanced-output configuration must be the same as those for the dual-input, balanced-output configuration (see Table 1-1). Thus

$$I_{CQ} = 0.988\text{ mA}$$
$$V_{CEQ} = 8.54\text{ V}$$
$$R_{i1} = R_{i2} = 5.06\text{ k}\Omega$$
$$R_o = 2.2\text{ k}\Omega$$

However, the voltage gain of the dual-input, unbalanced-output differential amplifier using Equation (1-21) is

$$A_d = \frac{R_C}{2r_e} = \frac{2.2\text{ k}\Omega}{(2)(25.3)} = 43.48$$

(b) Because the same biasing arrangement and same component values are used in both configurations, the results obtained in Example 1-1 for the dual-input, balanced-output configuration are also valid for the single-input, balanced-output configuration. That is,

$$I_{CQ} = 0.988\text{ mA}$$
$$V_{CEQ} = 8.54\text{ V}$$

$$A_d = 86.96$$

$$R_i = 5.06\ \text{k}\Omega$$

$$R_{o1} = R_{o2} = 2.2\ \text{k}\Omega$$

Example 1-3

For the dual-input, balanced-output differential amplifier of Example 1-1:
(a) Determine the output voltage (v_o) if $v_{\text{in}1} = 50$ mV peak to peak (pp) at 1 kHz and $v_{\text{in}2} = 20$ mV pp at 1 kHz.
(b) What is the maximum peak-to-peak output voltage without clipping?

Solution

(a) In Example 1-1 we have determined the voltage gain of the dual-input, balanced-output differential amplifier. Substituting this voltage gain ($A_d = 86.96$) and given values of input voltages in Equation (1-11), we get

$$\begin{aligned} v_o &= \frac{R_C}{r_e}(v_{\text{in}1} - v_{\text{in}2}) = 86.96(50\text{ mV} - 20\text{ mV}) \\ &= 2.61\text{ V pp} \end{aligned}$$

(b) Note that in Figure 1-2 the output voltage v_o is measured across the collectors. Therefore, to calculate the maximum peak-to-peak output voltage, we need to determine the voltage drop across each collector resistor:

$$V_{R_C} = R_C I_C$$

Substituting $I_C = I_{CQ} = 0.988$ mA (see Example 1-1), we get

$$V_{R_C} = (2.2\ \text{k}\Omega)(0.988\text{ mA}) = 2.17\text{ V} < V_{CE} = 8.54\text{ V} \quad \text{(see Example 1-1)}$$

This means that the maximum change in voltage across each collector resistor is ± 2.17 (ideally) or 4.34 V pp. In other words, the maximum peak-to-peak output voltage without clipping is $(2)(4.34) = 8.68$ V pp.

Remember that the maximum peak-to-peak output voltage without clipping would have been 4.34 V pp if the same circuit was used as the dual-input, unbalanced-output differential amplifier.

Example 1-4

Repeat Example 1-1 if 100-Ω resistors R'_E are placed externally in series with each emitter (see Figure 1-12).

Solution

(a) Substituting in Equations (1-34) and (1-4), we can determine the values of I_{CQ} and V_{CEQ}, respectively:

$$\begin{aligned} I_{CQ} &= \frac{V_{EE} - V_{BE}}{2R_E + R'_E + R_{\text{in}}/\beta_{\text{dc}}} = \frac{10 - 0.715}{9400 + 100 + 50/100} \\ &= 0.977\text{ mA} \end{aligned}$$

$$\begin{aligned} V_{CEQ} &= V_{CC} + V_{BE} - R_C I_{CQ} \\ &= 10 + 0.715 - (2.2\ \text{k}\Omega)(0.977\text{ mA}) \\ &= 8.57\text{ V} \end{aligned}$$

(b) The ac emitter resistance

$$r_e = \frac{25\text{ mV}}{I_E\text{ mA}} = \frac{25}{0.977} = 25.59\ \Omega$$

Therefore, substituting in Equation (1-35), the voltage gain is

$$A_d = \frac{R_C}{R'_E + r_e} = \frac{2.2\ \text{k}\Omega}{100 + 25.59} = 17.52$$

(c) The input resistance is given by Equation (1-36):

$$\begin{aligned} R_{i1} = R_{i2} &= 2\beta_{\text{ac}}(R'_E + r_e) \\ &= (2)(100)(100 + 25.59) \\ &= 25.12\ \text{k}\Omega \end{aligned}$$

The output resistance is independent of R'_E values and therefore is the same as in Example 1-1. That is,

$$R_{o1} = R_{o2} = 2.2\ \text{k}\Omega$$

Comparing the voltage gain and the input resistance values obtained in Examples 1-1 and 1-4, we can verify that the use of R'_E reduces the voltage gain but increases the input resistance significantly.

1-10 CONSTANT CURRENT BIAS

In the differential amplifiers discussed so far the combination of R_E and V_{EE} is used to set up the dc emitter current. We can also use constant current bias to set up the dc emitter current if desired. In fact, the constant current bias is better because it provides current stabilization and, in turn, assures a stable operating point for the differential amplifier. Figure 1-13 shows the dual-input, balanced-output differential amplifier using a resistive constant current bias. Notice that the resistor R_E is replaced by a constant current transistor (Q_3) circuit. The dc collector current in transistor Q_3 is established by resistors R_1, R_2, and R_E and can be determined as follows. Applying the voltage-divider rule, the voltage at the base of transistor Q_3 (neglecting base loading effect) is

$$V_{B3} = \frac{-R_2 V_{EE}}{R_1 + R_2}$$

$$V_{E3} = V_{B3} - V_{BE3} = -\frac{R_2 V_{EE}}{R_1 + R_2} - V_{BE3}$$

Therefore,

$$I_{E3} = I_{C3} = \frac{V_{E3} - (-V_{EE})}{R_E}$$

$$I_{C3} = \frac{V_{EE} - [R_2 V_{EE}/(R_1 + R_2)] - V_{BE3}}{R_E}$$

Because the two halves of the differential amplifier are symmetrical, each has half of the current I_{C3}. That is,

$$I_{E1} = I_{E2} = \frac{I_{C3}}{2} = \frac{V_{EE} - [R_2 V_{EE}/(R_1 + R_2)] - V_{BE3}}{2R_E} \qquad (1\text{-}37)$$

The collector current I_{C3} in transistor Q_3 is fixed and must be invariant because no signal is injected into either the emitter or the base of Q_3. Thus the transistor Q_3 is a source of constant emitter current for transistors Q_1 and Q_2 of the differential amplifier (see Figure 1-13).

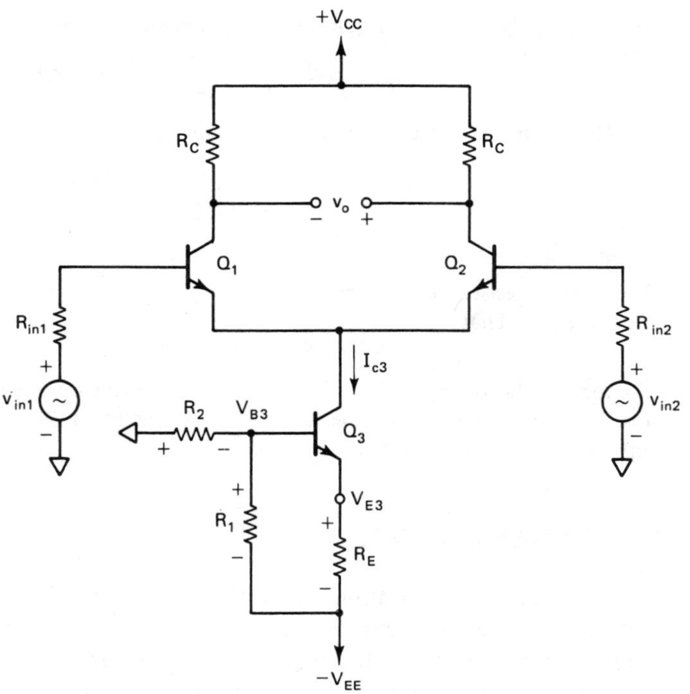

Figure 1-13 Dual-input, balanced-output differential amplifier using constant current bias.

Recall that in the analysis of the differential amplifier circuits with emitter bias, we required that $R_E \gg r_e$. Besides supplying constant emitter current, the constant current bias also provides a very high source resistance since the ac equivalent of the dc current source is ideally an open circuit. Therefore, all the performance equations obtained for the differential amplifier configurations using emitter bias are also applicable to differential amplifiers with constant current bias. For instance, the voltage-gain equation (1-12) is valid for the differential amplifier of Figure 1-2 as well as for that in Figure 1-13 with constant current bias.

To improve the thermal stability of the constant current transistor Q_3, resistor R_1 is replaced by diodes D_1 and D_2, as shown in Figure 1-14(a). The base of Q_3 is biased with the voltage divider containing components R_2, D_1, and D_2. Diodes D_1 and D_2 help to hold the emitter current I_{E3} constant even though the temperature changes. Note that I_2 flows to the node at the base of Q_3 and then divides into paths I_D and I_{B3}. If the temperature of Q_3 increases, its base–emitter voltage V_{BE3} decreases. In silicon units, V_{BE} decreases 2 mV/°C, and in germanium units, V_{BE} decreases 1.6 mV/°C. This reduced V_{BE3} tends to raise the voltage drop across R_E and, in turn, current I_{E3}. However, the voltage drops across D_1 and D_2 also decrease, causing a greater portion of I_2 to contribute to I_D, that is, causing I_D to increase. This causes I_{B3} to decrease, which prevents any significant increase in I_{E3}.

For better performance, a transistor array such as CA3086 may be used in Figure 1-14(a) as a constant current bias. Here an isolated transistor is used, and

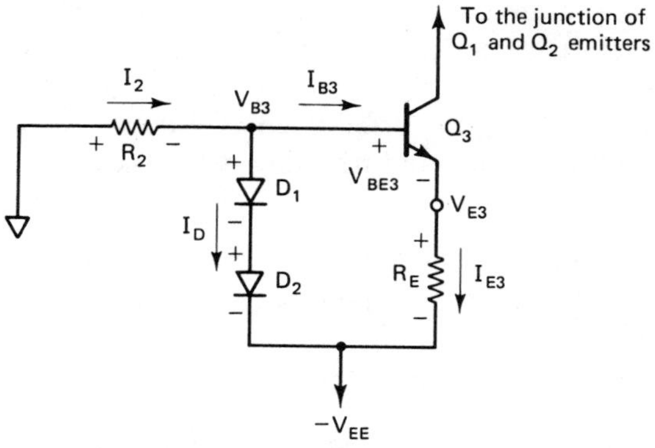

(a)

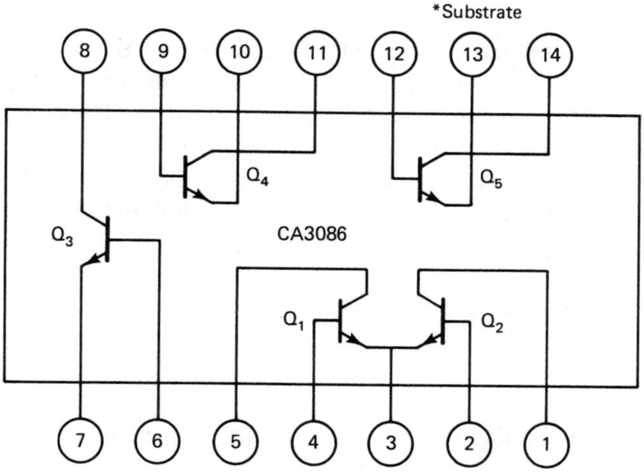

(b)

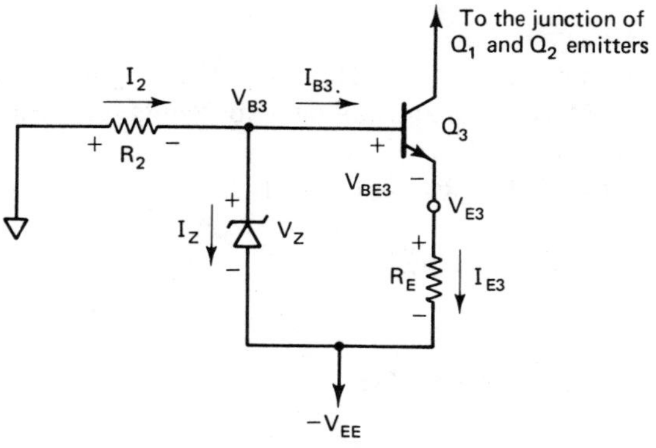

(c)

Figure 1-14 (a) Constant current bias with diode compensation for variations in V_{BE}. (b) Functional diagram of CA3086. (c) Constant current bias using zener.

needed diodes are formed by using transistors connected for diode operation, although discrete diodes will work as well. The functional diagram of the CA3086 is shown in Figure 1-14(b).

The emitter current I_{E3} in Figure 1-14(a) is determined as follows. Assuming that the voltage drop across the diodes is equal and denoting it by V_D, the voltage at the base of transistor Q_3 is

$$V_{B3} = -V_{EE} + 2V_D$$

$$V_{E3} = V_{B3} - V_{BE3} = -V_{EE} + 2V_D - V_{BE3}$$

$$I_{E3} = \frac{V_{E3} - (-V_{EE})}{R_E} = \frac{2V_D - V_{BE3}}{R_E}$$

Assuming that the transistor Q_3 has the same characteristics as diodes D_1 and D_2, that is, if $V_D = V_{BE3}$, then

$$I_{E3} = \frac{V_D}{R_E} \tag{1-38}$$

Thus for a given value of R_E the emitter current I_{E3} depends on the voltage drop across the diodes. But the voltage drop across the diodes is a function of the current (I_D) through them (according to the diode characteristic curve). However, current I_D is a part of current I_2, which is determined by the value of R_2. Obviously, the smaller the value of R_2, the larger is the drop across the diodes and, in turn, the greater is the emitter current I_{E3} for a given value of R_E. This means that we can change the value of I_{E3} by varying either the R_2 value or the R_E value [see Equation (1-38)].

To design the constant current bias circuit of Figure 1-14(a), we will use the following steps:

1. Choose a desired I_{E3} value.
2. Assuming that $V_D = 0.7$ V, determine the value of R_E using Equation (1-38).
3. Assuming that $I_2 = I_{E3}$ and $V_{D1} = V_{D2} = 0.7$ V, determine the value of R_2 from the following relation:

$$R_2 = \frac{V_{EE} - 1.4}{I_{E3}} \tag{1-39}$$

Very often a zener diode is used in place of the diodes as shown in Figure 1-14(c). zeners are available over a wide range of voltages and can have matching temperature coefficient of voltage to those of transistors.

Referring to the circuit of Figure 1-14(c), the voltage at the base of transistor Q_3 is

$$V_{B3} = -V_{EE} + V_Z$$

Then the voltage at the emitter is

$$V_{E3} = -V_{EE} + V_Z - V_{BE3}$$

Therefore, the current I_{E3} is

$$I_{E3} = \frac{V_{E3} - (-V_{EE})}{R_E}$$

Substituting and simplifying, we obtain

$$I_{E3} = \frac{V_Z - V_{BE3}}{R_E} \tag{1-40}$$

The value of R_2 should be selected so that $I_2 \cong 1.2I_{Zt}$, where I_{Zt} is the minimum current required to cause the zener diode to conduct in the reverse region, that is, to block the rated voltage V_Z. The value of I_{Zt} is specified on the data sheets of a zener diode. Thus the value of R_2 can be calculated using the equation

$$R_2 = \frac{V_{EE} - V_Z}{I_2} \tag{1-41}$$

where $I_2 \cong 1.2I_{Zt}$. The conclusion is quite simple: The zener diode is useful for maintaining a constant base voltage and in turn the constant emitter current in a constant current bias circuit.

1-11 CURRENT MIRROR

The circuit in which the output current is forced to equal the input current is said to be a current mirror circuit. Thus in a current mirror circuit the output current is a mirror image of the input current. The block diagram and the schematic circuit of the current mirror is shown in Figure 1-15(a) and (b), respectively. Once the current I_2 is set up, the current I_{C3} is automatically established to be nearly equal to I_2. The current mirror is a special case of constant current bias and therefore can be used to set up constant emitter currents in differential amplifier stages. Note that the current mirror circuit requires fewer components than constant current bias circuits. Because

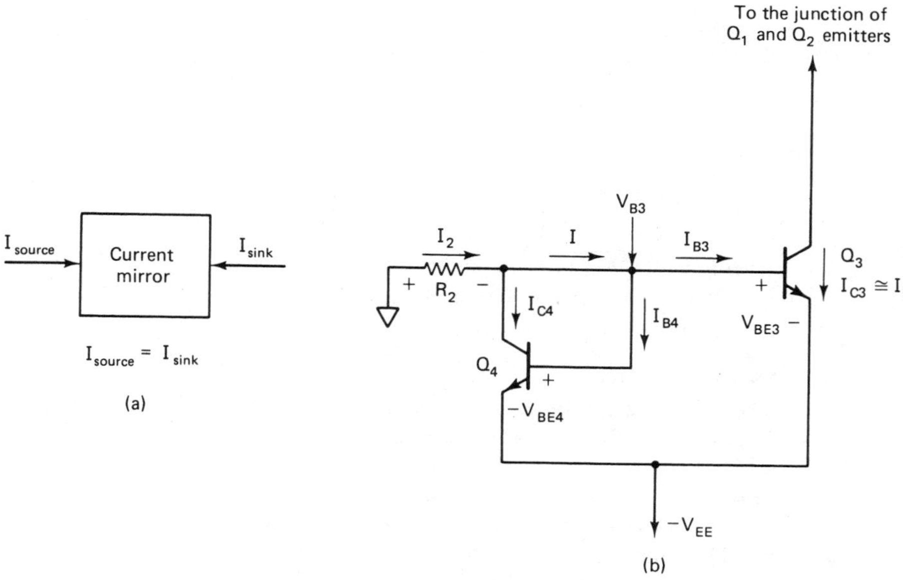

Figure 1-15 (a) Block diagram of current mirror. (b) Current mirror circuit.

of its simplicity and ease of fabrication, the current mirror circuit is most commonly used in integrated amplifiers such as differential and operational amplifiers.

Remember that for the proper operation of the circuit it is necessary to construct a current mirror using a transistor array such as the CA3086 in which a needed diode is formed by using an adjacent transistor [see Figure 1-15(b)]. When used this way, the transistor array helps to achieve the required collector current thermal stability in a current mirror circuit.

Since Q_3 and Q_4 are identical transistors, their base–emitter voltages must be the same, and their base and collector currents must also be approximately equal:

$$V_{BE3} = V_{BE4}$$
$$I_{C3} = I_{C4}$$
$$I_{B3} = I_{B4}$$

Summing currents at node V_{B3}, we obtain

$$\begin{aligned} I_2 &= I_{C4} + I \\ &= I_{C4} + 2I_{B4} = I_{C3} + 2I_{B3} \\ &= I_{C3} + 2\left(\frac{I_{C3}}{\beta_{dc}}\right) \\ &= I_{C3}\left(1 + \frac{2}{\beta_{dc}}\right) \end{aligned} \tag{1-42}$$

Generally, β_{dc} is large enough so that $2/\beta_{dc}$ is negligibly small. For instance, $\beta_{dc} =$ 100 for the CA3086. Thus the collector current of Q_3 is nearly equal to the current I_2.

$$I_2 \cong I_{C3} \tag{1-43}$$

The current I_2 is obtained by writing the Kirchhoff's voltage equation for the base–emitter loop of the transistor Q_3:

$$-R_2 I_2 - V_{BE3} + V_{EE} = 0$$

Rearranging, we get

$$I_2 = \frac{V_{EE} - V_{BE3}}{R_2} \tag{1-44}$$

Thus, using Equation (1-42), an appropriate value of R_2 can be chosen to set a desired value of collector current in a current mirror circuit. Remember that for satisfactory operation of the circuit, two identical transistors are necessary.

Example 1-5

The following specifications are given for the JFET differential amplifier shown in Figure 1-16: $R_D = 3.9$ kΩ, $R_G = 1$ MΩ, $R_E = 1.5$ kΩ, $R_2 = 82$ Ω, $R_{in1} = R_{in2} =$ 50 Ω, $V_{DD} = +10$ V, and $-V_{EE} = -10$ V. Assume that Q_1 and Q_2 are matched JFETs with $g_m = 8000$ μmhos, Q_3 is a BJT with $V_{BE} = 0.6$ V and $\beta_{ac} = \beta_{dc} = 100$, and D_1 is the 1N3826 zener with $V_Z = 5.1$ V and $I_{Zt} = 49$ mA. Determine:

(a) The emitter current in transistor Q_3.
(b) The voltage gain of the differential amplifier.
(c) The input resistance of the differential amplifier.

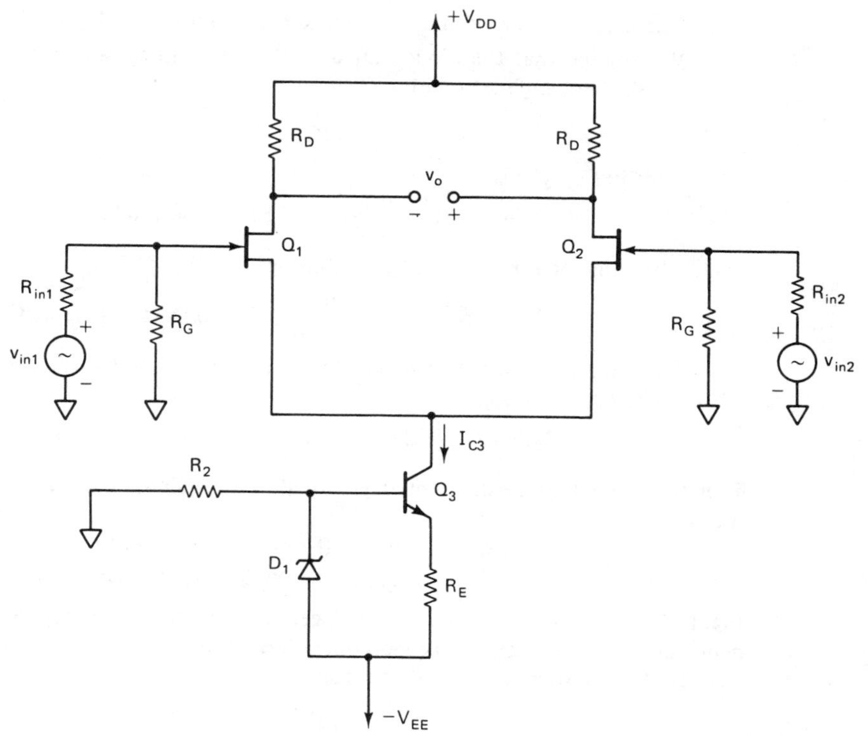

Figure 1-16 JFET differential amplifier with zener constant current bias.

Solution

(a) Using Equation (1-40), we can determine the value of emitter current in transistor Q_3.

$$I_{E3} = \frac{V_Z - V_{BE3}}{R_E} = \frac{5.1 - 0.6}{1.5\ \text{k}\Omega} = 3\ \text{mA}$$

(b) The voltage gain of the JFET dual-input, balanced-output differential amplifier is given by

$$A_d = g_m R_D = (8000\ \mu\text{mhos})(3.9\ \text{k}\Omega) = 31.2$$

(c) The input resistance seen from either input source is equal to the gate resistance R_G. That is,

$$R_{i1} = R_{i2} = 1\ \text{M}\Omega$$

Example 1-6

Design a dual-input, balanced-output differential amplifier with a constant current bias (using diodes) to satisfy the following requirements:

Differential voltage gain $A_d = 40 \pm 10$

Current supplied by the constant current bias circuit $= 4$ mA

Supply voltages $V_S = \pm 10$ V

Solution. First, we design a constant current bias circuit following the steps outlined in Section 1-10.

1. The desired value of emitter current $I_{E3} = 4$ mA (see Figure 1-17).
2. We assume that the diodes D_1 and D_2 are identical and that $V_{D1} = V_{D2} =$ 0.7 V. Rearranging Equation (1-38), we get

$$R_E = \frac{V_D}{I_{E3}}$$

Substituting yields

$$R_E = \frac{0.7 \text{ V}}{4 \text{ mA}} = 175\ \Omega \quad (\text{use } 150\ \Omega)$$

3. The value of R_2 can be calculated using Equation (1-39):

$$R_2 = \frac{V_{EE} - 1.4}{I_{E3}} = \frac{10 - 1.4}{4 \text{ mA}} = 2.15 \text{ k}\Omega \quad (\text{use } 2 \text{ k}\Omega)$$

Next, we determine the value of the collector resistor R_C using Equation (1-12). Since $I_{E3} = 4$ mA,

$$I_{E1} = I_{E2} = 2 \text{ mA} \qquad \text{and} \qquad r_e = \frac{25 \text{ mV}}{2 \text{ mA}} = 12.5\ \Omega$$

Therefore, to obtain the differential gain of 40, the required value of the collector resistor is

$$R_C = (A_d)r_e \qquad [\text{from Equation (1-12)}]$$
$$= (40)(12.5) = 500\ \Omega \quad (\text{use } 560\ \Omega)$$

Figure 1-17 shows the dual-input, balanced-output differential amplifier with component values. Note that for proper operation of the amplifier circuit the use of a transistor array such as CA3086 is recommended.

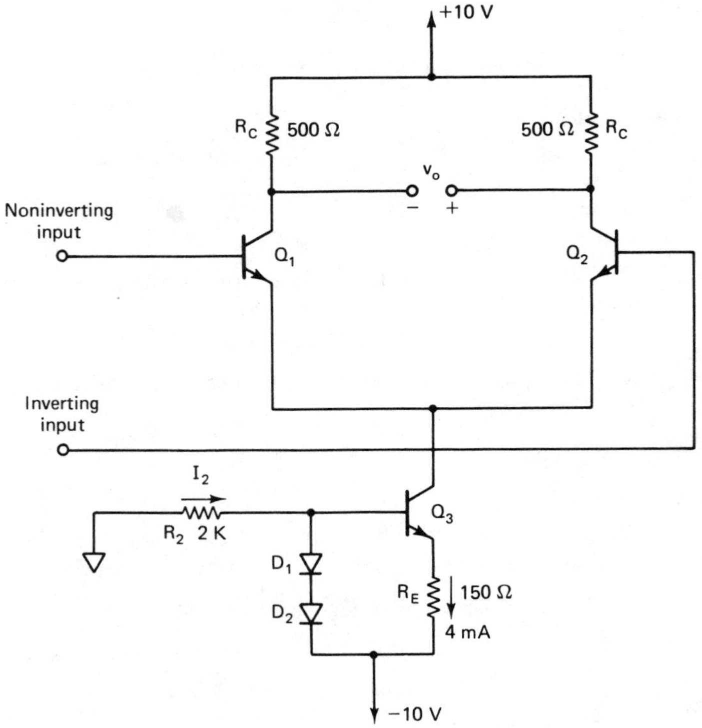

Figure 1-17 Differential amplifier of Example 1-6.

Example 1-7

Design the single-input, balanced-output differential amplifier shown in Figure 1-18(a) to meet the following specifications:

$$\text{Input resistance } R_i \geq 600 \text{ k}\Omega$$

$$\text{Peak-to-peak output voltage swing} \leq 5 \text{ V}$$

$$\text{Supply voltages } V_S = \pm 10 \text{ V}$$

Solution. The circuit shown in Figure 1-18(a) requires two Darlington transistor pairs. We will choose the CA3018 or SK3542 silicon transistor array, which consist of a Darlington-connected transistor pair and two isolated transistors, as shown in Figure 1-18(b). The isolated transistors can be used to form the current mirror circuit. Remember that we need two transistor arrays. The following specifications apply for each transistor in the CA3018 transistor array: $h_{FE} = 100$, $h_{fe} = 110$, and $V_{BE} = 0.715$ V. The desired input resistance for the single-input, balanced-output configuration is 600 kΩ. Therefore, using Equation (1-30), for the Darlington pair

$$R_i = 2\beta_{acT}(2r_{e3})$$

where β_{acT} = total current gain of the Darlington pair
$= (110)(110) = 12.1$ k
$2r_{e3}$ = ac emmiter resistance of the Darlington pair $Q_3Q_4 = Q_5Q_6$

Substituting yields

$$600 \text{ k}\Omega = (2)(12.1 \text{ k})(2r_{e3})$$

Therefore, the ac emitter resistance of transistor Q_3 ($= Q_5$) is

$$r_{e3} = \frac{600 \text{ k}\Omega}{48.2 \text{ k}} = 12.4 \ \Omega$$

But

$$r_{e3} = \frac{25 \text{ mV}}{I_{E3} \text{ mA}}$$

Therefore,

$$I_{E3} = \frac{25 \text{ mV}}{12.4\Omega} = 2.02 \text{ mA}$$

Let us choose $I_{E3} = I_{E5} = 2$ mA. This means that the current mirror circuit must supply $2I_{E3} = 4$ mA. To form the current mirror circuit we can use two isolated transistors in the CA3018, and the value of R_2 needed can be calculated by using Equation (1-44):

$$R_2 = \frac{V_{EE} - V_{BE3}}{I_2}$$

Substituting, we obtain

$$R_2 = \frac{10 - 0.715}{4 \text{ mA}} = 2.32 \text{ k}\Omega \quad (\text{use } 2.2 \text{ k}\Omega)$$

Assume that the voltage drop across collector resistor R_C determines the peak-to-peak output voltage swing. The expected output voltage swing in the circuit of Figure 1-18(a) is 5 V pp. Since the output is "differential," the voltage drop across each collector resistor is 2.5 V pp or ±1.25 V. In other words, $R_C I_C = 1.25$ V. Substituting, $I_C = 2$ mA we get

$$R_C = 625 \ \Omega \quad (\text{use } 620 \ \Omega)$$

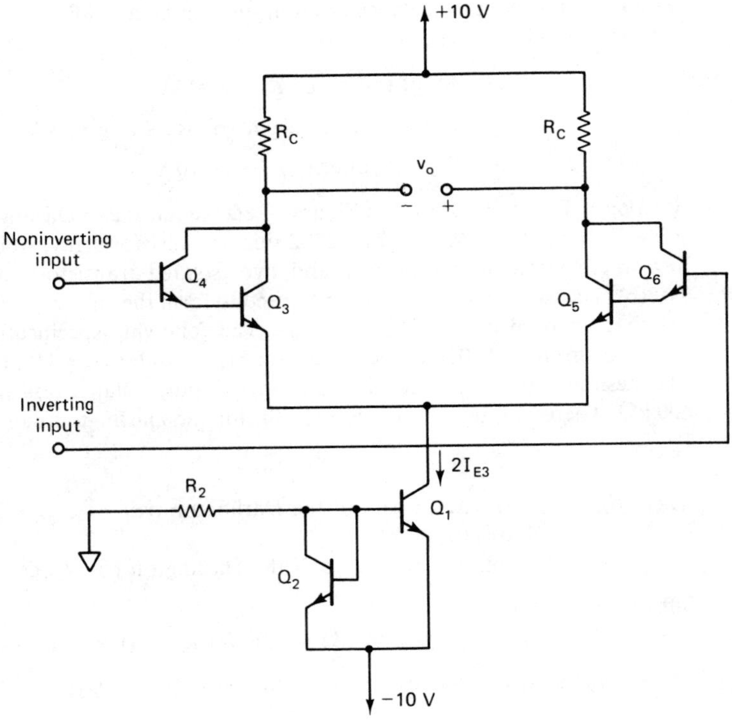

(a)

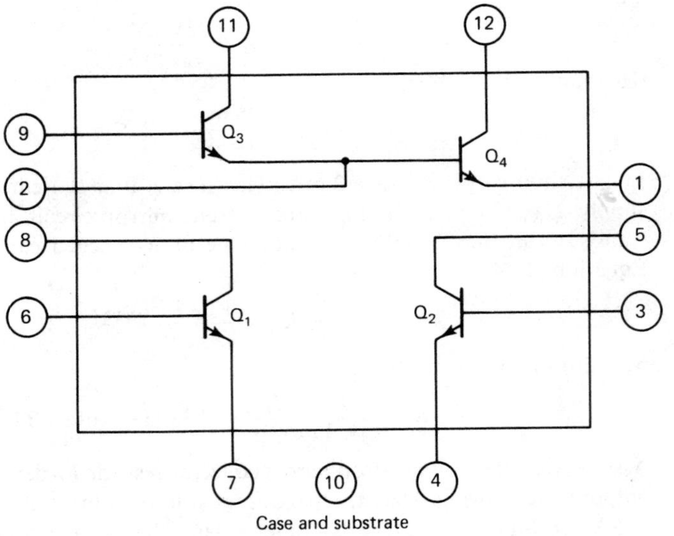

(b)

Figure 1-18 (a) Darlington-pair differential amplifier of Example 1-7. (b) CA3018 transistor array (metal can).

Thus for the single-input, balanced-output differential amplifier of Figure 1-18(a), we need the following components:

Two transistor arrays: CA3018

$R_C = 620\ \Omega$

$R_2 = 2.2\ k\Omega$

1-12 CASCADED DIFFERENTIAL AMPLIFIER STAGES

Having gone through an analysis of the differential amplifier circuit configurations, let us see how differential amplifiers can be cascaded (connected in series). Figure 1-19 shows a typical two-stage differential amplifier. The first stage is a dual-input, balanced-output differential amplifier. The second stage is another differential amplifier driven by the output of the first differential amplifier. A single-ended (unbalanced) output is taken from this second stage. Both stages use the emitter biasing technique (the combination of R_E and V_{EE}) to set up the emitter currents in the differential pairs.

In many cases the matching of transistors as well as resistor values is essential for proper operation of the differential stages. For this reason the use of transistor arrays such as CA3086 (or SK3543) is recommended. A transistor array such as CA3086 provides inherent advantages unique to integrated circuits: electrical and thermal matching, compactness, and ease of physical handling.

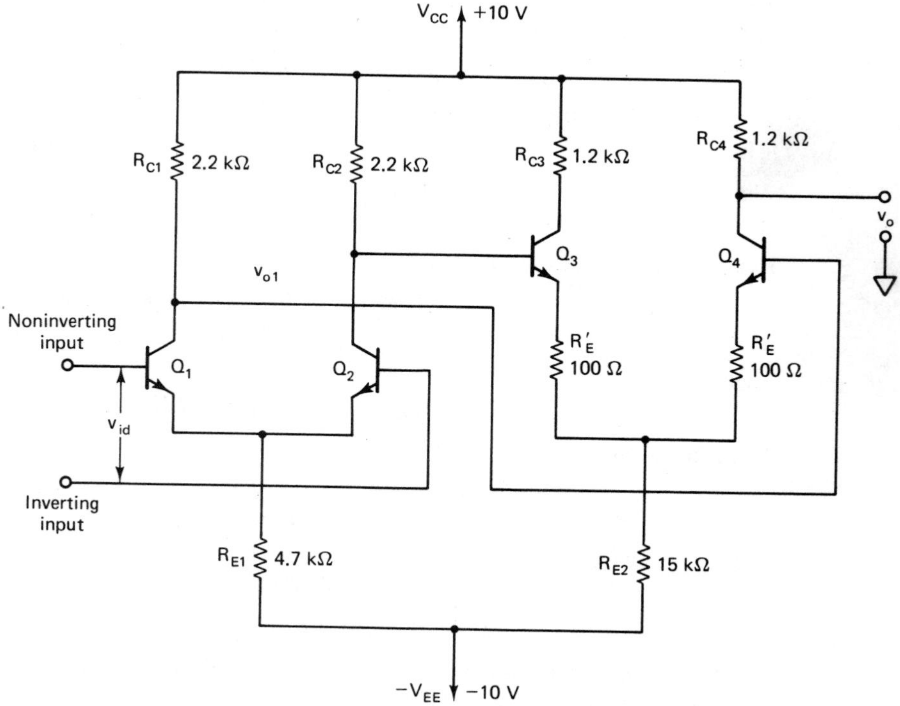

Figure 1-19 Cascaded differential amplifier of Example 1-8.

Example 1-8

For the cascaded differential amplifier shown in Figure 1-19, determine:
(a) The collector current and collector-to-emitter voltage for each transistor.
(b) The overall voltage gain.
(c) The input resistance.
(d) The output resistance.
Assume that the transistor array CA3086 is used (for which $h_{FE} = 100$ typical, $h_{fe} = 100$, and $V_{BE} = 0.715$ V). The pin diagram of the CA3086 is shown in Figure 1-14(b).

Solution

(a) To determine the collector current and collector-to-emitter voltage of transistors Q_1 and Q_2, we assume that the inverting and noninverting inputs are grounded. The collector current ($I_C \cong I_E$) in Q_1 and Q_2 is obtained by substituting the known values in Equation (1-2):

$$I_{E1} = \frac{V_{EE} - V_{BE}}{2R_{E1} + R_{\text{in}}/\beta_{\text{dc}}} = \frac{10 - 0.715}{2(4.7\text{ k}\Omega) + 0} = 0.988\text{ mA}$$

That is, $I_{C1} = I_{C2} = 0.988$ mA.

We can calculate the voltage between collector and emitter for Q_1 and Q_2 using the collector current as follows:

$$V_{C1} = V_{CC} - R_{C1}I_{C1} = 10 - (2.2\text{ k}\Omega)(0.988\text{ mA}) = 7.83\text{ V} = V_{C2}$$

Since the voltage at the emitter of Q_1 and Q_2 is -0.715 V,

$$V_{CE1} = V_{CE2} = V_{C1} - V_{E1} = 7.83 + 0.715 = 8.545\text{ V}$$

Next, we will determine the collector current in Q_3 and Q_4 by writing the Kirchhoff's voltage equation for the base–emitter loop of the transistor Q_3:

$$V_{CC} - R_{C2}I_{C2} - V_{BE3} - R'_E I_{E3} - R_{E2}(2I_{E3}) + V_{EE} = 0$$

$$10 - (2.2\text{ k}\Omega)(0.988\text{ mA}) - 0.715 - (100)(I_{E3}) - (30\text{ k}\Omega)I_{E3} + 10 = 0$$

$$10 - 2.17 - 0.715 + 10 - (30.1\text{ k}\Omega)I_{E3} = 0$$

$$I_{E3} = \frac{17.12}{30.1\text{ k}\Omega} = 0.569\text{ mA} = I_{E4}$$

Hence the voltage at the collector of Q_3 and Q_4 is

$$V_{C3} = V_{C4} = V_{CC} - R_{C3}I_{C3} = 10 - (1.2\text{ k}\Omega)(0.569\text{ mA})$$
$$= 9.32\text{ V}$$

Therefore,

$$V_{CE3} = V_{CE4} = V_{C3} - V_{E3} = 9.32 - 7.12 = 2.2\text{ V}$$

Thus, for Q_1 and Q_2:

$$I_{CQ} = 0.988\text{ mA}$$
$$V_{CEQ} = 8.545\text{ V}$$

and for Q_3 and Q_4:

$$I_{CQ} = 0.569\text{ mA}$$
$$V_{CEQ} = 2.2\text{ V}$$

[Note that the output terminal (V_{C4}) is at 9.32 V and not at zero volts.]
(b) First, we calculate the ac emitter resistance r_e of each stage and then its voltage gain.

$$r_{e1} = \frac{25 \text{ mV}}{I_{E1}} = \frac{25 \text{ mV}}{0.988 \text{ mA}} = 25.3 \ \Omega = r_{e2}$$

$$r_{e3} = \frac{25 \text{ mV}}{I_{E3}} = \frac{25 \text{ mV}}{0.569 \text{ mA}} = 43.94 \ \Omega = r_{e4}$$

The first stage is a dual-input, balanced-output differential amplifier; therefore, its voltage gain is

$$A_{d1} = \frac{v_{o1}}{v_{id}} = \frac{R_{C1} \parallel R_{i2}}{r_{e1}}$$

where

$$\begin{aligned} R_{i2} &= \text{input resistance of the second stage} \\ &= 2\beta_{ac}(r_{e3} + R'_E) \\ &= (200)(143.94) = 28.79 \text{ k}\Omega \end{aligned}$$

$$A_{d1} = \frac{2.2 \text{ k}\Omega \parallel 28.79 \text{ k}\Omega}{25.3} = 80.78$$

The second stage is a dual-input, unbalanced-output differential amplifier with swamping resistor R'_E, the voltage gain of which is

$$A_{d2} = \frac{v_o}{v_{o1}} = \frac{R_{C4}}{2(R'_E + r_{e4})} = \frac{1.2 \text{ k}\Omega}{287.88} = 4.17$$

Hence the overall voltage gain is

$$A_d = (A_{d1})(A_{d2}) = (80.78)(4.17) = 336.85$$

Thus we can obtain a higher voltage gain by cascading differential amplifier stages.
(c) The input resistance of the cascaded differential amplifier is the same as the input resistance of the first stage, that is,

$$R_i = 2\beta_{ac}(r_{e1}) = (200)(25.3) = 5.06 \text{ k}\Omega$$

(d) The output resistance of the cascaded differential amplifier is the same as the output resistance of the last stage. Hence, using Equation (1-17),

$$R_o = R_C = 1.2 \text{ k}\Omega$$

1-13 LEVEL TRANSLATOR

From the results of the cascaded differential amplifier of Example 1-8, the following observations can be made:

1. Because of the direct coupling, the dc level at the emitters rises from stage to stage. This increase in dc level tends to shift the operating point of the succeeding stages and, therefore, limits the output voltage swing and may even distort the output signal. For example, in the circuit of Figure 1-19, the voltage at the emitters of Q_1 and Q_2 of the first stage is -0.715, whereas that at the emitters of Q_3 and Q_4 of second stage is 7.12 V.

If used as a single stage, the voltage at the emitters of Q_3 and Q_4 would have been -0.715 V. Thus the increase in the emitter voltage of the second stage and in turn the change in its operating point are due to the cascading of differential amplifier stages.

2. The voltage at the output terminal of the second stage in Figure 1-19 is well above ground. This dc level is undesirable because it tends to limit the peak-to-peak

output voltage swing without distortion and also contributes to the error in the dc output signal.

Therefore, in Figure 1-19 a final stage should be included to shift the output dc level at the second stage down to about zero volts to ground. Such a stage is referred to as a *level translator* or *shifter*. Thus in the cascaded differential amplifier to shift the output dc level down to zero volts, the final stage must be followed by a level translator circuit. Although there are a variety of level translator circuits, an emitter follower with a voltage divider is the simplest among them [see Figure 1-20(a)]. The output of the second differential stage of Example 1-8 can work into the voltage-divider emitter-follower stage of Figure 1-20(a). Thus a positive 9.32 at the output terminal (V_{C4}) of the second stage in Figure 1-19 can produce zero volts at the junction

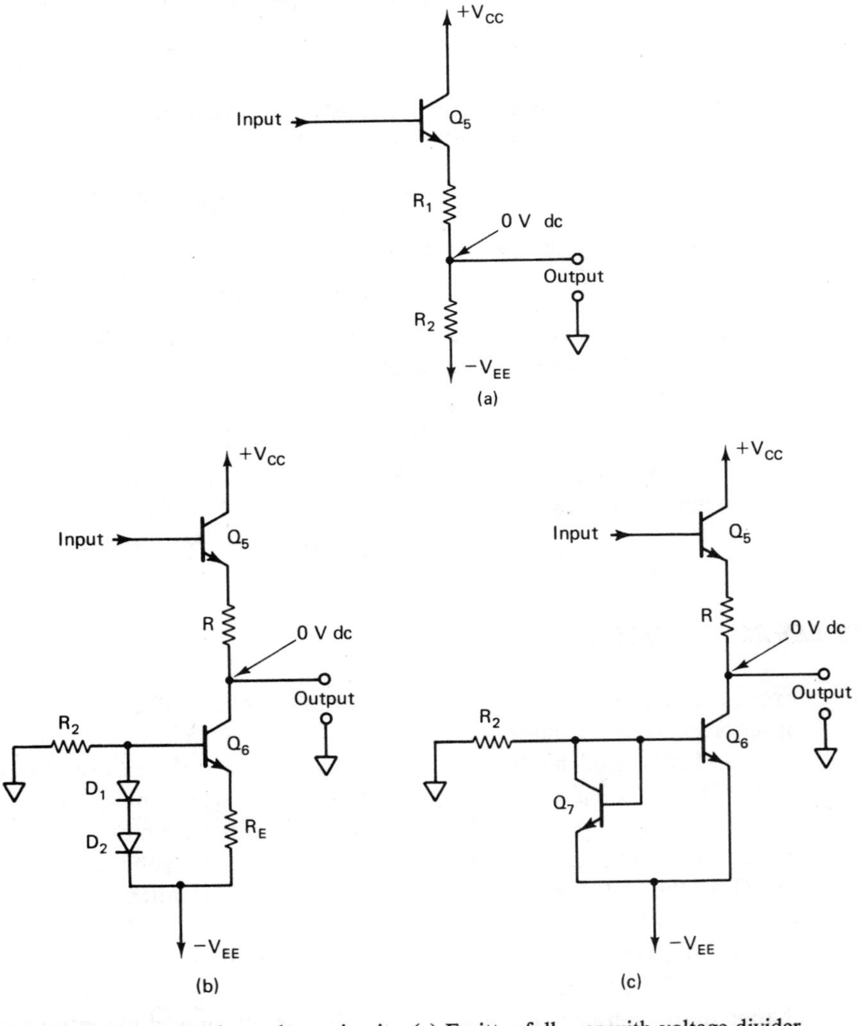

Figure 1-20 Level translator circuits. (a) Emitter follower with voltage divider. (b) Emitter follower with constant current bias. (c) Emitter follower with current mirror.

of R_1 and R_2 with proper selection of components. Better results are obtained by using an emitter follower either with a diode constant current bias or a current mirror instead of the voltage divider, as shown in Figure 1-20(b) and (c), respectively. Note that the resistive or zener constant current bias can also be used with the emitter follower in forming a level translator.

Example 1-9

Design a level translator circuit as shown in Figure 1-20(b) for the cascaded differential amplifier of Example 1-8.

Solution. We need to shift 9.32 V at the collector of Q_4 down to zero volts at the collector of Q_6 [see Figures 1-19 and 1-20(b)]. Following the design steps listed in Section 1-10, we first design the diode constant current bias circuit and then select an appropriate value for R so that the voltage at the collector of Q_6 is zero volts. Let us choose $I_{E6} = 3$ mA and $V_S = \pm 10$ V. Therefore, using Equation (1-38), the value of R_E is

$$R_E = \frac{V_D}{I_{E6}} = \frac{0.7}{3\text{ mA}} = 233.3\ \Omega \quad (\text{use } 270\ \Omega)$$

Substituting the known values in Equation (1-39) yields

$$R_2 = \frac{V_{EE} - 1.4}{I_{E6}} = \frac{10 - 1.4}{3\text{ mA}} = 2.87\text{ k}\Omega\ (\text{use } 2.7\text{ k}\Omega)$$

Assuming that $V_{BE5} = 0.7$ V, the voltage at the emitter of Q_5 is

$$V_{E5} = V_{B5} - V_{BE5} = 9.32 - 0.7 = 8.62\text{ V}$$

Since the voltage at the collector Q_6 is to be zero and $I_{E5} = I_{C6} = 3$ mA, the required value of R is

$$R = \frac{V_{E5}}{I_{E5}} = \frac{8.62}{3\text{ mA}} = 2.87\text{ k}\Omega \quad (\text{use a 3-k}\Omega\text{ potentiometer})$$

Thus the desired level translator circuit components are:

$$R = 3\text{-k}\Omega\text{ potentiometer}$$
$$R_E = 270\ \Omega$$
$$R_2 = 2.7\text{ k}\Omega$$
$$D_1, D_2, Q_5 \text{ and } Q_6 = \text{CA3086}$$

Example 1-10

For the cascaded differential amplifier of Figure 1-21, determine:
(a) The I_{CQ} and V_{CEQ} for each transistor.
(b) The overall voltage gain.
(c) The maximum peak-to-peak output voltage swing.
(Assume that the CA3086 transistor array is used for which $\beta_{ac} = \beta_{dc} = 100$ and $V_{BE} = 0.715$ V.)

Solution. The circuit in Figure 1-21 consists of three stages:

1. Dual-input, balanced-output differential amplifier
2. Dual-input, unbalanced-output differential amplifier
3. Level translator circuit

All the stages use current mirror biasing. Especially notice the way current mirror transistors Q_4, Q_7, and Q_9 are connected. The bases of these transistors are tied together to the base of transistor Q_3. For this reason the collector currents in these transistors (Q_3, Q_4, Q_7, and Q_9) are equal and are set up by a 2.2-kΩ base resistor (see Figure 1-21).

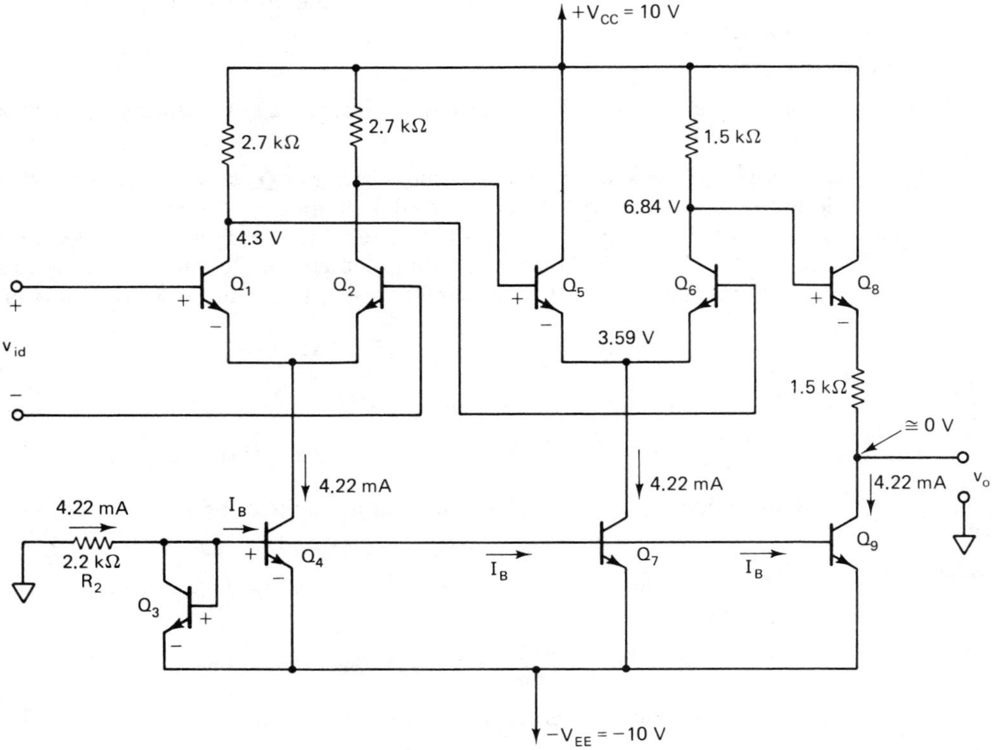

Figure 1-21 Cascaded differential amplifier of Example 1-10.

Another interesting point about this circuit is that there is no collector resistor used for Q_5 since the output is single-ended and is measured at the collector of transistor Q_6. To determine the dc conditions, we assume that the input terminals are shorted to ground.

(a) We can determine the collector currents in current mirror transistors Q_4, Q_7, and Q_9 using Equation (1-44):

$$I_2 = \frac{V_{EE} - V_{BE3}}{R_2}$$

Substituting gives us

$$I_2 = \frac{10 - 0.715}{2.2\text{ k}\Omega} = 4.22\text{ mA}$$

This means that

$$I_{CQ1} = I_{CQ2} = I_{CQ5} = I_{CQ6} = 2.11\text{ mA}$$

$$I_{CQ8} = 4.22\text{ mA} \qquad \text{since } I_{E8} = I_{C9}$$

Knowing these collector currents, we can calculate the collector-to-emitter voltages as follows:

$$V_{C1} = V_{C2} = V_{CC} - R_{C1}I_{C1} = 10 - (2.7\ \text{k}\Omega)(2.11\ \text{mA}) = 4.3\ \text{V}$$

Therefore,

$$V_{CEQ1} = V_{CEQ2} = 4.3 - (-0.715) = 5.015\ \text{V}$$
$$V_{E5} = V_{E6} = (4.3 - 0.715) = 3.59\ \text{V}$$
$$V_{C6} = V_{CC} - R_{C6}I_{C6} = 10 - (1.5\ \text{k}\Omega)(2.11\ \text{mA}) = 6.84\ \text{V}$$

Hence

$$V_{CE5} = 10 - 3.59 = 6.41\ \text{V}$$
$$V_{CE6} = 6.84 - 3.59 = 3.25\ \text{V}$$

Finally,

$$V_{E8} = 6.84 - 0.715 = 6.125\ \text{V} \qquad \text{and} \qquad V_{C8} = 10\ \text{V}$$

Therefore,

$$V_{CE8} = 10 - 6.125 = 3.875\ \text{V}$$

(b) The ac emitter resistances are

$$r_{e1} = r_{e6} = \frac{25\ \text{mV}}{I_{E1}} = \frac{25\ \text{mV}}{2.11\ \text{mA}} = 11.85\ \Omega$$
$$r_{e8} = \frac{25\ \text{mV}}{I_{E8}} = \frac{25\ \text{mV}}{4.22\ \text{mA}} = 5.92\ \Omega$$

and input resistance of the second stage is

$$R_{i2} = 2\beta_{ac}r_{e6} = (200)(11.85) = 2.37\ \text{k}\Omega$$

Therefore, the voltage gain of the first stage is

$$A_{d1} = \frac{R_{C1} \,||\, R_{i2}}{r_{e1}} = \frac{2.7\ \text{k}\Omega \,||\, 2.37\ \text{k}\Omega}{11.85} = 106.5$$

The input resistance of the third stage (looking into the base of Q_8) is

$$R_{i3} = \beta_{ac}(r_{e8} + R_{E8}) = (100)(5.92 + 1.5\ \text{k}\Omega) = 150.6\ \text{k}\Omega$$

Hence the voltage gain of the second stage is

$$A_{d2} = \frac{R_{C6} \,||\, R_{i3}}{2r_{e6}} = \frac{1.5\ \text{k}\Omega \,||\, 150.6\ \text{k}\Omega}{(2)(11.85)} = 62.66$$

The voltage gain of the third stage is

$$A_{d3} \cong 1$$

Thus the overall gain of the circuit in Figure 1-21 is

$$A_d = (A_{d1})(A_{d2})(A_{d3}) = (106.5)(62.66)(1) = 6673.3$$

(c) The third stage in Figure 1-21 is the level translator circuit and in addition, the voltage gain of this stage is unity. Therefore, the maximum peak-to-peak output voltage swing is determined by the voltage drop across the collector resistor of transistor Q_6. The voltage drop across collector resistor of Q_6 is

$$V_{1.5\text{k}\Omega} = (1.5\ \text{k}\Omega)(2.11\ \text{mA}) = 3.17\ \text{V}$$

Hence the maximum peak-to-peak output voltage swing is

$$v_o = \pm 3.17\ \text{V or } 6.34\ \text{V pp}$$

1-14 CASCODE OR CE-CB CONFIGURATION

Each of the small-signal amplifier configurations:—CE, CB, and CC has some unique advantages as well as drawbacks. The specific property of the amplifier configuration can be enhanced, a specific drawback can be removed, or a special feature can be added by connecting two configurations. Since there are three configurations, we can form nine possible combinations, each having two configurations. But unfortunately, out of these nine possible combinations, some do not improve the amplifier's performance in any way and are, therefore, not practical. The most commonly used combinational configurations are the following:

1. Common collector–common collector (CC-CC) connection
2. Common collector–common emitter (CC–CE) connection
3. Common collector–common base (CC–CB) connection
4. Common emitter–common emitter (CE–CE) connection
5. Common emitter–common base (CE–CB) connection

The Darlington pair can be used to obtain a CC-CC configuration. The Darlington pair is a composite two-transistor device in which the collectors of two transistors are tied together and the base of the second transistor is driven by the emitter of the first. The principal advantages of this configuration are that the input resistance and the current gain are increased by a factor β over those in the standard CC configuration.

Another example of the CC-CC configuration is the double follower circuit in which complementary transistors are connected in the CC-CC configuration. The main features of this circuit are that the output voltage is equal to the input voltage and that a very high current gain is provided.

To obtain the CC-CE configuration, we can use the Darlington pair in place of a single transistor in CE configuration. The principal advantage of this configuration is that its current gain and input resistance is improved by a factor β over that of the CE amplifier. Note that these increases are achieved at the sacrifice of a slight decrease in voltage gain.

The single-input, unbalanced-output differential amplifier of Figure 1-10 utilizes the CC-CB configuration. This configuration basically has the same input characteristics as the CC amplifier and the same output characteristics as the CB amplifier.

If the main objective is to obtain a very large voltage or power gain, we can use the CE-CE configuration. The complementary pair can also be utilized in forming the CE-CE configuration. Recall that in forming the dual-input, balanced-output differential amplifier, we also have to use two symmetrical CE sections (see Figure 1-2). Thus the CE-CE is the most commonly used configuration.

The CE-CB configuration (usually referred to as the *cascode amplifier*) is shown in Figure 1-22. Note that the CE and CB stages are direct-coupled. This configuration, of course, has basically the same input characteristics as the CE amplifier and the same output characteristics as the CB amplifier. More specifically, it has high output resistance and is inherently more stable. The high output resistance attainable is useful in achieving large amounts of voltage gain. Also, in this configuration no high-frequency feedback occurs from the output back to the input, and the input Miller capacitance effect is at a minimum because the voltage gain of the CE configuration is very low. Therefore, a CE-CB configuration is inherently more stable and hence ideally suited for high-frequency applications.

1-14.1 DC Analysis

In the cascode amplifier of Figure 1-22, R_1, R_2 and R_3 form a bias network for Q_1 and Q_2; C_{S1} and C_{S2} provide ac signal ground paths from the Q_1 emitter and Q_2 base, respectively; and C_i is the coupling capacitor. For the proper operation of the circuit we require that transistors Q_1 and Q_2 be identical. Keep in mind that for dc conditions all capacitors are assumed to be open-circuit.

Because of identical transistors and the direct coupling between the CE-CB stages,

$$I_{E1} = I_{E2} \quad \text{or} \quad I_{C1} = I_{C2} \qquad \text{and} \qquad I_{B1} = I_{B2}$$

This means that we need to determine the emitter current of Q_1 only. Applying the voltage-divider rule, we get

$$V_{R1} = \frac{R_1 V_{CC}}{R_1 + R_2 + R_3} \tag{1-45}$$

$$V_{E1} = V_{R1} - V_{BE1}$$

Therefore,

$$I_{E1} = \frac{V_{E1}}{R_E}$$

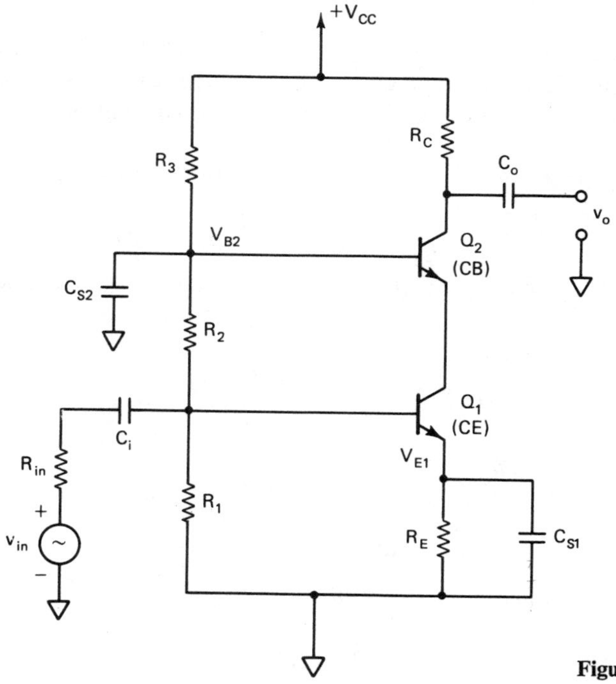

Figure 1-22 Cascode amplifier.

Substituting yields

$$I_{E1} = \frac{[(R_1)V_{CC}/(R_1 + R_2 + R_3)] - V_{BE1}}{R_E} \tag{1-46}$$

Thus once we know the emitter current of Q_1, we can determine all the currents in and voltages at the three transistor terminals for Q_1 and Q_2 if desired.

1-14.2 AC Analysis

The small-signal T-equivalent circuit for the cascode amplifier is shown in Figure 1-23. Since we are considering only the low-frequency performance, we assume all capacitors as short-circuit elements. We will determine the low-frequency, small-signal properties of the cascode circuit: the voltage gain, current gain, and input resistance.

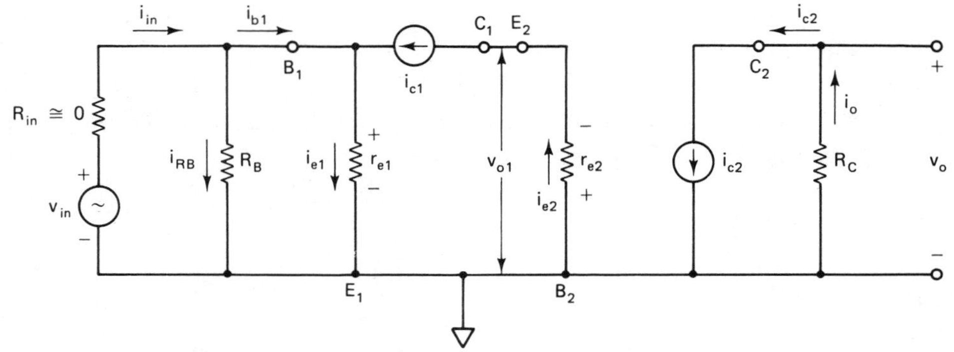

Figure 1-23 Small-signal T-equivalent circuit of cascode amplifier.

Note that in Figure 1-23, $R_B = R_1 \| R_2$; $r_{e1} \cong r_{e2}$ since $I_{E1} \cong I_{E2}$; and v_{o1} is the output voltage of the CE stage. Since Q_1 and Q_2 are identical transistors, $\beta_{dc1} = \beta_{dc2}$ and $\beta_{ac1} = \beta_{ac2}$; therefore, we will omit subscripts in the calculations that follow.

1-14.2(a) Voltage gain. For the input circuit in Figure 1-23,

$$v_{in} = (r_{e1})(i_{e1}) \qquad \text{since } R_B \gg r_{e1}$$

$$v_{o1} = -(r_{e2})(i_{e2})$$

Since $i_{e2} = i_{c1}$ and $i_{c1} \cong i_{e1}$, this implies that $i_{e2} \cong i_{e1}$. Therefore, the voltage gain A_v of the CE stage is

$$A_{v1} = \frac{v_{o1}}{v_{in}} = -\frac{(r_{e2})(i_{e2})}{(r_{e1})(i_{e1})} = -1 \qquad \text{since } r_{e1} = r_{e2} \tag{1-47}$$

The output voltage

$$v_o = -(R_C)(i_o)$$

but

$$i_o = i_{c2} = i_{e2} \qquad \text{and} \qquad i_{e2} = -\frac{v_{o1}}{r_{e2}}$$

Therefore, substituting the value of i_{e2} gives us

$$v_o = -(R_C)\left(-\frac{v_{o1}}{r_{e2}}\right)$$

Then the voltage gain of the CB stage is

$$\frac{v_o}{v_{o1}} = \frac{R_C}{r_{e2}} \tag{1-48}$$

Thus the overall voltage gain of the CE-CB configuration is

$$A_v = \frac{v_o}{v_{in}} = \frac{v_{o1}}{v_{in}}\frac{v_o}{v_{o1}} = -\frac{R_C}{r_{e2}} \tag{1-49}$$

Note that $r_{e1} = r_{e2}$.

1-14.2(b) Current gain. The current gain A_i of the CE-CB configuration can be expressed as

$$A_i = \frac{i_o}{i_{in}} = \frac{i_{b1}}{i_{in}}\left(\frac{i_{e1}}{i_{b1}}\right)\frac{i_{e2}}{i_{e1}}\left(\frac{i_{c2}}{i_{e2}}\right)\frac{i_o}{i_{c2}}$$

but

$$i_{e1} = i_{c1} = i_{e2} = i_{c2} = i_o$$

Therefore,

$$A_i = \frac{i_{b1}}{i_{in}}\frac{i_{e1}}{i_{b1}}$$

Applying the current-divider rule to the input circuit, we get

$$i_{b1} = \frac{R_B i_{in}}{R_B + \beta_{ac} r_{e1}}$$

where $R_B = R_1 \| R_2$

Also,

$$i_{e1} = \beta_{ac} i_{b1}$$

Therefore, the current-gain equation becomes

$$A_i = \frac{\beta_{ac} R_B}{R_B + \beta_{ac} r_{e1}} \tag{1-50}$$

1-14.2(c) Input resistance. For the input circuit

$$v_{in} = r_{e1} i_{e1} = \beta_{ac} r_{e1} i_{b1} \qquad \text{since } i_{e1} = \beta_{ac} i_{b1}$$

Substituting

$$i_{b1} = \frac{R_B i_{in}}{R_B + \beta_{ac} r_{e1}}$$

we have

$$v_{in} = \beta_{ac} r_{e1} \frac{R_B i_{in}}{R_B + \beta_{ac} r_{e1}}$$

Solving for the ratio v_{in}/i_{in}, we get for the input resistance:

$$R_i = \frac{v_{in}}{i_{in}} = \frac{\beta_{ac} r_{e1} R_B}{R_B + \beta_{ac} r_{e1}} \tag{1-51}$$

The cascode amplifier of Figure 1-22 is not encountered very often in discrete circuitry because of the following drawbacks:

1. The amplifier requires a large number of resistors and capacitors to provide bias and prevent undesirable feedback.
2. Unlike the integrated circuit (IC) cascode amplifier, it does not offer much

flexibility. For instance, the CA3095E is a differential cascode amplifier, which means that it can be used as a differential as well as a cascode amplifier. Some IC cascode amplifiers are even available with an automatic gain control (AGC) arrangement and a provision for thermal stability. Obviously, such amplifiers are usually a little more complex in structure than the basic circuit shown in Figure 1-22.

Example 1-11

The following specifications are given for the cascode amplifier of Figure 1-22.

$$R_1 = 27\text{ k}\Omega \qquad R_C = 1.8\text{ k}\Omega \qquad C_i = 10\ \mu\text{F}$$

$$R_2 = 39\text{ k}\Omega \qquad R_E = 1\text{ k}\Omega \qquad C_o = 10\ \mu\text{F}$$

$$R_3 = 47\text{ k}\Omega \qquad C_{s1} = 100\ \mu\text{F}$$

$$V_{CC} = 12\text{ V} \qquad C_{s2} = 100\ \mu\text{F}$$

Assume that the 2N3393 transistors are used with $\beta_{dc} = 140$, $\beta_{ac} = 125$, and $V_{BE} = 0.63$ V. Determine:

(a) The dc conditions for both transistors.
(b) The overall voltage gain.
(c) The overall current gain.
(d) The input resistance.

Solution

(a) Substituting in Equation (1-45), we get

$$V_{R1} = \frac{R_1 V_{CC}}{R_1 + R_2 + R_3} = \frac{(27\text{ k}\Omega)(12)}{27\text{ k}\Omega + 39\text{ k}\Omega + 47\text{ k}\Omega} = 2.87\text{ V}$$

Therefore,

$$V_{E1} = V_{R1} - V_{BE1} = 2.87 - 0.63 = 2.24\text{ V}$$

$$I_{E1} = \frac{V_{E1}}{R_E} = \frac{2.24\text{ V}}{1\text{ k}\Omega} = 2.24\text{ mA}$$

Thus,

$$I_{E1} = I_{C1} = I_{E2} = I_{C2} = 2.24\text{ mA}$$

$$V_{B2} = V_{CC} - \frac{R_3 V_{CC}}{R_1 + R_2 + R_3} = 12 - \frac{(47\text{ k}\Omega)(12)}{113\text{ k}\Omega} = 7.01\text{ V}$$

Hence,

$$V_{E2} = V_{C1} = V_{B2} - V_{BE2} = 7.01 - 0.63 = 6.38\text{ V}$$

$$V_{CE1} = V_{C1} - V_{E1} = 6.38 - 2.24 = 4.14\text{ V}$$

Since

$$V_{C2} = V_{CC} - R_C I_{C2} = 12 - (1.8\text{ k}\Omega)(2.24\text{ mA}) = 7.97\text{ V}$$

then

$$V_{CE2} = V_{C2} - V_{E2} = 7.97 - 6.38 = 1.59\text{ V}$$

(b) The ac emitter resistances of Q_1 and Q_2 are

$$r_{e1} = r_{e2} = \frac{25\text{ mV}}{I_{E1}} = \frac{25\text{ mV}}{2.24\text{ mA}} = 11.16\ \Omega$$

Substituting in Equation (1-49), we have the overall voltage gain

$$A_v = -\frac{R_C}{r_{e2}} = -\frac{1.8\text{ k}\Omega}{11.16} = 161.3$$

(c) $R_B = R_1 \| R_2 = 27\ \text{k}\Omega \| 39\ \text{k}\Omega = 15.95\ \text{k}\Omega$ and $\beta_{ac} = 125$
Substituting these values in Equation (1-50), we get

$$A_i = \frac{\beta_{ac} R_B}{R_B + \beta_{ac} R_B} = \frac{(125)(15.95\ \text{k}\Omega)}{15.95\ \text{k}\Omega + (125)(11.16\ \Omega)} = 114.9$$

(d) Input resistance is given by Equation (1-51):

$$R_i = \frac{\beta_{ac} r_{e1} R_B}{R_B + \beta_{ac} r_{e1}} = \frac{(125)(11.16)(15.95\ \text{k}\Omega)}{15.95\ \text{k}\Omega + (125)(11.16)} = 1.28\ \text{k}\Omega$$

SUMMARY

1. The differential amplifier consists of two symmetrical common emitter sections and is capable of amplifying the difference between two input signals. The differential amplifier can amplify ac as well as dc input signals because it employs direct coupling.
2. There are four types of differential amplifier configurations:
   (a) The dual input, balanced output
   (b) The dual input, unbalanced output
   (c) The single input, balanced output
   (d) The single input, unbalanced output
   These configurations are defined according to the number of input signals used and the way output is measured. When we use two inputs, the configuration is said to be dual input; otherwise, it is a single input. If the output voltage is measured between two collectors, the configuration is referred to as a balanced output. On the other hand, if the output is measured at either collector with respect to ground, the configuration is said to be unbalanced output.
3. For proper operation of the differential amplifier, a transistor array and matched components must be used. The differential amplifier can be biased by using emitter bias (a combination of R_E and V_{EE}), constant current bias, or current mirror techniques.
4. In a cascaded differential amplifier, the output of the first stage is used as an input for the second stage, the output of the second stage is applied as an input to the third stage, and so on. Because of direct coupling between the stages, the operating point of succeeding stages changes.
5. Generally, in a cascaded differential amplifier the last dual-input, unbalanced-output stage is followed by a level translator. The function of the level translator is to make the dc voltage at the output terminal zero.
6. The cascode amplifier is composed of direct-coupled common-emitter and common-base configurations. It provides a large amount of voltage gain at higher frequencies.

QUESTIONS

1-1. What is a differential amplifier?

1-2. What are the four differential amplifier configurations? Which one is not commonly used, and why?

1-3. What does the term "balanced output" mean?

1-4. What is the main advantage of constant current bias over emitter bias?

1-5. Explain the differences between constant current bias and current mirror.

1-6. What is a level translator circuit? Why is it used with the cascaded differential amplifier?

1-7. What is a cascode amplifier? List the characteristics of the cascode amplifier.

PROBLEMS

1-1. An emitter-biased dual-input, balanced-output differential amplifier has the following specifications: $|V_{CC}| = |-V_{EE}| = 10$ V, $R_{C1} = R_{C2} = 2.7$ kΩ, and $R_E = 3.9$ kΩ; the transistor array is CA3086 with $\beta_{ac} = \beta_{dc} = 100$ and $V_{BE} = 0.715$ V. Calculate:
(a) The operating current and voltage values for each transistor
(b) The voltage gain
(c) The input resistance of the circuit

1-2. Repeat Problem 1-1 for the dual-input, unbalanced-output differential amplifier.

1-3. Repeat Problem 1-1 with $R_E = 5.6$ kΩ.

1-4. Repeat Problem 1-1 with $R_{C1} = R_{C2} = 5.6$ kΩ.

1-5. Repeat Problem 1-3 for the single-input, balanced-output configuration.

1-6. Repeat Problem 1-4 for the dual-input, unbalanced-output configuration.

1-7. Repeat Problem 1-4 for the single-input, balanced-output configuration.

1-8. For the dual-input, balanced-output differential amplifier of Problem 1-1:
(a) Draw the output voltage waveform if $v_{in1} = 5$ mV pp and $v_{in2} = 2$ mV pp at 1 kHz. Assume that the collector C_1 is negative with respect to collector C_2 (see Figure 1-2).
(b) Determine the maximum peak-to-peak output voltage swing without clipping.

1-9. Repeat Problem 1-8 for the dual-input, unbalanced-output configuration.

1-10. Repeat Problem 1-8 if 180-Ω resistors R'_E are placed externally in series with each emitter (see Figure 1-12).

1-11. Design a zener constant current bias circuit as shown in Figure 1-14(c) according to the following specifications:
(a) Emitter current $I_{E3} = 5$ mA.
(b) 1N3825 Zener diode with $V_Z = 4.7$ V and $I_{Z_t} = 53$ mA.
(c) CA3086 transistor with $\beta_{ac} = \beta_{dc} = 100$ and $V_{BE} = 0.715$ V.
(d) Supply voltages $V_s = \pm 9$ V.

1-12. The JFET dual-input, balanced-output differential amplifier uses the Zener constant current bias circuit designed in Problem 1-11. Determine the voltage gain and the input resistance of the amplifier if $R_D = 3.9$ kΩ, $R_G = 1$ MΩ, and $g_m = 4000$ μmhos.

1-13. Design the dual-input, balanced-output differential amplifier using diode constant current bias to meet the following specifications:
(a) Supply voltages $(V_s) = \pm 12$ V.
(b) Emitter current (I_E) in each differential amplifier transistor $= 1.5$ mA.
(c) Voltage gain ≤ 60.

1-14. Design the dual-input, balanced-output differential amplifier with the current mirror bias according to the following specifications:
(a) Supply voltages $(V_s) = \pm 10$ V.
(b) Maximum output voltage swing $= 4$ V pp.

1-15. Design the single-input, balanced-output differential amplifier with diode constant current bias to meet the following specifications:

(a) Voltage gain ≤ 10.

(b) Input resistance $R_i \geq 20\ \text{k}\Omega$.

1-16. For the modified current mirror circuit shown in Figure 1-24, determine the emitter current in transistor Q_3. Assume that the CA3086 transistor array is used. For each transistor in the array, $\beta_{dc} = \beta_{ac} = 100$ and $V_{BE} = 0.715$ V.

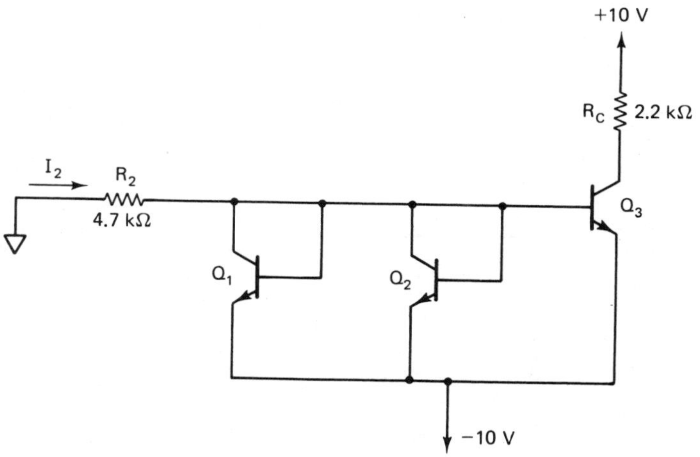

Figure 1-24 Modified current mirror circuit of Problem 1-16.

1-17. Referring to the circuit of Figure 1-25, determine:

(a) The current through collector resistor R_C.

(b) The collector current in each transistor.

Assume that the CA3086 transistor array is used.

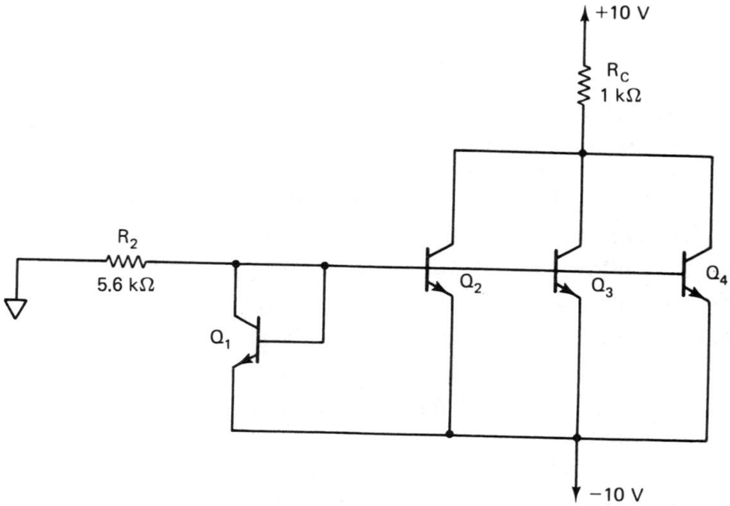

Figure 1-25 Modified current mirror circuit of Problem 1-17.

1-18. For the single-input, balanced-output differential amplifier shown in Figure 1-26, determine

(a) The operating point values (I_{CQ} and V_{CEQ}).

(b) The voltage gain.

(c) The input resistance.

(d) The maximum peak-to-peak output voltage.

Assume that $\beta_{ac} = \beta_{dc} = 100$, $V_{BE} = V_{D1} = 0.715$ V, $V_Z = 6.2$ V, and $I_{Zt} = 41$ mA.

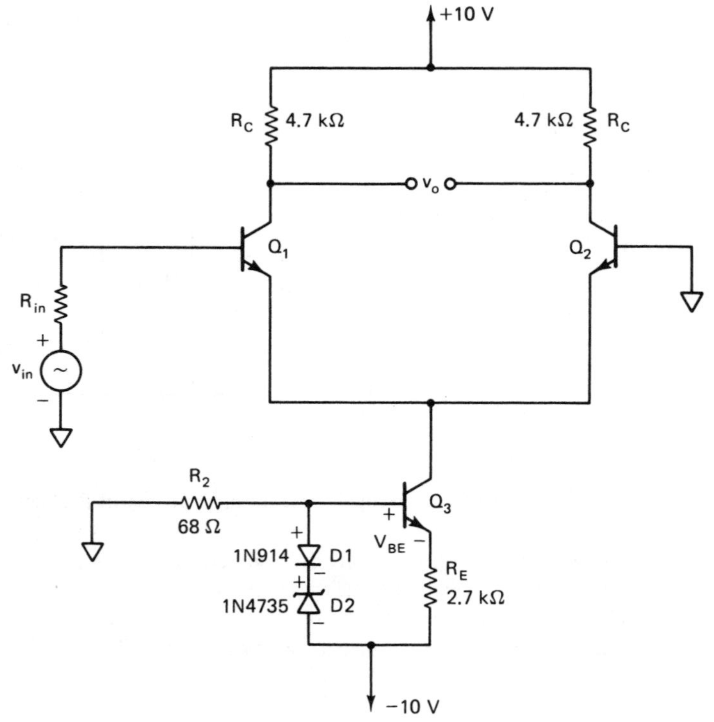

Figure 1-26 Differential amplifier of Problem 1-18.

1-19. For the cascaded differential amplifier of Figure 1-27, determine:

(a) The dc conditions for each stage.

(b) The overall voltage gain.

(c) The maximum peak-to-peak output voltage swing.

Assume that the CA3086 transistor array is used.

1-20. Repeat Problem 1-19 for the circuit shown in Figure 1-28.

1-21. Repeat Problem 1-19 for the circuit shown in Figure 1-29.

1-22. Repeat Problem 1-19 for the circuit shown in Figure 1-30.

1-23. Repeat Problem 1-19 for the circuit shown in Figure 1-31.

1-24. Repeat Problem 1-19 for the circuit shown in Figure 1-32. Assume that the input common mode voltage V_{CM} is 3.5 V.

1-25. Repeat Problem 1-19 for the circuit shown in Figure 1-33.

1-26. Design the level translator circuit using diode constant current bias for the circuit in Problem 1-19. Refer to Figure 1-20(b).

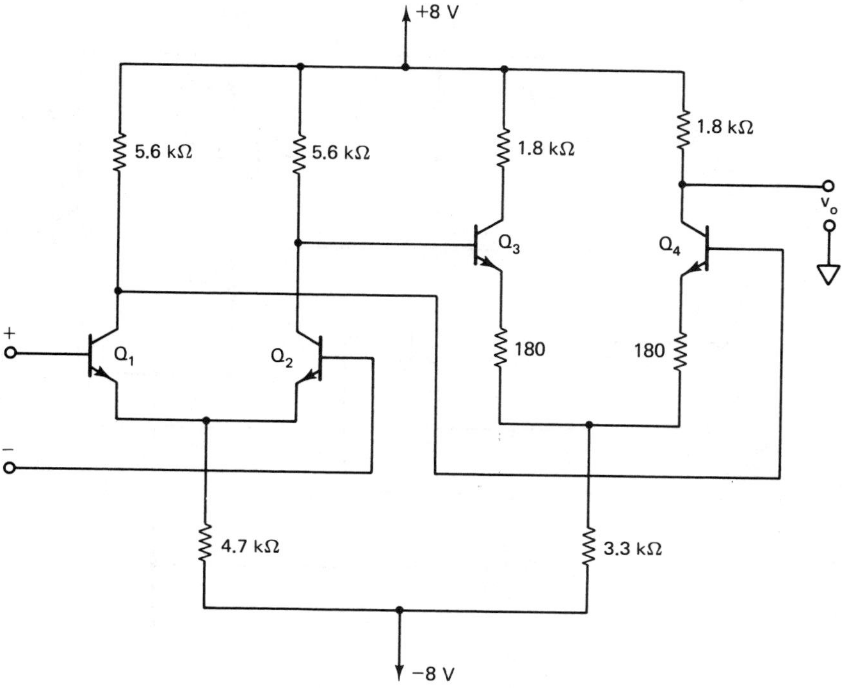

Figure 1-27 Cascaded differential amplifier of Problem 1-19.

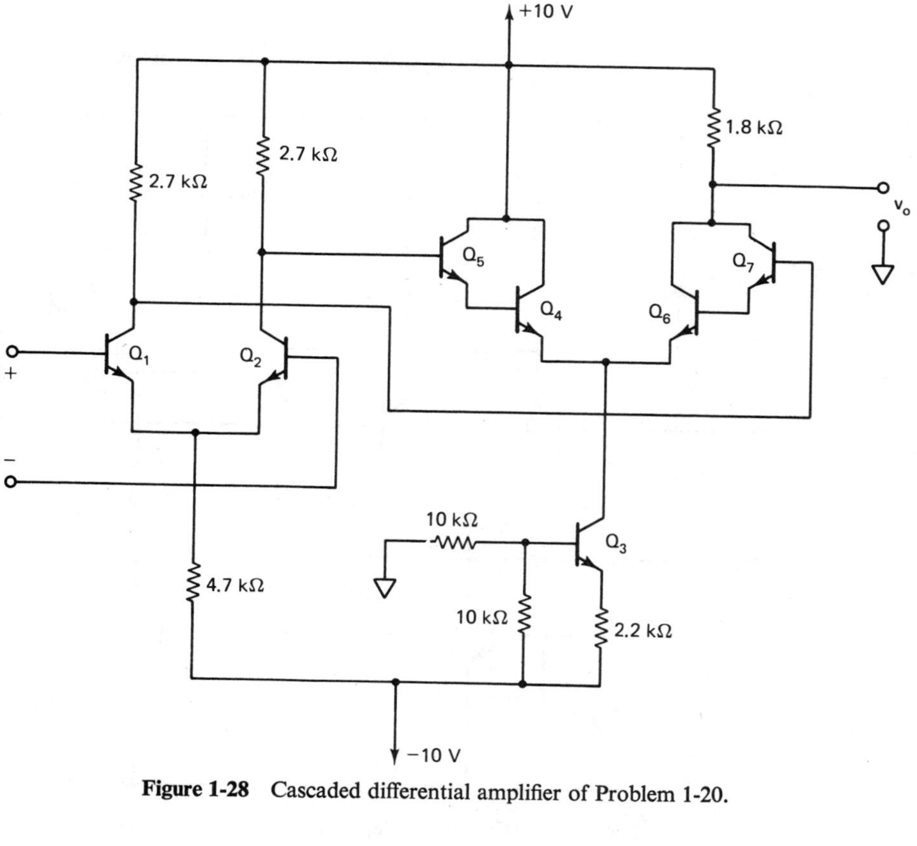

Figure 1-28 Cascaded differential amplifier of Problem 1-20.

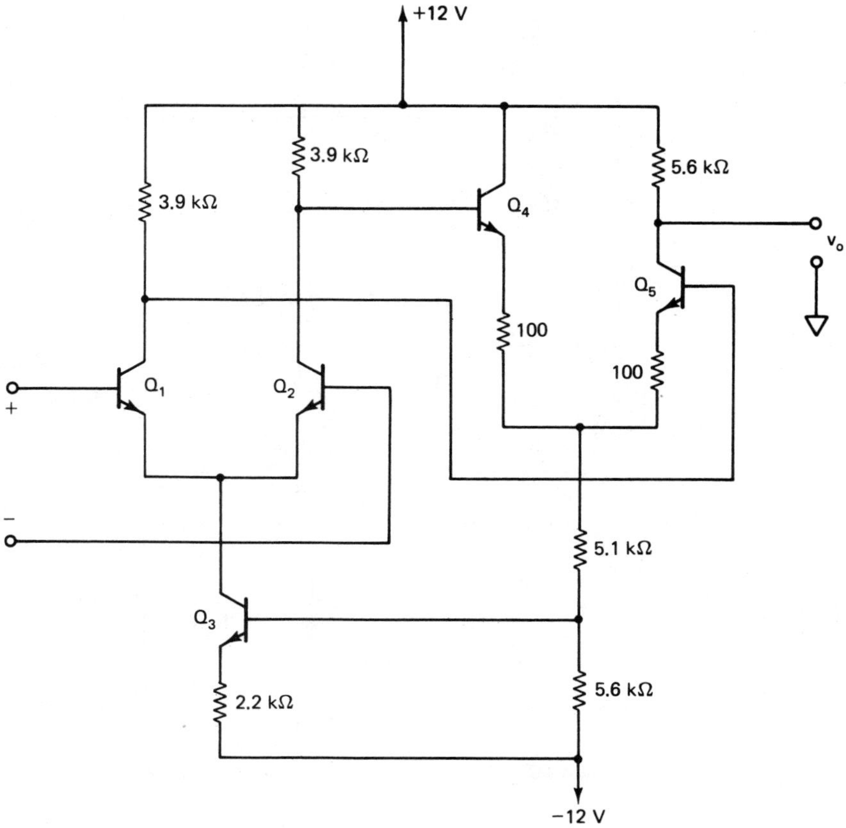

Figure 1-29 Cascaded differential amplifier of Problem 1-21.

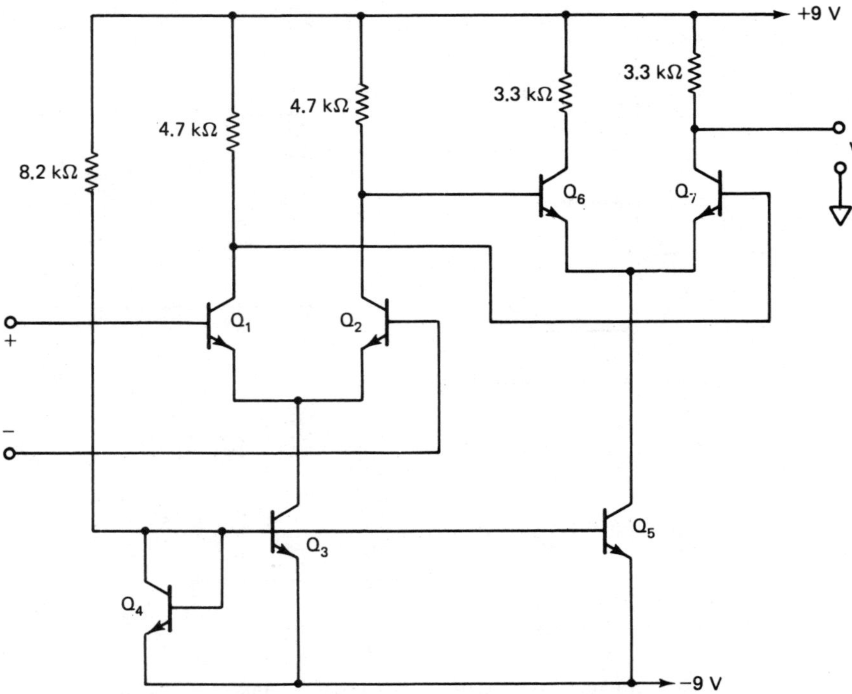

Figure 1-30 Cascaded differential amplifier of Problem 1-22.

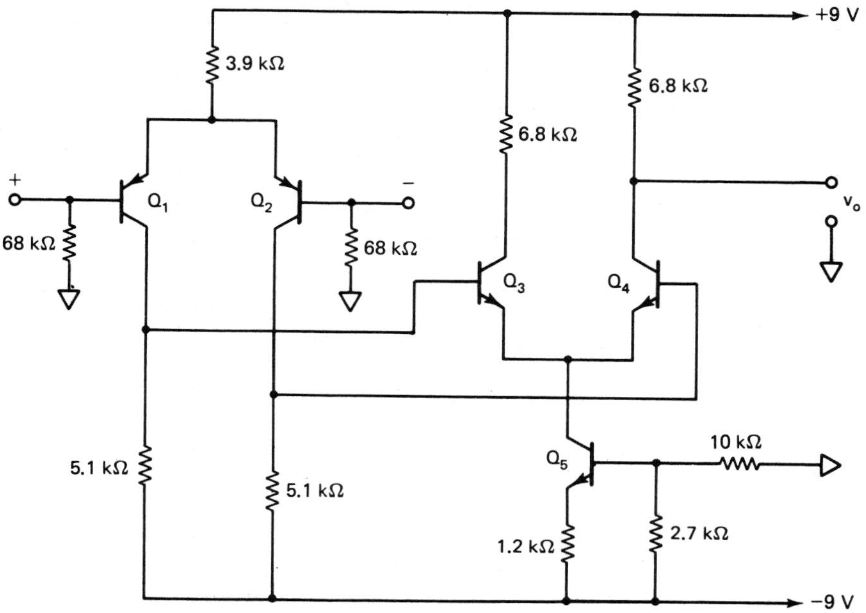

Figure 1-31 Cascaded differential amplifier of Problem 1-23.

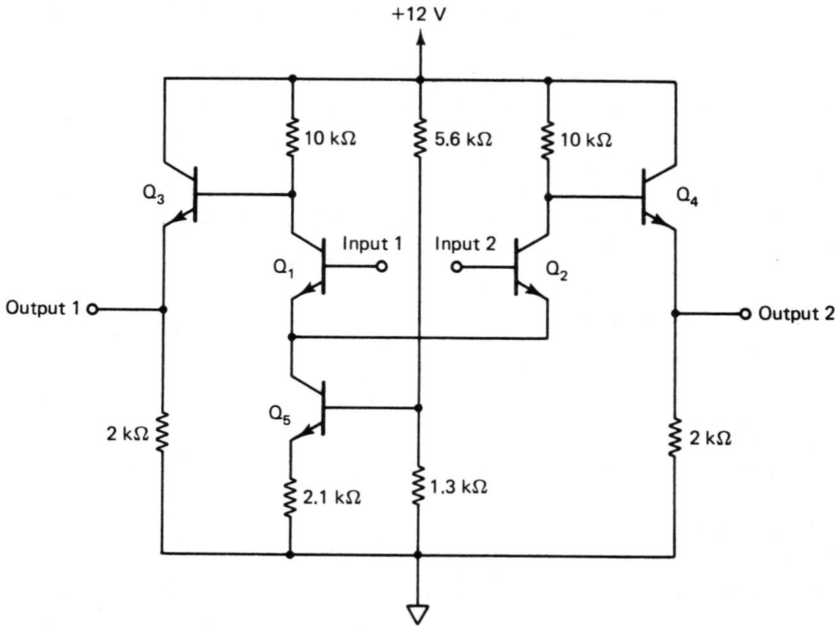

Figure 1-32 Schematic diagram of μA730 differential amplifier.

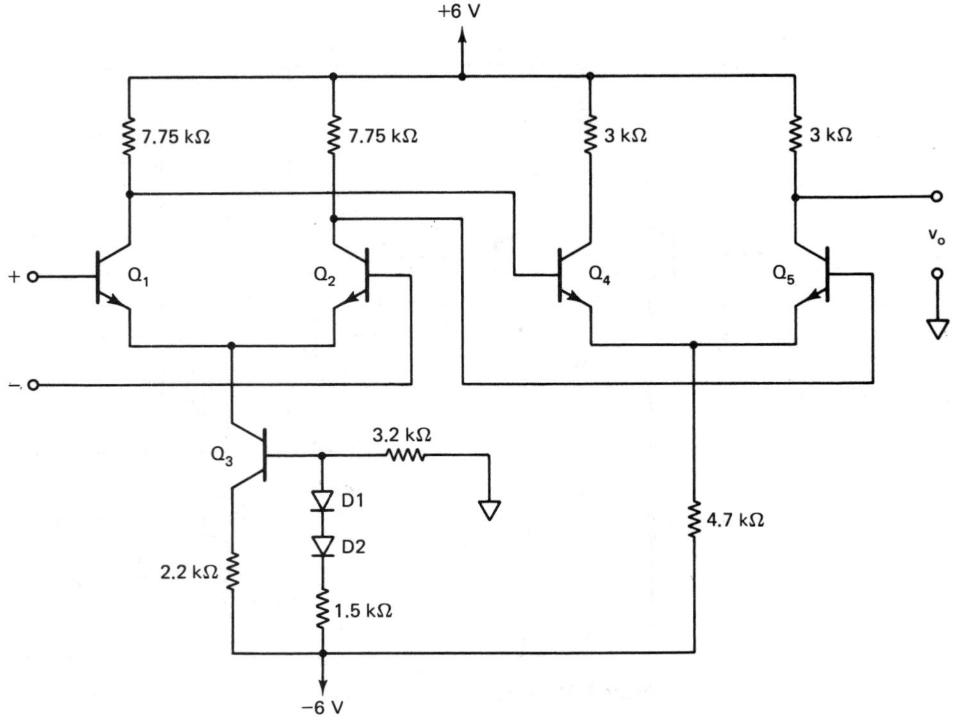

Figure 1-33 Cascaded differential amplifier of Problem 1-25.

1-27. Design the level translator circuit using zener constant current bias for the circuit in Problem 1-20. Refer to Figure 1-14(c).

1-28. For the circuit in Problem 1-21, design the level translator circuit using current mirror bias. Refer to Figure 1-20(c).

1-29. For the cascode amplifier shown in Figure 1-22, determine:

(a) The dc conditions for each transistor.

(b) The overall voltage gain.

(c) The overall current gain.

(d) The input resistance.

Assume that $R_1 = 12\ k\Omega$, $R_2 = 20\ k\Omega$, $R_3 = 24\ k\Omega$, $R_C = 1.8\ k\Omega$, $R_E = 1\ k\Omega$, $C_i = C_o = 10\ \mu F$, $C_{S1} = C_{S2} = 100\ \mu F$, and $V_{CC} = 12$ V. The transistors are 2N3393 with $\beta_{ac} = 125$ and $V_{BE} = 0.63$ V.

1-30. Repeat Example 1-11 with $R_E = 2.2\ k\Omega$.

1-31. Repeat Problem 1-29 with $R_C = 1\ k\Omega$.

EXPERIMENT 1-1
DIFFERENTIAL AMPLIFIER USING CONSTANT CURRENT BIAS

Objectives

In this experiment you will design a constant current source and current mirror networks and use them to bias the differential amplifier. At the end of this experiment you should be able to:

1. Measure and record the dc conditions in a differential amplifier using constant current bias and current mirror bias.
2. Explain the difference between a constant current bias and current mirror bias.

Equipment

1. Dual-trace oscilloscope
2. Audio-frequency generator
3. Low-voltage power supply
4. Multimeter

Materials

1. CA3086 transistor array or equivalent
2. 2.2-kΩ resistor
3. 3-kΩ potentiometer
4. Two 180-Ω matched resistors
5. Two small-signal diodes
6. Assorted components selected by the experimenter according to design

Procedure

1. Design a constant current bias circuit using diodes so that the emitter current in each differential amplifier transistor is 1 mA.
2. Measure the 2.2 kΩ resistor with ohmmeter and then adjust the 3 kΩ potentiometer to this value.
3. Connect the circuit of Figure E1-1.1(a) using the constant current bias network designed in step 1. However, do not turn on the signal generator at this time.
4. Using the voltmeter, measure the dc voltages at the emitter and collector of each transistor. Record these values in Data Table E1-1.1. Similarly, measure the dc emitter current in each transistor with the multimeter set to the dc current mode. Enter the measured values in Data Table E1-1.1.

DATA TABLE E1-1.1

DC measurements

	V_{C1}	V_{C2}	$V_{C3} = V_{E1} = V_{E2}$	V_{E3}	I_{C1}	I_{C2}	I_{C3}
Figure E1-1.1(a)							
Figure E1-1.1(b)							

5. Turn on the signal generator and set it to 1 kHz. Connect the oscilloscope probes of channels 1 and 2 to the collectors of transistors Q_1 and Q_2, respectively.

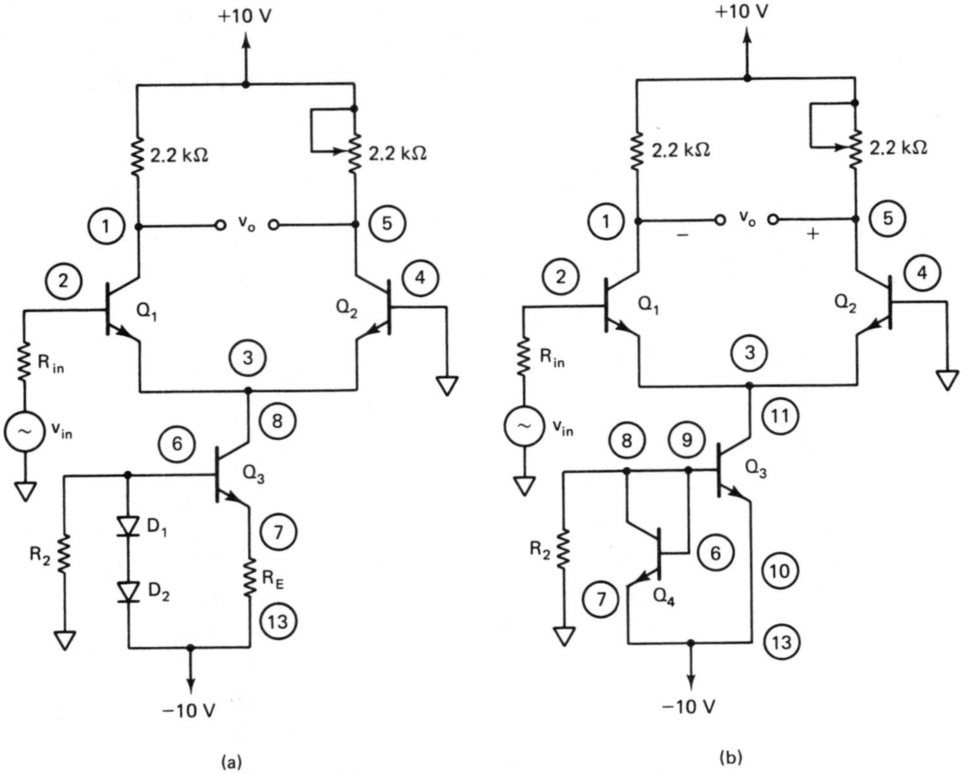

Figure E1-1.1 (a) Differential amplifier using constant current bias. (b) Differential amplifier using current mirror bias.

The ground leads of both channels must be connected to the ground point. Set the channel selector switch on the scope to ADD mode and channel 2 polarity switch on INVERT. Set the volts/cm switches of both channels to the same value. Adjust the amplitude of the signal applied to the differential amplifier so that the output waveform is a measurable sine wave. Measure the input and output voltages and record the measured values in Data Table E1-1.2.

DATA TABLE E1-1.2

AC measurements

	Input amplitude at 1 kHz, v_{in} pp	Output amplitude, v_o pp	Voltage gain	
			Measured	Calculated
Figure E1-1.1(a)				
Figure E1-1.1(b)				

6. Design a current mirror circuit so that the emitter current in each differential amplifier transistor is 1 mA.
7. Connect the circuit of Figure E1-1.1(b) using the current mirror circuit designed in step 6. However, do not turn on the signal generator at this time.
8. Repeat steps 4 and 5 for the circuit in Figure E1-1.1(b).

Questions

1. Describe the difference between a current mirror and constant current bias networks.
2. How does the experimental voltage gain compare with the calculated value? Give reasons if they differ by more than 20%.

EXPERIMENT 1-2
CASCADED DIFFERENTIAL AMPLIFIER WITH LEVEL TRANSLATOR

Objectives

In this experiment you will design a level translator circuit and use it to shift a dc output level in cascaded differential amplifier down to approximately zero volts. At the end of this experiment you should be able to:

1. Measure and record the dc conditions in a cascaded differential amplifier.
2. Explain the use of level translator circuit in a cascaded differential amplifier.
3. Explain the purpose of cascading differential amplifier stages.

Equipment

1. Dual-trace oscilloscope
2. Audio-frequency generator
3. Low-voltage power supply
4. Multimeter

Materials

1. Two CA3086 transistor arrays or equivalent
2. Two 2.2-kΩ matched resistors
3. 1.2-kΩ resistor
4. Two 180 Ω matched resistors
5. 4.7-Ω resistor
6. 1N3826 zener diode: $V_Z = 5.1$ V and $I_{Zt} = 49$ mA
7. 2.2-kΩ resistor
8. 82-Ω resistor
9. Components selected by the experimenter according to level translator design

Procedure

1. Connect the circuit shown in Figure E1-2.1(a) but do not turn on the signal generator at this time. Measure the voltages at the emitter and collector of each differential amplifier stage. Record the measured valuse in Data Table E1-2.1.

DATA TABLE E1-2.1

First differential amplifier stage					Second differential amplifier stage				
V_{C1}	V_{C2}	$V_{E1} = V_{E2}$	I_{E1}	I_{E2}	V_{C3}	V_{C4}	$V_{E3} = V_{E4}$	I_{E3}	I_{E4}

Similarly, measure the emitter currents in each differential amplifier stage. Enter these values in Data Table E1-2.1.

2. Turn on the signal generator and connect the scope probe to the collector of transistor Q_4. Set the signal generator frequency at 1 kHz. Adjust the amplitude of the input signal so that the output is undistorted but measurable sine wave. You may use the voltage divider arrangement across the signal generator in order to obtain a very small input amplitude. Measure and record the input and output voltages in Data Table E1-2.2.

DATA TABLE E1-2.2

Input amplitude, v_{in} pp	Output amplitude, v_o pp	Voltage gain	
		Measured	Calculated

3. Design a level translator circuit as shown in Figure E1-2.1(b). Connect the input terminal of the level translator circuit to the collector of transistor Q_4 and set the signal generator to 0 V. Measure the voltages at the emitter and collector of each transistor. Record the measured values in Data Table E1-2.3. Similarly, measure the emitter current in each transistor. Enter the measured values in Data Table E1-2.3.
4. Turn on the signal generator and connect the scope probe to the collector of transistor Q_7. Set the signal generator frequency at 1 kHz and adjust the ampli-

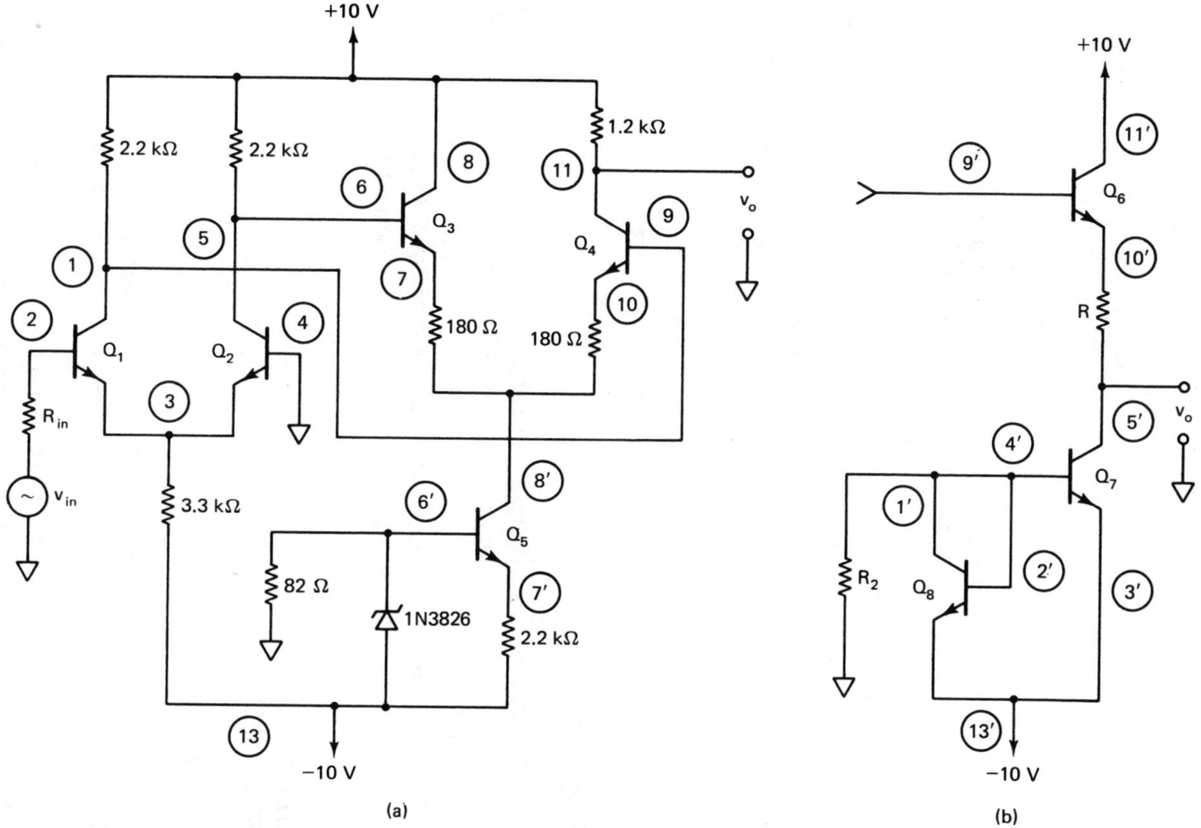

Figure E1-2.1 (a) Cascaded differential amplifier. (b) Level translator circuit.

DATA TABLE E1-2.3

V_{C1}	V_{C2}	$V_{E1} = V_{E2}$	V_{C3}	V_{C4}	V_{E3}	V_{E4}	V_{C5}	V_{C6}	V_{E6}	V_{C7}

I_{E1}	I_{E2}	I_{E3}	I_{E4}	I_{E5}	I_{E6}	I_{E7}	

tude of the input signal so that the output is measurable sinewave. Measure and record the input and output voltages in Data Table E1-2.4.

DATA TABLE E1-2.4

Input amplitude, v_{in} pp	Output amplitude, v_o pp	Voltage gain	
		Measured	Calculated

Questions

1. How does the experimental voltage gain compare to the calculated value, with and without level translator circuit? Give reasons if the values differ by more than 20%.
2. What is the main purpose of cascading differential amplifier stages?
3. Explain the use of level translator circuit in cascaded differential amplifier.

EXPERIMENT 1-3
CASCODE AMPLIFIER

Objectives

In this experiment you will investigate the dc conditions and ac operation of the cascode amplifier. At the end of this experiment you should be able to:

1. Measure and record the dc conditions in a cascode amplifier.
2. Measure and record input and output signals using an oscilloscope.
3. Construct a Bode plot of frequency versus voltage gain in dB.
4. Calculate the bandwidth of the amplifier.

Equipment

1. Oscilloscope
2. Audio-frequency generator
3. Low-voltage power supply
4. Multimeter

Materials

1. Two 2N3393 transistors
2. 47-kΩ resistor
3. 39-kΩ resistor
4. 27-kΩ resistor
5. 1.8-kΩ resistor
6. 1-kΩ resistor
7. Two 10-μF capacitors
8. Two 100-μF capacitors
9. Semilog paper

Procedure

1. Connect the circuit shown in Figure E1-3.1. With the signal generator set to 0 V, measure the voltages at and currents in both the transistor terminals. Record these measured values in Data Table E1-3.1.

DATA TABLE E1-3.1

V_{B1}	V_{C1}	V_{E1}	V_{B2}	V_{C2}	V_{E2}	I_{C1}	I_{E1}	I_{C2}	I_{E2}

2. Now turn on the signal generator, and set it to 50 Hz sine wave. Connect the scope across the output terminals and set the amplitude of the input sine wave so that the output waveform is a measurable sine wave. In order to obtain a very small input amplitude, a voltage-divider arrangement may be used across the signal generator. Measure the input and output voltages and record these values in Data Table E1-3.2.
3. Change the frequency setting of the signal generator to 70 Hz and again measure and record the input and output voltages in Data Table E1-3.2. Complete Data Table E1-3.2.

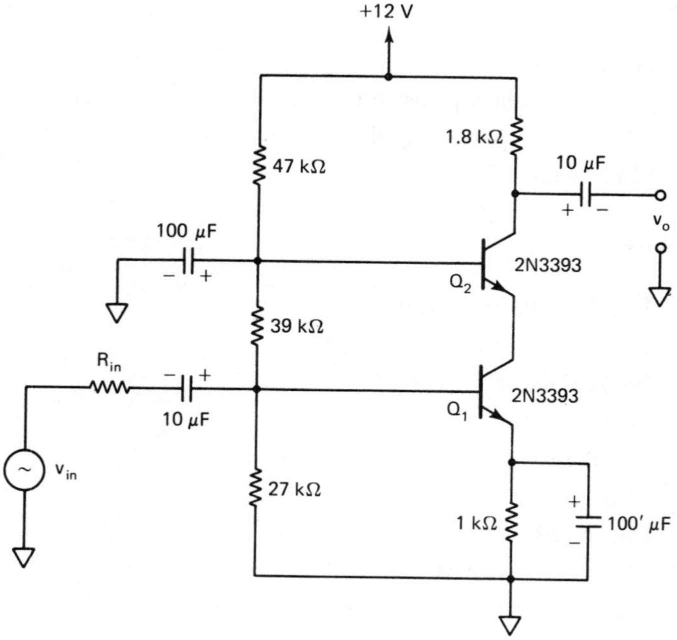

Figure E1-3.1 Cascode amplifier.

DATA TABLE E1-3.2

Input frequency (Hz)	Input amplitude, v_{in} pp	Output amplitude, v_o pp	Voltage gain, $A_v = v_o/v_{in}$	Voltage gain (dB)
1. 50				
2. 70				
3. 100				
4. 300				
5. 700				
6. 1,000				
7. 3,000				
8. 7,000				
9. 10,000				
10. 30,000				
11. 70,000				
12. 100,000				
13. 500,000				

4. Draw a Bode plot of frequency versus voltage gain in dB on semilog paper. Determine from the graph the ferquencies at which the midband frequency gain is down by 3 dB. Using these frequencies, calculate the bandwidth of the amplifier.

Questions

1. Compare the cascode amplifier with the *RC*-coupled CE amplifier for dc bias and ac (midband frequency) operation.
2. Why isn't the cascode amplifier of Figure E1-3.1 encountered very often in discrete circuitry?
3. Explain the use of cascode amplifier in electronic circuitry.

2

Introduction to Operational Amplifiers

2-1 INTRODUCTION

In Chapter 1 you studied the two important types of direct-coupled amplifiers—differential and cascode. You saw that when cascaded, the differential amplifier yields a very high voltage gain. However, in differential amplifiers the matching of resistors and transistors is essential for the proper operation of the circuit. For this reason the differential amplifier and the cascode amplifier are manufactured as individual integrated circuits, and the differential amplifiers, in particular, are used as "building blocks" for more elaborate circuits called operational amplifiers. The cost of integrated amplifiers is currently so low that they have effectively displaced the discrete amplifiers.

In this chapter we study the basic structure of the operational amplifier (op-amp) and then analyze some typical operational amplifier circuits. We will also investigate the op-amp symbol, types of op-amps, different grades of op-amps, and op-amp development—among other things.

2-2 THE OPERATIONAL AMPLIFIER

An operational amplifier is a direct-coupled high-gain amplifier usually consisting of one or more differential amplifiers and usually followed by a level translator and an output stage. The output stage is generally a push-pull or push-pull complementary symmetry pair. An operational amplifier is available as a single integrated circuit package.

The operational amplifier is a versatile device that can be used to amplify dc as well as ac input signals and was originally designed for computing such mathematical functions as addition, subtraction, multiplication, and integration. Thus the name *operational amplifier* stems from its original use for these mathematical operations and is abbreviated to *op-amp*. With the addition of suitable external feedback components, the modern day op-amp can be used for a variety of applications, such as ac and dc signal amplification, active filters, oscillators, comparators, regulators, and others.

2-3 BLOCK DIAGRAM REPRESENTATION OF A TYPICAL OP-AMP

Since an op-amp is a multistage amplifier, it can be represented by a block diagram as shown in Figure 2-1.

The input stage is the dual-input, balanced-output differential amplifier. This stage generally provides most of the voltage gain of the amplifier and also establishes the input resistance of the op-amp. The intermediate stage is usually another differential amplifier, which is driven by the output of the first stage. In most amplifiers the intermediate stage is dual input, unbalanced (single-ended) output. Because direct coupling is used, the dc voltage at the output of the intermediate stage is well above ground potential. Therefore, the level translator (shifting) circuit is used after the intermediate stage to shift the dc level at the output of the intermediate stage downward to zero volts with respect to ground. The final stage is usually a push-pull complementary amplifier output stage. The output stage increases the output voltage swing and raises the current supplying capability of the op-amp. A well-designed output stage also provides low output resistance.

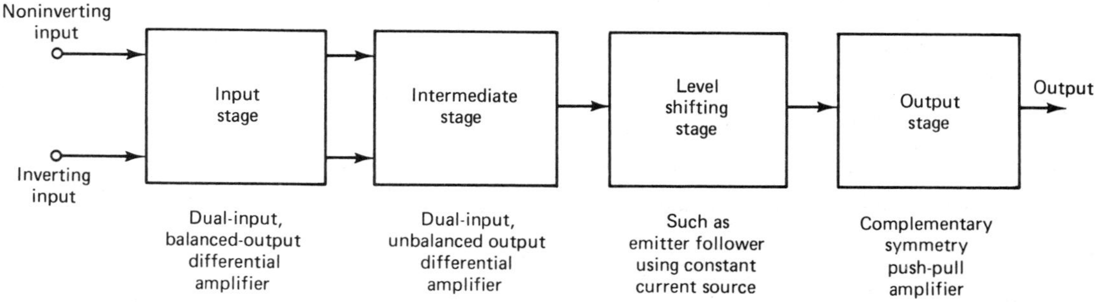

Figure 2-1 Block diagram of a typical op-amp.

2-4 ANALYSIS OF TYPICAL OP-AMP EQUIVALENT CIRCUITS

First we consider a specific op-amp with a simpler equivalent circuit and perform the analysis of this circuit to determine the input resistance and voltage gain. Then the op-amp with a more complex equivalent circuit will be presented. However, instead of a quantitative analysis, a qualitative analysis will be used for this complex op-amp equivalent circuit.

While there are a variety of op-amps, each with specific inner design features such as internal frequency compensation, FET inputs, Darlington inputs, current sources as active loads, input voltage and output current limiter, and many others, the analysis of specific op-amp equivalent circuits will provide a good basis for understanding the inner operation and construction of the op-amp and aid in selecting a proper op-amp for a desired application.

Example 2-1

The equivalent circuit of the Motorola op-amp MC 1435 is shown in Figure 2-2.

(a) Determine the collector current in each transistor and the dc voltage at the output terminal.

(b) Calculate the voltage gain of the op-amp.

(c) Determine the input resistance of the op-amp.

Assume that the $\beta_{ac} = \beta_{dc} = 150$ and $V_{BE} = 0.7$ V for each transistor.

Solution. Figure 2-2 should give you an idea of the internal structure of the op-amp. A close examination of this equivalent circuit reveals that it consists of four stages, as shown in Section 2-3. These stages are labeled in Figure 2-2 and are separated by dashed lines. Transistors Q_1 and Q_2 form the first differential amplifier stage, which uses a constant current bias provided by transistor Q_3 and associated resistors. This first stage has inverting and noninverting inputs and hence can be driven with two inputs or a single input. The second differential amplifier formed by transistors Q_4 and Q_5 uses an emitter bias and is driven by the outputs of the first differential amplifier. The single-ended output of the second differential amplifier drives a level translator (emitter follower), which is composed of transistor Q_6 and a 15-kΩ resistor. The output of the level translator drives a complementary transistor pair Q_7 and Q_8, which produces the final output voltage.

(a) To determine the collector current and the collector-to-emitter voltage of the first differential amplifier stage, the inverting and noninverting input terminals are assumed to be connected to ground. Referring to Figure 2-2, the collector current in transistor Q_1 and Q_2 is then determined as follows:

$$V_{C1} = V_{CC} - (6.7\text{ k}\Omega)I_{C1}$$
$$= 6 - (6.7\text{ k}\Omega)I_{C1}$$
$$V_{E4} = V_{C1} - V_{BE5}$$
$$= 5.3 - (6.7\text{ k}\Omega)I_{C1}$$

Since I_{B3} is negligibly small,

$$I_4 = \frac{V_{E4} + V_{EE}}{9.1\text{ k}\Omega + 5.5\text{ k}\Omega}$$
$$= \frac{11.3 - (6.7\text{ k}\Omega)I_{C1}}{14.6\text{ k}\Omega}$$
$$V_{B3} = (5.5\text{ k}\Omega)I_4 - V_{EE}$$
$$= 5.5\text{ k}\Omega\left[\frac{11.3 - (6.7\text{ k}\Omega)I_{C1}}{14.6\text{ k}\Omega}\right] - 6$$
$$V_{E3} = V_{B3} - V_{BE3}$$
$$= \frac{5.5\text{ k}\Omega}{14.6\text{ k}\Omega}[11.3 - (6.7\text{ k}\Omega)I_{C1}] - 6.7$$
$$I_{E3} = \frac{V_{E3} + V_{EE}}{3.3\text{ k}\Omega}$$
$$= \frac{(5.5\text{ k}\Omega/14.6\text{ k}\Omega)[11.3 - (6.7\text{ k}\Omega)I_{C1}] - 6.7 + 6}{3.3\text{ k}\Omega}$$

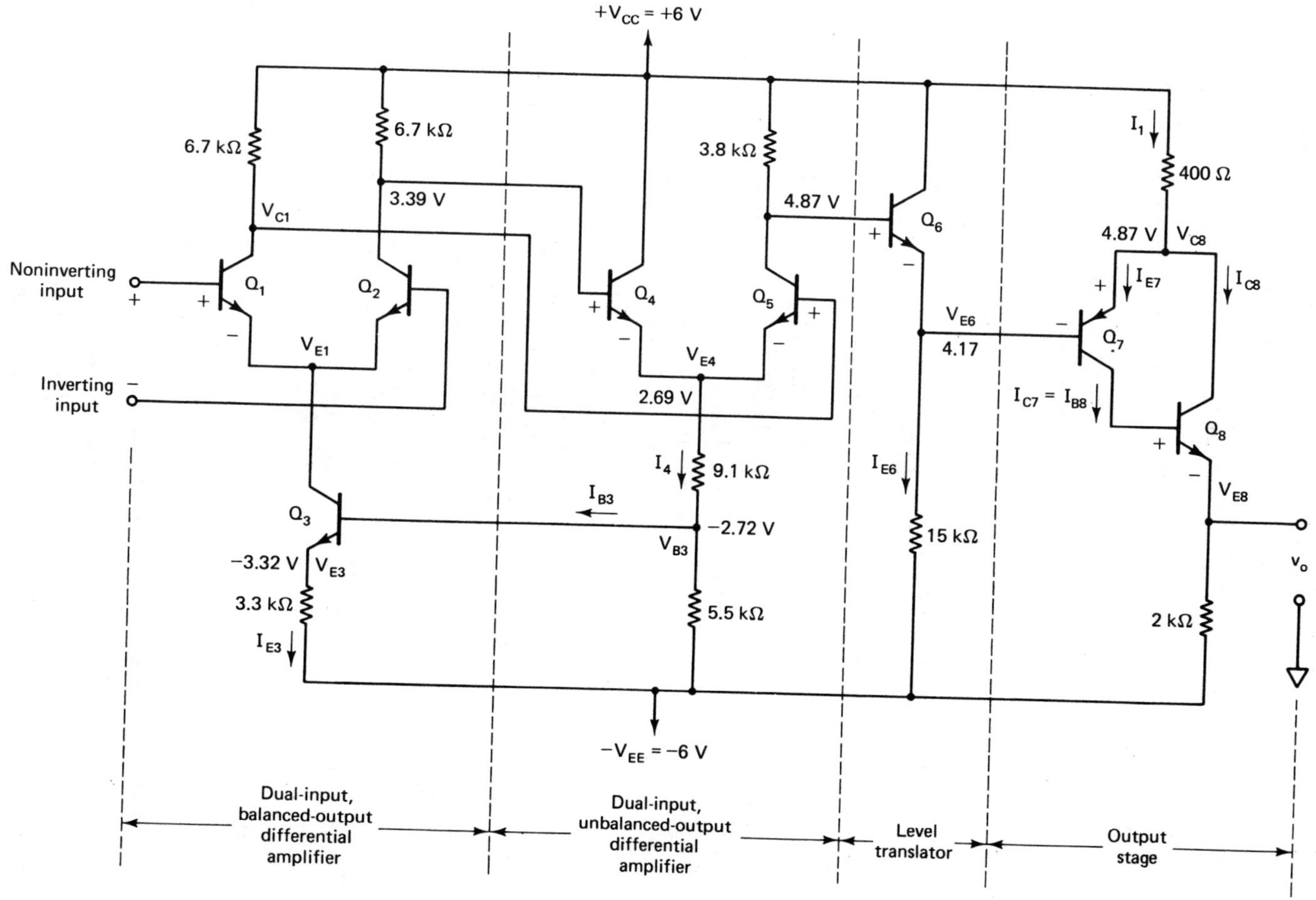

Figure 2-2 Equivalent circuit of the MC1435 op-amp. (Courtesy of Motorola Semiconductor Inc.)

But $I_{E3} = 2I_{C1}$. Therefore,

$$2I_{C1} = \frac{(5.5\text{ k}\Omega)(11.3)}{(14.6\text{ k}\Omega)(3.3\text{ k}\Omega)} - \frac{(5.5\text{ k}\Omega)(6.7\text{ k}\Omega)I_{C1}}{(14.6\text{ k}\Omega)(3.3\text{ k}\Omega)} - \frac{0.7}{3.3\text{ k}\Omega}$$

$$2I_{C1} + 0.765I_{C1} = 1.29\text{ mA} - 0.212\text{ mA}$$

$$I_{C1} = \frac{1.08\text{ mA}}{2.765} \cong 0.39\text{ mA}$$

This collector current can be used to compute the collector currents of the remaining stages as follows:

$$V_{C1} = V_{CC} - (6.7\text{ k}\Omega)I_{C1}$$

$$= 6 - (6.7\text{ k}\Omega)(0.39\text{ mA}) = 3.39\text{ V}$$

$$V_{E4} = V_{C1} - V_{BE5}$$

$$= 2.69\text{ V}$$

Hence

$$2I_{E4} = \frac{V_{E4} + V_{EE}}{14.6\text{ k}\Omega}$$

$$I_{E4} = \frac{2.69 + 6}{29.2\text{ k}\Omega} = 0.298\text{ mA}$$

$$V_{C5} = V_{CC} - (3.8\text{ k}\Omega)I_{C5}$$

$$= 6 - (3.8\text{ k}\Omega)(0.298\text{ mA}) = 4.87\text{ V}$$

$$V_{E6} = V_{C5} - V_{BE6}$$

$$= 4.87 - 0.7 = 4.17\text{ V}$$

Therefore,

$$I_{E6} = \frac{V_{E6} + V_{EE}}{15\text{ k}\Omega}$$

$$= \frac{4.17 + 6}{15\text{ k}\Omega} = 0.678\text{ mA}$$

$$V_{E7} = V_{C8} = V_{B7} + V_{BE7}$$

$$= 4.17 + 0.7 = 4.87\text{ V}$$

Hence

$$I_1 = \frac{V_{CC} - V_{E7}}{400}$$

$$= \frac{6 - 4.87}{400} = 2.83\text{ mA}$$

Since $I_{C7} = I_{B8}$ and $I_{C7} \cong I_{E7}$,

$$I_1 = I_{C8} + I_{B8}$$

$$= I_{E8} = 2.83\text{ mA}$$

The voltage V_{E8} at the output terminal is

$$V_{E8} = -V_{EE} + (2\text{ k}\Omega)I_{E8}$$

$$= -6 + (2\text{ k}\Omega)(2.83\text{ mA})$$

$$= -0.34\text{ V} \cong 0\text{ V} \qquad \text{as expected}$$

Thus

$$I_{C1} = I_{C2} \cong 0.39\text{ mA}$$

$$I_{C3} \cong 0.78 \text{ mA}$$
$$I_{C4} = I_{C5} \cong 0.298 \text{ mA}$$
$$I_{C6} \cong 0.678 \text{ mA}$$
$$I_{C7} = I_{B8} = 18.8 \ \mu\text{A}$$
$$I_{E8} = 2.83 \text{ mA}$$
$$V_{E8} = -0.34 \text{ V}$$

(b) Voltage gain of the level translator as well as that of the output stage is approximately 1. To calculate the voltage gains of the differential amplifier stages, we need to know their ac emitter resistances. The ac emitter resistance of each stage can be calculated using the collector current obtained in part (a):

$$r_{e1} = r_{e2} = \frac{25 \text{ mV}}{I_{E1}} = \frac{25 \text{ mV}}{0.39 \text{ mA}} \cong 64.1 \ \Omega$$

$$r_{e4} = r_{e5} = \frac{25 \text{ mV}}{I_{E4}} = \frac{25 \text{ mV}}{0.298 \text{ mA}} \cong 83.89 \ \Omega$$

$$r_{e6} = \frac{25 \text{ mV}}{I_{E6}} = \frac{25 \text{ mV}}{0.678 \text{ mA}} \cong 36.87 \ \Omega$$

Therefore, the voltage gain (A_{d1}) of the dual-input, balanced-output differential amplifier is

$$A_{d1} = \frac{R_{C1} \,||\, 2\beta_{\text{ac}}(r_{e4})}{r_{e1}} = \frac{6.7 \text{ k}\Omega \,||\, (2)(150)(83.89)}{64.1}$$
$$= 82.55$$

And the voltage gain (A_{d2}) of the dual-input, unbalanced output-differential amplifier is

$$A_{d2} = \frac{R_{C5} \,||\, \beta_{\text{ac}}(r_{e6} + 15 \text{ k}\Omega)}{2r_{e5}} = \frac{3.8 \text{ k}\Omega \,||\, (150)(15.04 \text{ k}\Omega)}{(2)(83.89)}$$
$$= 22.6$$

Thus the overall voltage gain of the op-amp is

$$A_d = (A_{d1})(A_{d2}) = (82.55)(22.6)$$
$$= 1865.63$$

(c) The input resistance of the op-amp is the same as the input resistance of the first differential amplifier stage, and is given by Equation (1-14):

$$R_i = 2\beta_{\text{ac}} r_{e1}$$
$$= (2)(150)(64.1) = 19.23 \text{ k}\Omega$$

This completes the analysis of MC1435 op-amp.

Figure 2-3 shows the schematic diagram of the 741 op-amp that has become an industry standard. This unit is a high-gain, stable amplifier that features internal compensation. A desired phase control at high frequencies is provided by a 30-pF capacitor that is fabricated on the chip. Since the equivalent circuit of the 741 op-amp is more complex than that of the 1435, we will perform a qualitative (rather than quantitative) analysis of the circuit.

The input differential amplifier consists of transistors Q_1–Q_3 and Q_2–Q_4. A differential input signal is applied between the bases of transistors Q_1 and Q_2. The

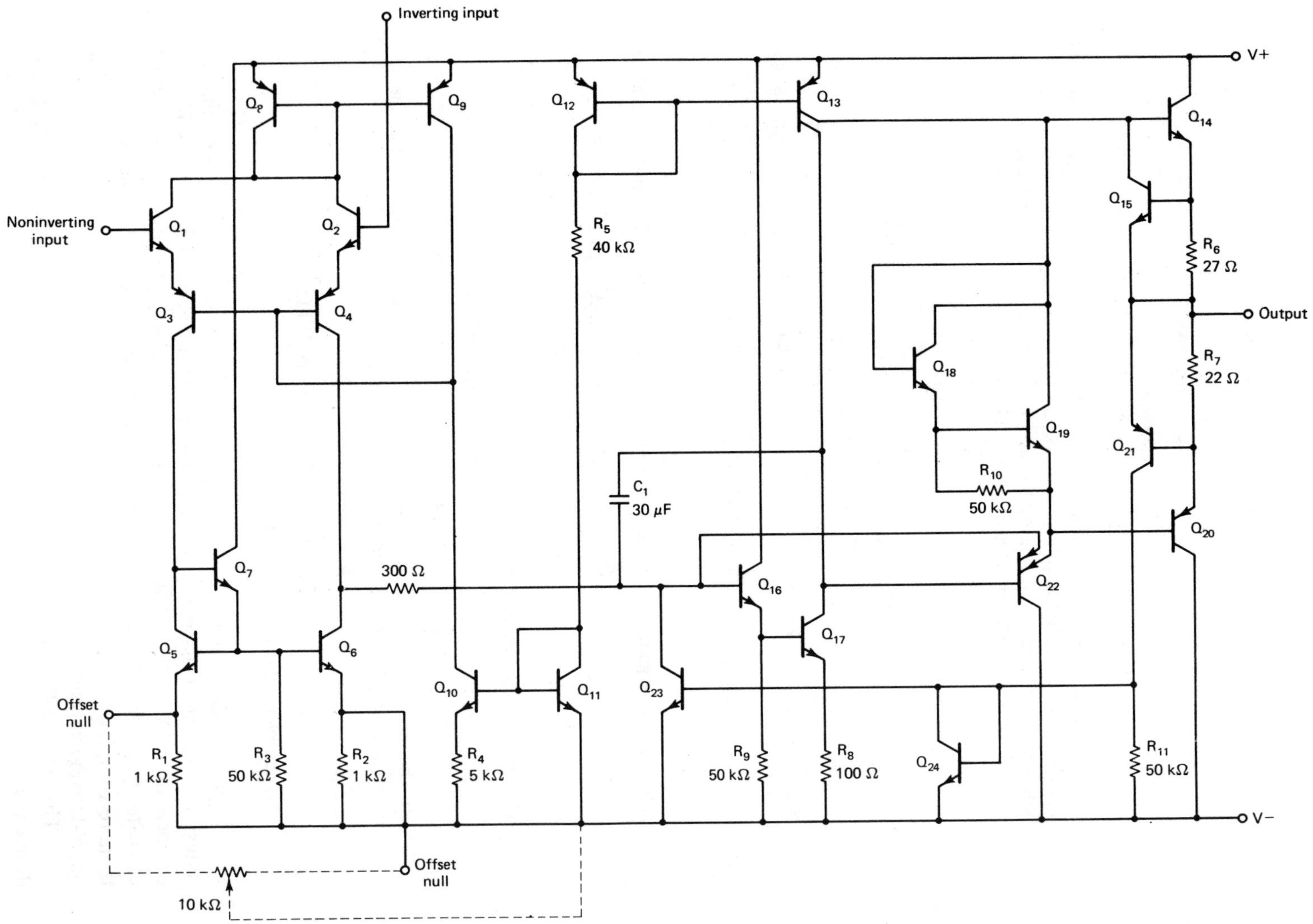

Figure 2-3 Equivalent circuit of the 741 op-amp. (Courtesy of Fairchild Camera and Instrument Corporation.)

maximum input signal capacity of the input differential stage is increased by PNP transistors Q_3 and Q_4, since the base–emitter junction of Q_3 is in series with the base–emitter junction of Q_1 and the base–emitter junction of Q_4 is in series with that of Q_2. Since the signal levels normally encountered do not exceed the voltage across both of the series junctions, the input differential stage is practically indestructible.

The constant base current in Q_3 and Q_4 is supplied by a constant current source consisting of transistors Q_9 and Q_{10} and diode-connected transistors Q_8 and Q_{11}. This constant base current in transistors Q_3 and Q_4 in turn establishes the operating current for the input differential stage. The diode-connected transistors Q_8 and Q_{11} stabilize the base potentials of transistors Q_9 and Q_{10}, respectively.

A controlled current source is formed by transistors Q_5 and Q_6 and resistors R_1, R_2, and R_3. The emitter current of transistors Q_5 and Q_6 can, however, be controlled by varying a 10-kΩ-potentiometer that is externally connected between the offset null terminals and by connecting the wiper to the negative supply. (For more information on offset null capabilities of the 741, refer to Section 5-2). Transistor Q_7 supplies the base currents to transistors Q_5 and Q_6. Since transistors Q_5 and Q_6 are identical (and $R_1 = R_2$), their base currents must be equal, and in turn their collector currents must also be equal at all times. The collector voltage of transistor Q_5 controls the operation of transistor Q_7; that is, any change in Q_5 collector voltage produces identical changes in transistor Q_5 and Q_6 collector currents via transistor Q_7.

The output of the differential stage is taken at the junction of transistors Q_4 and Q_6, which act as a complementary-symmetry amplifier. For example, suppose that a differential input signal applied between Q_1 and Q_2 causes the collector currents of Q_1, Q_3, and Q_5 to increase, and the collector currents of Q_2 and Q_4 to decrease. This differential input signal also causes the collector voltage of Q_5 to increase, which in turn causes the collector currents of Q_5 and Q_6 to increase. Thus at the junction of transistors Q_4 and Q_6, the net change in current is twice that of Q_4 alone. This action verifies that transistors Q_4 and Q_6 act as a complementary-symmetry amplifier. The output at the junction of Q_4 and Q_6 is proportional to the differential input signal.

The output of the differential stage is applied to the common collector amplifier formed by Q_{16} and R_9. The output of Q_{16} drives the common emitter amplifier, which is a bias source for transistors Q_{18} and Q_{19}. The CE amplifier is composed of Q_{17}, R_8 and constant current load formed by Q_{13}. Transistors Q_{13} and Q_{12} form a current mirror which establishes the current through transistors Q_{17}, Q_{18}, and Q_{19}. In fact, the currents in collectors of Q_{13} is equal to the current through R_5. The network consisting of transistor Q_{18}, Q_{19}, and R_{10} is designed to bias the output complementary symmetry stage formed by transistors Q_{14}, Q_{20} and resistors R_6, R_7 in the linear region. These transistors also separate the bases of Q_{14} and Q_{20} by two diode drops and in turn temperature-compensate currents in Q_{14} and Q_{20}. The output is measured at the junction of R_6 and R_7, which are in the emitter circuits of Q_{14} and Q_{20}, respectively.

Transistor Q_{22} performs two functions. It serves as a buffer between Q_{17} and Q_{20} and also provides a negative feedback to Q_{16}. The current limiting for the output complementary-symmetry stage is provided by transistors Q_{15}, Q_{21}, and Q_{23}. If the output (load) current exceeds the safe limit, the voltage drops across R_6 and R_7 increase, causing Q_{15} and Q_{21} to turn on, which in turn causes Q_{23} to turn on. This action however, turns off Q_{16} and in turn Q_{17}, which reduces the emitter currents in Q_{22}. The reduction in the currents of Q_{22} causes the currents in Q_{18} and Q_{19} to decrease, which in turn lowers the currents in Q_{14} and Q_{20}. The diode-connected transistor Q_{24} is a temperature-compensating diode for transistor Q_{23}.

2-5 SCHEMATIC SYMBOL

Given an op-amp schematic diagram like the one in Figure 2-2 or 2-3, we can save time by using a schematic symbol for the entire circuit. Figure 2-4 shows the most widely used of such symbols for a circuit with two inputs and one output. For simplicity, power supply and other pin connections are omitted. Since the input differential amplifier stage of the op-amp is designed to be operated in the differential mode, the differential inputs are designated by the (+) and (−) notations. The (+) input is the noninverting input. An ac signal (or dc voltage) applied to this input produces an in-phase (or same polarity) signal at the output. On the other hand, the (−) input is the inverting input because an ac signal (or dc voltage) applied to this input produces an 180° out-of-phase (or opposite polarity) signal at the output.

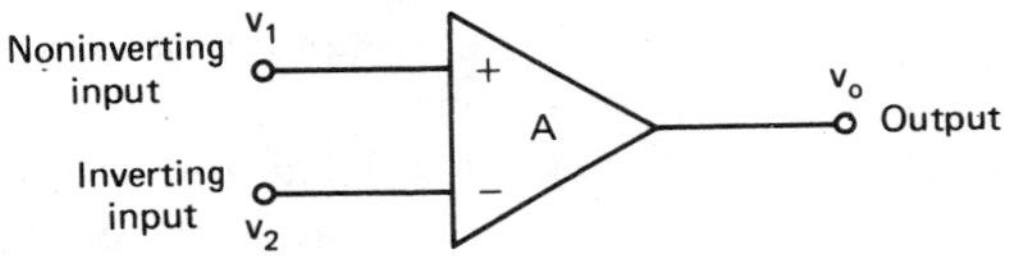

Figure 2-4 Schematic symbol of op-amp.

In Figure 2-4,

$$v_1 = \text{voltage at the noninverting input (volts)}$$

$$v_2 = \text{voltage at the inverting input (volts)}$$

$$v_o = \text{output voltage (volts)}$$

All these voltages are measured with respect to ground

$$A = \text{large-signal voltage gain that is specified on the data sheets for an op-amp}$$

2-6 INTEGRATED CIRCUITS

The circuits shown in Figure 2-2 and 2-3 are *integrated* circuits (ICs), meaning that all the components in each circuit are fabricated on the same "chip." ICs have become a vital part of modern electronic circuit design. They are used in the computer industry, automobile industry, home appliances, communication, and control systems, where they permit miniaturization and superior performance not possible with discrete components. ICs are now being used in all types of electronic equipment because of the long, trouble-free service they provide. In addition, they are economical because they are mass produced.

Classified according to their mode of operation, ICs are of two basic types: digital or linear.

Digital ICs are complete functioning logic networks that are equivalents of basic transistor logic circuits. They are used to form such circuits as gates, counters, multiplexers, demultiplexers, shift registers, and others. Since a digital IC is a complete predesigned package, it usually requires nothing more than a power supply, input, and output.

Digital circuits are primarily concerned with only two levels of voltage (or current): "high" and "low." Therefore, accurate control of operating-region characteristics is not required in digital circuits, unlike in linear circuits. For this reason, digital circuits are easy to design and are produced in large quantities as low-cost devices.

Linear ICs are equivalents of discrete transistor networks, such as amplifiers, filters, frequency multipliers, and modulators, that often require additional external components for satisfactory operation. For example, external resistors are necessary to control the voltage gain and frequency response of an op-amp. In linear circuits the output electrical signals vary in proportion to the input signals applied or the physical quantities (parameters) they represent. Since the electrical signals are *analogous* to the physical quantities, linear circuits are also referred to as *analog circuits*.

Of all presently available linear ICs, the majority are operational amplifiers. With suitable external components, the op-amp is used in amplifiers, active filters, integrators, differentiators, and in countless other applications. A wide variety of special-purpose linear ICs is available for uses in comparators, voltage regulators, digital-interface circuits, and in radio-frequency and power amplifiers.

In addition to the op-amp and special-purpose linear ICs, component arrays are also available in IC form. Such arrays may consist of groups of isolated transistors, diodes, and resistors as well as of Darlington pairs or individual stages such as differential and cascode amplifiers.

Op-amps are further classified into two groups: general-purpose and special-purpose. *General-purpose* op-amps may be used for a variety of applications, such as integrator, differentiator, summing amplifier, and others. An example of a widely used general-purpose op-amp is the 741/351. On the other hand, *special-purpose* op-amps are used only for the specific applications they are designed for. For example, the LM380 op-amp can be used *only* for audio power applications.

2-7 TYPES OF INTEGRATED CIRCUITS

Integrated circuits may be classified as either monolithic or hybrid. Most linear ICs are produced by the monolithic process in that all transistors and passive elements (resistors and capacitors) are fabricated on a single piece of semiconductor material, usually silicon. "Monolithic" is a Greek-based word meaning "one stone."

In *monolithic ICs* all components (active and passive) are formed simultaneously by a diffusion process. Then a metallization process is used in interconnecting these components to form the desired circuit. Electrical isolation between the components in monolithic ICs can be achieved by any one of the three isolation techniques: dielectric, beam-lead, or PN-junction. However, the PN-junction isolation is most economical and is, therefore, commonly used.

The monolithic process makes low-cost mass production of ICs possible. Also, monolithic ICs exhibit good thermal stability because all the components are integrated on the same chip very close to each other.

However, the large values of resistance and capacitance that are required in some linear circuits cannot be formed using the monolithic process. Moreover, there is no method available to fabricate transformers or to form large values of inductors in integrated-circuit form. However, if these components are required in a given application, external discrete components can be used with the IC.

In *hybrid ICs*, passive components (such as resistors and capacitors) and the interconnections between them are formed on an insulating substrate. The substrate is used as a chassis for the integrated components. Active components such as transistors and diodes, as well as monolithic integrated circuits, are then connected to form a complete circuit. For this reason, low-volume production methods are best suited to hybrid IC technology.

Hybrid ICs are further classed as thin-film or thick-film, depending on the method used to form the resistors, capacitors, and related interconnections on the substrate. When a suitable material is evaporated on a substrate in forming resistors, capacitors, and interconnections, a *thin-film* hybrid IC is obtained. On the other hand, in a *thick-film* hybrid IC the resistors, capacitors, and interconnections are etched on the substrate by silk screening.

2-8 MANUFACTURERS' DESIGNATIONS FOR INTEGRATED CIRCUITS

In the United States alone there are well over 30 IC manufacturers producing millions of ICs per year. Each manufacturer uses a specific code and assigns a specific type number to the ICs it produces. That is, each manufacturer uses its own identifying initials followed by its own type number. For example, the 741 type of internally compensated op-amp was originally manufactured by Fairchild and is sold as the μA741, where "μA" represents the identifying initials used by Fairchild. Initials used by some of the well-known manufacturers of *linear* ICs are as follows:

Fairchild	μA
	μAF
National Semiconductor	LM
	LH
	LF
	TBA
Motorola	MC
	MFC
RCA	CA
	CD
Texas Instruments	SN

Sprague	ULN
	ULS
	ULX
Intersil	ICL
	IH
Siliconix, Inc.	L
Signetics	N/S
	NE/SE
	SU
Burr-Brown	BB

Remember that the initials used by manufacturers in designating digital ICs may differ from those used for linear ICs. For example, "DM" and "CD" are the initials used for digital monolithic and CMOS (complementary metal-oxide semiconductor) digital ICs, respectively, by National Semiconductor.

In addition to producing their own ICs, a number of manufacturers also produce one another's popular ICs. In second-sourcing such ICs, the manufacturers usually retain the original type number of the IC in their own IC designation. For example, Fairchild's original μA741 is also manufactured by various other manufacturers under their own designations, as follows:

National Semiconductor	LM 741
Motorola	MC 1741
RCA	CA 3741
Texas Instruments	SN 52741
Signetics	N 5741

Note that the last three digits in each manufacturer's designation are 741. All these op-amps have the same specifications and, therefore, behave the same.

More information is available in the linear industry's cross reference guides on the types of ICs manufactured by different manufacturers. An industry cross-reference guide is generally included in the manufacturer's data book.

Since a number of manufacturers produce the same IC, for convenience we shall refer to such ICs by their type numbers and delete manufacturers' identifying initials. For example, instead of referring to an op-amp as a μA741 or MC1741, we shall refer to it simply as a 741.

Some linear ICs are available in different classes, such as A, C, E, S, and SC. For example, the 741, 741A, 741C, 741E, 741S, and 741SC are different versions of the same op-amp. The 741 is a military-grade op-amp (operating temperature range: -55 to 125°C) and the 741C is a commercial-grade op-amp (operating temperature range: 0 to 70°C/75°C). On the other hand, the 741A and 741E are improved versions of the 741 and 741C, respectively, in that they have improved electrical specifications over their counterparts. The 741C and 741E are identical to the 741 and 741A except that the former have their performance guaranteed over a 0 to 70°/75°C temperature range, instead of −55 to 125°C. The 741S and 741SC are military- and commercial-grade op-amps, respectively, with higher slew rate (rate of change of output voltage per unit of time) than the 741 and 741C.

2-9 DEVELOPMENT OF INTEGRATED CIRCUITS

The development of linear ICs can be traced back to early 1960s, when arrays were first fabricated on a single silicon chip. The arrays are combinations of isolated components such as diodes and transistors or individual stages such as differential amplifiers and Darlington pairs. The use of such IC arrays helped minimize the temperature-drift problem inherent in discrete transistor and diode circuits. It also greatly reduced the size of discrete electronic circuits.

In 1963 Fairchild Semiconductor introduced the first IC op-amp, its μA702, which set the stage for the development of other IC op-amps. The unequal supply voltages, such as $+V_{CC} = -12$ V and $-V_{EE} = -6$ V; relatively low input resistance (40 kΩ typically); and low voltage gain (3600 V/V) were the major drawbacks of the μA702 op-amp. For these reasons the μA702 was not universally accepted.

In 1965 Fairchild introduced the μA709, an improved op-amp compared to the μA702. More specifically, the μA709 had symmetrical supply voltages such as $+V_{CC} = +15$ V and $-V_{EE} = -15$ V; much higher input resistance (400 kΩ, typically); and a voltage gain of 45,000 V/V. The μA709 was the first quality op-amp and is, therefore, remembered for its historical significance. The μA709 is also regarded as a *first-generation* op-amp. Another example of a first-generation IC is the MC1537. The disadvantages of first-generation ICs are as follows:

1. *No short-circuit protection.* The op-amp is susceptible to burnout if output is accidentally shorted to ground.
2. *A possible latch-up problem.* Output voltage can be latched up to some value and then fails to respond to changes in input signal applied.
3. *Requires an external frequency-compensating network* (*two capacitors and a resistor*) *for stable operation.* This means that when an external compensating network is not used, the op-amp may oscillate. Ideally, these compensating components should be internally integrated so that no extra work of connecting them to the op-amp is required.

The next major advancement in IC op-amp technology came in 1968 with the introduction of the Fairchild μA741, an internally compensated op-amp. Unlike the μA709, it has short-circuit protection, has no latch-up problem, and is inherently stable. Besides that, it has very high input resistance (2 MΩ, typically), extremely high voltage gain (200,000 V/V), and offset null capability. It is regarded as a classic among IC op-amps because of its performance, versatility, and economy. It is one of the most widely used general-purpose op-amps in industry even a decade after its introduction. The 741 is an example of *second-generation* op-amps. Other examples of second-generation op-amps are the LM101, LM307, μA748, and MC1558. General-purpose op-amps such as the 741 and 307 are used in the greatest percentage of applications.

The design philosophy and technology used in the 741 was implemented in later-generation op-amps to enhance and optimize their performance characteristics. This trend gave birth to a variety of special-purpose op-amps. These devices had one or more of the following distinguishing features: high slew rate, that is, high rate of change of output voltage per unit of time; wide bandwidth; high input resistance; very low input currents and voltage offsets or drifts; programmability, and other characteristics. For example, the LM318 is a high-speed op-amp with 15-MHz small-signal bandwidth and guaranteed 50-V/μs slew rate. The key features of the μA771 op-amp are a low input bias current of 200 pA and a high slew rate of 13 V/μs, typically. The μA714 is a precision op-amp with low offset voltage (75 μV), low input offset current (2.8 nA), low offset voltage drift (1.3 μV/1°C), and low noise. The μA791 op-amp is a power amplifier with output current capability of 1 A, whereas the μA776 is a multi-purpose programmable op-amp. Thus most later-generation ICs have the improved electrical specifications and hence exhibit superior performance over preceding-generation ICs.

New directions in linear ICs are aimed at the development of building-block circuits which not only include consumer circuits (audio, radio, and TV), but also industrial circuits (timers, regulators, etc.) The prime example of this concept is the emergence of cost-saving quad devices such as the LM324 and MC4741. The trend is to increase component density and yield while keeping the cost down through improved technology. Today's state of the art developments are concerned with BI-FET and laser trimming technology. National Semiconductor is the developer of BI-FET technology, which has introduced wide bandwidths and fast slew rates, together with low bias currents and low input offset currents. For example, the LF351 and LF353 BI-FET op-amps have low input bias and offset currents (200 and 100 pA, respectively), a large unity gain bandwidth (4 MHz), and a fast slew rate (13 V/μs). In addition, these op-amps have the same pin-outs as the 741 and 1458 (dual 741), respectively. This means that designers can immediately upgrade the overall performance of existing designs by simply placing LF351/LF353 op-amps into sockets presently hosting the 741 types. Laser trimming technology, at the same time, has achieved lower input offset voltages. *Trim*ming and BI-*FET* technologies are combined (TRIMFET, for short) by Motorola to provide precision input characteristics at a low cost. Motorola's MC34001/MC34002/MC34004 single, dual, and quad op-amps are the best examples of TRIMFET technologies. These devices are pin-compatible with the 741, 1458, and 324 (quad) op-amps and have nearly identical electrical characteristics to those of the LF351/LF353.

The growth of IC technology and the implementation of new techniques is more evident in the development of digital ICs. Advances in semiconductor technology have had a remarkable effect on the development of digital ICs. The best current example of such developments is the MC68000 microprocessor, which has 70,000 active components integrated on a single chip. The fabrication of such a complex and powerful digital IC was made possible through dry plasma etching, projection printing, and HMOS (high-density, short-channel MOS) circuit design techniques. In short, advances in technology have made ICs easy to use, more reliable, and more flexible in application, while maximizing performance.

2-9.1 SSI, MSI, LSI, and VLSI Packages

ICs are classified according to the number of components (or gates, in the case of digital ICs) integrated on the same chip, as listed below:

Small-scale integration	SSI
Medium-scale integration	MSI
Large-scale integration	LSI
Very large scale integration	VLSI

In the SSI package, the number of components integrated on the same chip is typically < 10. Most of the arrays fall into this group. In the MSI IC, the number of components is < 100, whereas the LSI package includes > 100 components. Almost all the linear integrated circuits and integrated combination logic circuits are MSI packages. Most of the sequential logic circuits are of the LSI type. In the VLSI package, the number of components formed on the same chip is typically > 1000. A classic example of a VLSI package is Motorola's MC68000 microprocessor IC, which houses some 70,000 components on the same chip.

As IC technology progresses and more efficient manufacturing techniques allow the manufacture of more complex yet better quality ICs, the price of an IC package continues to fall. A typical example would be the hand-held calculator. Current calculator prices are remarkably below the levels established when these devices were introduced in the late 1960s.

2-10 INTEGRATED CIRCUIT PACKAGE TYPES, PIN IDENTIFICATION, AND TEMPERATURE RANGES

2-10.1 Package Types

Figure 2-5 shows the three basic types of linear IC packages:

1. The flat pack
2. The metal can or transistor pack
3. The dual-in-line package (for short DIP)

In the *flat pack*, the chip is enclosed in a rectangular ceramic case with terminal leads extending through the sides and ends as shown in Figure 2-5(a). The flat pack comes with 8, 10, 14, or 16 leads. These leads accommodate the power supplies, inputs, outputs, and several special connections required to complete the circuit.

In the *metal can* or *transistor pack*, the chip is encapsulated in a metal or plastic case [see Figure 2-5(b)–(d)]. The transistor pack is available with 3, 5, 8, 10, or 12 pins. Most of the voltage-regulator ICs, such as the LM117, have 3 pins. Power op-amps and audio power amplifiers are usually available in 5-pin packages. The metal can

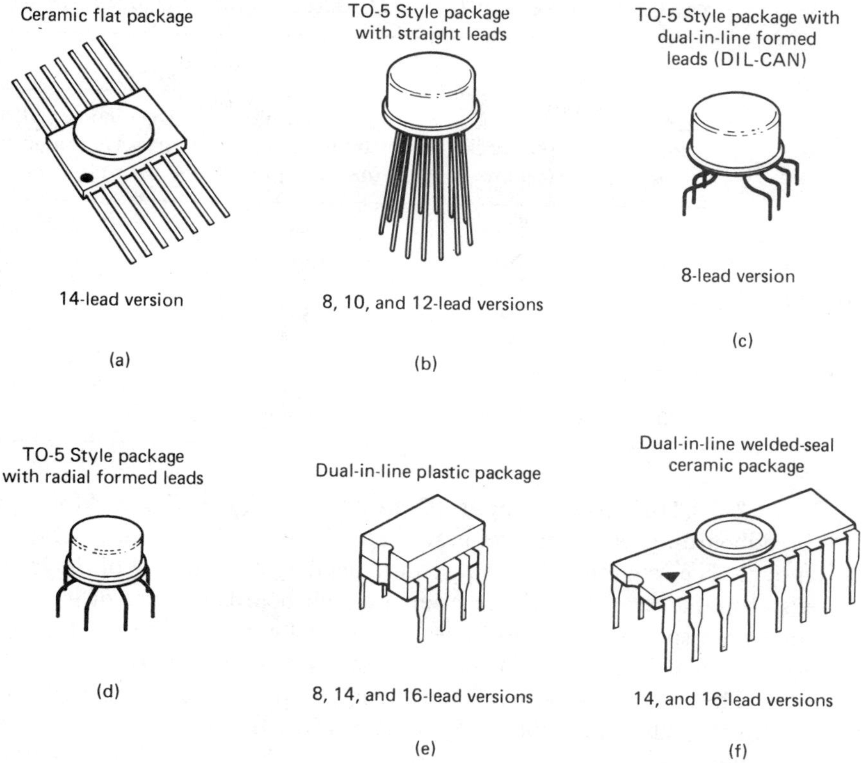

Figure 2-5 Types of IC packages: (a) flat pack; (b–d) metal can; (e, f) dual-in-line package. (Courtesy of RCA Corporation.)

package is best suited for power amplifiers because metal is a good heat conductor and consequently has better dissipation capability than the flat-pack or dual-in-line package. In addition, the metal can package permits the use of external heat sinks. Most of the general-purpose op-amps come in 8, 10, or 12-pin packages.

In the *dual-in-line package* (DIP), the chip is mounted inside a plastic or ceramic case, as shown in Figure 2-5(e) and (f). The DIP is the most widely used package type because it can be mounted easily. The 8-pin dual-in-line packages are referred to as mini DIPs. Dual-in-line packages are also available with 12, 14, 16, and 20 pins. In general, as the density of components integrated on the same chip increases, the number of pins also goes up. This is especially true in digital ICs. For example, there are 64 pins on the MC68000 microprocessor chip, compared to 40 pins on MC6800 microprocessor.

Table 2-1 is a summary of the number of pins and types of packages in which linear integrated circuits are presently available. On the other hand, almost all digital ICs are DIP packages. Metal can packages are also available with dual-in-line formed leads (DIL-CAN) and with radial formed leads as shown in Figure 2-5(c) and (d). Different outlines exist within each package style to accommodate various die sizes and number of pins (leads). For example, TO-99, TO-100, and TO-101 are some of the outlines available in a transistor pack.

TABLE 2-1 LINEAR INTEGRATED CIRCUIT PACKAGES AND PIN NUMBERS

	Type of package		
Number of pins	Flat pack (ceramic)	Metal can or transistor pack (metal or plastic)	Dual-in-line (DIP) package (ceramic or plastic)
3	No	Yes	No
5	No	Yes	No
8	Yes	Yes	Yes (mini DIP)
10	Yes	Yes	No
12	No	Yes	Yes
14	Yes	No	Yes
16	No	No	Yes
20	No	No	Yes

2-10.1 (a) Selecting an IC package. If the IC is used for experimentation/breadboarding purposes, the best choice is the DIP package, because it is easy to mount. The mounting does not require bending or soldering of the leads. The DIP is also suitable for mounting on printed circuit boards because of its lead construction and more spacing between the leads. Generally, the ceramic DIP is more expensive than the plastic DIP, but the ceramic mainly dissipates more heat.

The flat pack is more reliable and lighter than a comparable DIP package and is, therefore, suited for airborne applications. On the other hand, the metal can is the best choice, if the IC is to be operated at relatively high power and expected to dissipate considerable heat.

When all three packages are available for a specific application, the choice can be made based on the relative cost and ease of fabrication/bread boarding the IC. Some applications may not allow a choice of package because the IC is available in only one package style.

2-10.2 Pin Identification

Although there has been a tendency to standardize IC terminal connections, the various manufacturers still use their own systems. Most manufacturers use the pin designations depicted in Figure 2-6 for the three types of packages.

2-10.3 Temperature Ranges

All ICs manufactured fall into one of the three basic temperature grades:

1. Military temperature range: -55 to $+125°C$ (or -55 to $+85°C$)
2. Industrial temperature range: -20 to $+85°C$ (or -40 to $+85°C$)
3. Commercial temperature range: 0 to $+70°C$ (or 0 to $+75°C$)

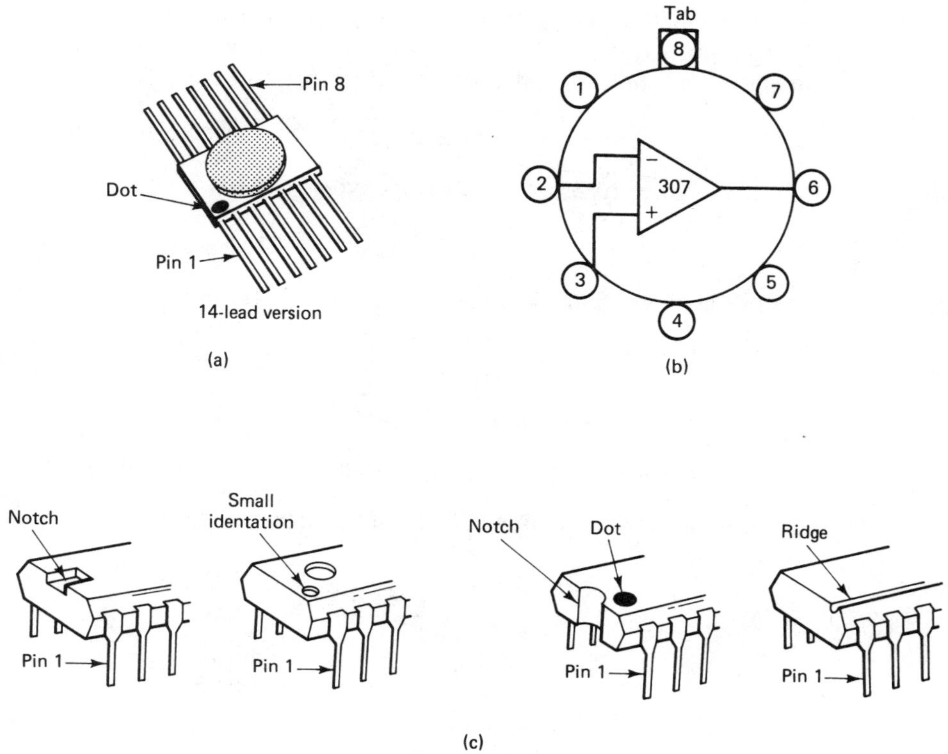

Figure 2-6 Typical pin designations for different packages: (a) flat pack, (b) metal can, (c) DIP.

The individual data sheets specify exact values of the IC parameters and conditions under which these parameters are determined. The standard practice is to specify IC parameters at room temperature, that is, 25°C.

The military- and commercial-grade ICs differ in specifications for supply voltages, input current and voltage offsets and drifts, voltage gains, and others. The military grade devices are almost always of superior quality, with tightly controlled parameters, and consequently cost more. Commercial-grade ICs have the worst tolerances among the three types but are the cheapest. In short, performance and cost are the important factors in selecting an IC.

2-11 ORDERING INFORMATION

Generally, in ordering an IC the following information must be specified: device type, package type, and temperature range. The device type is a group of alphanumeric characters such as μA741, LM741, and MC1741. These characters refer to the data sheet that specifies the device's functional and electrical characteristics. The basic package type—flat pack, transistor pack, or DIP—is represented by one letter. The military, industrial, or commercial temperature range is either numerically specified or included in the device type number or represented by a letter.

Unfortunately, the manufacturers do not have a consistent format by which the ordering information should be specified. For example, the ordering information for Fairchild's 741 mini DIP with commercial temperature range is as follows:

$\underline{\mu A741}$ $\underline{T}$ $\underline{C}$

- $\mu A741$ → Device type (op-amp)
- T → Package type (mini DIP)
- C → Commercial temperature range (0 to 70°C)

On the other hand, the ordering information format for a typical Motorola IC is as follows:

$\underline{MC34001}$ $\underline{P}$ $\underline{0° \text{ to } 70°C}$

- MC34001 → Device type (op-amp)
- P → Package type (plastic DIP)
- 0° to 70°C → Temperature range (commercial)

In National Semiconductor ICs the temperature range is denoted in the device number itself:

$\underline{LM101A}$ $\underline{F}$

- LM101A → Device type and temperature range
- F → Package type (flat pack)

In National's linear ICs, a 1–2–3 numbering system is used to represent the temperature range. For example, in LM101/LM201/LM301 the 1 denotes a military temperature range device, the 2 denotes an industrial temperature range device, and the 3 denotes a commercial temperature range device.

Note that the package letter designations and temperature range nomenclatures are independently defined by each manufacturer. Therefore, the best source of information for ordering an IC is the data book.

2-12 DEVICE IDENTIFICATION

The IC is identified by marking the device type number on the face of the IC. This number is usually accompanied by the date code, indicating the year and week the device was manufactured. Figure 2-7 shows an example of Fairchild's device identification method.

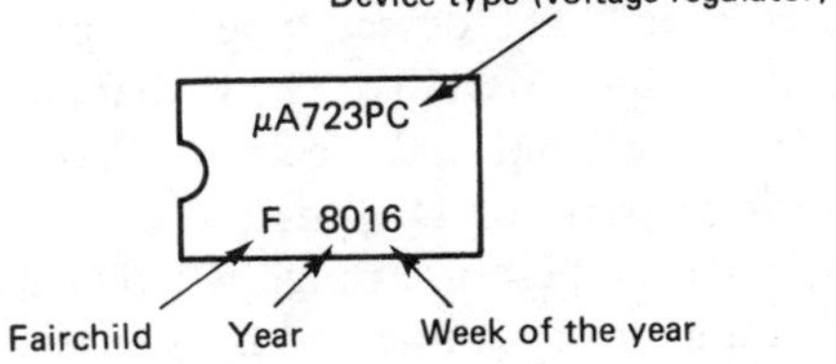

Figure 2-7 Device identification.

2-13 POWER SUPPLIES FOR INTEGRATED CIRCUITS

Most linear ICs (particularly op-amps) use one or more differential amplifier stages, and differential amplifiers require both a positive and negative power supply for proper operation of the circuit. This means that most linear ICs need both a positive and a negative power supply. A few linear ICs (especially earlier op-amps) use unequal power supplies, and some ICs require only a positive supply. For example, the 702 op-amp requires unequal power supplies, whereas the 324 requires only a positive supply. When a single supply is used, it is normally necessary to connect an extra circuit to the IC. Some dual-supply op-amp ICs can also be operated from a single supply voltage, provided that a special external circuit is used with it. Digital ICs, on the other hand, generally require only one positive supply voltage. [An exception is the emitter-coupled logic (ECL)IC.]

The two power supplies required for a linear IC are usually equal in magnitude, +15 V and −15 V, for example. These power supply voltages must be referenced to a common point or ground. Unfortunately, as in the case of ordering information, manufacturers do not agree on power supply labeling. For example, Fairchild uses $V+$ to indicate the positive voltage and $V-$ to indicate the negative voltage or V_S to indicate both positive and negative voltages. On the other hand, Motorola uses the symbols $+V_{CC}$ and $-V_{EE}$ (as in discrete transistor circuits) to represent positive and negative voltages, respectively. We will follow Motorola's notation of the power supplies.

Figure 2-8 shows power supply connections for the 741/351 op-amp. The numerals adjacent to terminals are pin numbers. Thus for the 741/351 op-amp, pin 7 is a positive supply pin and pin 4 is a negative supply pin. The remaining pins on the 741/351 are omitted for the sake of simplicity.

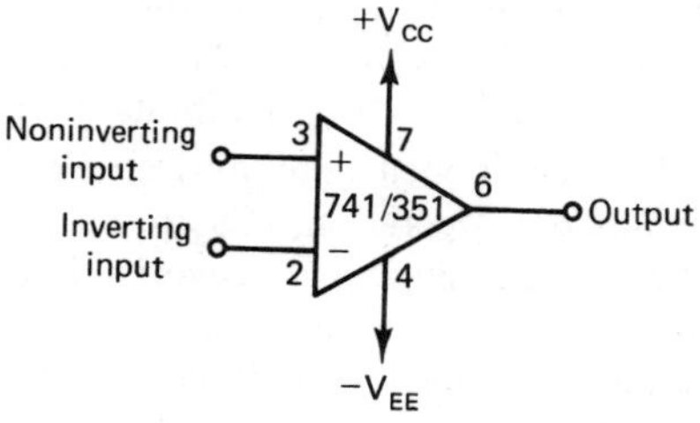

Figure 2-8 Typical linear IC op-amp power supply connections.

Instead of using two separate power supplies, we can use a single power supply to obtain $+V_{CC}$ and $-V_{EE}$, as shown in Figure 2-9. In Figure 2-9(a) the value of R_R should be $> 10\ \text{k}\Omega$ so that it does not draw much current from the supply V_S. The two capacitors provide for decoupling (bypass) of the power supply; they range in value from 0.01 to 10 μF. In Figure 2-9(b) Zener diodes are used to obtain symmetrical supply voltages. The value of R_S should be chosen such that it supplies sufficient current for the diodes to operate in the avalanche mode. The potentiometer is used in Figure 2-9(c) to assure equality between $+V_{CC}$ and $-V_{EE}$ values. Diodes D_1 and D_2 are intended to protect the IC if the positive and negative leads of the supply voltage V_S are accidentally reversed. Note that these diodes could also be connected in Figure 2-9(a) and (b).

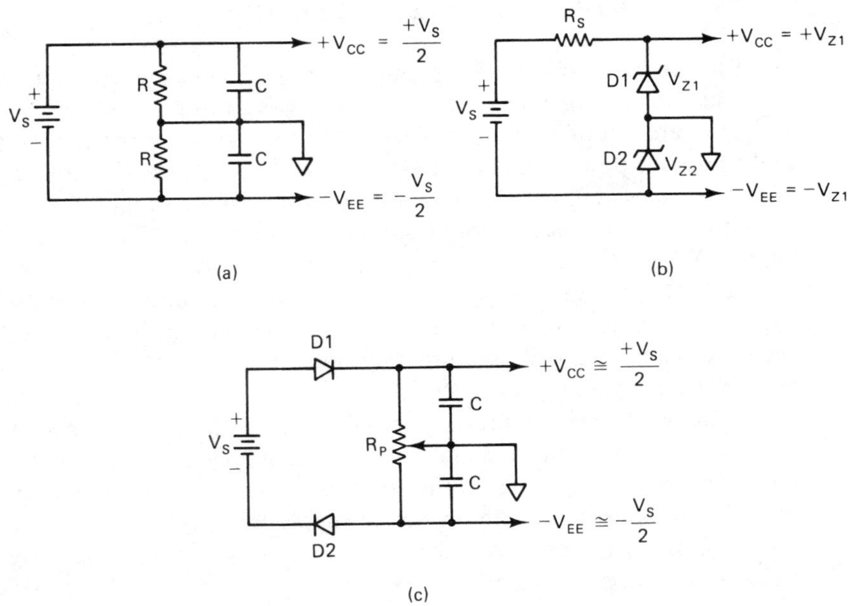

Figure 2-9 Different arrangements of obtaining positive and negative supply voltages for an op-amp.

SUMMARY

1. An operational amplifier is a direct-coupled high-gain amplifier consisting of one or more differential amplifiers, usually followed by a level translator and an output stage. The block diagram representation of the op-amp helps to visualize its internal construction.
2. In integrated circuits (ICs) all the components are fabricated on the same chip (chassis). Classified according to their mode of operation, ICs are of two basic types: digital or linear. Digital ICs are complete functioning logic networks that are equivalents of basic transistor logic circuits. On the other hand, linear ICs are equivalents of discrete transistor networks such as amplifiers, filters, frequency multipliers, and modulators. Most linear ICs are produced by a monolithic process in that all transistors and passive elements (R and C) are fabricated on a single piece of semiconductor material. In hybrid ICs, passive components and the interconnections between them are formed on an insultating substrate. Active components and monolithic ICs are then connected to form a complete circuit.
3. Each manufacturer uses a specific code and assigns a specific type number to the ICs it produces. Since the invention of the op-amp in early 1960s, there have been a number of op-amp (IC) generations. Later-generation ICs have improved specifications and hence exhibit superior performance.
4. Linear ICs are available in three basic types: the flat pack, the metal can, and the dual-in-line (DIP) package. Almost all digital ICs are DIPs.
5. Most linear ICs require both a positive and a negative supply. The standard supply voltages for op-amps are ± 15 V. On the other hand, most digital ICs require only one positive supply voltage.

QUESTIONS

2-1. What is an op-amp?

2-2. List four basic building blocks of an op-amp.

2-3. What is the advantage of using a schematic symbol for an op-amp?

2-4. Explain briefly the difference between digital and linear ICs.

2-5. What is the difference between monolithic and hybrid ICs?

2-6. List the disadvantages of first-generation op-amps, such as the 709.

2-7. What is the major difference between SSI, MSI, LSI, and VLSI ICs?

2-8. List three types of linear IC packages.

2-9. What are the three operating temperature ranges of the IC?

2-10. What is the major difference between the power supply requirements of linear and digital ICs?

2-11. Name two recent op-amps (1980 technology) that are direct replacements for the 741 op-amp (1960 technology).

PROBLEMS

2-1. Referring to the LH0005 op-amp equivalent circuit shown in Figure 2-10, determine (a) the collector current through each transistor and (b) the dc voltage at the output terminal. (Assume that each transistor has $\beta_{dc} = \beta_{ac} = 100$ and $V_{BE} = 0.7$ V.) (c) Draw

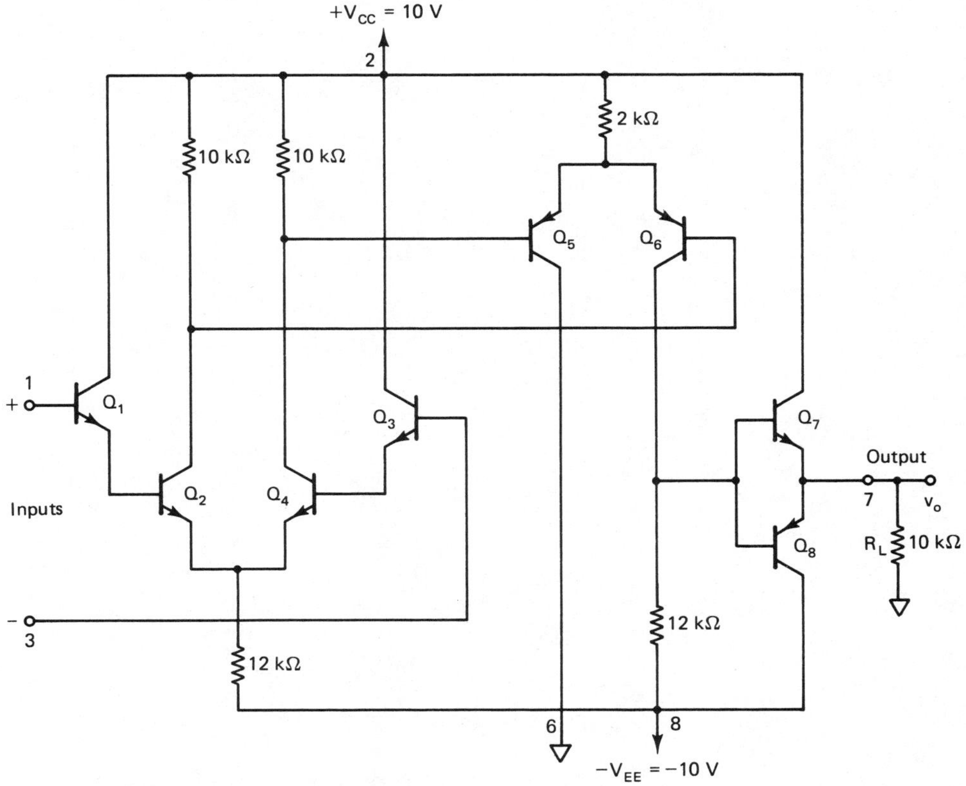

Figure 2-10 Equivalent circuit of the LH0005 for Problem 2-1. (Courtesy of National Semiconductor.)

the block diagram for the circuit shown in Figure 2-10 and explain the function of each block.

2-2. For the op-amp of Problem 2-1, determine:

(a) The overall voltage gain
(b) The input resistance
(c) The maximum output voltage swing

3

Interpretation of Data Sheets and Characteristics of an Op-Amp

3-1 INTRODUCTION

In present-day electronic systems a basic building block is the op-amp. Therefore, a better understanding of the characteristics of op-amps and of their importance to overall circuit operation is essential. The manufacturers supply data sheets for the ICs they produce. These data sheets provide a wealth of information: absolute maximum ratings, intended applications, electrical characteristics, performance limitations, pin diagrams, equivalent circuits of the devices, and more. To get the most use out of these data sheets, we must be able to interpret properly the information presented in them. The lack of standardization of the IC specifications, which includes package type designations, device nomenclature, pin configuration, and ordering information, also makes it imperative to refer to the data sheets.

The purpose of this chapter is to explain how to read the typical op-amp data sheet, to define most electrical parameters given there, and to evaluate their significance. Proper interpretation of data sheet should not only help to understand the characteristics of the op-amp but should also help to select a proper op-amp for a desired application. After these considerations, the chapter discusses equivalent circuit and open-loop op-amp configurations.

3-2 INTERPRETING A TYPICAL SET OF DATA SHEETS

Figure 3-1 shows data sheets for a Fairchild μA741 op-amp. Although the series includes the 741, 741A, 741C, and 741E models, the schematic diagrams and electrical parameters for all of them are the same, with only the values of the parameters differing

μA741

FREQUENCY-COMPENSATED OPERATIONAL AMPLIFIER

FAIRCHILD LINEAR INTEGRATED CIRCUITS

GENERAL DESCRIPTION – The μA741 is a high performance monolithic Operational Amplifier constructed using the Fairchild Planar* epitaxial process. It is intended for a wide range of analog applications. High common mode voltage range and absence of latch-up tendencies make the μA741 ideal for use as a voltage follower. The high gain and wide range of operating voltage provides superior performance in integrator, summing amplifier, and general feedback applications.

- **NO FREQUENCY COMPENSATION REQUIRED**
- **SHORT CIRCUIT PROTECTION**
- **OFFSET VOLTAGE NULL CAPABILITY**
- **LARGE COMMON MODE AND DIFFERENTIAL VOLTAGE RANGES**
- **LOW POWER CONSUMPTION**
- **NO LATCH-UP**

ABSOLUTE MAXIMUM RATINGS

Supply Voltage	
μA741A, μA741, μA741E	±22 V
μA741C	±18 V
Internal Power Dissipation (Note 1)	
Metal Can	500 mW
Molded and Hermetic DIP	670 mW
Mini DIP	310 mW
Flatpak	570 mW
Differential Input Voltage	±30 V
Input Voltage (Note 2)	±15 V
Storage Temperature Range	
Metal Can, Hermetic DIP, and Flatpak	–65°C to +150°C
Mini DIP, Molded DIP	–55°C to +125°C
Operating Temperature Range	
Military (μA741A, μA741)	–55°C to +125°C
Commercial (μA741E, μA741C)	0°C to +70°C
Pin Temperature (Soldering)	
Metal Can, Hermetic DIPs, and Flatpak (60 s)	300°C
Molded DIPs (10 s)	260°C
Output Short Circuit Duration (Note 3)	Indefinite

CONNECTION DIAGRAMS

8-PIN METAL CAN
(TOP VIEW)
PACKAGE OUTLINE 5B
PACKAGE CODE H

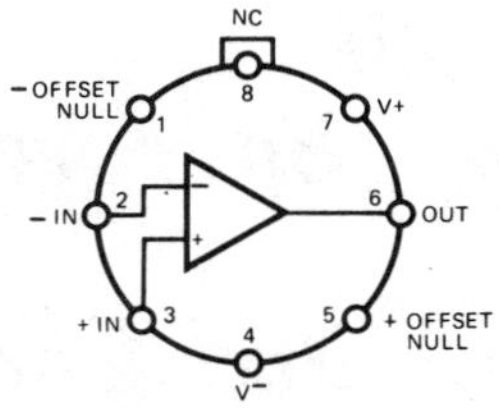

Note: Pin 4 connected to case

ORDER INFORMATION

TYPE	PART NO.
μA741A	μA741AHM
μA741	μA741HM
μA741E	μA741EHC
μA741C	μA741HC

14-PIN DIP
(TOP VIEW)
PACKAGE OUTLINES 6A, 9A
PACKAGE CODES D P

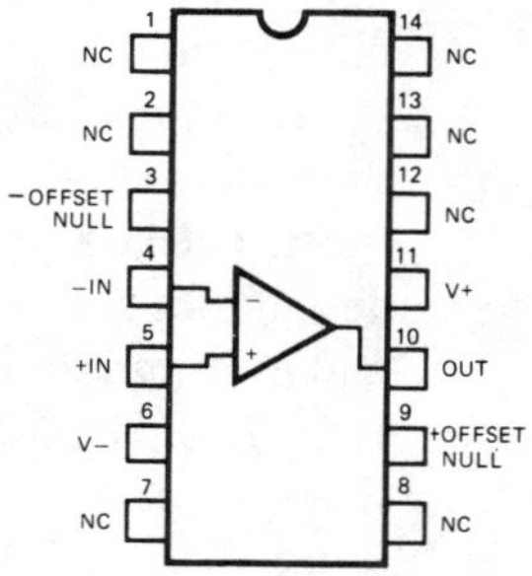

ORDER INFORMATION

TYPE	PART NO.
μA741A	μA741ADM
μA741	μA741DM
μA741E	μA741EDC
μA741C	μA741DC
μA741C	μA741PC

8-PIN MINI DIP
(TOP VIEW)
PACKAGE OUTLINES 6T 9T
PACKAGE CODES R T

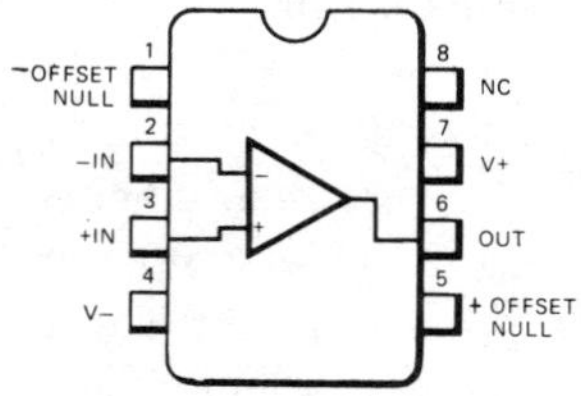

ORDER INFORMATION

TYPE	PART NO.
μA741C	μA741TC
μA741C	μA741RC

10-PIN FLATPAK
(TOP VIEW)
PACKAGE OUTLINE 3F
PACKAGE CODE F

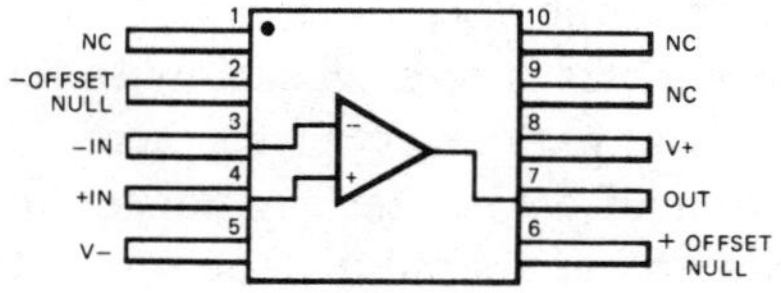

ORDER INFORMATION

TYPE	PART NO.
μA741A	μA741AFM
μA741	μA741FM

Notes on following pages.

*Planar is a patented Fairchild process.

Figure 3-1 Data sheets for the μA741. (Courtesy of Fairchild Camera and Instrument Corporation.)

μA741A

ELECTRICAL CHARACTERISTICS: $V_S = \pm 15$ V, $T_A = 25°C$ unless otherwise specified.

CHARACTERISTICS (see definitions)		CONDITIONS		MIN	TYP	MAX	UNITS
Input Offset Voltage		$R_S \leqslant 50\Omega$			0.8	3.0	mV
Average Input Offset Voltage Drift						15	μV/°C
Input Offset Current					3.0	30	nA
Average Input Offset Current Drift						0.5	nA/°C
Input Bias Current					30	80	nA
Power Supply Rejection Ratio		$V_S = +20, -20; V_S = -20, +10V, R_S = 50\Omega$			15	50	μV/V
Output Short Circuit Current				10	25	40	mA
Power Dissipation		$V_S = \pm 20V$			80	150	mW
Input Impedance		$V_S = \pm 20V$		1.0	6.0		MΩ
Large Signal Voltage Gain		$V_S = \pm 20V, R_L = 2k\Omega, V_{OUT} = \pm 15V$		50			V/mV
Transient Response (Unity Gain)	Rise Time				0.25	0.8	μs
	Overshoot				6.0	20	%
Bandwidth (Note 4)				.437	1.5		MHz
Slew Rate (Unity Gain)		$V_{IN} = \pm 10V$		0.3	0.7		V/μs
The following specifications apply for $-55°C \leqslant T_A \leqslant +125°C$							
Input Offset Voltage						4.0	mV
Input Offset Current						70	nA
Input Bias Current						210	nA
Common Mode Rejection Ratio		$V_S = \pm 20V, V_{IN} = \pm 15V, R_S = 50\Omega$		80	95		dB
Adjustment For Input Offset Voltage		$V_S = \pm 20V$		10			mV
Output Short Circuit Current				10		40	mA
Power Dissipation		$V_S = \pm 20V$	$-55°C$			165	mW
			$+125°C$			135	mW
Input Impedance		$V_S = \pm 20V$		0.5			MΩ
Output Voltage Swing		$V_S = \pm 20V$,	$R_L = 10k\Omega$	±16			V
			$R_L = 2k\Omega$	±15			V
Large Signal Voltage Gain		$V_S = \pm 20V, R_L = 2k\Omega, V_{OUT} = \pm 15V$		32			V/mV
		$V_S = \pm 5V, R_L = 2k\Omega, V_{OUT} = \pm 2$ V		10			V/mV

NOTES

1. Rating applies to ambient temperatures up to 70°C. Above 70°C ambient derate linearly at 6.3mW/°C for the metal can, 8.3mW/°C for the DIP and 7.1mW/°C for the Flatpak.
2. For supply voltages less than ±15V, the absolute maximum input voltage is equal to the supply voltage.
3. Short circuit may be to ground or either supply. Rating applies to +125°C case temperature or 75°C ambient temperature.
4. Calculated value from: $BW(MHz) = \dfrac{0.35}{\text{Rise Time } (\mu s)}$

Figure 3-1 (Continued)

μA741

ELECTRICAL CHARACTERISTICS: $V_S = \pm 15$ V, $T_A = 25°$C unless otherwise specified.

CHARACTERISTICS (see definitions)	CONDITIONS	MIN	TYP	MAX	UNITS
Input Offset Voltage	$R_S \leqslant 10$ kΩ		1.0	5.0	mV
Input Offset Current			20	200	nA
Input Bias Current			80	500	nA
Input Resistance		0.3	2.0		MΩ
Input Capacitance			1.4		pF
Offset Voltage Adjustment Range			±15		mV
Large Signal Voltage Gain	$R_L \geqslant 2$ kΩ, $V_{OUT} = \pm 10$ V	50,000	200,000		
Output Resistance			75		Ω
Output Short Circuit Current			25		mA
Supply Current			1.7	2.8	mA
Power Consumption			50	85	mW
Transient Response (Unity Gain) — Rise time	$V_{IN} = 20$ mV, $R_L = 2$ kΩ, $C_L \leqslant 100$ pF		0.3		μs
Transient Response (Unity Gain) — Overshoot	$V_{IN} = 20$ mV, $R_L = 2$ kΩ, $C_L \leqslant 100$ pF		5.0		%
Slew Rate	$R_L \geqslant 2$ kΩ		0.5		V/μs

The following specifications apply for $-55°C \leqslant T_A \leqslant +125°C$:

CHARACTERISTICS	CONDITIONS	MIN	TYP	MAX	UNITS
Input Offset Voltage	$R_S \leqslant 10$ kΩ		1.0	6.0	mV
Input Offset Current	$T_A = +125°C$		7.0	200	nA
	$T_A = -55°C$		85	500	nA
Input Bias Current	$T_A = +125°C$		0.03	0.5	μA
	$T_A = -55°C$		0.3	1.5	μA
Input Voltage Range		±12	±13		V
Common Mode Rejection Ratio	$R_S \leqslant 10$ kΩ	70	90		dB
Supply Voltage Rejection Ratio	$R_S \leqslant 10$ kΩ		30	150	μV/V
Large Signal Voltage Gain	$R_L \geqslant 2$ kΩ, $V_{OUT} = \pm 10$ V	25,000			
Output Voltage Swing	$R_L \geqslant 10$ kΩ	±12	±14		V
	$R_L \geqslant 2$ kΩ	±10	±13		V
Supply Current	$T_A = +125°C$		1.5	2.5	mA
	$T_A = -55°C$		2.0	3.3	mA
Power Consumption	$T_A = +125°C$		45	75	mW
	$T_A = -55°C$		60	100	mW

TYPICAL PERFORMANCE CURVES FOR μA741A AND μA741

OPEN LOOP VOLTAGE GAIN AS A FUNCTION OF SUPPLY VOLTAGE

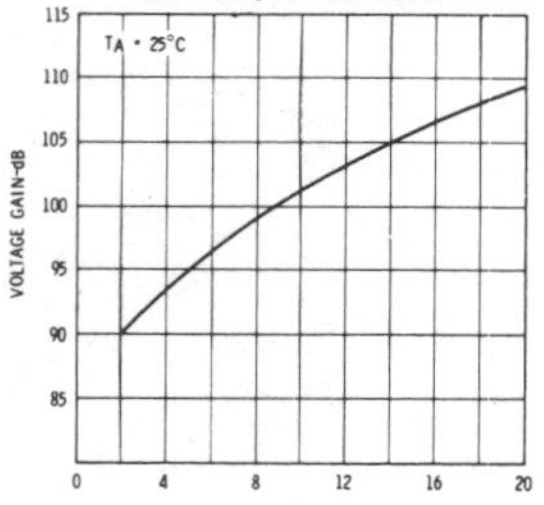

OUTPUT VOLTAGE SWING AS A FUNCTION OF SUPPLY VOLTAGE

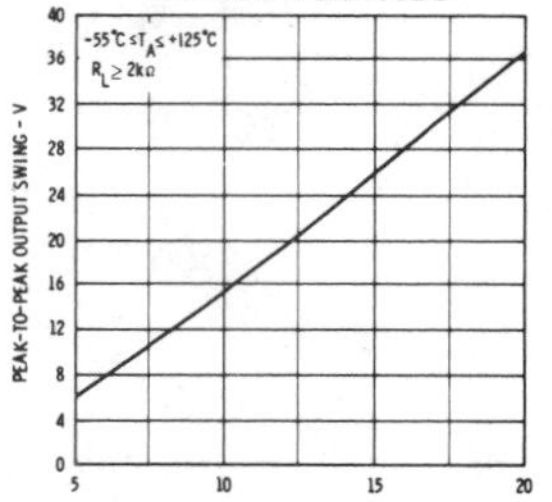

INPUT COMMON MODE VOLTAGE RANGE AS A FUNCTION OF SUPPLY VOLTAGE

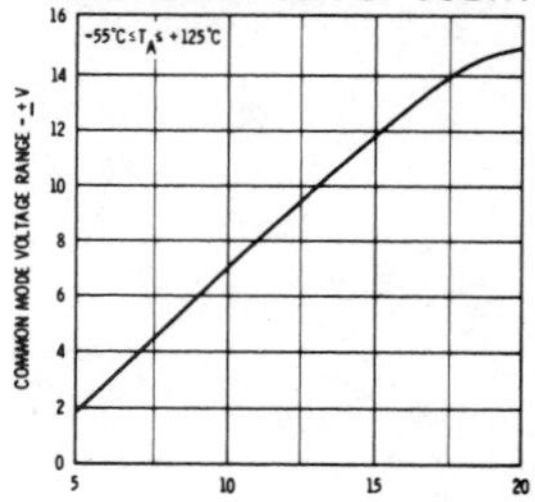

Figure 3-1 (Continued)

μA741E

ELECTRICAL CHARACTERISTICS: $V_S = \pm 15$ V, $T_A = 25°C$ unless otherwise specified.

CHARACTERISTICS (see definitions)		CONDITIONS		MIN	TYP	MAX	UNITS
Input Offset Voltage		$R_S \leq 50\Omega$			0.8	3.0	mV
Average Input Offset Voltage Drift						15	μV/°C
Input Offset Current					3.0	30	nA
Average Input Offset Current Drift						0.5	nA/°C
Input Bias Current					30	80	nA
Power Supply Rejection Ratio		$V_S = +10, -20$; $V_S = +20, -10V$, $R_S = 50\Omega$			15	50	μV/V
Output Short Circuit Current				10	25	40	mA
Power Dissipation		$V_S = \pm 20V$			80	150	mW
Input Impedance		$V_S = \pm 20V$		1.0	6.0		MΩ
Large Signal Voltage Gain		$V_S = \pm 20V$, $R_L = 2k\Omega$, $V_{OUT} = \pm 15V$		50			V/mV
Transient Response	Rise Time				0.25	0.8	μs
(Unity Gain)	Overshoot				6.0	20	%
Bandwidth (Note 4)				.437	1.5		MHz
Slew Rate (Unity Gain)		$V_{IN} = \pm 10V$		0.3	0.7		V/μs
The following specifications apply for $0°C \leq T_A \leq 70°C$							
Input Offset Voltage						4.0	mV
Input Offset Current						70	nA
Input Bias Current						210	nA
Common Mode Rejection Ratio		$V_S = \pm 20V$, $V_{IN} = \pm 15V$, $R_S = 50\Omega$		80	95		dB
Adjustment For Input Offset Voltage		$V_S = \pm 20V$		10			mV
Output Short Circuit Current				10		40	mA
Power Dissipation		$V_S = \pm 20V$				150	mW
Input Impedance		$V_S = \pm 20V$		0.5			MΩ
Output Voltage Swing		$V_S = \pm 20V$,	$R_L = 10k\Omega$	±16			V
			$R_L = 2k\Omega$	±15			V
Large Signal Voltage Gain		$V_S = \pm 20V$, $R_L = 2k\Omega$, $V_{OUT} = \pm 15V$		32			V/mV
		$V_S = \pm 5V$, $R_L = 2k\Omega$, $V_{OUT} = \pm 2$ V		10			V/mV

EQUIVALENT CIRCUIT

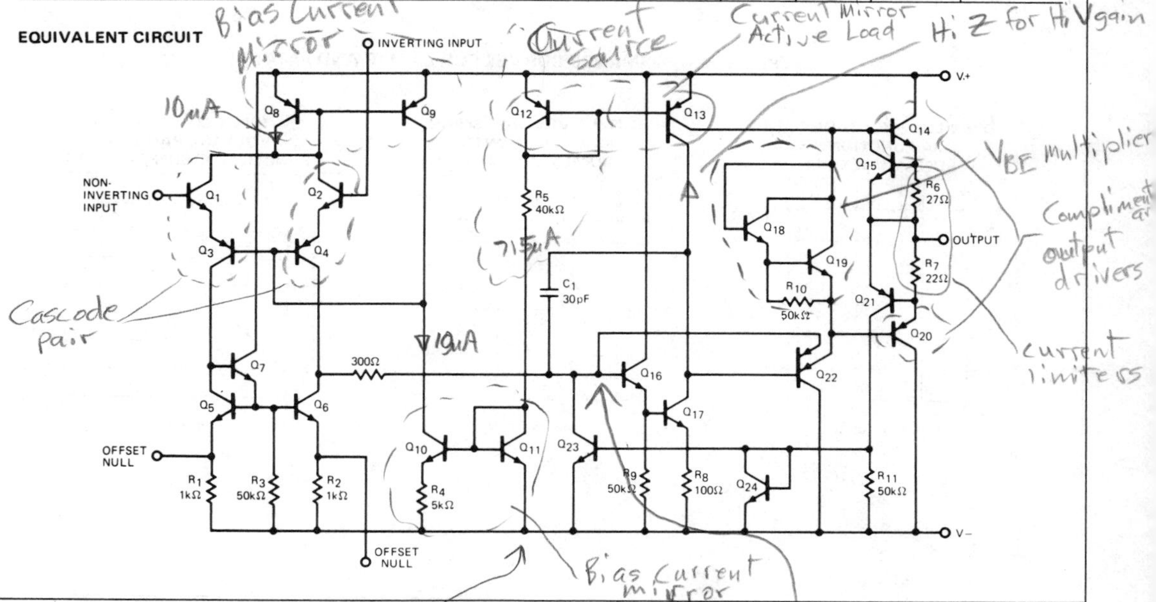

Figure 3-1 (Continued)

Current set up by this branch $= I_{ref} = \frac{30-1.4}{40K} = 715\mu A$

μA741C

ELECTRICAL CHARACTERISTICS: $V_S = \pm 15$ V, $T_A = 25°C$ unless otherwise specified.

CHARACTERISTICS (see definitions)		CONDITIONS	MIN	TYP	MAX	UNITS
Input Offset Voltage		$R_S \leq 10$ kΩ		2.0	6.0	mV
Input Offset Current				20	200	nA
Input Bias Current				80	500	nA
Input Resistance			0.3	2.0		MΩ
Input Capacitance				1.4		pF
Offset Voltage Adjustment Range				±15		mV
Input Voltage Range			±12	±13		V
Common Mode Rejection Ratio		$R_S \leq 10$ kΩ	70	90		dB
Supply Voltage Rejection Ratio		$R_S \leq 10$ kΩ		30	150	μV/V
Large Signal Voltage Gain		$R_L \geq 2$ kΩ, $V_{OUT} = \pm 10$ V	20,000	200,000		
Output Voltage Swing		$R_L \geq 10$ kΩ	±12	±14		V
		$R_L \geq 2$ kΩ	±10	±13		V
Output Resistance				75		Ω
Output Short Circuit Current				25		mA
Supply Current				1.7	2.8	mA
Power Consumption				50	85	mW
Transient Response (Unity Gain)	Rise time	$V_{IN} = 20$ mV, $R_L = 2$ kΩ, $C_L \leq 100$ pF		0.3		μs
	Overshoot			5.0		%
Slew Rate		$R_L \geq 2$ kΩ		0.5		V/μs
The following specifications apply for $0°C \leq T_A \leq +70°C$:						
Input Offset Voltage					7.5	mV
Input Offset Current					300	nA
Input Bias Current					800	nA
Large Signal Voltage Gain		$R_L \geq 2$ kΩ, $V_{OUT} = \pm 10$ V	15,000			
Output Voltage Swing		$R_L \geq 2$ kΩ	±10	±13		V

TYPICAL PERFORMANCE CURVES FOR μA741E AND μA741C

OPEN LOOP VOLTAGE GAIN AS A FUNCTION OF SUPPLY VOLTAGE

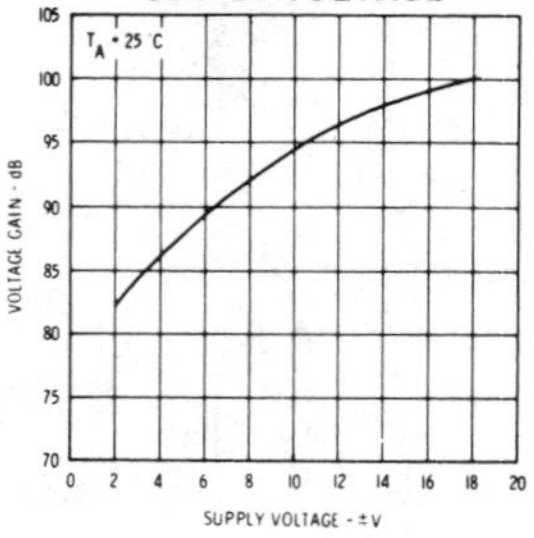

OUTPUT VOLTAGE SWING AS A FUNCTION OF SUPPLY VOLTAGE

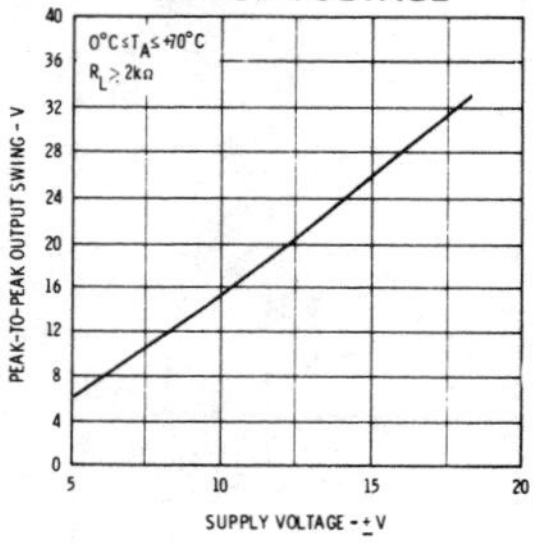

INPUT COMMON MODE VOLTAGE RANGE AS A FUNCTION OF SUPPLY VOLTAGE

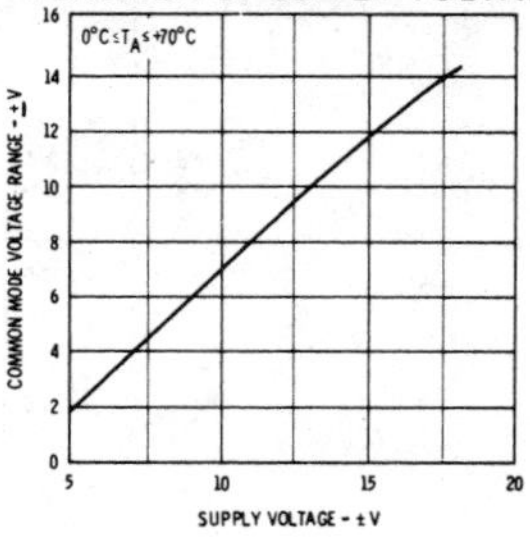

Figure 3-1 (Continued)

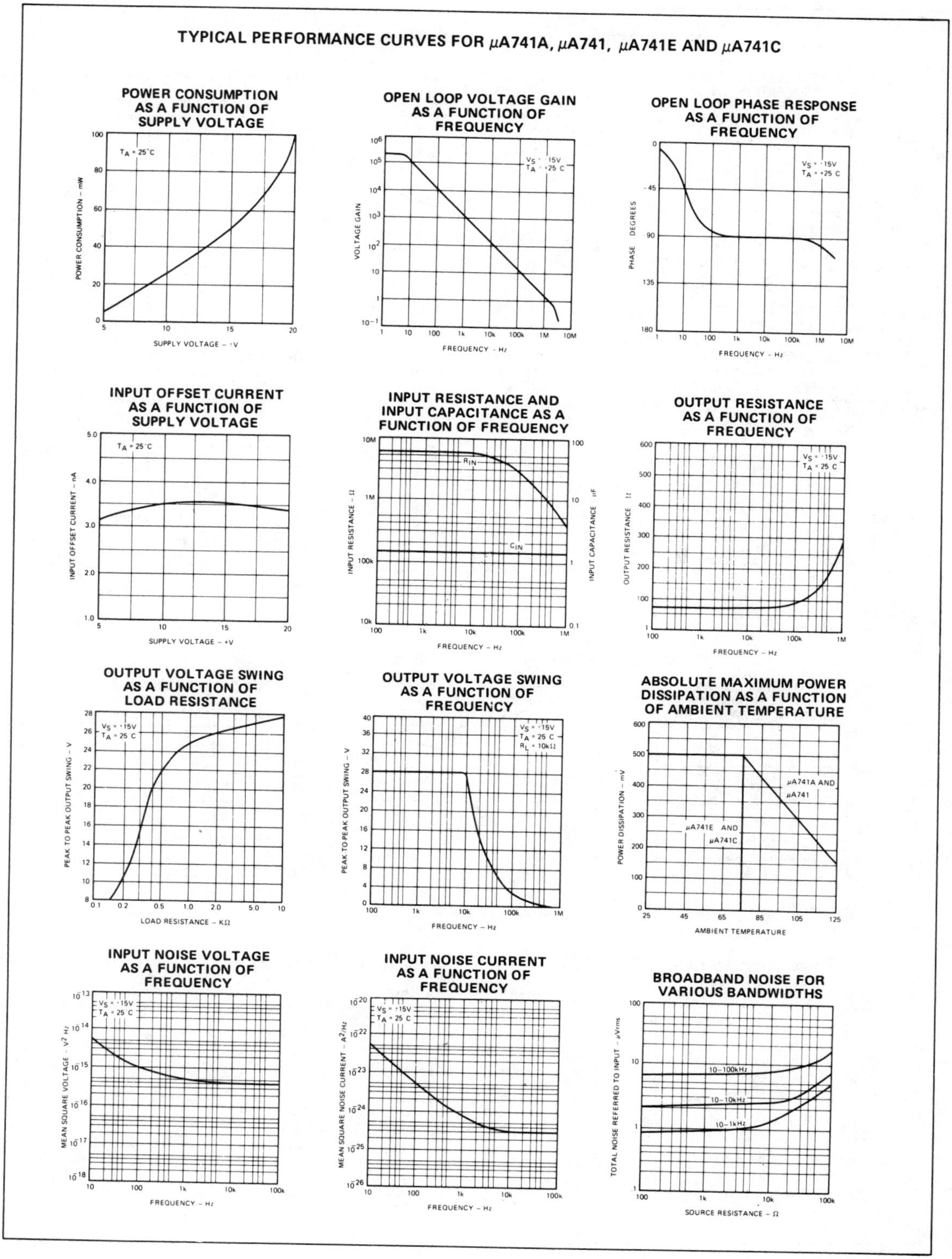

Figure 3-1 (Continued)

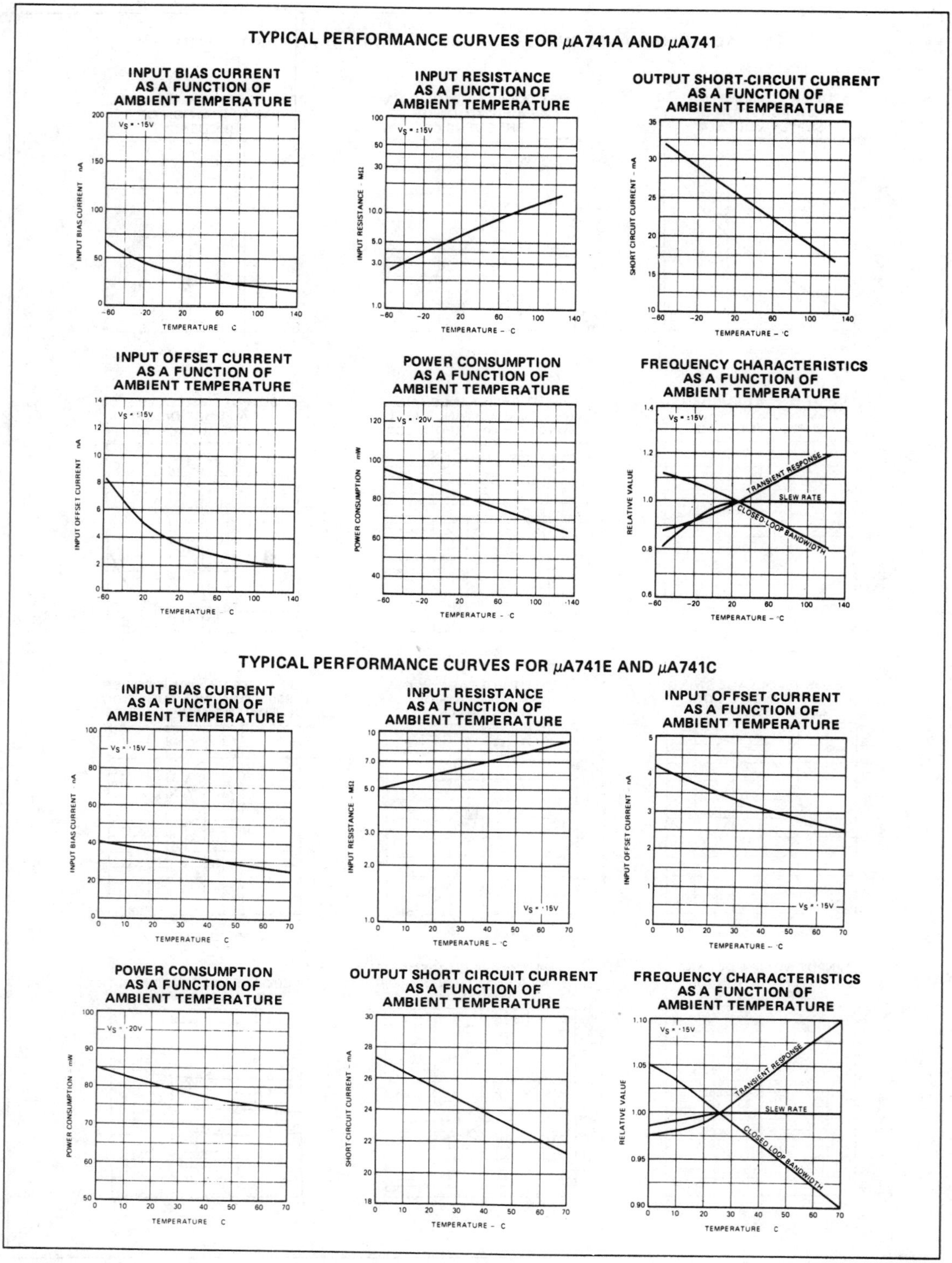

Figure 3-1 (Continued)

TRANSIENT RESPONSE

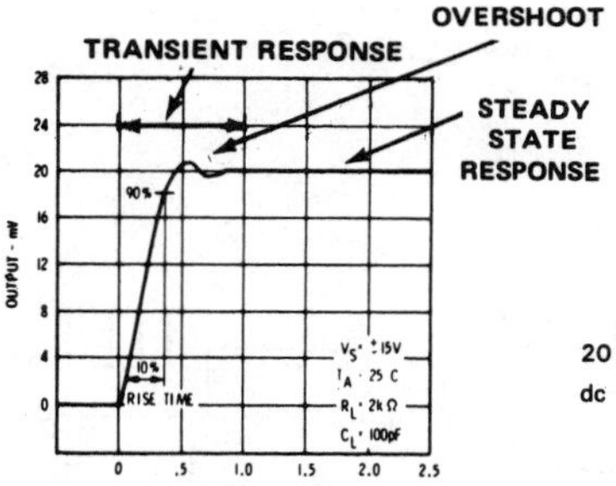

TRANSIENT RESPONSE TEST CIRCUIT

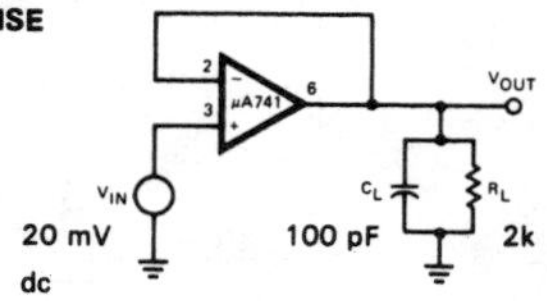

COMMON MODE REJECTION RATIO AS A FUNCTION OF FREQUENCY

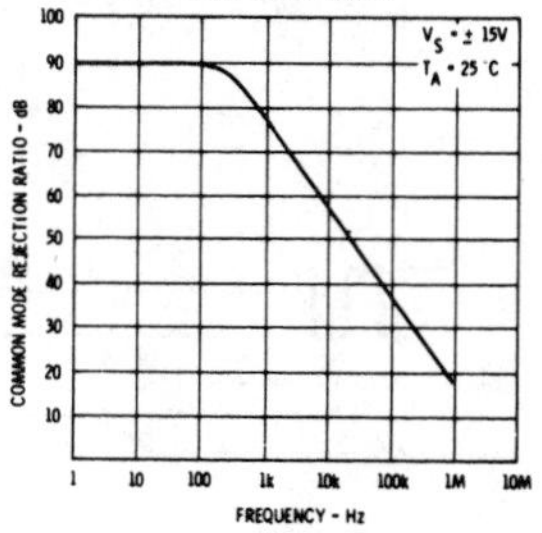

FREQUENCY CHARACTERISTICS AS A FUNCTION OF SUPPLY VOLTAGE

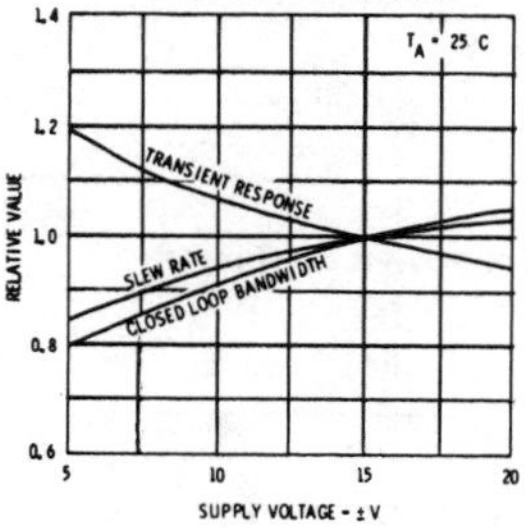

VOLTAGE OFFSET NULL CIRCUIT

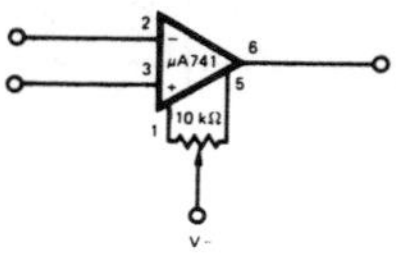

VOLTAGE FOLLOWER LARGE SIGNAL PULSE RESPONSE

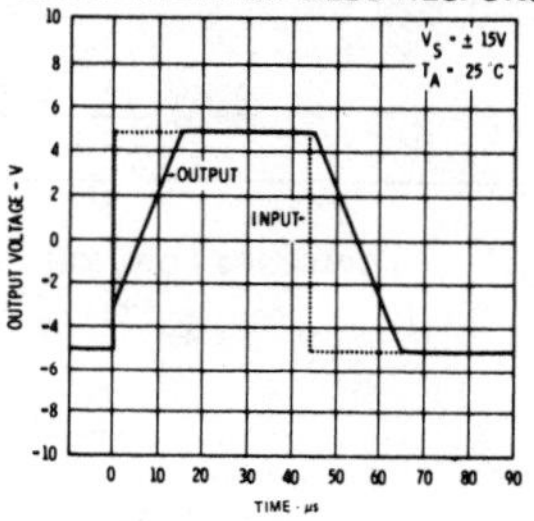

TYPICAL APPLICATIONS

UNITY-GAIN VOLTAGE FOLLOWER

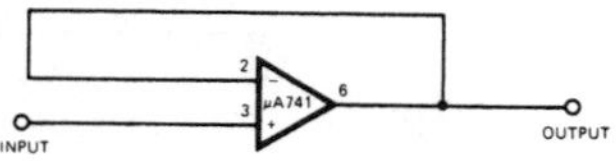

$R_{IN} = 400\ M\Omega$

$C_{IN} = 1\ pF$

$R_{OUT} << 1\ \Omega$

B.W. = 1 MHz

NON-INVERTING AMPLIFIER

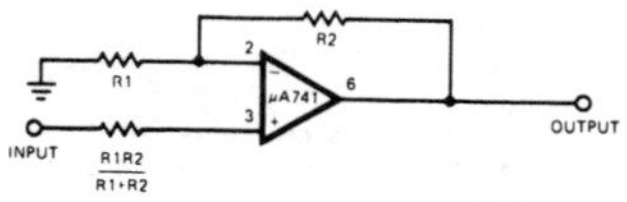

GAIN	R1	R2	B W	R_{IN}
10	1 kΩ	9 kΩ	100 kHz	400 MΩ
100	100 Ω	9.9 kΩ	10 kHz	280 MΩ
1000	100 Ω	99.9 kΩ	1 kHz	80 MΩ

INVERTING AMPLIFIER

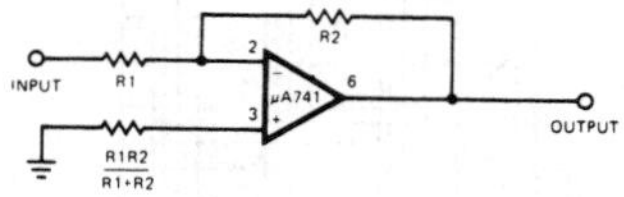

GAIN	R1	R2	B W	R_{IN}
1	10 kΩ	10 kΩ	1 MHz	10 kΩ
10	1 kΩ	10 kΩ	100 kHz	1 kΩ
100	1 kΩ	100 kΩ	10 kHz	1 kΩ
1000	100 Ω	100 kΩ	1 kHz	100 Ω

CLIPPING AMPLIFIER

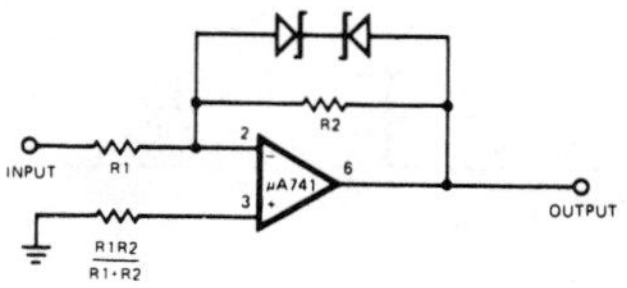

$$\frac{E_{OUT}}{E_{IN}} = \frac{R2}{R1} \text{ if } |E_{OUT}| \leq V_Z + 0.7\ V$$

where V_Z = Zener breakdown voltage

Figure 3-1 (Continued)

TYPICAL APPLICATIONS (Cont'd)

SIMPLE INTEGRATOR

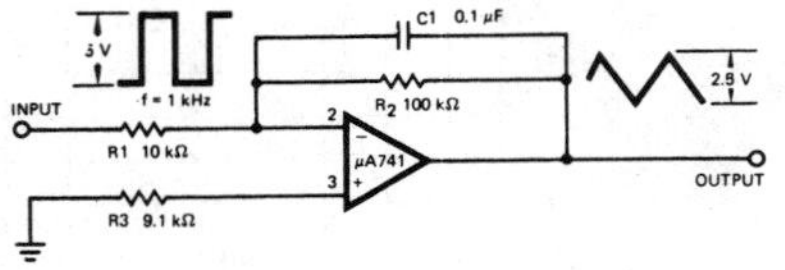

$$E_{OUT} = -\frac{1}{R_1C_1}\int E_{IN}dt$$

SIMPLE DIFFERENTIATOR

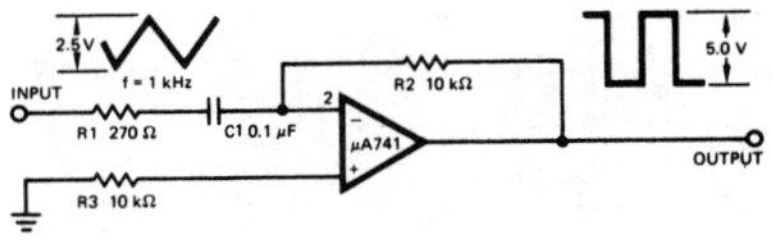

$$E_{OUT} = -R2C\frac{dE_{IN}}{dt}$$

LOW DRIFT LOW NOISE AMPLIFIER

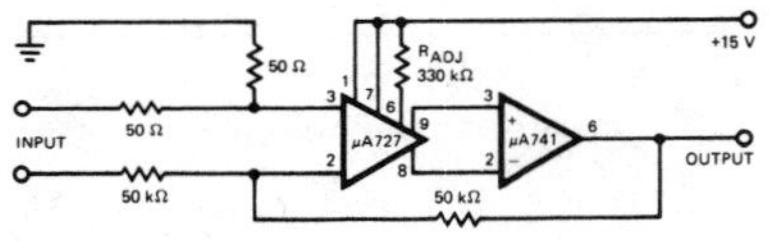

Voltage Gain = 10^3
Input Offset Voltage Drift = 0.6 μV/°C
Input Offset Current Drift = 2.0 pA/°C

HIGH SLEW RATE POWER AMPLIFIER

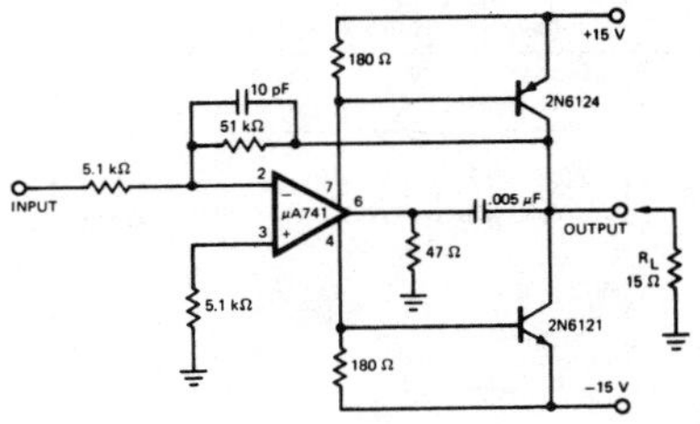

NOTCH FILTER USING THE μA741 AS A GYRATOR

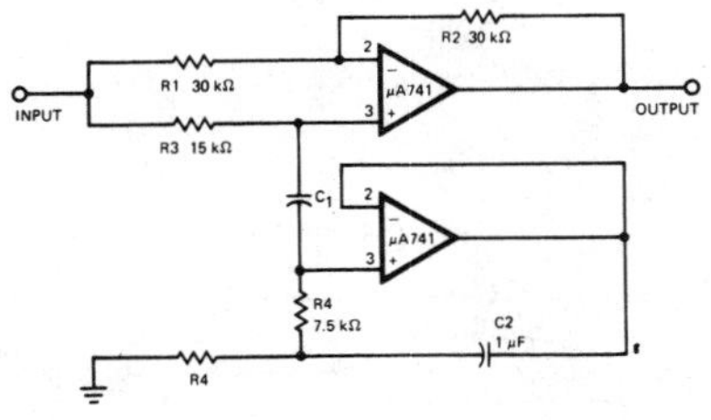

Trim R3 such that

$$\frac{R1}{R2} = \frac{R3}{2\ R4}$$

NOTCH FREQUENCY AS A FUNCTION OF C_1

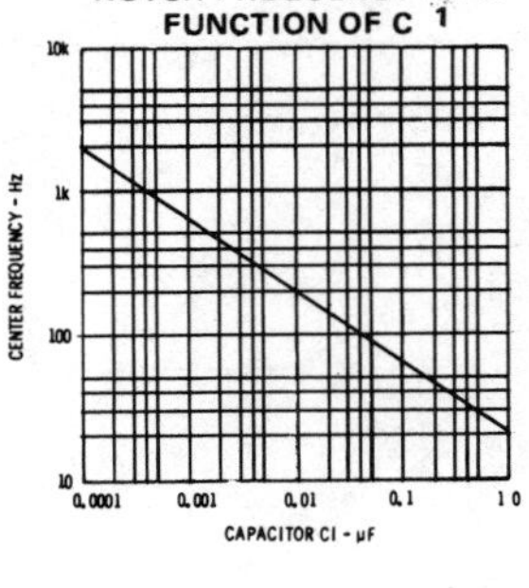

Figure 3-1 (Continued)

from one model to another. For instance, the 741A has the best tolerance (tightly controlled parameters) and costs the most. At the other extreme, the 741C has the worst tolerance and consequently costs the least. In this section we consider only the 741C op-amp specifications.

Generally, the information found on data sheets can be broken down into the following groups:

1. At the top of the data sheet is a device number and a brief description of the basic type of the device, such as *frequency-compensated op-amp*, *low-power op-amp*, or *low-cost programmable op-amp*.
2. Next, a general description is given that includes the construction process of the device, intended applications, and a list of the main features.
3. Absolute maximum ratings for the proper operation of the device are then specified. These ratings are limiting values of operating and environmental conditions applicable to the device and should not be exceeded.
4. The pin configuration (connection diagram), package types, and order information is given.
5. The internal schematic diagram (equivalent circuit) is shown.
6. Electrical characteristics and parameter values under specific conditions are also given.
7. Typical performance curves such as voltage gain versus supply voltage, output voltage swing as a function of frequency, and power consumption as a function of temperature are provided.
8. Finally, typical applications and test circuits for the device are illustrated.

Let us now take a close look at the μA741C data sheets with reference to the information outlined above.

1. The Fairchild 741 is an internally frequency compensated operational amplifier.

2. The 741 is a monolithic IC constructed by a special process called "planer epitaxial." It is suited for integrator, summing amplifier, voltage follower, and other general feedback applications. The features of the 741 are as follows:

a. No external frequency compensation required
b. Short-circuit protection
c. Offset null capability
d. Large common-mode and differential voltage ranges
e. Low power consumption
f. No latch-up problem

3. Absolute maximum ratings are specified for supply voltage, input and differential input voltages, storage and operating temperature ranges, soldering pin temperature, and output short-circuit duration. Supply voltage and operating temperature range are given for Models 741, 741A, 741C, and 741E, whereas internal

power dissipation, storage temperature range, and soldering pin temperature are listed for all package types: flat pack, metal can, and DIP. For safe and proper operation of the device, these ratings should not be exceeded even under the worst operating conditions.

4. The 741 is available in all *three* package types: 8-pin metal can, 10-pin flat pack, and 8- or 14-pin DIP. The pin diagram for all these packages is shown on the data sheets. Order information is included for each package as well.

5. The equivalent circuit diagram illustrates the internal structure of the 741 op-amp. This diagram helps to clarify the capabilities and limitations of the op-amp. Note that all the 741 models have the same equivalent circuit.

6. For the 741C, two sets of electrical specifications are given. One set of specifications applies at room temperature (T_A) of 25°C, whereas the other set applies to the commercial temperature range 0 to +70°C. Since our main interest at this point is in the definition and significance of such parameters rather than their values, we shall discuss the parameters at 25°C. These parameters are applicable at supply voltages $+V_{CC} = +15$ V and $-V_{EE} = -15$ V. The electrical parameters are defined in the following paragraphs.

Input Offset Voltage. Input offset voltage is the voltage that must be applied between the two input terminals of an op-amp to null the output, as shown in Figure 3-2. In the figure V_{dc1} and V_{dc2} are dc voltages and R_s represents the source resistance. We denote input offset voltage by V_{io}. This voltage V_{io} could be positive or negative; therefore, its absolute value is listed on the data sheet. For a 741C the maximum value of V_{io} is 6 mV dc. The smaller the value of V_{io}, the better the input terminals are matched. For instance, the 714C *precision op-amp* has $V_{io} = 150\ \mu$V maximum.

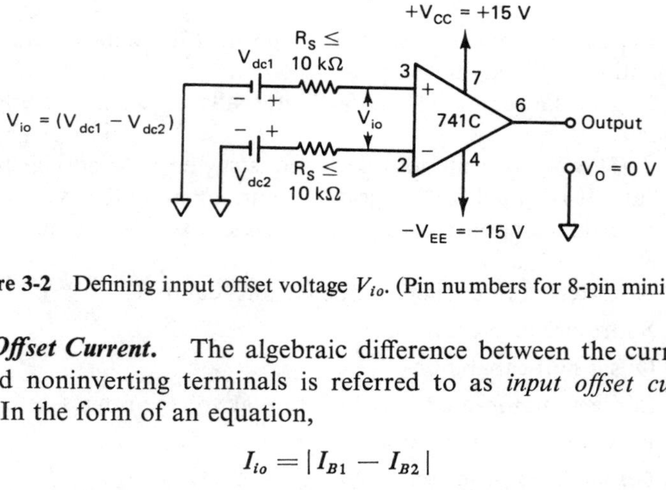

Figure 3-2 Defining input offset voltage V_{io}. (Pin numbers for 8-pin mini DIP.)

Input Offset Current. The algebraic difference between the currents into the inverting and noninverting terminals is referred to as *input offset current* I_{io} (see Figure 3-3). In the form of an equation,

$$I_{io} = |I_{B1} - I_{B2}| \tag{3-1}$$

where I_{B1} is the current into the noninverting input, I_{B2} the current into the inverting input.

The input offset current for the 741C is 200 nA maximum. As the matching between two input terminals is improved, the difference between I_{B1} and I_{B2} becomes smaller; that is, the I_{io} value decreases further. For instance, the precision op-amp

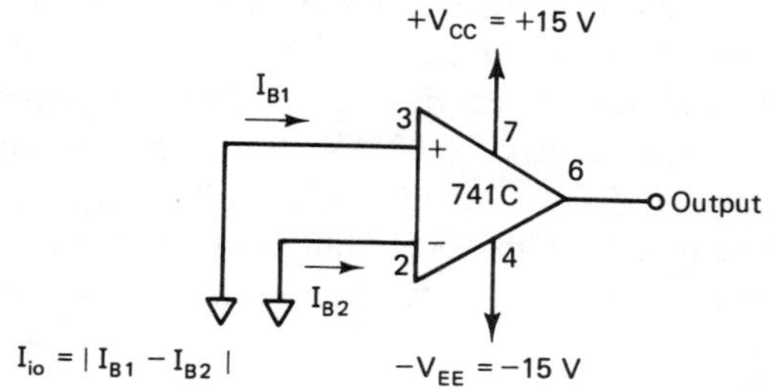

Figure 3-3 Defining input offset current I_{io}.

714C has a maximum value of I_{io} equal to 6 nA, a dramatic improvement over older technology.

Input Bias Current. Input bias current I_B is the average of the currents that flow into the inverting and noninverting input terminals of the op-amp. In equation form,

$$I_B = \frac{I_{B1} + I_{B2}}{2} \tag{3-2}$$

$I_B = 500$ nA maximum for the 741C, whereas I_B for the precision 714C is ± 7 nA.

Note that the two input currents I_{B1} and I_{B2} are actually the base currents of the first differential amplifier stage.

Differential Input Resistance. Differential input resistance R_i (often referred to as input resistance) is the equivalent resistance that can be measured at either the inverting or noninverting input terminal with the other terminal connected to ground. For the 741C the input resistance is a relatively high 2 MΩ. However, for FET input op-amps this value is amazingly large. For example, $R_i = 1000$ GΩ (10^{12} Ω) for the μAF771 FET input op-amp.

Input Capacitance. Input capacitance C_i is the equivalent capacitance that can be measured at either the inverting or noninverting terminal with the other terminal connected to ground. A typical value of C_i is 1.4 pF for the 741C. This parameter is not listed on all op-amp data sheets.

Offset Voltage Adjustment Range. One of the features of the 741 family op-amps is an offset voltage null capability. The 741 op-amps have pins 1 and 5 marked as *offset null* for this purpose. As shown in Figure 3-4, a 10-kΩ potentiometer can be connected between offset null pins 1 and 5, and the wiper of the potentiometer can be connected to the negative supply $-V_{EE}$. By varying the potentiometer, the

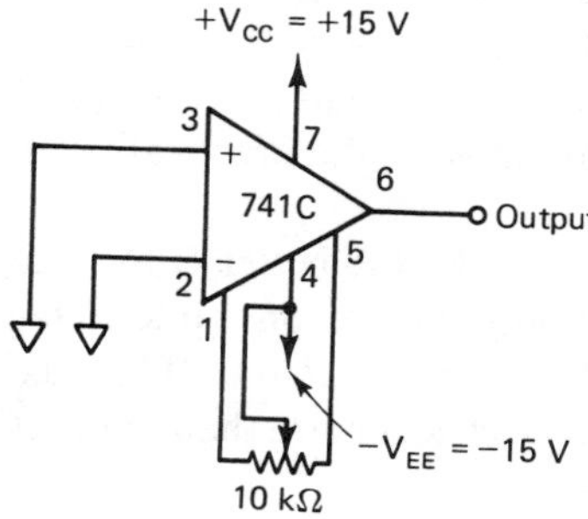

Figure 3-4 Offset voltage adjustment range.

output offset voltage (output voltage without any input applied) can be reduced to zero volts. Thus the offset voltage adjustment range is the range through which the input offset voltage can be adjusted by varying the 10-kΩ potentiometer. For the 741C the offset voltage adjustment range is ±15 mV. Very few op-amps have the offset voltage null capability, some of these being the 301, 748, and 777. This means that for most op-amps we have to design an offset voltage compensating network in order to reduce the output offset voltage to zero. The design of such a network is presented in Chapter 5.

Input Voltage Range. When the same voltage is applied to both input terminals, the voltage is called a common-mode voltage V_{cm} and the op-amp is said to be operating in the common-mode configuration (see Figure 3-5). For the 741C the range of the input common-mode voltage is ±13 V maximum. This means that the common-mode voltage applied to both input terminals can be as high as +13 V or as low as −13 V without disturbing proper functioning of the op-amp. In other words, the input voltage range is the range of common-mode voltages over which the offset specifications apply. Obviously, the common-mode configuration is used only for test purposes to determine the degree of matching between the inverting and non-inverting input terminals.

Common-Mode Rejection Ratio. The common-mode rejection ratio (CMRR) is defined in several essentially equivalent ways by the various manufacturers. Generally, it can be defined as the ratio of the differential voltage gain A_d to the common-mode voltage gain A_{cm}; that is,

$$\text{CMRR} = \frac{A_d}{A_{cm}} \tag{3-3}$$

The differential voltage gain A_d is the same as the large-signal voltage gain A which is specified on the data sheets; however, the common-mode voltage gain can be determined from the circuit of Figure 3-5 using the equation

$$A_{cm} = \frac{V_{ocm}}{V_{cm}} \tag{3-4}$$

where V_{ocm} = output common-mode voltage
V_{cm} = input common-mode voltage
A_{cm} = common-mode voltage gain

Generally, the A_{cm} is very small and $A_d = A$ is very large; therefore, the CMRR is very large. Being a large value, CMRR is most often expressed in decibels (dB). For the 741C, CMRR is 90 dB typically. Note that this value of CMRR is determined under the test condition that the input source resistance $R_s \leq 10$ kΩ. In Figure 3-5 R_s is assumed to be zero because most of the practical voltage sources have negligible source resistances.

The higher the value of CMRR, the better is the matching between two input terminals and the smaller is the output common-mode voltage. For the 714C precision op-amp CMRR = 120 dB. This means that the 714C has a better ability to reject common-mode voltages such as electrical noise than the 741C and is preferred in noise environments.

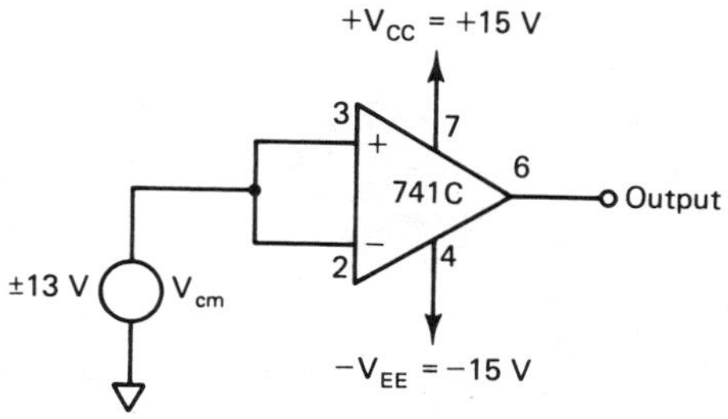

Figure 3-5 Common-mode configuration

Supply Voltage Rejection Ratio. The change in an op-amp's input offset voltage V_{io} caused by variations in supply voltages is called the *supply voltage rejection ratio* (SVRR). A variety of terms equivalent to SVRR are used by different manufacturers, such as the *power supply rejection ratio* (PSRR) and the *power supply sensitivity* (PSS). These terms are expressed either in microvolts per volt or in decibels. If we denote the change in supply voltages by ΔV and the corresponding change in input offset voltage by ΔV_{io}, SVRR can be defined as follows:

$$\text{SVRR} = \frac{\Delta V_{io}}{\Delta V} \tag{3-5}$$

For the 741C, SVRR = 150 μV/V. On the other hand, for the 714C,

$$\text{SVRR} = 20 \log \left(\frac{\Delta V}{\Delta V_{io}}\right) = 104 \text{ dB} \tag{3-6}$$

or, equivalently,

$$\text{SVRR} = 6.31\ \mu\text{V/V}$$

This means that the lower the value of SVRR in microvolts/volt, the better for op-amp performance.

Note that for the 741C (and most other op-amps) SVRR is measured for both supply magnitudes increasing or decreasing simultaneously, with $R_s \leq 10$ kΩ. However, there are some op-amps for which SVRR is separately specified as *positive SVRR* for positive voltage and *negative SVRR* for negative supply voltage.

Large-Signal Voltage Gain. Since the op-amp amplifies difference voltage between two input terminals, the voltage gain of the amplifier is defined as

$$\text{voltage gain} = \frac{\text{output voltage}}{\text{differential input voltage}}$$

that is,

$$A = \frac{V_o}{V_{id}} \tag{3-7}$$

Because output signal amplitude is much larger than the input signal, the voltage gain is commonly called *large-signal voltage gain.* Under the test conditions $R_L \geq 2$ kΩ and $V_o = \pm 10$ V (or 20 V peak to peak), the large-signal voltage gain of the 741C is 200,000 typically (see Figure 3-6).

Output Voltage Swing. The output voltage swing V_{omax} of the 741C is guaranteed to be between -13 and $+13$ V for $R_L \geq 2$ kΩ, that is, giving a 26-V peak-to-peak undistorted sine wave for ac input signals. In fact, the output voltage swing

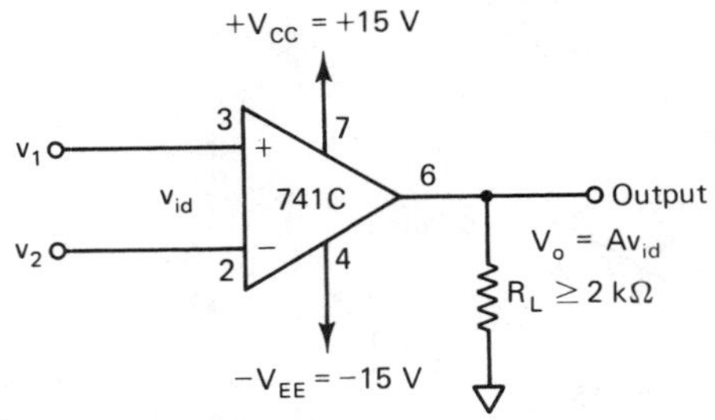

Figure 3-6 Determining voltage gain.

indicates the values of positive and negative saturation voltages of the op-amp. The output voltage never exceeds these limits for given supply voltages $+V_{CC}$ and $-V_{EE}$.

Output Resistance. Output resistance R_o is the equivalent resistance that can be measured between the output terminal of the op-amp and the ground (or common point). It is 75 Ω for the 741C op-amp.

Output Short-Circuit Current. Nobody would deliberately want to short the output terminal of the op-amp to ground, but if such an event were to happen accidentally, the current through the short would certainly be much higher in value than either I_B or I_{io}. This high current may damage the op-amp if it does not have output short-circuit protection. Recall, however, that the 741 family op-amps do have short-circuit protection circuitry built in.

The short-circuit current $I_{sc} = 25$ mA for the 741C op-amp. This means that the built-in short-circuit protection is guaranteed to withstand 25 mA of current in protecting the op-amp.

Supply Current. Supply current I_s is the current drawn by the op-amp from the power supply. This parameter is not given on most of the op-amp data sheets. For the 741C op-amp the supply current $I_s = 2.8$ mA.

Power Consumption. Power consumption P_c is the amount of quiescent power ($v_{in} = 0$ V) that must be consumed by the op-amp in order to operate properly. The amount of power consumed by the 741C is 85 mW.

Transient Response. The response of any practically useful network to a given input is composed of two parts: the *transient* and *steady-state* response. The transient response is that portion of the complete response before the output attains some fixed value. Once reached, this fixed value remains at that level and is, therefore, referred to as a *steady-state value*. The response of the network after it attains a fixed value is independent of time and is called the *steady-state response*. Unlike the steady-state response, the transient response is time-variant. The rise time and the percent of overshoot are the characteristics of the transient response. The time required by the output to go from zero to 90% of its final value is called the *rise time*. Conversely, *overshoot* is the maximum amount by which the output deviates from the steady-state value. Overshoot is generally expressed as a percentage.

The transient response test circuit for the 741C as well as the response of this test circuit for $V_{in} = 20$ mV dc is included in the data sheets. The rise time is 0.3 μs and overshoot is 5% for the 741C op-amp.

The transient response is one of the important considerations in selecting an op-amp in ac applications. In fact, the rise time is inversely proportional to the unity gain bandwidth of the op-amp. This means that the smaller the value of rise time, the higher is the bandwidth.

Slew Rate. Slew rate (SR) is defined as the maximum rate of change of output voltage per unit of time and is expressed in volts per microseconds. In equation form,

$$SR = \left.\frac{dV_o}{dt}\right|_{\text{maximum}} \qquad V/\mu s \tag{3-8}$$

Slew rate indicates how rapidly the output of an op-amp can change in response to changes in the input frequency with input amplitude constant. The slew rate changes with change in voltage gain and is normally specified at unity (+1) gain. The slew rate of an op-amp is fixed, therefore, if the slope requirements of the output signal are greater than the slew rate, then distortion occurs. Thus slew rate is one of the important factors in selecting the op-amp for ac applications, particularly at relatively high frequencies. One of the drawbacks of the 741C is its low slew rate (0.5 V/μs), which limits its use in relatively high frequency applications, especially in oscillators, comparators, and filters. The newer op-amps—LF351, μAF771, and MC34001—which are direct replacements for 741, have a slew rate of 13 V/μs. In high-speed op-amps especially the slew rate is significantly improved. For instance, the 318 has a slew rate of 70 V/μs.

On the data sheets of some op-amps you may find additional electrical parameters that are related to their intended applications. These electrical parameters include gain–bandwidth product, average temperature coefficients of input offset voltage and current, long-term input offset voltage and current stability, and equivalent noise voltage and current. Besides that, there is one more parameter that is applicable to only dual and quad op-amps: channel separation.

Gain–Bandwidth Product. The gain–bandwidth product (BW) is the bandwidth of the op-amp when the voltage gain is 1. Although for the 741 op-amp it is not listed under electrical characteristics, from the open-loop voltage gain versus frequency graph it can be found to be approximately 1 MHz. Equivalent terms for gain–bandwidth product are *closed-loop bandwidth, unity gain bandwidth,* and *small-signal bandwidth.* The newer op-amps LF351 and MC34001, which are direct replacements for the 741, have a gain–bandwidth product of 4 MHz.

Average Temperature Coefficient of Input Offset Voltage (and Current). These parameters are also referred to as *average input offset voltage* or *current drift.* The average temperature coefficient of input offset voltage is the average rate of change in input offset voltage per unit change in temperature expressed as μV/°C. Similarly, the average temperature coefficient of input offset current is the average rate of change in input offset current per unit change in temperature and is usually expressed as pA/°C. Both of these parameters are generally given for the instrumentation and precision type op-amps. For example, for the precision op-amp 714C, the average temperature coefficient of input offset voltage $\Delta V_{io}/\Delta T = 0.5\ \mu V/°C$ typically, and the average temperature coefficient of input offset current $\Delta I_{io}/\Delta T = 12\ pA/°C$.

Long-Term Input Offset Voltage (and Current) Stability. $\Delta V_{io}/\Delta t$ is the average rate of change in input offset voltage per unit of time and is generally expressed as

μV/week. It is also referred to as *input offset voltage drift with time.* Similarly, $\Delta I_{io}/\Delta t$ is the average rate of change in input offset current per unit of time in pA/week. Again, these parameters are normally given for instrumentation and precision-type op-amps. For instance, the precision-type op-amp 714C has $\Delta V_{io}/\Delta t = 0.1\ \mu$V/week typically. The value of $\Delta I_{io}/\Delta t$ is negligible for the 714C; hence it is not listed on the data sheet.

Equivalent Input Noise Voltage and Current. Since electrical noise is random in nature, it is expressed as a root-mean-square value. Standard industry practice is to express the noise as a power density. The equivalent input noise voltage is, therefore, expressed as square voltage (V^2/Hz) and the equivalent input noise current as square noise current (A^2/Hz). Using input noise voltage and input noise current versus frequency curves given on the data sheets (see Figure 3-1), the minimum amount of signal power that is necessary to overcome the noise signal and produce a measurable output signal can be determined. As a general rule, the signal-to-noise ratio must be at least larger than by a factor of 10.

Channel Separation. This parameter is specified in the data sheets of dual and quad (four) op-amps such as the μAF772 and μAF774, respectively. It is a measure of the amount of electrical coupling between op-amps that are integrated on the same chip. Because of the physical closeness of op-amps in dual and quad packages, when a signal is applied to the input of only one op-amp, some signal will appear at the output of other op-amps. The amplitude of these output signals is approximately the same and can be calculated using channel separation and a given input signal. Channel separation is also called *amplifier-to-amplifier coupling.*

The 774/348 is a true quad 741 and has a channel separation of -120 dB. This means that if a signal is applied to one of the op-amps, the signal at the outputs of the undriven op-amps will be at least 120 dB (equivalent to a ratio of 10^6) below the signal output of the driven op-amp.

Thus as more and more data are made available on op-amp characteristics, it becomes easier to select a proper op-amp for a desired application.

A list of parameters that should be considered for ac and dc applications is given in Table 3-1. Depending on the application, a designer sets priorities among these parameters and then selects a suitable op-amp for his or her needs.

7. Typical performance curves included on the data sheets of the 741C are the graphs of electrical parameters that are affected by mainly three factors:

a. Change in supply voltage
b. Change in operating frequency
c. Change in temperature

The graphs of parameters that are affected by each of these factors are as listed below:

Supply-voltage-dependent parameters. Voltage gain, output voltage swing, input common-mode voltage range, power consumption, and input offset current.

Frequency-dependent parameters. Voltage gain, input resistance, output resistance, output voltage swing, input noise voltage and noise current, and CMRR.

TABLE 3-1 PARAMETER EVALUATION FOR AC AND DC APPLICATIONS

For ac applications, consider:	For dc applications, consider:
Input resistance Output resistance Large-signal voltage gain Output voltage swing Transient response Slew rate Gain–bandwidth product Average input offset voltage and current drifts Long-term input offset voltage stability Equivalent input noise voltage and current	Input offset voltage and input offset current Input resistance Output resistance Power supply rejection ratio Large-signal voltage gain Output voltage swing Average input offset voltage and current drifts Long-term input offset voltage stability

Temperature-dependent parameters. Absolute maximum power dissipation, input bias current, input offset current, power consumption, input resistance, output short-circuit current, transient response, and gain–bandwidth product.

In short, the information obtained from the performance curves can be used to improve the op-amp's performance.

8. Finally, a collection of amplifier applications with circuit diagrams is included on the data sheets. These are the applications in which the op-amp is guaranteed to perform satisfactorily. Such applications of the 741C include voltage follower, inverting and noninverting amplifiers, clipping amplifier, simple integrator, and differentiator, and others.

We have seen that a wide range of important information, including a pin diagram, electrical characteristics, and applications, is included on the data sheets of an op-amp. To get the most out of this device, it is essential to refer to the data sheets.

3-3 THE IDEAL OP-AMP

An *ideal* op-amp would exhibit the following electrical characteristics:

1. Infinite voltage gain A.
2. Infinite input resistance R_i, so that almost any signal source can drive it and there is no loading of the preceding stage.
3. Zero output resistance R_o, so that output can drive an infinite number of other devices.

4. Zero output voltage when input voltage is zero.
5. Infinite bandwidth, so that any frequency signal from 0 to ∞ Hz can be amplified without attenuation.
6. Infinite common-mode rejection ratio, so that the output common-mode noise voltage is zero.
7. Infinite slew rate, so that output voltage changes occur simultaneously with input voltage changes.

There are practical op-amps that can be made to approximate some of these characteristics using a negative feedback arrangement. In particular, the input resistance, output resistance, and bandwidth can be brought close to ideal values by this method. Chapter 4 includes a discussion of negative feedback configurations.

3-4 EQUIVALENT CIRCUIT OF AN OP-AMP

Figure 3-7 shows an equivalent circuit of an op-amp. This circuit includes important values from the data sheets, namely A, R_i, and R_o. Note that Av_{id} is an equivalent Thévenin voltage source and R_o is the Thévenin equivalent resistance looking back into the output terminal of an op-amp.

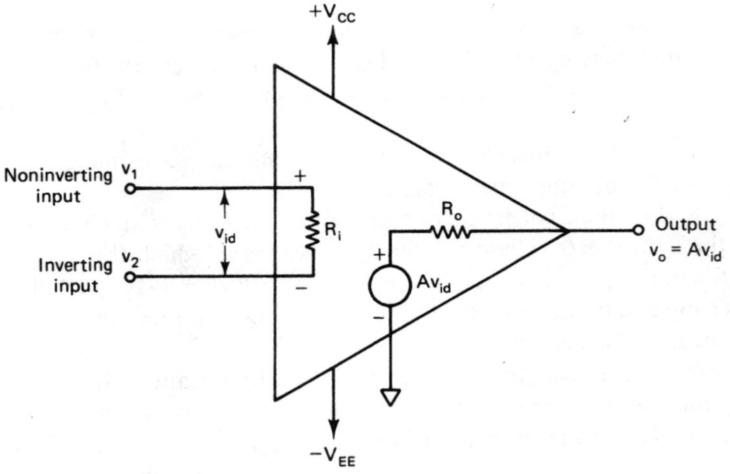

Figure 3-7 Equivalent circuit of an op-amp.

The equivalent circuit is useful in analyzing the basic operating principles of op-amps and in observing the effects of standard feedback arrangements. For the circuit shown in Figure 3-7, the output voltage is

$$v_o = Av_{id} = A(v_1 - v_2) \tag{3-9}$$

where A = large-signal voltage gain
v_{id} = difference (differential) input voltage
v_1 = voltage at the noninverting input terminal with respect to ground
v_2 = voltage at the inverting terminal with respect to ground

Equation (3-9) indicates that the output voltage v_o is directly proportional to the algebraic difference between the two input voltages. In other words, the op-amp amplifies the difference between the two input voltages; it does not amplify the input voltages themselves. For this reason the polarity of the output voltage depends on the polarity of the difference voltage.

3-5 IDEAL VOLTAGE TRANSFER CURVE

Equation (3-9) is the basic op-amp equation, in which the output offset voltage is assumed to be zero. This equation is useful in studying the op-amp's characteristics and in analyzing different circuit configurations that employ feedback. The graphic representation of this equation is shown in Figure 3-8, in that the output voltage v_o is plotted against input difference voltage v_{id}, keeping gain A constant.

Note, however, that the output voltage cannot exceed the positive and negative saturation voltages. These saturation voltages are specified by an output voltage swing rating of the op-amp for given values of supply voltages. This means that the output voltage is directly proportional to the input difference voltage only until it reaches the saturation voltages and that thereafter output voltage remains constant, as shown in Figure 3-8.

The curve shown in Figure 3-8 is called an ideal voltage transfer curve, ideal because output offset voltage is assumed to be zero. In normal op-amp use (with negative feedback), this voltage is near zero and is ignored for simplicity of calculation. Notice that the curve is not drawn to scale. If drawn to scale, the curve would be almost vertical because of the very large values of A.

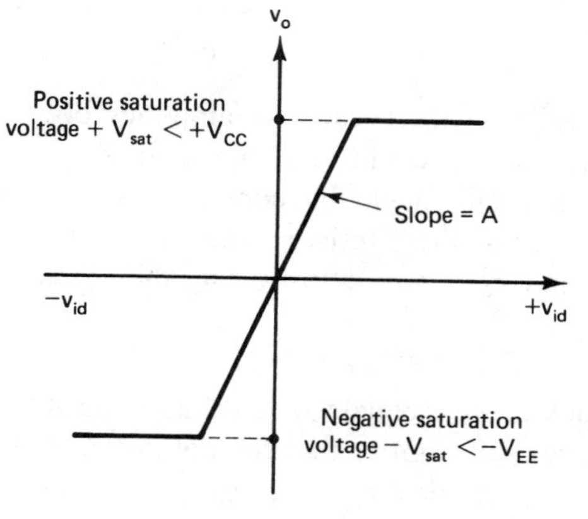

Figure 3-8 Ideal voltage transfer curve.

3-6 OPEN-LOOP OP-AMP CONFIGURATIONS

In the case of amplifiers the term "open-loop" indicates that no connection, either direct or via another network, exists between the output and input terminals. That is, the output signal is not fed back in any form as part of the input signal, and the "loop" that would have been formed with feedback is open.

When connected in open-loop configuration, the op-amp simply functions as a high-gain amplifier. There are three open-loop op-amp configurations:

1. The differential amplifier
2. The inverting amplifier
3. The noninverting amplifier

These configurations are classed according to the number of inputs used and the terminal to which the input is applied when a single input is used.

3-6.1 The Differential Amplifier

Figure 3-9 shows the open-loop differential amplifier in which input signals $v_{\text{in}1}$ and $v_{\text{in}2}$ are applied to the positive and negative input terminals. Since the op-amp amplifies the difference between the two-input signals, this configuration is called the *differential amplifier*.

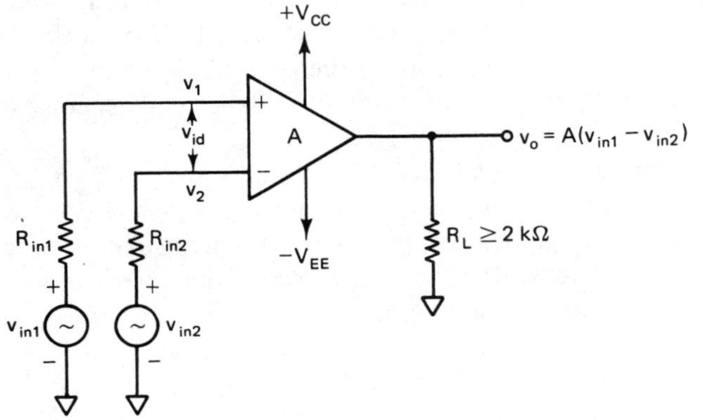

Figure 3-9 Open-loop differential amplifier.

The op-amp is a versatile device because it amplifies both ac and dc input signals. This means that $v_{\text{in}1}$ and $v_{\text{in}2}$ could be either ac or dc voltages. The source resistances $R_{\text{in}1}$ and $R_{\text{in}2}$ are normally negligible compared to the input resistance R_i. Therefore, the voltage drops across these resistors can be assumed to be zero, which then implies that $v_1 = v_{\text{in}1}$ and $v_2 = v_{\text{in}2}$. Substituting these values of v_1 and v_2 in Equation (3-9), we get

$$v_o = A(v_{\text{in}1} - v_{\text{in}2})$$

Thus, as expected, the output voltage is equal to the voltage gain A times the difference between the two input voltages. Also notice that the polarity of the output voltage is dependent on the polarity of the input difference voltage $(v_{\text{in}1} - v_{\text{in}2})$. In open-loop configurations, gain A is commonly referred to as *open-loop gain*.

3-6.2 The Inverting Amplifier

In the inverting amplifier only one input is applied and that to the inverting input terminal. The noninverting input terminal is grounded (refer to Figure 3-10). Since $v_1 = 0$ V, and $v_2 = v_{\text{in}}$ from Equation (3-9),

$$v_o = -Av_{\text{in}}$$

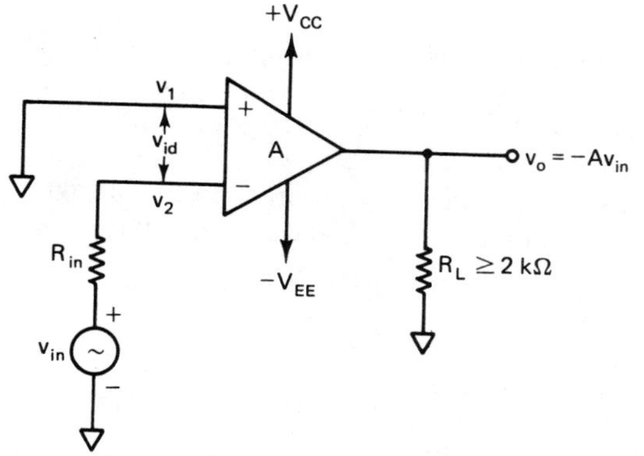

Figure 3-10 Inverting amplifier.

The negative sign indicates that the output voltage is out of phase with respect to input by 180° or is of opposite polarity. Thus in the inverting amplifier the input signal is amplified by gain A and is also inverted at the output.

3-6.3 The Noninverting Amplifier

Figure 3-11 shows the open-loop noninverting amplifier. In this configuration the input is applied to the noninverting input terminal, and the inverting terminal is connected to ground.

In the circuit of Figure 3-11, $v_1 = v_{\text{in}}$ and $v_2 = 0$ V. Therefore, according to Equation (3-9),

$$v_o = Av_{\text{in}}$$

This means that the output voltage is larger than the input voltage by gain A and is in phase with the input signal.

In all three open-loop configurations any input signal (differential or single) that is only slightly greater than zero drives the output to saturation level. This results from the very high gain (A) of the op-amp. Thus, when operated open-loop, the output of the op-amp is either negative or positive saturation or switches between positive and negative saturation levels. For this reason the open-loop op-amp configurations are not used in linear applications.

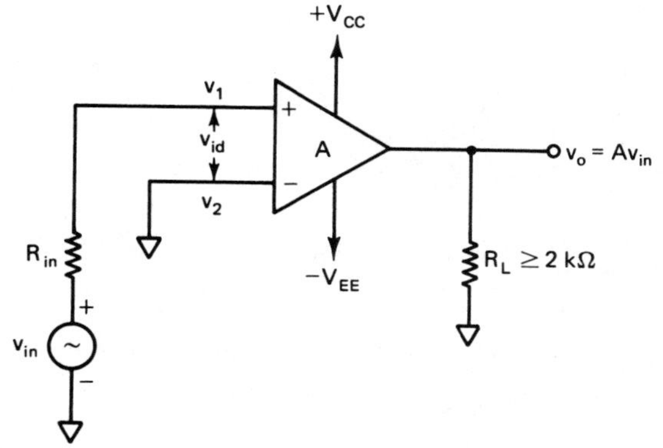

Figure 3-11 Noninverting amplifier.

Example 3-1

Determine the output voltage in each of the following cases for the open-loop differential amplifier of Figure 3-9:

(a) $v_{in1} = 5\ \mu V$ dc, $v_{in2} = -7\ \mu V$ dc

(b) $v_{in1} = 10$ mV rms, $v_{in2} = 20$ mV rms

The op-amp is a 741 with the following specifications: $A = 200{,}000$, $R_i = 2\ M\Omega$, $R_o = 75\ \Omega$, $+V_{CC} = +15$ V, $-V_{EE} = -15$ V, and output voltage swing $= \pm 14$V.

Solution

(a) By Equation (3-9):

$$v_o = 200{,}000[(5)(10^{-6}) - (-7)(10^{-6})] = 2.4 \text{ V dc}$$

Remember that $v_o = 2.4$ V dc with the assumption that the dc output voltage is zero when the input signals are zero.

(b) Equation (3-9) is valid for both ac and dc input signals. However, the restriction on ac input signals is that they must be of the same frequency (see Figure 3-12). By Equation (3-9):

$$v_o = 200{,}000[(10)(10^{-3}) - (20)(10^{-3})] = -2000 \text{ V rms}$$

Thus the theoretical value of output voltage $v_0 = -2000$ V rms. However, the op-amp saturates at ± 14 V. Therefore, the actual output waveform will be *clipped* as shown in Figure 3-12. This nonsinusoidal waveform is unacceptable in amplifier applications.

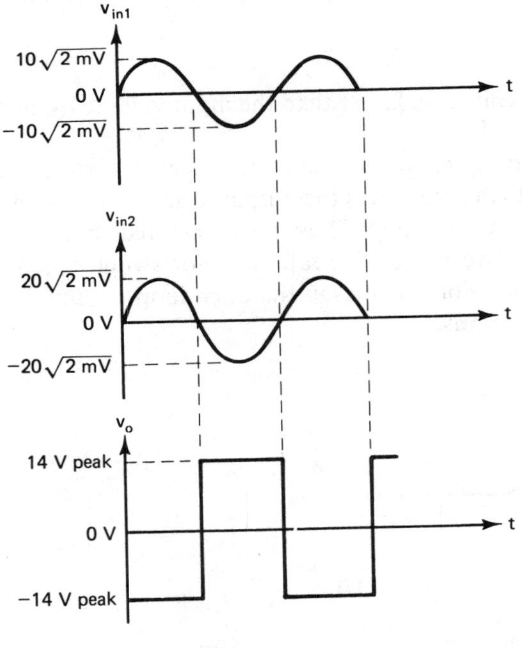

Figure 3-12 Waveforms for Example 3-1.

The normal solution to this problem is to use a negative feedback which is discussed in Chapter 4.

Example 3-2

Determine the output voltage for the inverting amplifier shown in Figure 3-10, if
(a) $v_{in} = 20$ mV dc
(b) $v_{in} = -50\ \mu$V peak sine wave
Assume that the op-amp is a 741.

Solution. By Equation (3-9):

$$\text{(a)}\quad v_o = -Av_{in} = -(2)(10^5)(20)(10^{-3}) = -4000 \text{ V}$$

This is the theoretical value; the actual value will be a negative saturation voltage of −14 V.

$$\text{(b)}\quad v_o = -Av_{in} = -(2)(10^5)(-50)(10^{-6}) = 10 \text{ V peak}$$

This means that the output is a positive sine wave, since it is less than the output voltage swing of ±14 V or 28 V peak to peak.

SUMMARY

1. The manufacturers supply data sheets for the ICs they produce. These data sheets include absolute maximum ratings, electrical characteristics, pin diagrams, equivalent circuits, intended applications of the devices, and more. To get most use out of these sheets, a proper interpretation of them is essential.
2. Generally, the information found on the op-amp data sheets can be broken down into the following groups: device type, absolute maximum ratings, connection diagram, equivalent circuit, electrical characteristics, and typical applications.
3. An ideal op-amp has infinity voltage gain, input resistance, CMRR, and slew rate, together with zero output resistance and output offset voltage.
4. The equivalent circuit is useful in analyzing the basic operating principles of an op-amp and in observing the effects of feedback arrangements.
5. The voltage transfer characteristic curve of an op-amp is the graph of output voltage versus differential input voltage.
6. Differential, inverting, and noninverting amplifiers are the three open-loop op-amp configurations in which the output signal is not fed back in any form as part of the input signal. When operated open-loop, generally the op-amp's output is either positive or negative saturation or switches between positive and negative saturation levels. This action is undesirable in linear applications; hence open-loop configurations are rarely used in linear applications.

QUESTIONS

3-1. What information is contained in the typical op-amp data sheet?

3-2. Explain why proper interpretation of op-amp data sheets is important.

3-3. Define the following electrical parameters: input offset voltage, input resistance, CMRR, output voltage swing, slew rate.

3-4. List the parameters that should be considered for ac and dc applications.

3-5. What are the three factors that affect electrical parameters of an op-amp?

3-6. What are the characteristics of an ideal op-amp?

3-7. What is a voltage transfer curve of an op-amp?

3-8. List three open-loop op-amp configurations.

3-9. Explain why open-loop op-amp configurations are not used in linear applications.

PROBLEMS

3-1. For the μAF771 op-amp, PSRR = 70 dB minimum. What is the numerical value of PSRR?

3-2. For a given op-amp, CMRR = 10^5 and differential gain $A_d = 10^5$. Determine the common-mode gain A_{cm} of the op-amp.

3-3. The output voltage of a certain op-amp circuit changes by 20 V in 4 μs. What is its slew rate?

3-4. In the differential amplifier of Figure 3-9, v_{in1} = 2.1 V dc and v_{in2} = 2.0 V dc. Determine the output voltage v_o. Assume that the op-amp is a 741, with supply voltages = ±15 V.

3-5. Repeat Problem 3-4 for the following values of inputs: v_{in1} = −25 μV rms and v_{in2} = 20 μV rms.

3-6. Referring to the circuit of Figure 3-10, determine v_o if v_{in} = −15 μV dc. Assume that the op-amp is a 741 and that supply voltages = ±15 V.

3-7. Repeat Problem 3-6 for v_{in} = −10 μV rms.

4

An Op-Amp with Negative Feedback

4-1 INTRODUCTION

In Chapter 3 the characteristics of the op-amp and interpretation of its parameters are discussed. Recall that clipping occurs in open-loop configurations when the output attempts to exceed the saturation levels of the op-amp. In other words, because the open-loop gain of the op-amp is very high, only the smaller signals (of the order of microvolts or less) having very low frequency may be amplified accurately without distortion. However, signals this small are very susceptible to noise and almost impossible to obtain in the laboratory.

Besides being large, the open-loop voltage gain of the op-amp is not a constant. The voltage gain varies with changes in temperature and power supply as well as mass production techniques. The variations in voltage gain are relatively large in open-loop op-amps in particular, which makes the open-loop op-amp unsuitable for many linear applications. In most linear applications the output is proportional to the input and is of the same type.

In addition, the *bandwidth* (band of frequencies for which the gain remains constant) of most open-loop op-amps is negligibly small—almost zero. For this reason the open-loop op-amp is impractical in ac applications. For instance, the open-loop bandwidth of the 741C is approximately 5 Hz. (Refer to the voltage gain versus frequency graph supplied on the data sheets.) However, in almost all ac applications a bandwidth larger than 5 Hz is needed.

For the reasons stated above the open-loop op-amp is generally not used in linear applications. Nevertheless, in certain applications the open-loop op-amp is purposely used as a nonlinear device; that is, a square-wave output is obtained by deliberately applying a relatively large input signal. Open-loop op-amp configurations are most suitable in such applications.

We can select as well as control the gain of the op-amp if we introduce a modification in the basic circuit. This modification involves the use of *feedback*; that is, an output signal is fed back to the input either directly or via another network. If the signal fed back is of opposite polarity or out of phase by 180° (or odd integer multiples of 180°) with respect to the input signal, the feedback is called *negative feedback*. An amplifier with negative feedback has a self-correcting ability against any change in output voltage caused by changes in environmental conditions. Negative feedback is also known as *degenerative* feedback because when used it degenerates (reduces) the output voltage amplitude and in turn reduces the voltage gain.

On the other hand, if the signal fed back is in phase with the input signal, the feedback is called *positive feedback*. In positive feedback the feedback signal aids the input signal. For this reason it is also referred to as *regenerative* feedback. Positive feedback is necessary in oscillator circuits.

When used in amplifiers, negative feedback stabilizes the gain, increases the bandwidth, and changes the input and output resistances. Of course, the price paid for these improvements is reduced voltage gain. Other benefits of negative feedback include a decrease in harmonic or nonlinear distortion and reduction in the effect of input offset voltage at the output. Negative feedback also reduces the effect of variations in temperature and supply voltages on the output of the op-amp.

Negative feedback and its effect on the performance of an op-amp are the subject of this chapter. First the block diagrams for different op-amp configurations using negative feedback are presented. Then specific arrangements of these feedback configurations are studied in more depth.

4-2 BLOCK DIAGRAM REPRESENTATION OF FEEDBACK CONFIGURATIONS

An op-amp that uses feedback is called a *feedback amplifier*. A feedback amplifier is sometimes referred to as a *closed-loop amplifier* because the feedback forms a closed loop between the input and the output. A feedback amplifier essentially consists of two parts: an op-amp and a *feedback circuit*. The feedback circuit can take any form whatsoever, depending on the intended application of the amplifier. This means that the feedback circuit may be made up of either passive components, active components, or combinations of both. This chapter, in order to develop the basic feedback concepts, presents only purely resistive feedback circuits.

A closed-loop amplifier can be represented by using two blocks, one for an op-amp and another for a feedback circuit. There are four ways to connect these two blocks. These connections are classified according to whether the voltage or current is fed back to the input in *series* or in *parallel*, as follows:

1. Voltage—series feedback
2. Voltage—shunt feedback

3. Current—series feedback
4. Current—shunt feedback

The four types of configurations are illustrated in Figure 4-1. In Figure 4-1(a) and (b) the voltage across load resistor R_L is the input voltage to the feedback circuit. The feedback quantity (either voltage or current) is the output of the feedback circuit and is proportional to the output voltage. On the other hand, in the current-series and current-shunt feedback circuits of Figure 4-1(c) and (d), the load current i_L (output current) flows into the feedback circuit. The output of the feedback circuit is the feedback quantity (either voltage or current), which is proportional to the load current i_L.

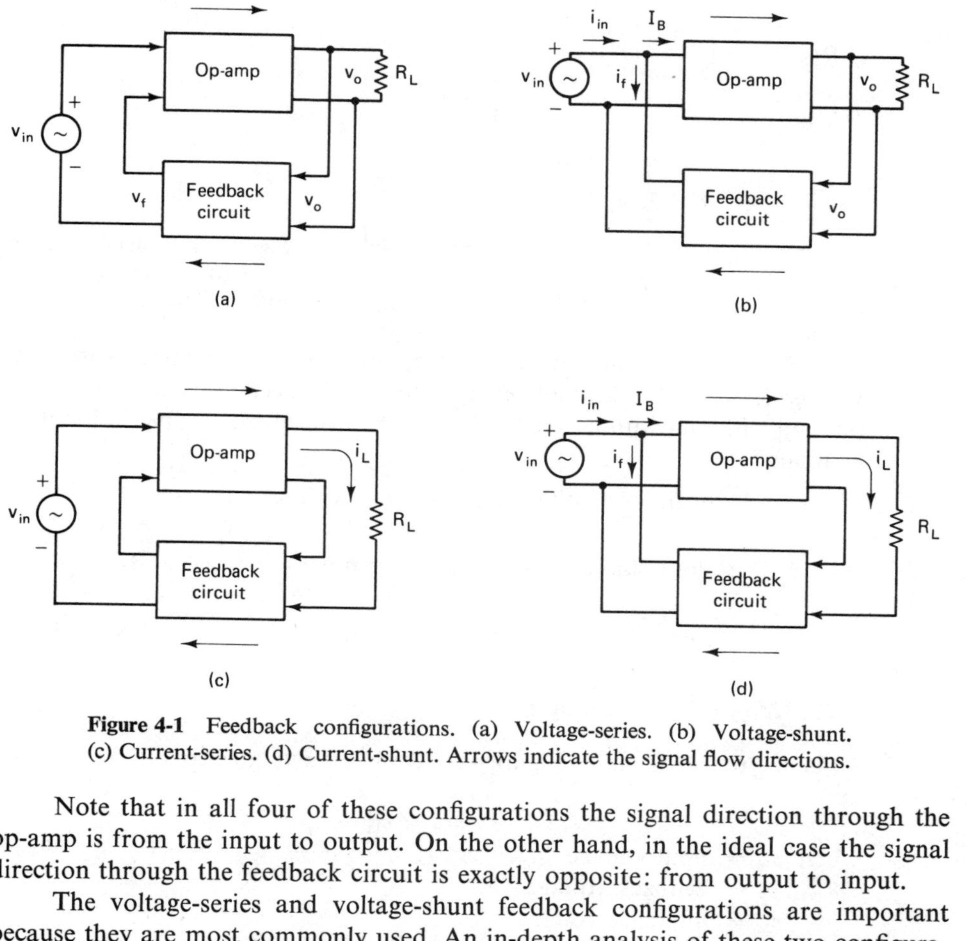

Figure 4-1 Feedback configurations. (a) Voltage-series. (b) Voltage-shunt. (c) Current-series. (d) Current-shunt. Arrows indicate the signal flow directions.

Note that in all four of these configurations the signal direction through the op-amp is from the input to output. On the other hand, in the ideal case the signal direction through the feedback circuit is exactly opposite: from output to input.

The voltage-series and voltage-shunt feedback configurations are important because they are most commonly used. An in-depth analysis of these two configurations is presented here, computing voltage gain, input resistance, output resistance, and bandwidth for each. The other two configurations are not discussed in this chapter. However, the interested reader may analyze the latter using the same procedures as illustrated for the first two configurations.

4-3 VOLTAGE-SERIES FEEDBACK AMPLIFIER

The schematic diagram of the voltage-series feedback amplifier is shown in Figure 4-2. The op-amp is represented by its schematic symbol, including its large-signal voltage gain A, and the feedback circuit is composed of two resistors R_1 and R_F.

The circuit shown in Figure 4-2 is commonly known as a *noninverting amplifier with feedback* (or *closed-loop noninverting amplifier*) because it uses feedback, and the input signal is applied to the noninverting input terminal of the op-amp.

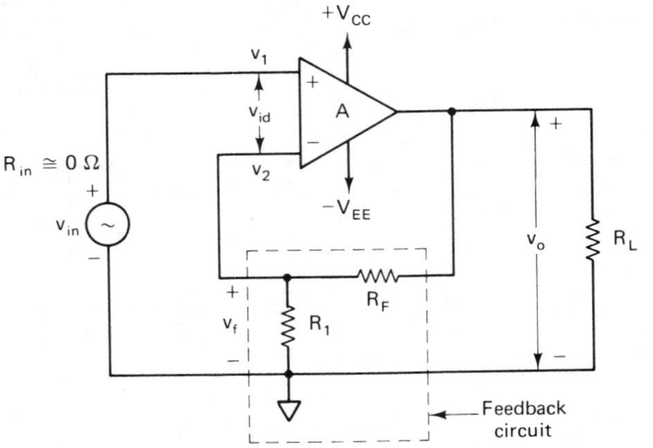

Figure 4-2 Voltage-series feedback amplifier (or noninverting amplifier with feedback).

Before proceeding, it is necessary to define some important terms for the voltage-series feedback amplifier of Figure 4-2. Specifically, the voltage gain of the op-amp with and without feedback, and the gain of the feedback circuit are defined as follows:

$$\text{open-loop voltage gain (or gain without feedback)} \quad A = \frac{v_o}{v_{id}}$$

$$\text{closed-loop voltage gain (or gain with feedback)} \quad A_F = \frac{v_o}{v_{\text{in}}}$$

$$\text{gain of the feedback circuit} \quad B = \frac{v_f}{v_o}$$

4-3.1 Negative Feedback

Referring to the circuit of Figure 4-2, Kirchhoff's voltage equation for the input loop is

$$v_{id} = v_{\text{in}} - v_f \tag{4-1}$$

where v_{in} = input voltage
v_f = feedback voltage
v_{id} = difference voltage

Recall, however, that an op-amp always amplifies the difference input voltage v_{id}. From Equation (4-1), this difference voltage is equal to the input voltage v_{in} minus the feedback voltage v_f. In other words, the feedback voltage always opposes the input voltage (or is out of phase by 180° with respect to the input voltage); hence the feedback is said to be *negative*.

Returning now to the analysis of the voltage-series feedback amplifier, we should note that it will be performed by computing closed-loop voltage gain, input and output resistances, and the bandwidth.

4-3.2 Closed-Loop Voltage Gain

As defined previously, the closed-loop voltage gain

$$A_F = \frac{v_o}{v_{in}}$$

However, by Equation (3-9),

$$v_o = A(v_1 - v_2)$$

Referring to Figure 4-2, we see that

$$v_1 = v_{in}$$

$$v_2 = v_f = \frac{R_1 v_o}{R_1 + R_F} \qquad \text{since } R_i \gg R_1$$

Therefore,

$$v_o = A\left(v_{in} - \frac{R_1 v_o}{R_1 + R_F}\right)$$

Rearranging, we get

$$v_o = \frac{A(R_1 + R_F)v_{in}}{R_1 + R_F + AR_1}$$

Thus

$$A_F = \frac{v_o}{v_{in}} = \frac{A(R_1 + R_F)}{R_1 + R_F + AR_1} \quad \text{(exact)} \tag{4-2}$$

Generally, A is very large (typically 10^5). Therefore,

$$AR_1 \gg (R_1 + R_F) \qquad \text{and} \qquad (R_1 + R_F + AR_1) \cong AR_1$$

Thus

$$A_F = \frac{v_o}{v_{in}} = 1 + \frac{R_F}{R_1} \quad \text{(ideal)} \tag{4-3}$$

Equation (4-3) is important because it shows that the gain of the voltage-series feedback amplifier is determined by the ratio of two resistors, R_1 and R_F. For instance, if a gain of 11 is desired, we can then choose $R_1 = 1\ \text{k}\Omega$ and $R_F = 10\ \text{k}\Omega$ or $R_1 = 100\ \Omega$ and $R_F = 1\ \text{k}\Omega$. In other words, in setting the gain the ratio of R_1 and R_F is important and not the absolute values of these resistors. As a general rule, however, all external component values should be less than 1 MΩ so that they do not adversely affect the internal circuitry of the op-amp. This is especially true for older-generation ICs such as the 709 and 741.

Another interesting result can be obtained from Equation (4-3). As defined previously, the gain of the feedback circuit (B) is the ratio of v_f to v_o. Referring to Figure 4-2, this gain is

$$B = \frac{v_f}{v_o}$$

$$= \frac{R_1}{R_1 + R_F} \tag{4-4}$$

Comparing Equations (4-3) and (4-4), we can conclude that

$$A_F = \frac{1}{B} \qquad \text{(ideal)} \tag{4-5}$$

This means that the gain of the feedback circuit is the reciprocal of the closed-loop voltage gain. In other words, for given R_1 and R_F the values of A_F and B are fixed. Besides that, Equation (4-5) is an alternative to Equation (4-3), and its simpler form makes it easier to remember.

Finally, the closed-loop voltage gain A_F can be expressed in terms of open-loop gain A and feedback circuit gain B as follows. Rearranging Equation (4-2), we get

$$A_F = \frac{A\left(\dfrac{R_1 + R_F}{R_1 + R_F}\right)}{\dfrac{R_1 + R_F}{R_1 + R_F} + \dfrac{AR_1}{R_1 + R_F}}$$

Using Equation (4-4) yields

$$A_F = \frac{A}{1 + AB} \tag{4-6}$$

where A_F = closed-loop gain
A = open-loop gain
B = gain of the feedback circuit
AB = loop gain

A *one-line* block diagram of Equation (4-6) is shown in Figure 4-3. This block diagram illustrates a standard form for representing a system with feedback and also indicates the relationship between different variables of the system. In short, the block diagram approach helps to simplify the analysis of complex closed-loop networks, particularly if they are composed of nonresistive feedback circuits.

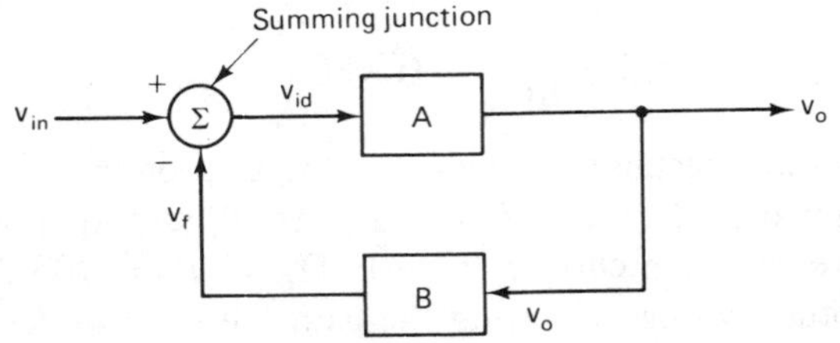

Figure 4-3 Block diagram representation of noninverting amplifier with feedback.

4-3.3 Difference Input Voltage Ideally Zero

Let us reconsider Equation (3-9), which can be rewritten as

$$v_{id} = \frac{v_o}{A}$$

since A is very large (ideally infinite),

$$v_{id} \cong 0 \tag{4-7a}$$

That is,

$$v_1 \cong v_2 \qquad \text{(ideal)} \tag{4-7b}$$

Equation (4-7b) says that the voltage at the noninverting input terminal of an op-amp is approximately equal to that at the inverting input terminal provided that A is very large. This concept is useful in the analysis of *closed loop* op-amp circuits. For example, ideal closed-loop voltage gain [Equation (4-3)] can be obtained using the results above as follows. In the circuit of Figure 4-2,

$$v_1 = v_{\text{in}}$$

$$v_2 = v_f = \frac{R_1 v_o}{R_1 + R_F}$$

Substituting these values of v_1 and v_2 in Equation (4-7b), we get

$$v_{\text{in}} = \frac{R_1 v_o}{R_1 + R_F}$$

That is,

$$A_F = \frac{v_o}{v_{\text{in}}} = 1 + \frac{R_F}{R_1}$$

4-3.4 Input Resistance with Feedback

Figure 4-4 shows a voltage-series feedback amplifier with the op-amp equivalent circuit. In this circuit R_i is the input resistance (open-loop) of the op-amp, and R_{iF} is the input resistance of the feedback amplifier. The input resistance with feedback is defined as

$$R_{iF} = \frac{v_{\text{in}}}{i_{\text{in}}} = \frac{v_{\text{in}}}{v_{id}/R_i}$$

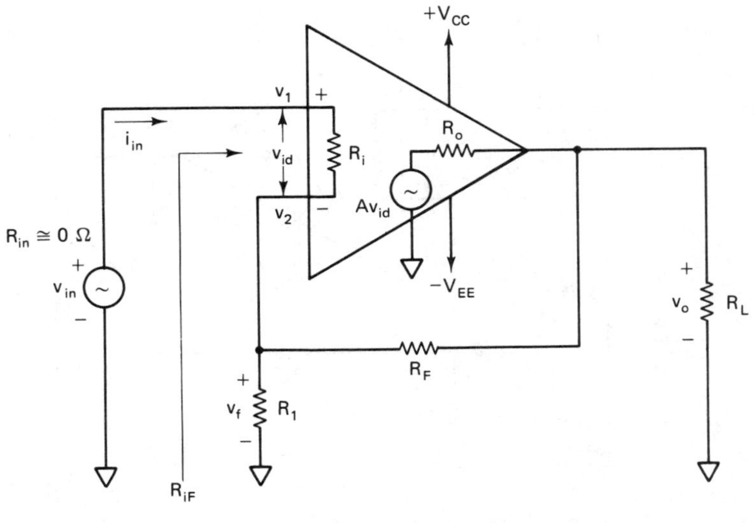

Figure 4-4 Derivation of input resistance with feedback.

However,

$$v_{id} = \frac{v_o}{A} \quad \text{and} \quad v_o = \frac{A}{1 + AB} v_{\text{in}}$$

Therefore,

$$\begin{aligned} R_{iF} &= R_i \frac{v_{\text{in}}}{v_o/A} \\ &= AR_i \frac{v_{\text{in}}}{Av_{\text{in}}/(1 + AB)} \\ &= R_i(1 + AB) \end{aligned} \tag{4-8}$$

This means that the input resistance of the op-amp with feedback is $(1 + AB)$ times that without feedback.

4-3.5 Output Resistance with Feedback

Output resistance is the resistance determined looking back into the feedback amplifier from the output terminal as shown in Figure 4-5. This resistance can be obtained by using Thévenin's theorem for *dependent* sources. Specifically, to find output resistance with feedback R_{oF}, reduce independent source v_{in} to zero, apply an external voltage v_o, and then calculate the resulting current i_o. In short, the R_{oF} is defined as follows:

$$R_{oF} = \frac{v_o}{i_o} \tag{4-9a}$$

Writing Kirchhoff's current equation at output node N, we get

$$i_o = i_a + i_b$$

since $R_F + R_1 \| R_i \gg R_o$, $i_a \gg i_b$. Therefore,

$$i_o \cong i_a$$

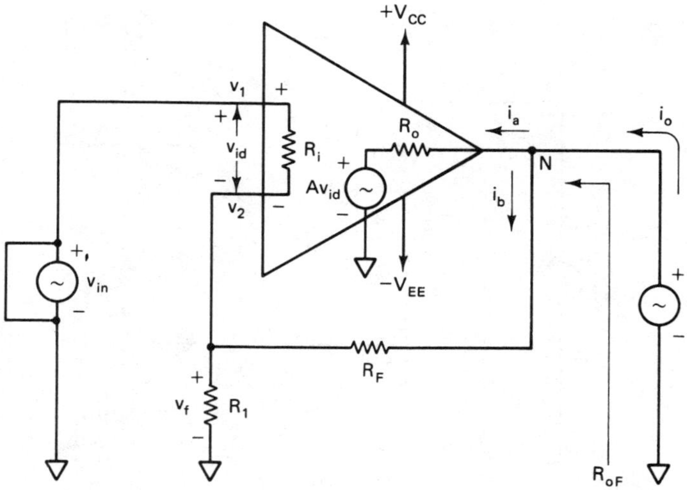

Figure 4-5 Derivation of output resistance with feedback.

The current i_o can be found by writing Kirchhoff's voltage equation for the output loop:

$$v_o - R_o i_o - Av_{id} = 0$$

$$i_o = \frac{v_o - Av_{id}}{R_o}$$

However,

$$v_{id} = v_1 - v_2$$

$$= 0 - v_f$$

$$= -\frac{R_1 v_o}{R_1 + R_F} = -Bv_o$$

Therefore,

$$i_o = \frac{v_o + ABv_o}{R_o}$$

Substituting the value of i_o in Equation (4-9a), we get

$$R_{oF} = \frac{v_o}{(v_o + ABv_o)/R_o}$$

$$= \frac{R_o}{1 + AB} \tag{4-9b}$$

This result shows that the output resistance of the voltage series feedback amplifier is $1/(1 + AB)$ times the output resistance R_o of the op-amp. That is, the output resistance of the op-amp with feedback is much smaller than the output resistance without feedback.

4-3.6 Bandwidth with Feedback

The bandwidth of an amplifier is defined as the band (range) of frequencies for which the gain remains constant. Manufacturers generally specify either the gain–bandwidth product or supply open-loop gain versus frequency curve for the op-amp. For the 741 op-amp the latter is typical.

Figure 4-6 shows the open-loop gain versus frequency curve of the 741C op-amp. From this curve for a gain of 200,000, the bandwidth is approximately 5 Hz. On the other extreme, the bandwidth is approximately 1 MHz when the gain is unity. The frequency at which the gain equals 1 is known as the *unity gain–bandwidth.* It is the maximum frequency the op-amps can be used for. Furthermore, the *gain–bandwidth product* obtained from the open-loop gain versus frequency curve of Figure 4-6 is equal to the unity gain–bandwidth of an op-amp. However, this holds true only for those op-amps which, like the 741, have just one break frequency below unity gain–bandwidth.

Since the gain–bandwidth product is constant, obviously the higher the gain, the smaller the bandwidth, and vice versa. As we have seen, if negative feedback is used, gain A decreases to $A/(1 + AB)$. Therefore, to obtain the closed-loop bandwidth, the open-loop bandwidth must be multiplied by the same factor by which the gain is divided, that is, by the factor $(1 + AB)$. In short,

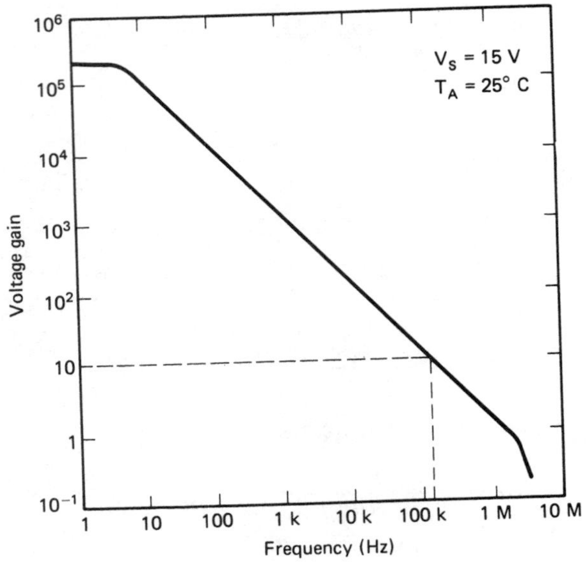

Figure 4-6 Open-loop gain versus frequency curve of the 741C.

$$\text{bandwidth with feedback} = (\text{bandwidth without feedback})(1 + AB)$$

$$f_F = f_o(1 + AB) \tag{4-10a}$$

or alternatively,

$$f_F = \frac{\text{unity gain–bandwidth}}{\text{closed-loop gain}}$$

$$= \frac{\text{UBW}}{A_F} \tag{4-10b}$$

For instance, let us assume that the 741C is used in the circuit of Figure 4-2 and that the desired voltage gain is 10; then the closed-loop bandwidth f_F [using Equation (4-10b)] will be approximately 1 MHz/10 = 100 kHz.

The closed-loop bandwidth can also be determined from the open-loop gain versus frequency plot. To do this we locate the closed-loop voltage gain value on the gain axis and draw a parallel line through this value to the frequency axis. Then we project the point of intersection of the line with the curve on the frequency axis and read the value of the closed-loop bandwidth. Using this procedure in Figure 4-6, the bandwidth is approximately 100 kHz for a closed-loop gain of 10.

4-3.7 Total Output Offset Voltage with Feedback

In an op-amp when the input is zero, the output is also expected to be zero. However, because of the effect of input offset voltage and current, the output is significantly larger, a result in large part of very high open-loop gain. That is, the high gain aggravates the effect of input offset voltage at the output. We call this enhanced output voltage the *total output offset voltage* V_{ooT}. In an open-loop op-amp the total output offset voltage is equal to either the positive or negative saturation voltage. The saturation voltages are specified on the data sheets as *output voltage swing.*

Since with feedback the gain of the noninverting amplifier changes from A to $A/(1 + AB)$ [Equation (4-6)], the total output offset voltage with feedback must also be $1/(1 + AB)$ times the voltage without feedback. That is,

$$\begin{matrix}\text{total output offset} \\ \text{voltage with feedback}\end{matrix} = \frac{\text{total output offset voltage without feedback}}{1 + AB}$$

or

$$V_{ooT} = \frac{\pm V_{sat}}{1 + AB} \tag{4-11}$$

where $1/(1 + AB)$ is always less than 1 and $\pm V_{sat}$ = saturation voltages. Note that V_{ooT} is either positive or negative voltage.

Negative feedback can also be used to reduce significantly the effect of noise, variations in supply voltages, and of changes in temperature on the output voltage of a noninverting amplifier. In fact, the higher the value of $(1 + AB)$, the smaller is the effect of noise and of variations in supply voltages and changes in temperature on the output voltage of a noninverting amplifier.

From the analysis above it is clear that the noninverting amplifier with feedback exhibits the characteristics of the *perfect voltage amplifier.* That is, it has very high input resistance, very low output resistance, stable voltage gain, large bandwidth, and very little (ideally zero) output offset voltage.

4-3.8 Voltage Follower

The lowest gain that can be obtained from a noninverting amplifier with feedback is 1. When the noninverting amplifier is configured for unity gain, it is called a *voltage follower* because the output voltage is equal to and in phase with the input. In other words, in the voltage follower the output follows the input.

Although it is similar to the discrete emitter follower, the voltage follower is preferred because it has much higher input resistance, and the output amplitude is exactly equal to the input.

To obtain the voltage follower from the noninverting amplifier of Figure 4-2, simply *open* R_1 and *short* R_F. The resulting circuit is shown in Figure 4-7. In this

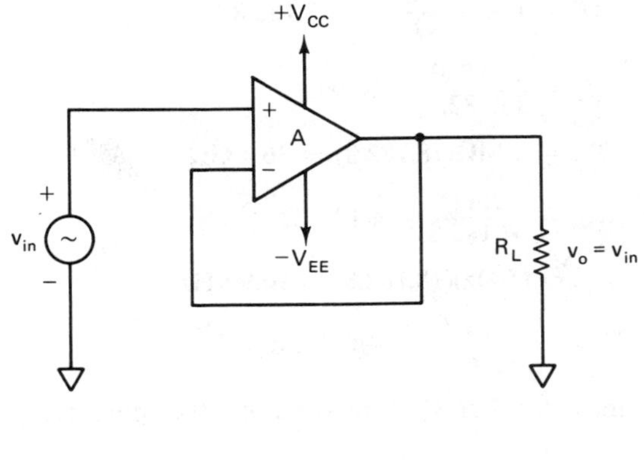

Figure 4-7 Voltage follower.

figure all the output voltage is fed back into the inverting terminal of the op-amp; consequently, the gain of the feedback circuit is 1 ($B = A_F = 1$).

Since the voltage follower is a special case of the noninverting amplifier, all the formulas developed for the latter are indeed applicable to the former except that the gain of the feedback circuit is 1 ($B = 1$). The applicable formulas are:

$$A_F = 1$$

$$R_{iF} = AR_i$$

$$R_{oF} = \frac{R_o}{A}$$

$$f_F = Af_o$$

$$V_{ooT} = \frac{\pm V_{sat}}{A}$$

since $(1 + A) \cong A$.

The voltage follower is also called a *noninverting buffer* because when placed between two networks, it removes the loading on the first network.

Example 4-1

The 741C op-amp having the following parameters is connected as a noninverting amplifier (Figure 4-2) with $R_1 = 1\ \text{k}\Omega$ and $R_F = 10\ \text{k}\Omega$:

$$A = 200{,}000$$

$$R_i = 2\ \text{M}\Omega$$

$$R_o = 75\ \Omega$$

$$f_o \cong 5\ \text{Hz}$$

$$\text{supply voltages} = \pm 15\ \text{V}$$

$$\text{output voltage swing} = \pm 13\ \text{V}$$

Compute the values of A_F, R_{iF}, R_{oF}, f_F, and V_{ooT}.

Solution. Let us first calculate the value of B. Then the closed-loop parameters A_F, R_{iF}, R_{oF}, f_F, and V_{ooT} can be obtained by using Equations (4-6), (4-8), (4-9b), (4-10a), and (4-11), respectively.

$$B = \frac{R_1}{R_1 + R_F} = \frac{1\ \text{k}\Omega}{1\ \text{k}\Omega + 10\ \text{k}\Omega} = \frac{1}{11}$$

$$1 + AB = 1 + \frac{200{,}000}{11} = 18{,}182.8$$

$$A_F = \frac{200{,}000}{18{,}182.8} = 10.99$$

$$R_{iF} = 2\ \text{M}\Omega(18{,}182.8) = 36.4\ \text{G}\Omega$$

$$R_{oF} = \frac{75\Omega}{18{,}182.8} = 4.12\ \text{m}\Omega$$

$$f_F = (5\ \text{Hz})(18{,}182.8) = 90.9\ \text{kHz}$$

$$V_{ooT} = \frac{\pm 13\text{V}}{18{,}182.8} = \pm 0.715\ \text{mV}$$

Note that the ($\pm$) sign indicates that V_{ooT} could be of either polarity.

In the example above, the voltage gain calculated using the exact equation [Equation (4-2)] is 10.99. The gain would have been 11 if we had used the ideal voltage-gain equation [Equation (4-3)]. Thus the difference error is very small (0.09%) and can be ignored. That is, for all practical purposes we may use the ideal voltage-gain equation, provided that $A \gg A_F$. Remember that as A_F approaches A, the difference error also increases.

Example 4-2

Repeat Example 4-1 for the voltage follower of Figure 4-7.

Solution. For the voltage follower, $B = 1$; therefore, $1 + AB = 200{,}000$. To compute the closed-loop parameters we merely substitute the known values into Equations (4-6), (4-8), (4-9b), (4-10a), and (4-11).

$$A_F = 1$$

$$R_{iF} = 2\ \text{M}\Omega(200{,}000) = 400\ \text{G}\Omega$$

$$R_{oF} = \frac{75\Omega}{200{,}000} = 0.375\ \text{m}\Omega$$

$$f_F = (5\ \text{Hz})(200{,}000) = 1\ \text{MHz}$$

$$V_{ooT} = \frac{\pm 13\ \text{V}}{200{,}000} = \pm 65\ \mu\text{V}$$

Thus the input and output resistances of the voltage follower approach ideal values, and the bandwidth is equal to the maximum operating frequency of the op-amp. In addition, since $(1 + AB) = A$, the smallest possible value for V_{ooT} is possible.

4-4 VOLTAGE-SHUNT FEEDBACK AMPLIFIER

Figure 4-8 shows the voltage-shunt feedback amplifier using an op-amp. The input voltage drives the inverting terminal, and the amplified as well as inverted output signal is also applied to the inverting input via the feedback resistor R_F. This arrange-

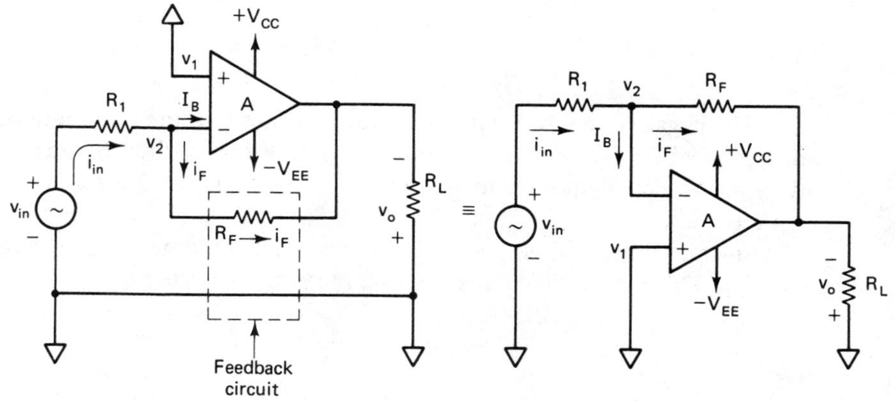

Figure 4-8 Voltage-shunt feedback amplifier (or inverting amplifier with feedback).

ment forms a negative feedback because any increase in the output signal results in a feedback signal into the inverting input, causing a decrease in the output signal.

Note that the noninverting terminal is grounded, and the feedback circuit has only one resistor R_F. However, an extra resistor R_1 is connected in series with the input signal source v_{in}.

First we derive the formulas for the voltage gain, input and output resistances, bandwidth, and the total output offset voltage. Then we study the special cases of this configuration.

4-4.1 Closed-Loop Voltage Gain

The closed-loop voltage gain A_F of the voltage-shunt feedback amplifier can be obtained by writing Kirchhoff's current equation at the input node v_2 [see Figure 4-8] as follows:

$$i_{in} = i_F + I_B \tag{4-12a}$$

Since R_i is very large, the input bias current I_B is negligibly small. For instance, $R_i = 2\ \text{M}\Omega$ and $I_B = 0.5\ \mu\text{A}$ for the 741C. Therefore,

$$i_{in} \cong i_F$$

That is,

$$\frac{v_{in} - v_2}{R_1} = \frac{v_2 - v_o}{R_F} \tag{4-12b}$$

However, from Equation (3-9),

$$v_1 - v_2 = \frac{v_o}{A}$$

Since $v_1 = 0$ V,

$$v_2 = -\frac{v_o}{A}$$

Substituting this value of v_2 in Equation (4-12b) and rearranging, we get

$$\frac{v_{in} + v_o/A}{R_1} = \frac{-(v_o/A) - v_o}{R_F}$$

$$A_F = \frac{v_o}{v_{in}} = -\frac{AR_F}{R_1 + R_F + AR_1} \qquad \text{(exact)} \tag{4-13}$$

The negative sign in Equation (4-13) indicates that the input and output signals are out of phase by 180° (or of opposite polarities). In fact, because of this phase inversion, the configuration in Figure 4-8 is commonly called an *inverting amplifier* with feedback.

Since the internal gain A of the op-amp is very large (ideally infinity), $AR_1 \gg R_1 + R_F$. This means that Equation (4-13) can be rewritten as

$$A_F = \frac{v_o}{v_{in}} = -\frac{R_F}{R_1} \qquad \text{(ideal)} \tag{4-14}$$

This equation shows that the gain of the inverting amplifier is set by selecting a ratio of feedback resistance R_F to the input resistance R_1. In fact, the ratio R_F/R_1 can be set to any value whatsoever, even to less than 1. Because of this property of the gain equation, the inverting amplifier configuration with feedback lends itself to a majority of applications as against those of the noninverting amplifier.

Let us now rewrite Equation (4-13) in the feedback form of Equation (4-6), for a couple of reasons. First, it facilitates analysis of the inverting amplifier with feedback. Second, it helps compare and contrast inverting and noninverting amplifier configurations, as we shall soon see. However, to express Equation (4-13) in the form of Equation (4-6), we must represent the current-summing junction at the input terminals of an amplifier as a voltage-summing junction (see Figures 4-8 and 4-9).

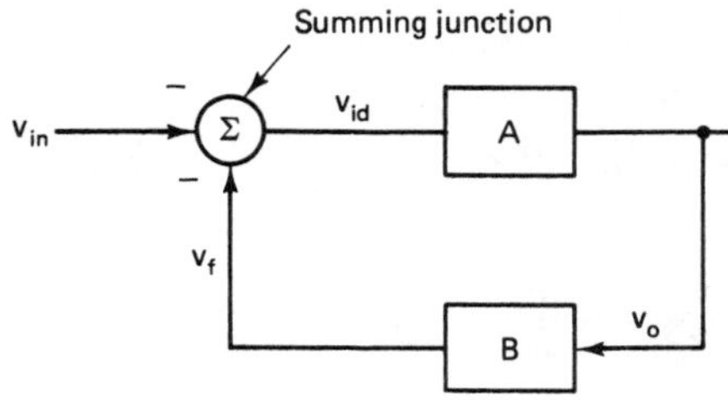

Figure 4-9 Block diagram of inverting amplifier with feedback using a voltage-summing junction as a model for current-summing.

To begin with, we divide both numerator and denominator of Equation (4-13) by R_F:

$$A_F = -\frac{A}{1 + (R_1/R_F)(1 + A)}$$

Since A is very large, $1 + A \cong A$; hence

$$A_F = -\frac{A}{1 + A(R_1/R_F)} \tag{4-15}$$

A comparison of Equation (4-15) with the feedback Equation (4-6) indicates that we can *equate*

$$B = \frac{R_1}{R_F} \tag{4-16}$$

Note that in actuality $B = 1/R_F$; however, the representation $B = R_1/R_F$ is possible because the voltage-shunt case is again modeled as a voltage amplifier. Thus the magnitude of the ideal closed-loop gain is

$$|A_F| = \frac{1}{B} = \frac{R_F}{R_1} \tag{4-17}$$

The one-line block diagram of the inverting amplifier with feedback is shown in Figure 4-9. However, remember that here a voltage-summing junction is merely being used as a model for what is actually current-summing. This approach facilitates the analysis of the inverting amplifier and also enables us to express the performance equations in the same form as those for the noninverting amplifier.

4-4.2 Inverting Input Terminal at Virtual Ground

Refer again to the inverting amplifier of Figure 4-8. In this figure, the noninverting terminal is grounded, and the input signal is applied to the inverting terminal via resistor R_1. However, as discussed in Section 4-3.3, the difference input voltage is ideally zero; that is, the voltage at the inverting terminal (v_2) is approximately equal to that at the noninverting terminal (v_1). In other words, the inverting terminal voltage v_2 is approximately at ground potential. Therefore, the inverting terminal is said to be at *virtual ground.* This concept is extremely useful in the analysis of closed-loop inverting amplifier circuits. For example, ideal closed-loop gain [Equation (4-14)] can be obtained using the *virtual ground* concept, as follows:

In the circuit of Figure 4-8,

$$i_{in} \cong i_F$$

That is,

$$\frac{v_{in} - v_2}{R_1} = \frac{v_2 - v_o}{R_F}$$

However,

$$v_1 = v_2 = 0 \text{ V}$$

Therefore,

$$\frac{v_{in}}{R_1} = -\frac{v_o}{R_F}$$

or

$$A_F = \frac{v_o}{v_{in}} = -\frac{R_F}{R_1}$$

This is the same result obtained in Equation (4-14).

4-4.3 Input Resistance with Feedback

The easiest method of finding the input resistance is to Millerize the feedback resistor R_F, that is, split R_F into its two Miller components, as shown in Figure 4-10.

In the circuit of Figure 4-10, the input resistance with feedback R_{iF} is then

$$R_{iF} = R_1 + \frac{R_F}{1 + A} \Big\| (R_i) \qquad \text{(exact)} \tag{4-18}$$

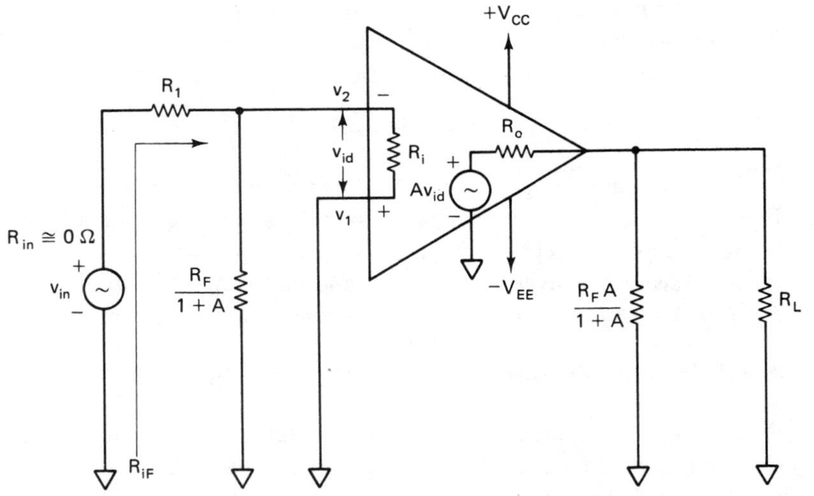

Figure 4-10 Inverting amplifier with Millerized feedback resistor.

Since R_i and A are very large,

$$\frac{R_F}{1+A}\bigg\|R_i \cong 0\,\Omega$$

Hence

$$R_{iF} = R_1 \qquad \text{(ideal)} \tag{4-19}$$

4-4.4 Output Resistance with Feedback

The output resistance with feedback R_{oF} is the resistance measured at the output terminal of the feedback amplifier. The output resistance of the noninverting amplifier was obtained by using Thévenin's theorem, and we can do the same for the inverting amplifier. Thévenin's equivalent circuit for R_{oF} of the inverting amplifier is shown in Figure 4-11. Note that this Thévenin's equivalent circuit is exactly the same as that for noninverting amplifier (Figure 4-5) because the output connections in both amplifiers are the same. This means that the equation for the output resistance R_{oF} of the inverting amplifier must be identical to that of the noninverting amplifier [Equation (4-9b)]. Specifically,

$$R_{oF} = \frac{R_o}{1+AB} \tag{4-20}$$

However, note that $B = R_1/R_F$ for the inverting amplifier.

4-4.5 Bandwidth with Feedback

As mentioned previously, the gain bandwidth product of a single break frequency op-amp is always constant. We also saw that the gain of the amplifier with feedback is always less than the gain without feedback. Therefore, the bandwidth of the amplifier

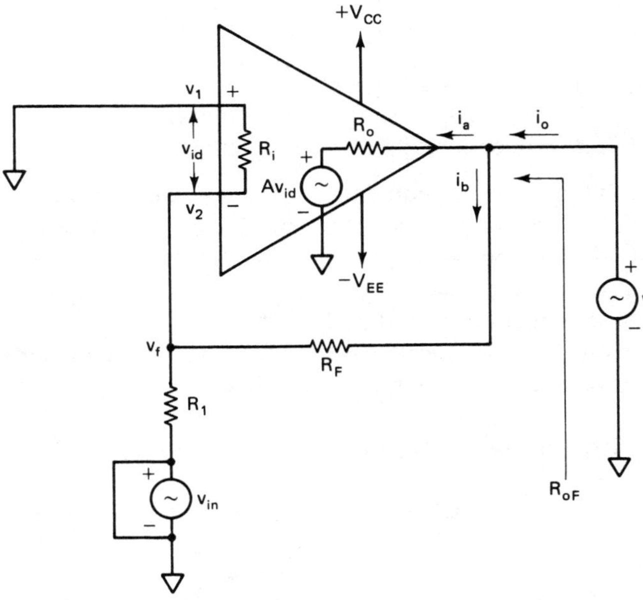

Figure 4-11 Thévenin's equivalent circuit for R_{oF} of the inverting amplifier.

with feedback f_F must be larger than that without feedback. This means that the equation for the bandwidth of the inverting amplifier can be written as

$$f_F = f_o(1 + AB) \tag{4-21a}$$

or, equivalently,

$$f_F = \frac{\text{unity gain bandwidth}}{\text{closed-loop gain}} \tag{4-21b}$$

where f_o is the break frequency of the op-amp and $B = R_1/R_F$. Note that the equation for the bandwidth of the inverting and noninverting amplifier is the same except the B value.

To find the closed-loop bandwidth of the inverting amplifier, use Equation (4-21a) if f_o is known and use Equation (4-21b) if unity gain bandwidth is given. However, to calculate the bandwidth graphically from the gain versus frequency curve, we can use the same procedure that was outlined for the noninverting amplifier.

4-4.6 Total Output Offset Voltage with Feedback

When the temperature and power supply voltages are fixed, the output offset voltage is a function of the gain of an op-amp. However, we saw that the gain of the op-amp with feedback is always less than that without feedback. Therefore, the output offset voltage with feedback V_{ooT} must always be smaller than that without feedback. Specifically,

$$\text{total output offset voltage with feedback} = \frac{\text{total output offset voltage without feedback}}{1 + AB}$$

That is,

$$V_{ooT} = \frac{\pm V_{\text{sat}}}{1 + AB} \tag{4-22}$$

where $\pm V_{\text{sat}}$ = saturation voltages
$B = R_1/R_F$ for the inverting amplifier
A = open-loop gain of the op-amp

Note again that the V_{ooT} equation for the inverting amplifier is the same as that for the noninverting amplifier except for the B value.

In addition, because of the negative feedback the effect of noise, variations in supply voltages, and changes in temperature on the output voltage of the inverting amplifiers are significantly reduced.

Finally, the two special cases of the inverting amplifier with feedback are the current-to-voltage converter and the inverter.

4-4.7 Current-to-Voltage Converter

Let us reconsider the ideal voltage-gain Equation (4-14) of the inverting amplifier,

$$\frac{v_o}{v_{\text{in}}} = -\frac{R_F}{R_1}$$

Therefore,

$$v_o = -\left(\frac{v_{in}}{R_1}\right)R_F$$

However, since $v_1 = 0$ V and $v_1 = v_2$,

$$\frac{v_{in}}{R_1} = i_{in}$$

$$v_o = -i_{in}R_F \tag{4-23}$$

This means that if we replace the v_{in} and R_1 combination by a current source i_{in} as shown in Figure 4-12, the output voltage v_o becomes proportional to the input current i_{in}. In other words, the circuit of Figure 4-12 *converts* the input current into a proportional output voltage.

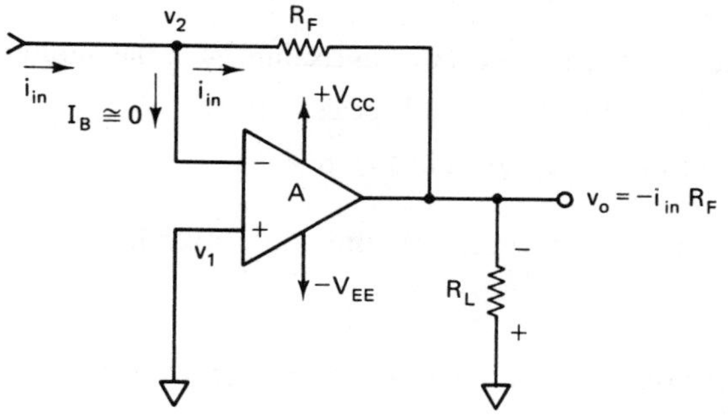

Figure 4-12 Current-to-voltage converter.

One of the most common uses of the current-to-voltage converter is in sensing current from photodetectors and digital-to-analog converter applications, which are discussed in more depth in Chapter 7.

4-4.8 Inverter

If we need an output signal equal in amplitude but opposite in phase to that of the input signal, we can use the inverter. The inverting amplifier of Figure 4-8 works as an inverter if $R_1 = R_F$. Since the inverter is a special case of the inverting amplifier, all the equations developed for the inverting amplifier are also applicable here. The equations can be applied by merely substituting A for $(1 + AB)$, since $B = 1$.

Example 4-3

For the inverting amplifier of Figure 4-8, $R_1 = 470\ \Omega$ and $R_F = 4.7$ kΩ. Assume that the op-amp is the 741 having the same specifications as those given in Example 4-1. Calculate the values of A_F, R_{iF}, R_{oF}, f_F, and V_{ooT}.

Solution. From Equation (4-16),

$$B = \frac{470}{4.7\text{ k}\Omega} = \frac{1}{10}$$

Then $1 + AB \cong 20{,}000$. Therefore, using Equations (4-15), (4-18), (4-20), (4-21a), and (4-22), the values of the closed-loop parameters are:

$$A_F = -\frac{200{,}000}{20{,}000} = -10$$

$$R_{iF} = 470\ \Omega + \left(\frac{4700\ \Omega}{200000} \middle\| 2\ \text{M}\Omega\right) = 470\ \Omega$$

$$R_{oF} = \frac{75\ \Omega}{20{,}000} = 3.75\ \text{m}\Omega$$

$$f_F = (5\ \text{Hz})(20{,}000) = 100\ \text{kHz}$$

$$V_{ooT} = \frac{\pm 13\ \text{V}}{20{,}000} = \pm 650\ \mu\text{V}$$

Example 4-4

For the inverting amplifier of Example 4-3, determine the value of output voltage if the input is a 1-V peak-to-peak sine wave at 1 kHz. Also sketch the output waveform. Assume that V_{ooT} is zero.

Solution. Using the value of the gain calculated in Example 4-3, the output voltage is

$$v_o = -(10)(1) = -10\ \text{V peak to peak}$$

The input and output waveforms are shown in Figure 4-13.

For quick review, the results of noninverting and inverting amplifiers are summarized in Table 4-1.

TABLE 4-1 SUMMARY OF RESULTS OBTAINED FOR NONINVERTING AND INVERTING AMPLIFIERS

Parameter	Noninverting amplifier	Inverting amplifier
1. Voltage gain	$A_F = \frac{A(R_1 + R_F)}{R_1 + R_F + AR_1}$ (exact) $= \frac{A}{1 + AB}$ $= 1 + \frac{R_F}{R_1}$ (ideal)	$A_F = \frac{-AR_F}{R_1 + R_F + AR_1}$ (exact) $= \frac{-A}{1 + AB}$ $= -\frac{R_F}{R_1}$ (ideal)
2. Gain of the feedback circuit	$B = \frac{R_1}{R_1 + R_F}$	$B = \frac{R_1}{R_F}$
3. Input resistance	$R_{iF} = R_i(1 + AB)$	$R_{iF} = R_1 + \left(\frac{R_F}{1 + A} \middle\| R_i\right)$
4. Output resistance	$R_{oF} = \frac{R_o}{1 + AB}$	$R_{oF} = \frac{R_o}{1 + AB}$
5. Bandwidth	$f_F = f_o(1 + AB)$ $= \frac{UBW}{A_F}$	$f_F = f_o(1 + AB)$ $= \frac{UBW}{A_F}$
6. Total output offset voltage	$V_{ooT} = \frac{\pm V_{sat}}{1 + AB}$	$V_{ooT} = \frac{\pm V_{sat}}{1 + AB}$

As mentioned earlier, current-shunt and current-series feedback configurations seldom find use in practice; therefore, they will not be discussed here. However, interested readers may analyze them following the same procedure as used for voltage-series and voltage-shunt feedback configurations.

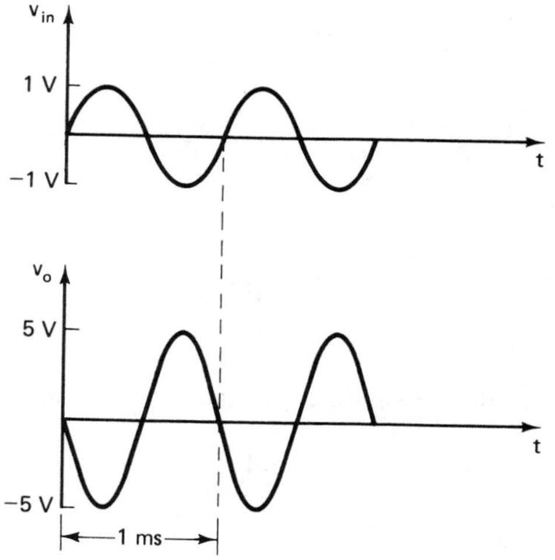

Figure 4-13 Waveforms for Example 4-4.

4-5 DIFFERENTIAL AMPLIFIERS

We will now study the closed-loop differential amplifier configurations. Specifically, we evaluate three different arrangements of the differential amplifier with negative feedback. We classify these arrangements according to the number of op-amps used. That is,

1. Differential amplifier with one op-amp
2. Differential amplifier with two op-amps
3. Differential amplifier with three op-amps

Generally, the differential amplifiers are used in instrumentation and industrial applications to amplify differences between two input signals, such as the outputs of the *Wheatstone bridge* circuits. Differential amplifiers are preferred in these applications because they are better able to reject common-mode (noise) voltages than single-input circuits such as inverting and noninverting amplifiers. They also present a balanced input impedance.

4-5.1 Differential Amplifier with One Op-Amp

Figure 4-14 shows the differential amplifier with one op-amp. We will analyze this circuit by deriving voltage gain, and input resistance. A close examination of Figure 4-14 reveals that the differential amplifier is a combination of inverting and noninverting amplifiers. That is, when v_x is reduced to zero the circuit is a noninverting amplifier, whereas the circuit is an inverting amplifier when input v_y is reduced to zero.

4-5.1(a) Voltage gain. The circuit in Figure 4-14 has two inputs, v_x and v_y; we will, therefore, use the superposition theorem in order to establish the relationship between inputs and output. When $v_y = 0$ V, the configuration becomes an

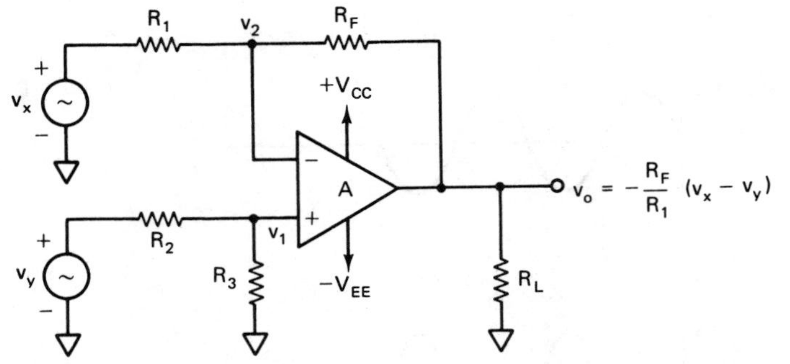

Figure 4-14 Differential amplifier with one op-amp. $R_1 = R_2$ and $R_F = R_3$.

inverting amplifier; hence the output due to v_x only is

$$v_{ox} = -\frac{R_F(v_x)}{R_1} \tag{4-24a}$$

Similarly, when $v_x = 0$ V, the configuration is a noninverting amplifier having a voltage-divider network composed of R_2 and R_3 at the noninverting input. Therefore,

$$v_1 = \frac{R_3(v_y)}{R_2 + R_3}$$

and the output due to v_y then is

$$v_{oy} = \left(1 + \frac{R_F}{R_1}\right)v_1$$

That is,

$$v_{oy} = \frac{R_3}{R_2 + R_3}\left(\frac{R_1 + R_F}{R_1}\right)v_y$$

Since $R_1 = R_2$ and $R_F = R_3$,

$$v_{oy} = \frac{R_F(v_y)}{R_1} \tag{4-24b}$$

Thus, from Equations (4-24a) and (4-24b), the net output voltage is

$$v_o = v_{ox} + v_{oy}$$

$$v_o = -\frac{R_F}{R_1}(v_x - v_y) = -\frac{R_F(v_{xy})}{R_1}$$

or the voltage gain

$$A_D = \frac{v_o}{v_{xy}} = -\frac{R_F}{R_1} \tag{4-25}$$

Note that the gain of the differential amplifier is the same as that of the inverting amplifier.

4-5.1(b) Input resistance. Assume that when both input signals are zero, the output voltage is also zero. That is, the total output offset voltage V_{ooT} is zero. Therefore, with $V_{ooT} = 0$ V, $R_1 = R_2$, and $R_F = R_3$, the input resistance between either input terminal v_x or v_y and ground is

$$R_{iF} = R_1 + R_F \tag{4-26}$$

If we need a variable gain, we can use the differential amplifier of Figure 4-15. In this circuit $R_1 = R_2$, $R_F = R_3$, and the potentiometer $R_P = R_4$. Therefore, depending on the position of the wiper in R_P, voltage gain can be varied from the closed-loop gain of $-2R_F/R_1$ to the open-loop gain of A.

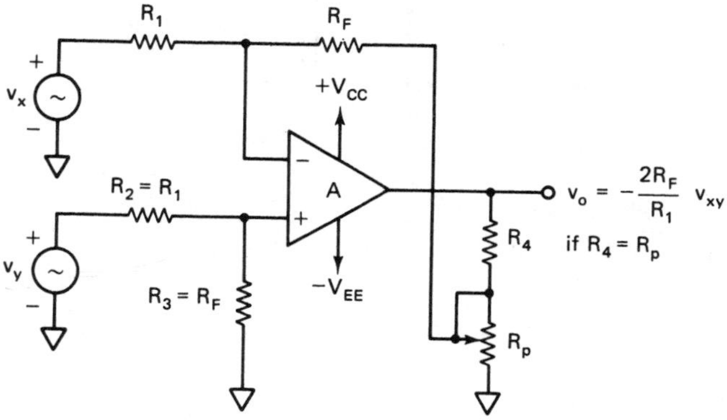

Figure 4-15 Differential amplifier with variable gain.

Example 4-5

In the circuit of Figure 4-14, $R_1 = R_2 = 1\ \text{k}\Omega$, $R_F = R_3 = 10\ \text{k}\Omega$, and the op-amp is a 741C.

(a) What are the gain and input resistance of the amplifier?

(b) Calculate the output voltage v_o if $v_x = 2.7$ V pp and $v_y = 3$ V pp sine waves at 100 Hz.

Solution

(a) Substituting known values in Equations (4-25) and (4-26), we obtain A_D and R_{iF}:

$$A_D = -\frac{10\ \text{k}\Omega}{1\ \text{k}\Omega} = -10$$

$$R_{iF} = 1\ \text{k}\Omega + 10\ \text{k}\Omega = 11\ \text{k}\Omega$$

(b) Rearranging Equation (4-25) yields

$$\begin{aligned} v_o &= -A_D v_{xy} \\ &= -(10)(2.7 - 3) = 3 \text{ V pp sine wave at 100 Hz} \end{aligned}$$

4-5.2 Differential Amplifier with Two Op-Amps

Recall that the gain expression for the differential amplifier of Figure 4-14 is the same as that of the inverting amplifier of Figure 4-8 [see Equations (4-14) and (4-25)]. We can increase the gain of the differential amplifier and also increase the input resistance R_{iF} if we use two op-amps. Such a circuit is shown in Figure 4-16. In fact, the characteristics of this amplifier are identical to those of the noninverting amplifier.

4-5.2(a) Voltage gain. A close examination of the circuit of Figure 4-16 shows that it is composed of two stages: first, the noninverting amplifier, and second, the differential amplifier with unequal gains. By finding the gain of these two stages, we can obtain the overall gain of the circuit as follows:

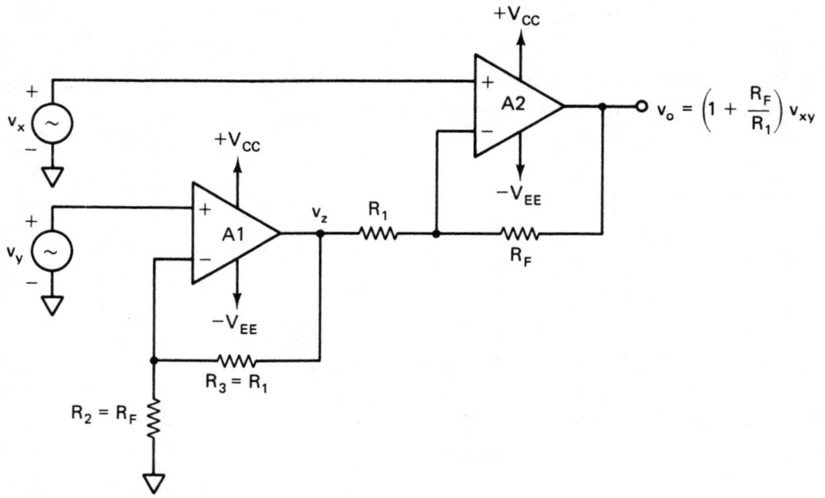

Figure 4-16 Differential amplifier with two op-amps.

The output v_z of the first stage is

$$v_z = \left(1 + \frac{R_3}{R_2}\right)v_y \tag{4-27a}$$

By applying the superposition theorem to the second stage, we can obtain the output voltage:

$$v_o = -\frac{R_F(v_z)}{R_1} + \left(1 + \frac{R_F}{R_1}\right)v_x \tag{4-27b}$$

Substituting the value of v_z from Equation (4-27a), we get

$$v_o = -\left(\frac{R_F}{R_1}\right)\left(1 + \frac{R_3}{R_2}\right)v_y + \left(1 + \frac{R_F}{R_1}\right)v_x$$

Since $R_1 = R_3$ and $R_F = R_2$,

$$v_o = \left(1 + \frac{R_F}{R_1}\right)(v_x - v_y)$$

Therefore,

$$A_D = \frac{v_o}{v_{xy}} = 1 + \frac{R_F}{R_1} \tag{4-28}$$

where $v_{xy} = v_x - v_y$.

4-5.2(b) Input resistance. The input resistance R_{iF} of the differential amplifier is the resistance determined looking into either one of the two noninverting input terminals with the other grounded (see Figure 4-16). Note, however, that the first stage (A_1) is a noninverting amplifier; therefore [from Equation (4-8)], its input resistance is

$$R_{iF1} = R_i(1 + AB) \tag{4-29a}$$

where R_i = open-loop input resistance of the op-amp

$$B = \frac{R_2}{R_2 + R_3}$$

Similarly, with v_y shorted to ground ($v_y = 0$ V), the second stage (A_2) also becomes a noninverting amplifier whose input resistance can then be written as

$$R_{iF2} = R_i(1 + AB) \tag{4-29b}$$

where R_i = open-loop input resistance of the op-amp

$$B = \frac{R_1}{R_1 + R_F}$$

However, since $R_1 = R_3$ and $R_F = R_2$, the $R_{iF1} \neq R_{iF2}$. This is the drawback of the differential amplifier of Figure 4-16. Nevertheless, with proper selection of components, both R_{iF1} and R_{iF2} can be made much larger than the source resistances, so that the loading of the input sources does not occur.

Example 4-6

The following specifications are given for the differential amplifier of Figure 4-16: $R_1 = R_3 = 680\ \Omega$, $R_F = R_2 = 6.8\ \text{k}\Omega$, $v_x = -1.5$ V pp, and $v_y = -2$ V pp sine waves at 1 kHz. The op-amp is a 741C. Calculate (a) the voltage gain and the input resistance and (b) the output voltage of the amplifier. Assume that the output is initially nulled ($V_{ooT} = 0$ V).

Solution

(a) From Equation (4-28),

$$A_D = 1 + \frac{6.8\ \text{k}\Omega}{680\ \Omega} = 11$$

The input resistance can be calculated using Equations (4-29a) and (4-29b).

$$R_{iF1} = (2\ \text{M}\Omega)\left[1 + \frac{(2)(10^5)(6.8\ \text{k}\Omega)}{6.8\ \text{k}\Omega + 680\ \Omega}\right] = 364\ \text{G}\Omega$$

$$R_{iF2} = (2\ \text{M}\Omega)\left[1 + \frac{(2)(10^5)(680)\ \Omega}{6.8\ \text{k}\Omega + 680\ \Omega}\right] = 36.4\ \text{G}\Omega$$

(b) Output voltage can be calculated by rearranging Equation (4-28):

$$v_o = \left(1 + \frac{R_F}{R_1}\right)v_{xy}$$

$$= (11)(-1.5 + 2) = 5.5 \text{ V pp sine wave at 1 kHz}$$

4-5.3 Differential Amplifier with Three Op-Amps

The advantages of the two differential amplifiers discussed above can be combined in a single differential amplifier having variable voltage gain and very high input resistance. However, to achieve this desirable combination, three op-amps must be used, as shown in Figure 4-17.

4-5.3(a) Voltage gain. The differential op-amp of Figure 4-17 consists of two stages. The first stage is composed of op-amps A_1 and A_2, while the second stage is formed by op-amp A_3. Therefore, to find the overall voltage gain A_D of the amplifier, the voltage gain of each stage must be determined. To begin with, the first stage can be viewed as two separate differential amplifiers, as shown in Figure 4-18. The output voltages of these differential amplifiers can be found by applying the superposition theorem. For Figure 4-18(a),

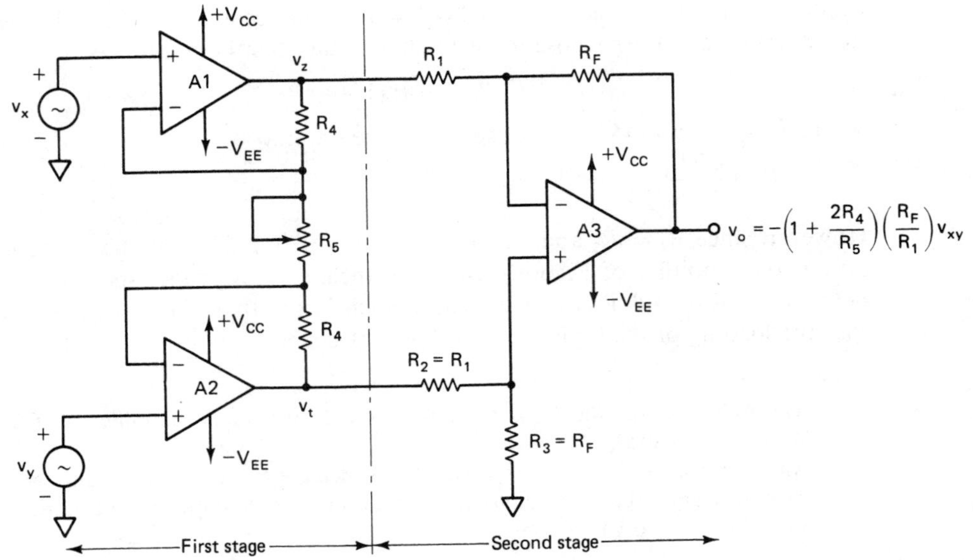

Figure 4-17 Differential amplifier with three op-amps.

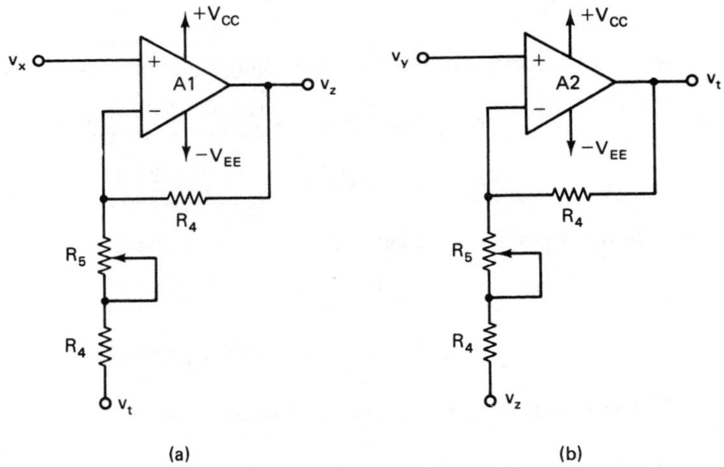

Figure 4-18 Deriving the voltage gain of the first stage of the circuit in Figure 4-17.

$$v_z = \frac{2R_4 + R_5}{R_4 + R_5} v_x - \frac{R_4}{R_4 + R_5} v_t \tag{4-30a}$$

and for Figure 4-18(b),

$$v_t = \frac{2R_4 + R_5}{R_4 + R_5} v_y - \frac{R_4}{R_4 + R_5} v_z \tag{4-30b}$$

However, the output voltage of the first stage is

$$v_{zt} = v_z - v_t$$

Therefore, from Equations (4-30a) and (4-30b),

$$v_{zt} = \frac{2R_4 + R_5}{R_4 + R_5}(v_x - v_y) + \frac{R_4}{R_4 + R_5}(v_z - v_t)$$

$$= \frac{2R_4 + R_5}{R_4 + R_5}v_{xy} + \frac{R_4}{R_4 + R_5}v_{zt}$$

Simplifying and rearranging the voltage gain of the first stage gives

$$\frac{v_{zt}}{v_{xy}} = \frac{2R_4 + R_5}{R_5} \tag{4-30c}$$

Next, using Equation (4-25), the gain of the second stage is

$$\frac{v_o}{v_{zt}} = -\frac{R_F}{R_1} \tag{4-30d}$$

Thus, from Equations (4-30c) and (4-30d), the overall voltage gain is

$$A_D = \frac{v_{zt}}{v_{xy}}\frac{v_o}{v_{zt}} = -\left(\frac{2R_4 + R_5}{R_5}\right)\frac{R_F}{R_1}$$

or

$$A_D = \frac{v_o}{v_{xy}} = -\left(1 + \frac{2R_4}{R_5}\right)\frac{R_F}{R_1} \tag{4-31}$$

Remember that the gain can be changed by varying potentiometer R_5. Of course, R_5 should never be set to zero or infinity. If R_5 is set to zero, voltage gain Equation (4-31) becomes indeterminent; whereas if R_5 is set to infinity (open R_5), the first stage acts as a voltage follower.

4-5.3(b) Input resistance. The input resistance R_{iF} of the differential amplifier in Figure 4-17 is the same as the input resistance of the first stage, that is, the resistance determined at input v_x and v_y looking into the circuit with the other terminal grounded.

In Figure 4-18a, for instance, when v_t is reduced to zero, that is, when v_y is grounded, the circuit is a noninverting amplifier. Applying the concepts developed for the noninverting amplifier, the input resistance determined at input v_x is

$$R_{iF} = R_i\left(1 + A\frac{R_4 + R_5}{2R_4 + R_5}\right) \tag{4-32}$$

Similarly, the input resistance determined at input v_y will be the same as that given in Equation (4-32).

A simple modification in the differential amplifier of Figure 4-17 is to replace the first stage by voltage followers, as shown in Figure 4-19. The main advantage of this circuit is that it has the highest input resistance of the three amplifiers; this resistance is given by

$$R_{iF} = R_i(1 + A) \tag{4-33a}$$

where R_i = input resistance of the op-amp
A = large-signal voltage gain of the op-amp

However, the voltage gain of the circuit in Figure 4-19 is equal to the voltage gain of

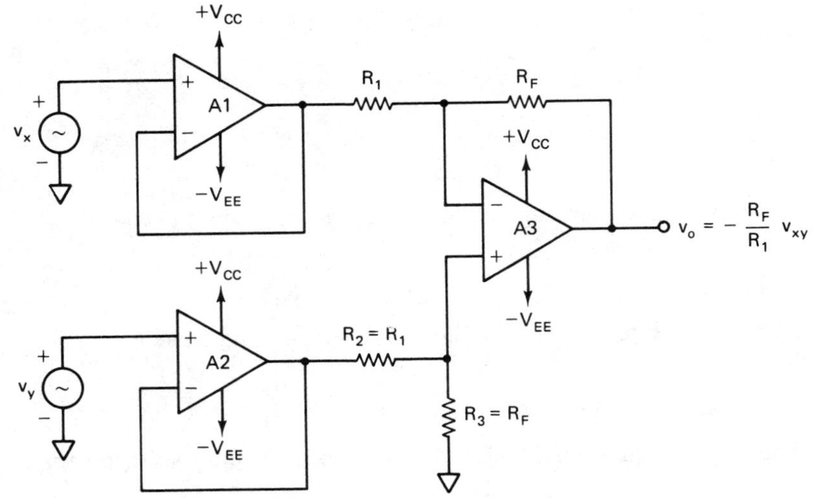

Figure 4-19 Special case of differential amplifier with three op-amps.

the differential amplifier with one op-amp; that is,

$$A_D = \frac{v_o}{v_{xy}} = -\frac{R_F}{R_1} \tag{4-33b}$$

4-5.4 Output Resistance and Bandwidth of Differential Amplifiers

Recall that the closed-loop op-amp configurations—noninverting, inverting, and differential amplifier—employ negative feedback and use the same type of output connections. Therefore, the output resistance of all three configurations must be identical. In other words, the output resistance of the differential amplifier should be the same as that of the noninverting or inverting amplifier, except that $B = 1/A_D$. That is,

$$R_{oF} = \frac{R_o}{1 + A/A_D} \tag{4-34}$$

where A_D = closed-loop gain of the differential amplifier
R_o = output resistance of the op-amp
A = large-signal voltage gain of the op-amp

Remember that A_D is different for each of the differential amplifier configurations.

As in the case of noninverting or inverting amplifiers, the bandwidth of differential amplifiers also depends on the closed-loop gain of the amplifier, and is given by

$$f_F = \frac{\text{unity gain bandwidth}}{\text{closed-loop gain } A_D} \tag{4-35a}$$

or alternatively,

$$f_F = f_o\left(1 + \frac{A}{A_D}\right) \tag{4-35b}$$

where f_o is the open-loop break frequency of the op-amp. For example, the output resistance and bandwidth of the differential amplifier with one op-amp will be as follows:

$$R_{oF} = \frac{R_0}{1 + AR_1/R_F}$$

$$f_F = f_o\left(1 + \frac{AR_1}{R_F}\right)$$

Example 4-7

The following specifications are given for the differential amplifier of Figure 4-17: $R_4 = 3.9\text{ k}\Omega$, R_5 is a 5-kΩ potentiometer, $R_1 = R_2 = 1\text{ k}\Omega$, and $R_F = R_3 = 4.7\text{ k}\Omega$. The op-amps are 741C's and the supply voltages are ± 15 V. If R_5 is set at 3 kΩ, calculate (a) the voltage gain, (b) the input resistance, (c) the output resistance, and (d) the bandwidth of the differential amplifier.

Solution

(a) Substituting the given values in Equation (4-31), the voltage gain is

$$A_D = -\left[1 + \frac{(2)(3{,}9\text{ k}\Omega)}{3\text{ k}\Omega}\right]\frac{4.7\text{ k}\Omega}{1\text{ k}\Omega} = -16.92$$

Note that the negative sign indicates the phase reversal between differential input and output.

(b) Using Equation (4-32),

$$R_{iF} = 200{,}0000\left[1 + \frac{(200{,}000)(3.9\text{ k}\Omega + 3\text{ k}\Omega)}{(2)(3.9\text{ k}\Omega) + 3\text{ k}\Omega}\right] = 256\text{ G}\Omega$$

(c) Output resistance can be obtained substituting the values in Equation (4-34):

$$R_{oF} = \frac{75\Omega}{1 + 200{,}000/16.92} = 6.3\text{ m}\Omega$$

(d) Finally, the bandwidth of the amplifier is given by Equation (4-35b):

$$f_F = 5\left(1 + \frac{200{,}000}{16.92}\right) = 59.12\text{ kHz}$$

SUMMARY

1. The very high voltage gain of the open-loop op-amp as well as its variation with temperature, power supply, and production yield makes the open-loop op-amp configuration unsuitable for linear applications.
2. The introduction of negative feedback stabilizes the gain; however, it is also smaller than the open-loop gain. On the other hand, positive feedback is necessary in oscillator circuits.
3. There are four closed-loop configurations using negative feedback: voltage-series, voltage-shunt, current-series, and current-shunt. These configurations are labeled according to whether the voltage or current is fed back to the input in series or in parallel.
4. The voltage-series negative feedback configuration is commonly called a noninverting amplifier with feedback, while the voltage-shunt feedback configuration is called an inverting amplifier with feedback. The ideal closed-loop gain of these two configurations depends

only on the feedback network (components) and is independent of the internal gain A of the op-amp.

5. The use of negative feedback in a noninverting amplifier increases the input impedance and bandwidth, and decreases the output resistance, total output-offset voltage, and the effect of varying environmental conditions on the gain. The same results are obtained for the inverting amplifier using negative feedback, except for input resistance. The input resistance of the inverting amplifier is relatively smaller because it depends on the resistance connected in series with the input signal source.
6. A special case of the noninverting amplifier is the voltage follower, while the current-to-voltage converter and the inverter are two special cases of the inverting amplifier. The voltage follower has a gain of unity. The current-to-voltage converter produces output voltage proportional to input current, and the inverter has an output equal in amplitude but opposite in phase to that of the input.
7. Negative feedback is also used in differential amplifiers. According to the number of op-amps used, the differential amplifier configurations are classified as differential amplifier with one op-amp, differential amplifier with two op-amps, and differential amplifier with three op-amps. The differential amplifier with one op-amp has the same characteristics as the inverting amplifier, while the amplifier with two op-amps has the same characteristics as the noninverting amplifier. Very high input resistance and variable gain are the features of the differential amplifier with three op-amps.

QUESTIONS

4-1. Give two reasons why an open-loop op-amp is unsuitable for linear applications.

4-2. What is feedback? List two types of feedback. Which type is used in linear applications?

4-3. List the four negative feedback configurations. Which two configurations are most commonly used?

4-4. Explain briefly why negative feedback is desirable in amplifier applications.

4-5. What is the effect of negative feedback in noninverting amplifiers?

4-6. How does negative feedback affect the performance of an inverting amplifier?

4-7. In what way is the voltage follower a special case of the noninverting amplifier?

4-8. List two special cases of inverting amplifiers.

4-9. What are the three differential amplifier configurations? Briefly compare and contrast these configurations.

PROBLEMS

4-1. The open-loop gain A of a particular op-amp varies from 15,000 to 20,000 over the operating temperature range 25° to 35°C. This op-amp is used in the circuit of Figure 4-2 with $R_1 = 470\ \Omega$ and $R_F = 47\ k\Omega$.

(a) Calculate the closed-loop gain at both temperature extremes.

(b) Compute the percentage variation in the open-loop gain and in the closed-loop gain; explain the results.

4-2. The following specifications apply to the noninverting amplifier of Figure 4-2:

Op-amp is the 741C with:	Feedback circuit
$A = 200{,}000$	$R_1 = 1.8\ \text{k}\Omega$
$R_i = 2\ \text{M}\Omega$	$R_F = 18\ \text{k}\Omega$
$R_o = 75\ \Omega$	
$f_o \cong 5\ \text{Hz}$	
UBW $\cong 1$ MHz	
Supply voltages $= \pm 15$ V	
Maximum output voltage swing $= \pm 13$ V	

(a) Calculate the *exact* closed-loop gain.
(b) Calculate the *ideal* closed-loop gain.
(c) Explain the results obtained in parts (a) and (b).

4-3. Repeat Problem 4-2 with $R_1 = 470\ \Omega$ and $R_F = 47\ \text{k}\Omega$.

4-4. The 714C is configured as a noninverting amplifier as shown in Figure 4-2. The following data are given for the circuit:

$A = 400{,}000 \qquad R_1 = 470\ \Omega$

$R_i = 33\ \text{M}\Omega \qquad R_F = 4.7\ \text{k}\Omega$

$R_o = 60\ \Omega$

Supply voltages $= \pm 15$ V

Maximum output voltage swing $= \pm 13$ V

Unity gain bandwidth $= 0.6$ MHz

Compute the closed-loop parameters A_F, R_{iF}, R_{oF}, f_F, and V_{ooT}.

4-5. Repeat Problem 4-4 with $R_1 = 100\ \Omega$.

4-6. For the noninverting amplifier of Problem 4-4, compute the output voltage v_o and then sketch it if $v_{in} = 100$ mV pp sine wave at 1 kHz. Assume that the op-amp is initially nulled.

4-7. The op-amp of Problem 4-4 is used in a voltage follower. Compute the closed-loop parameters A_F, R_{iF}, R_{oF}, f_F, and V_{ooT}.

4-8. The 741C is configured as an inverting amplifier as shown in Figure 4-8, with $R_1 = 1\ \text{k}\Omega$ and $R_F = 10\ \text{k}\Omega$.
(a) Calculate the *exact* closed-loop gain.
(b) Calculate the *ideal* closed-loop gain.
(c) Compare and explain the results of parts (a) and (b).
(For op-amp specifications, refer to Problem 4-2.)

4-9. Repeat Problem 4-8 with $R_1 = 100\ \Omega$.

4-10. The 714C is connected as an inverting amplifier with $R_1 = 1\ \text{k}\Omega$ and $R_F = 4.7\ \text{k}\Omega$. Compute the closed-loop parameters A_F, R_{iF}, R_{oF}, f_F, and V_{ooT}. (Refer to Problem 4-4 for the op-amp parameters.)

4-11. Repeat Problem 4-10 with $R_F = 1\ \text{k}\Omega$. What is the name of this circuit?

4-12. For the inverting amplifier of Problem 4-10, $v_{in} = 1$ V pp sine wave at 100 Hz.
(a) Compute the output voltage v_o.
(b) Draw the output voltage waveform, assuming that the output is initially nulled.

4-13. The input current into the current-to-voltage converter of Figure 4-12 varies from 500 μA to 1 mA. Calculate the variations in the output voltage if $R_F = 4.7\ \text{k}\Omega$. Assume that the op-amp is a 714C with its output initially nulled.

4-14. The 741C is used for the differential amplifier of Figure 4-14 with $R_1 = R_2 = 100\ \Omega$ and $R_F = R_3 = 3.9\ \text{k}\Omega$.
(a) Calculate the closed-loop voltage gain.
(b) Calculate the input resistance R_{iF} assuming that the circuit is initially nulled.
(The 741C parameters are given in Problem 4-2.)

4-15. For the differential amplifier of Problem 4-14, compute the output voltage if $v_x = -3.2$ V and $v_y = -3.0$ V. Assume that the circuit is initially nulled.

4-16. In the variable-gain differential amplifier of Figure 4-15, $R_1 = R_2 = 100\ \Omega$ $R_F = R_3 = 2.2\ \text{k}\Omega$, $R_4 = 10\ \text{k}\Omega$, and R_P is a 20-kΩ potentiometer. The op-amp is a 714C. Compute the gain of the differential amplifier if (a) $R_P = 10\ \text{k}\Omega$ and (b) $R_P = 5\ \text{k}\Omega$.

4-17. The following specifications apply to the differential amplifier of Figure 4-16: $R_2 = R_F = 2.2\ \text{k}\Omega$, $R_3 = R_1 = 1\ \text{k}\Omega$, $v_x = 700$ mV pp, and $v_y = 500$ mV pp sine waves at 500 Hz; the op-amp is a 741C. Assume that the op-amp is initially nulled. Calculate:
(a) The voltage gain.
(b) The bandwidth.
(c) The input resistance seen by each signal source.
(d) The output resistance.
(e) The output voltage of the amplifier.
(For the electrical parameters of the 741C, refer to Problem 4-2).

4-18. For the differential amplifier of Figure 4-17, $R_4 = 1\ \text{k}\Omega$, R_5 is a 10-kΩ potentiometer, $R_1 = R_2 = 1.5\ \text{k}\Omega$, and $R_F = R_3 = 15\ \text{k}\Omega$. The op-amps are 741Cs with supply voltages $= \pm 15$ V. If R_5 is set at 5 kΩ, calculate:
(a) The voltage gain.
(b) The input resistance seen by each signal source.
(c) The output resistance.
(d) The bandwidth of the differential amplifier.

EXPERIMENT 4-1
NONINVERTING AMPLIFIER WITH FEEDBACK

Objectives

In this experiment you will study the effect of negative feedback on the performance of the noninverting amplifier. At the end of this experiment you should be able to:

1. Verify that the gain of the noninverting amplifier depends on the feedback components and not on the internal gain of an op-amp.
2. Confirm that the use of negative feedback increases the bandwidth and decreases the output offset voltage of the op-amp.
3. Understand that the voltage follower is a special case of the noninverting amplifier.
4. Discuss the differences between the noninverting amplifier and the voltage follower.

Equipment

1. Dual-trace oscilloscope
2. Audio-frequency generator
3. ±15-V power supply
4. Multimeter

Materials

1. 741 or 351 op-amp
2. Two 1-kΩ resistors
3. Two 10-kΩ resistors
4. 100-kΩ resistor

Procedure

1. Connect the 741/351 op-amp as shown in Figure E4-1.1(a). The pin numbers indicated in this figure refer to an 8-pin mini DIP.
2. With the voltmeter, measure the output voltage and enter the measured value in Table E4-1.1(a).

TABLE E4-1.1

(a) dc measurements (with $v_{in} = 0$ V)

Configuration	Output offset voltage
Noninverting amplifier without feedback	
Noninverting amplifier with feedback	
Voltage follower	

(b) ac measurements

Input voltage, v_{in}	Input frequency, f_{in}	Output voltage, v_o			Voltage gain, v_o/v_{in}		
		Noninverting amplifier with feedback		Voltage follower	Noninverting amplifier with freedback		Voltage follower
		$R_F = 10$ kΩ	$R_F = 100$ kΩ		$R_F = 10$ kΩ	$R_F = 100$ kΩ	
100 mV 50 mV 50 mV	1 kHz 1 kHz 100 Hz						

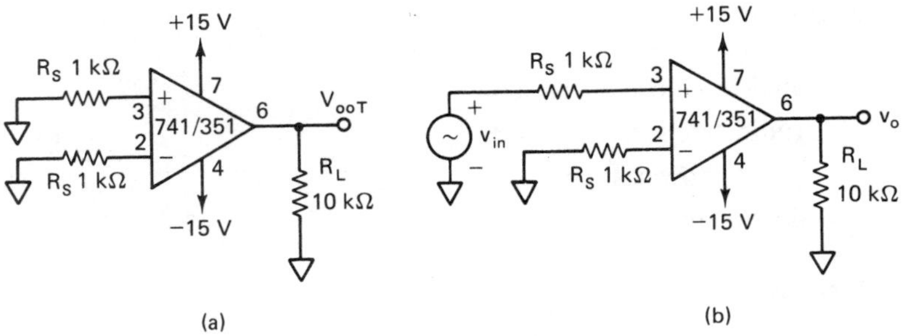

Figure E4-1.1 (a) Open-loop op-amp with input terminals connected to ground. (b) Noninverting open-loop op-amp.

3. Disconnect the resistor R_S connected to the noninverting input from ground and apply a 100-mV pp sine wave at 1 kHz through R_S to the noninverting terminal, as shown in Figure E4-1.1(b). Using the dual-trace oscilloscope, observe the input and output waveforms simultaneously. Change the input amplitude to 50 mV pp and observe the input and output waveforms. Then change the input frequency to 100 Hz and again observe the input and output waveforms.
4. Connect the circuit as shown in Figure E4-1.2(a). With $v_{in} = 0$ V, measure the output voltage using the voltmeter and enter the measured value in Table E4-1.1(a).

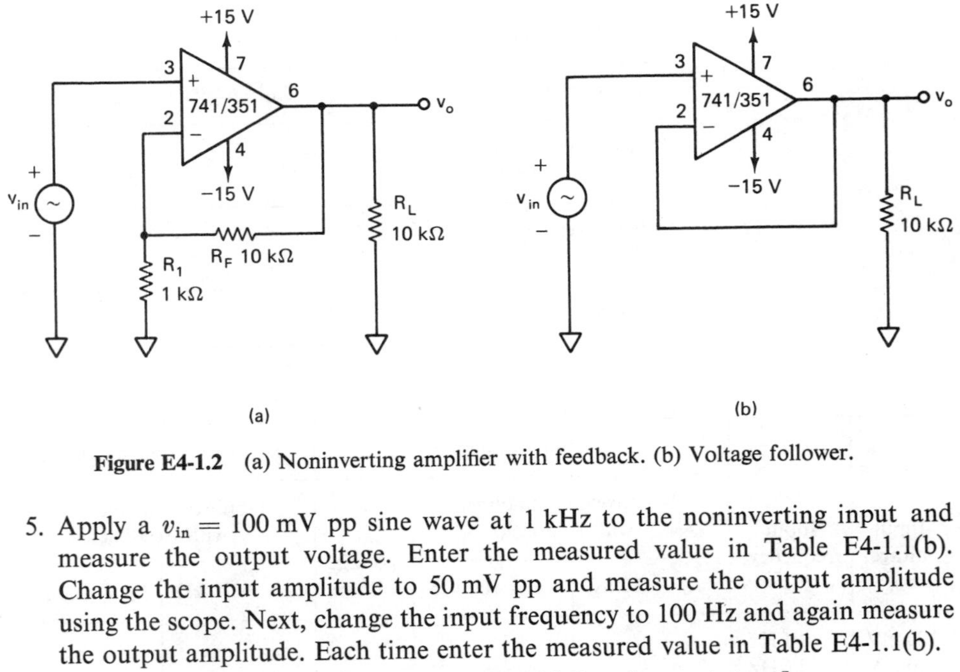

Figure E4-1.2 (a) Noninverting amplifier with feedback. (b) Voltage follower.

5. Apply a $v_{in} = 100$ mV pp sine wave at 1 kHz to the noninverting input and measure the output voltage. Enter the measured value in Table E4-1.1(b). Change the input amplitude to 50 mV pp and measure the output amplitude using the scope. Next, change the input frequency to 100 Hz and again measure the output amplitude. Each time enter the measured value in Table E4-1.1(b).
6. Change the value of R_F from 10 kΩ to 100 kΩ and repeat step 5.

7. Connect the circuit of Figure E4-1.2(b). With $v_{in} = 0$ V, measure the output voltage using the voltmeter and enter the measured value in Table E4-1.1(a).
8. Repeat step 5 for the circuit of Figure E4-1.2(b).
9. Complete Table E4-1.1(b).

Questions

1. List the amplifiers with the highest and lowest output offset voltage; then give the reason for the difference.
2. What is the input offset voltage for the voltage follower?
3. What type of output waveform is obtained in step 3? Explain why.
4. What is the gain of the noninverting amplifier when $R_F = 10\ \text{k}\Omega$ and $R_F = 100\ \text{k}\Omega$? What conclusion can be drawn from these results?
5. Referring to the data in Table E4-1.1(b), does the gain of the noninverting amplifier change when the input amplitude or frequency is changed? Explain your answer.
6. For the voltage follower, what is the relationship between the input and output voltages?
7. Explain how a voltage follower is a special case of the noninverting amplifier with feedback.

EXPERIMENT 4-2
INVERTING AMPLIFIER WITH FEEDBACK

Objectives

In this experiment you will study the effect of negative feedback on the performance of the inverting amplifier. At the end of this experiment you should be able to:

1. Verify that the input and output voltages of the inverting amplifier are out of phase by 180°.
2. Confirm that the gain of the inverting amplifier with feedback depends on external components and not on the internal gain of the op-amp.
3. Verify that the closed-loop bandwidth of an inverting amplifier depends on its closed-loop gain.
4. Understand that the inverter is a special case of the inverting amplifier.

Equipment

1. Dual-trace oscilloscope
2. Audio-frequency generator
3. $\pm$15-V power supply

Materials

1. 741 or 351 op-amp
2. 1-kΩ resistor
3. Three 10-kΩ resistors
4. 100-kΩ resistor

Procedure

1. Connect the 741/351 op-amp as shown in Figure E4-2.1(a). The pin numbers indicated refer to an 8-pin mini DIP.
2. Set v_{in} = 100 mV at 100 Hz sine wave and observe the input and output waveforms on the scope simultaneously. Make sure that the scope is used in the proper triggered mode. Enter the amplitudes of the input and output voltages in Table E4-2.1. Change the input frequency to 500 Hz and again observe the input and output waveforms. Enter the input and output amplitudes in Table E4-2.1.

TABLE E4-2.1

Input voltage, v_{in}	Input frequency, f_{in}	Output voltage, v_o			Voltage gain, v_o/v_{in}		
		Inverting amplifier with feedback			Inverting amplifier with feedback		
		R_F = 10 kΩ	R_F = 100 kΩ	Inverter	R_F = 10 kΩ	R_F = 100 kΩ	Inverter
100 mV	100 Hz						
100 mV	500 Hz						
100 mV	1 kHz						

Configuration	Frequency at which output starts decreasing
Inverting amplifier with feedback (i) R_F = 10 kΩ (ii) R_F = 100 kΩ Inverter	

3. Change the frequency of the input to 1 kHz, keeping the amplitude the same (v_{in} = 100 mV pp). Measure the input and output amplitudes and enter the results in Table E4-2.1.
4. Slowly increase the frequency of the input until the output amplitude just starts decreasing. Record this value of frequency in Table E4-2.1.
5. Change R_F from 10 kΩ to 100 kΩ and repeated steps 2, 3, and 4.
6. Connect the circuit of Figure E4-2.1(b) and again repeat steps 2, 3, and 4.
7. Complete Table E4-2.1.

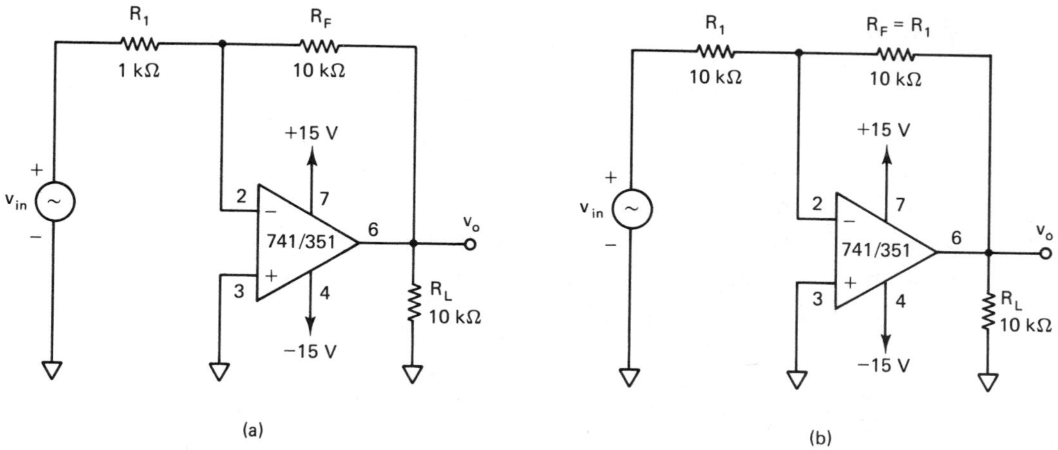

Figure E4-2.1 (a) Inverting amplifier with feedback. (b) Inverter.

Questions

1. What is the phase shift between the input and output voltages of an inverting amplifier?
2. Does the gain of the inverting amplifier with $R_F = 10\ \text{k}\Omega$ remain constant over 100 Hz to 1 kHz? Explain why or why not.
3. What determines the gain of the amplifier?
4. Do the gains of the inverting amplifier and inverter start decreasing at the same frequency? Explain your answer.
5. What is the difference between an inverting amplifier and an inverter?

EXPERIMENT 4-3
DIFFERENTIAL AMPLIFIER WITH FEEDBACK

Objectives

In this experiment you will learn how the differential amplifier works. At the end of this experiment you should be able to:

1. Verify that the differential amplifier amplifies the algebraic difference between two input voltages.
2. Verify that the gain of the differential amplifier with feedback depends on external components and not on the internal gain of the op-amp.
3. Confirm that the bandwidth of the differential amplifier depends on its gain.

Equipment

1. Dual-trace oscilloscope
2. Audio-frequency generator
3. ±15-V power supply

Materials

1. 741 or 351 op-amp
2. 2N3391 transistor
3. 1-kΩ resistor
4. 2-kΩ resistor
5. 10-kΩ resistor
6. 22-kΩ resistor
7. Three 47-kΩ resistors
8. Two 100-kΩ resistors
9. Two 470-kΩ resistors
10. 3-kΩ potentiometer
11. Three 5-μF capacitors

Procedure

1. Construct the circuit of Figure E4-3.1. Connect channels 1 and 2 of the scope to points x and y, set $v_{in} = 100$ mV at 1 kHz sine wave, and adjust the 3-kΩ potentiometer so that the voltages at points x and y are equal in amplitude.

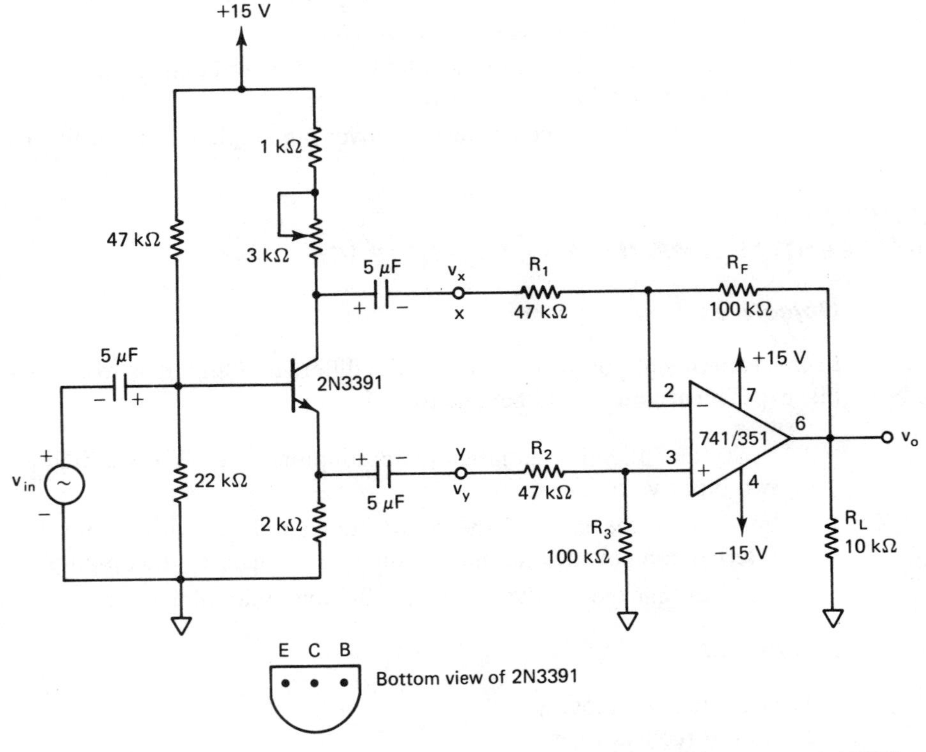

Figure E4-3.1 Differential amplifier with feedback. (Pin numbers refer to 8-pin mini DIP.)

2. Measure the input voltages v_x and v_y and output voltage v_o. Enter the measured values in Table E4-3.1(a).

TABLE E4-3.1

(a) If the voltage at the noninverting input terminal leads that at the inverting terminal by 180°:

v_x pp (V)	v_y pp (V)	v_{xy} pp (V)	v_o pp (V)	Voltage gain, v_o/v_{xy}	Calculated voltage gain, R_F/R_1

(b) If the voltage at the inverting input terminal leads that at the noninverting input terminal by 180°:

v_x pp (V)	v_y pp (V)	v_{xy} pp (V)	v_o pp (V)	Voltage gain, v_o/v_{xy}	Calculated voltage gain, R_F/R_1

(c) 100-kΩ resistors replaced by 470-kΩ resistors:

v_x pp (V)	v_y pp (V)	v_{xy} pp (V)	v_o pp (V)	Voltage gain, v_o/v_{xy}	Calculated voltage gain, R_F/R_1

(d) The frequency at which the output voltage of the differential amplifier starts decreasing: ______________Hz.

3. Switch off the input voltage v_{in} and power supply voltages ±15 V. Then reconnect the noninverting input terminal to point x and the inverting input terminal to point y, respectively. Turn the input and power supply voltages on and measure the v_x, v_y, and v_o voltages using the scope. Enter the results in Table E4-3.1(b).
4. Switch off the input voltage and power supply voltages and replace the 100-kΩ resistors by the 470-kΩ resistors. Turn on the v_{in} and ±15-V supply voltages. Measure v_x, v_y, and v_o voltages using the scope and enter the results in Table E4-3.1(c).
5. Increase the frequency of the input signal v_{in} until the amplitude of the output voltage v_o starts decreasing. Enter this value in Table E4-3.1(d).
6. Complete Table E4-3.1.

Questions

1. What conclusion can be drawn from the results of Table E4-3.1(a) and (b)?
2. How is the gain of the differential amplifier set?
3. What is the significance of the frequency at which the output voltage of the differential amplifier starts decreasing?

5

The Practical Op-Amp

5-1 INTRODUCTION

In preceding chapters we have seen that the op-amp may be used as an inverting, noninverting, or differential amplifier, and that the negative feedback can be used to stabilize the voltage gain and increase the bandwidth of the op-amp circuit. However, up to this point in our discussion we have treated the op-amp as an ideal device that gives us the desired results, such as high input impedance, low output impedance, high voltage gain, and broader bandwidth, if we use appropriate external components. We have also treated the op-amp as though it responds equally to both ac and dc input voltages. However, these ideal characteristics are not fully present in the practical op-amp circuits. The practical op-amp has some dc output voltage, called *output offset voltage*, even though both inverting and noninverting input terminals are grounded. Such an output offset voltage is an error voltage and is therefore undesirable. In this chapter we discuss the properties of the practical op-amp that produce the output offset voltage. We also see that it is necessary to incorporate additional circuitry to increase the effectiveness of the op-amp, especially if it is used as a dc amplifier.

5-2 INPUT OFFSET VOLTAGE

Input offset voltage V_{io} is the differential input voltage that exists between two input terminals of an op-amp without any external inputs applied. In other words, it is the amount of the input voltage that should be applied between two input terminals in order to force the output voltage to zero [see Figure 5-1(a)]. Let us denote the output

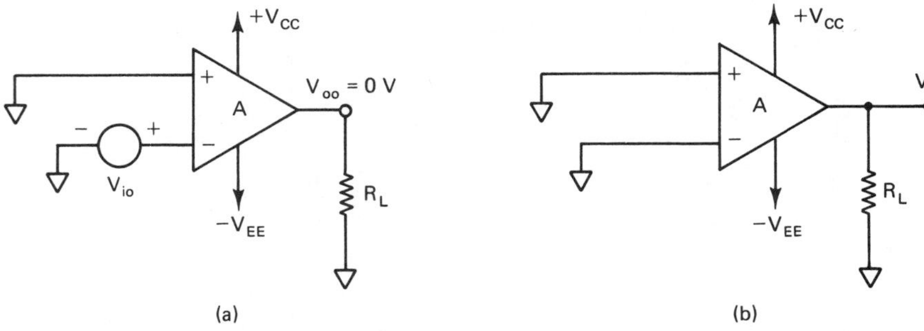

Figure 5-1 (a) Input offset voltage in an op-amp. (b) Output offset voltage in an op-amp.

offset voltage due to input offset voltage V_{io} as V_{oo}. The output offset voltage V_{oo} is caused by mismatching between two input terminals. Even though all the components are integrated on the same chip, it is not possible to have two transistors in the input differential amplifier stage with exactly the same characteristics. This means that the collector currents in these two transistors are not equal, which causes a differential output voltage from the first stage. The output of first stage is amplified by following stages and possibly aggravated by more mismatching in them. Thus the output voltage caused by mismatching between two input terminals is the output offset V_{oo}. Figure 5-1(b) shows the output offset voltage in an op-amp without feedback.

The output offset voltage V_{oo} is a dc voltage; it may be positive or negative in polarity depending on whether the potential difference between two input terminals is positive or negative. It is impossible to predict the polarity of the input offset voltage since it is dependent on mismatching between two input terminals. Therefore, on the data sheets the value of the input offset voltage V_{io} is listed as an absolute maximum value. For example, for a 741, $V_{io} = 6$ mV maximum, whereas for a 740, $V_{io} = 20$ mV maximum, which means that the maximum potential difference between two input terminals in a 741 op-amp can be as large as 6 mV dc; that is, voltage at the noninverting input terminal may differ from that at the inverting input terminal by as much as 6 mV dc, or vice versa. This input offset voltage gives rise to an output offset voltage V_{oo}. Thus we need to apply a differential input voltage of specific amplitude and correct polarity in order to reduce the output offset voltage V_{oo} to zero. This voltage is referred to as input offset voltage V_{io}. The V_{io} value for op-amps of the same type may not be the same in amplitude and polarity because of mass production, but it will always be less than the maximum value given on the specification sheets. For example, if we take three different 741s and use them one at a time as in Figure 5-1(b) and measure the corresponding output offset voltage for each of them, we find that the output offset voltage in these three op-amps is not of the same amplitude and polarity, which means that input offset V_{io} is not of the same amplitude and polarity even though the op-amps are of the same type. To reduce V_{oo} to zero, we need to have some circuit at the input terminals of the op-amp which will give us the flexibility of obtaining V_{io} of proper amplitude and polarity. Such a circuit is called an input offset voltage compensating network. Before we apply external input to the op-amp, with the help of an offset voltage compensating network we reduce the output offset voltage V_{oo} to zero;

the op-amp is then said to be *nulled* or *balanced.* Before we get into the design of a compensating network, it is worth noting that the compensating network is not needed for those op-amps that have offset null pins, such as 741, 748, 777, and 201. For the 741-type op-amp the manufacturer recommends that a 10-kΩ potentiometer be placed across offset null pins 1 and 5 and a wiper be connected to the negative supply pin 4 as shown in Figure 5-2. Adjustment of this pot will null the output.

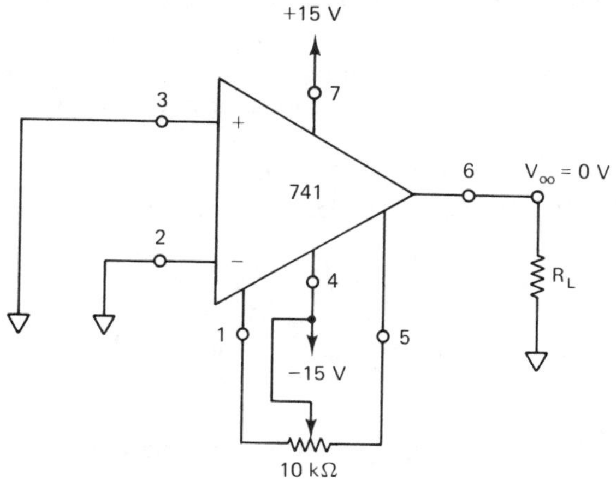

Figure 5-2 Voltage offset null circuit for 741.

Refer to the 741 equivalent circuit of Section 2-4. By varying the position of the wiper on the 10-kΩ potentiometer, we are trying to remove the mismatch between inverting and noninverting input terminals of the op-amp. Adjust the wiper until the output offset voltage is reduced to zero. See the analysis of the 741 op-amp circuit in Section 2-4.

5-2.1 Offset-Voltage Compensating Network Design

The op-amp with offset-voltage compensating network is shown in Figure 5-3. The compensating network consists of potentiometer R_a and resistors R_b and R_c. If we are planning to make use of the op-amp as an inverting amplifier, the compensating network should be connected to the noninverting input terminal of the op-amp. The circuit in Figure 5-3 can be used as a noninverting amplifier since the compensating network is connected to the inverting input terminal of the op-amp.

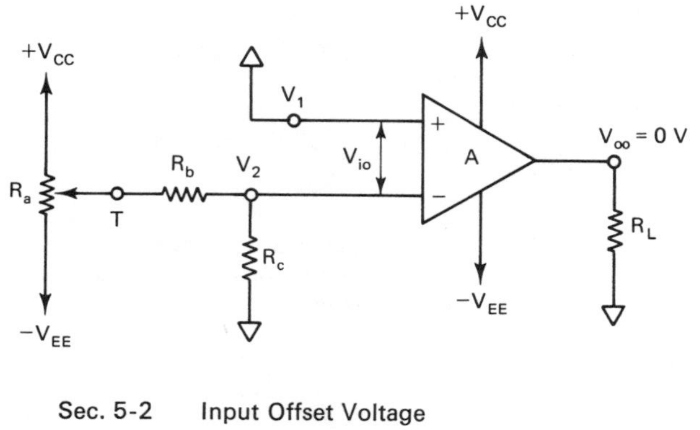

Figure 5-3 Op-amp with offset voltage compensating network.

Note that the potentiometer R_a is connected to $+V_{CC}$ and $-V_{EE}$ and the voltage across R_c is the voltage at the inverting input terminal V_2. By adjusting the wiper on R_a, voltage V_2 can be made to equal voltage V_1: that is, V_{io} can be made zero, which forces V_{oo} to go to zero. Say, for example, that V_{oo} is positive, which implies that $V_1 > V_2$. This means that V_2 should be increased until it equals V_1; the increase can be accomplished by moving the wiper toward V_{CC}. On the other hand, if V_{oo} is negative, that is, $V_2 > V_1$, the wiper can be moved toward $-V_{EE}$ until V_{oo} is reduced to zero.

To establish a relationship between V_{io}, supply voltages and the compensating components, we first have to Thévenize the circuit, looking back into R_a from point T.

The maximum Thévenin's equivalent resistance R_{max} occurs when the wiper is at the center of the potentiometer, as shown in Figure 5-4(a). Thus

$$R_{max} = \frac{R_a}{2} \left\| \frac{R_a}{2} \right. = \frac{R_a}{4}.$$

The maximum Thévenin's equivalent voltage V_{max} is equal to either V_{CC} or $-V_{EE}$ when the wiper is uppermost in the potentiometer or lowest in the potentiometer [see Figure 5-4(b) and (c)].

Supply voltages V_{CC} and $-V_{EE}$ are equal in magnitude; therefore, let us denote their magnitude by voltage V. Thus $V_{max} = V$.

Next we redraw the compensating network using the maximum Thévenin's voltage and resistance as shown in Figure 5-5. Applying the voltage-divider rule to

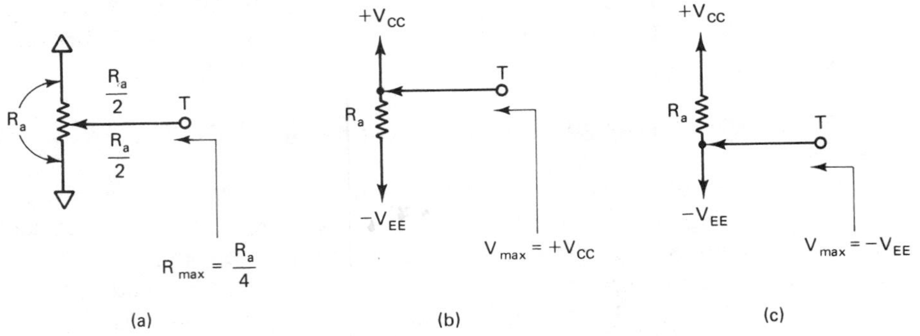

Figure 5-4 (a) Thévenin's equivalent for resistance R_{max}. (b, c) Thévenin's equivalent for voltage V_{max}.

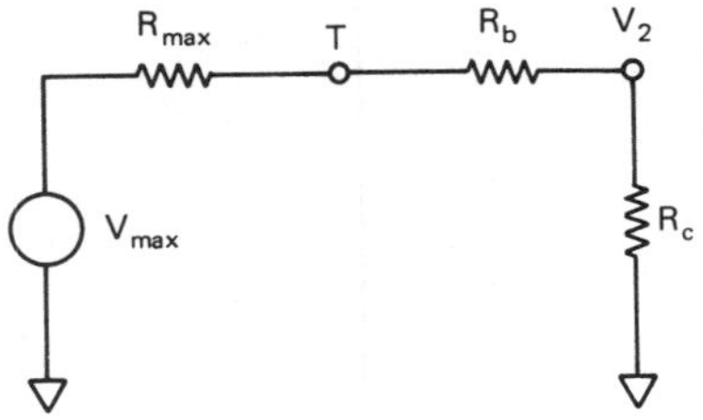

Figure 5-5 Compensating network with maximum Thévenin's equivalent resistance and voltage.

the circuit in Figure 5-5, we get

$$V_2 = \frac{R_c}{R_{\max} + R_b + R_c} V_{\max} \tag{5-1}$$

where V_2 has been expressed as a function of maximum Thévenin's voltage $V_{\max}$ and maximum Thévenin's resistance $R_{\max}$. But the maximum value of V_2 can be equal to V_{io}, since $|V_1 - V_2| = V_{io}$. Thus Equation (5-1) becomes

$$V_{io} = \frac{R_c}{R_{\max} + R_b + R_c} V_{\max} \tag{5-2}$$

There are too many unknowns in Equation (5-2). To simplify Equation (5-2), let us make the assumption that $R_b > R_{\max} > R_c$, where $R_{\max} = R_a/4$. An explanation for this assumption is in order. The bias currents in the op-amp are fixed and very small, in the range of nanoamperes. Therefore, the bias current in an inverting terminal of the op-amp in Figure 5-3 will also be very small and fixed. In no case will the bias current be larger than the value listed on the data sheets. To null the op-amp the required variation in the voltage across R_c is also small, on the order of millivolts; in the case of some op-amps it is even on the order of microvolts. This means that not only the current through R_c should be small, but the value of R_c should also be very small. The bias current in an inverting terminal and the current through R_c are derived from the current through R_b. Since two of these currents will be very small, the current through R_b will be significantly smaller than the current through R_a. To accomplish this we needed $R_b > R_a$, which implies that $R_b > R_{\max}$ since $R_{\max} = R_a/4$. Thus the assumption $R_b > R_{\max} > R_c$ is valid. Using this assumption, we can say that $R_{\max} + R_b + R_c \cong R_b$. Therefore, Equation (5-2) can be rewritten as

$$V_{io} = \frac{R_c V_{\max}}{R_b} \tag{5-3}$$

where

$$V_{\max} = V = |V_{CC}| = |-V_{EE}|$$

$$V_{io} = \frac{R_c V}{R_b} \tag{5-4}$$

Note that V_{io} depends on the magnitude of the supply voltages $+V_{CC}$ and $-V_{EE}$.

Equation (5-4) will be used to design the compensating network. We will obtain the value of V_{io} from the data sheet of a given op-amp, whereas the value of V will be fixed according to the supply voltages chosen; once these two values are known, we will establish the relationship between R_b and R_c. Then we will select the value for R_c to compute the value of R_b. The value for R_c will be selected to be less than 100 Ω so that R_b and R_a values will not be too large.

Example 5-1

Design a compensating network for the LM307 op-amp. The op-amp uses ±10-V supply voltages.

Solution. The value of V_{io} specified on the data sheets of the LM307 is 10 mV maximum. The value of $V = |V_{CC}| = |-V_{EE}| = 10$ V. Substituting these values in Equation (5-4), we get

$$10 \text{ mV} = \frac{R_c(10 \text{ V})}{R_b}$$

$$R_b = \frac{10 \text{ V}}{10 \text{ mV}} R_c$$

If we select $R_c = 10\ \Omega$, the value of R_b should be

$$R_b = (1000)R_c = 10{,}000\ \Omega$$

Since $R_b > R_{\max}$, let us choose $R_b = 10R_{\max}$, where

$$R_{\max} = \frac{R_a}{4}$$

$$R_b = (10)\frac{R_a}{4}$$

$$R_a = \frac{R_b}{2.5} = \frac{10\text{ k}\Omega}{2.5} = 4\text{-k}\Omega \text{ potentiometer}$$

If a 4 kΩ potentiometer is not available, we may prefer to use the next lower value available, such as 3 kΩ, so that the value of R_a will be larger than R_b by a factor of 10. If we select a 3-kΩ potentiometer as the R_a value, R_b is 13.3 times larger than R_a. Thus

$$R_a = 3\text{-k}\Omega \text{ potentiometer}$$

$$R_b = 10\text{ k}\Omega$$

$$R_c = 10\ \Omega$$

The final circuit, which also includes the pin connections for the LM307, is shown in Figure 5-6.

After the circuit in Figure 5-6 is breadboarded, we will adjust the potentiometer R_a until the output is reduced to zero. This nulled op-amp circuit now can be used as a noninverting amplifier by applying an external voltage to the noninverting input terminal. But the magnitude of the input voltage applied has to be very very small to avoid any distortion and clipping of the output signal. A small input voltage is needed

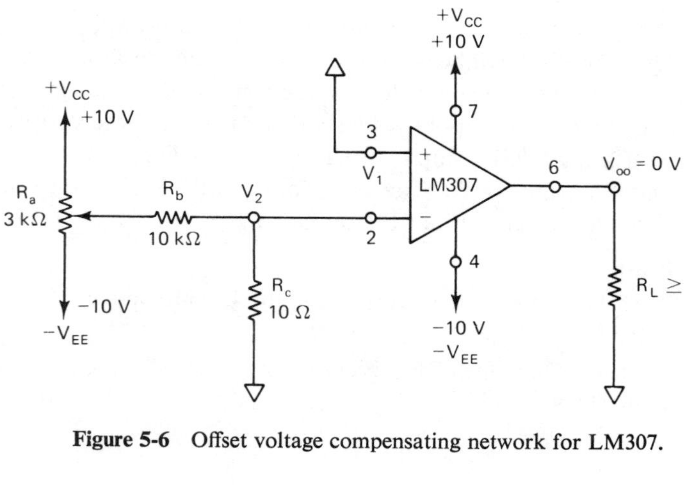

Figure 5-6 Offset voltage compensating network for LM307.

because the voltage gain of the op-amp without feedback is extremely high, ideally infinity. Because of the high risk of distortion and clipping of the output signal, op-amp in open-loop configuration is not used in linear circuit applications.

Let us now examine the effect of V_{io} in amplifiers with feedback. The noninverting and inverting amplifiers with feedback are shown in Figure 5-7. To determine the effect of V_{io} in each case, we have to reduce the input voltage v_{in} to zero.

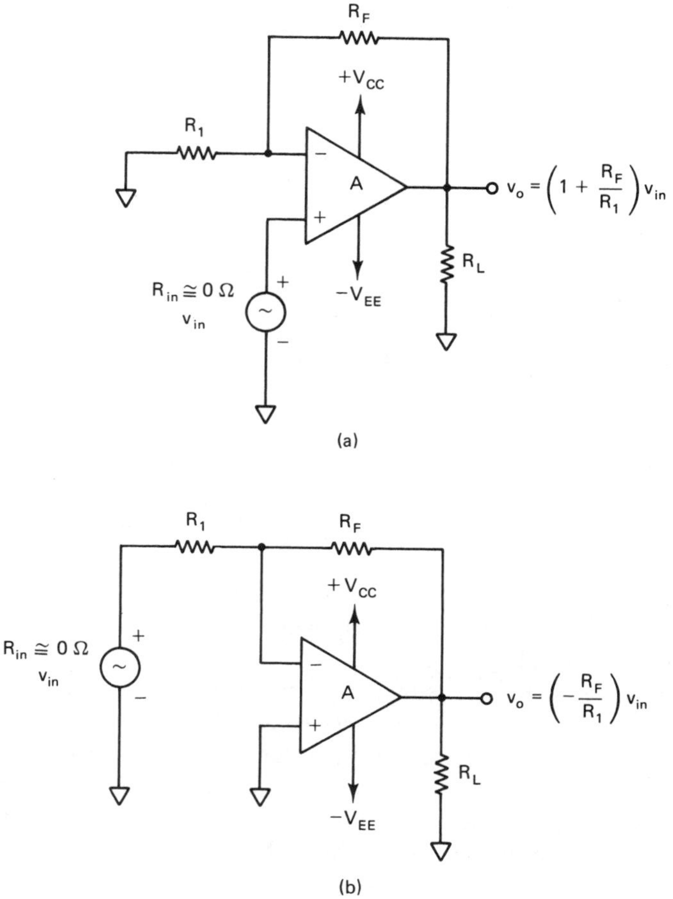

Figure 5-7 (a) Noninverting amplifier with feedback. (b) Inverting amplifier with feedback.

With v_{in} reduced to zero, the circuits of both noninverting and inverting amplifiers are the same as the circuit in Figure 5-8. We assume here and in future cases that the internal resistance R_{in} of the input signal voltage is negligibly small.

We wish to express V_{oo} in terms of external components R_1, R_F, and the specified input offset voltage V_{io} for a given op-amp. In the figure, the noninverting input terminal is connected to ground; therefore, we assume voltage V_1 at the noninverting input terminal to be zero. The voltage V_2 at the inverting input terminal can be determined by applying the voltage-divider rule:

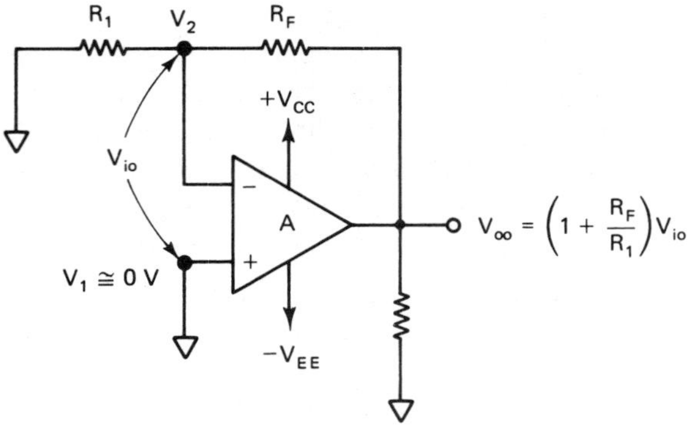

Figure 5-8 Closed-loop noninverting or inverting amplifier with $v_{\text{in}} = 0$ V.

$$V_2 = \frac{R_1 V_{oo}}{R_1 + R_F} \tag{5-5}$$

Therefore,

$$V_{oo} = \frac{R_1 + R_F}{R_1} V_2 \tag{5-6}$$

Since $V_{io} = |V_1 - V_2|$ and $V_1 = 0$ V,

$$V_{io} = |0 - V_2| = V_2 \tag{5-7}$$

Therefore,

$$V_{oo} = \left(1 + \frac{R_F}{R_1}\right) V_{io} = (A_{oo}) V_{io} \tag{5-8}$$

where $A_{oo} = 1 + R_F/R_1$. Thus when an op-amp is used in a closed-loop configuration as either a noninverting or inverting amplifier, we can compute the maximum possible output offset voltage V_{oo} caused by the input offset voltage V_{io}, by using Equation (5-8). According to this equation, for a given op-amp the amount of V_{oo} depends on the values of external components R_1 and R_F, that is, A_{oo}. Therefore, the smaller the value of A_{oo}, the smaller will be the value of V_{oo} for a given value of V_{io}. In the extreme case when $R_1 \gg R_F$, $A_{oo} \cong 1$ and $V_{oo} \cong V_{io}$. Thus, in practice all op-amp circuits will have some output offset voltage V_{oo}. To null the output offset voltage V_{oo}, we can use a compensating network identical to the one that was used to null the open-loop op-amp circuit. The closed-loop noninverting and inverting amplifiers with their compensating networks are shown in Figure 5-9.

The offset-voltage compensating networks in Figure 5-9 will be designed by using Equation (5-4). The compensating network is connected in the noninverting terminal for the inverting amplifier and in the inverting terminal for the noninverting amplifier. Note that the voltage gain of the noninverting amplifier with compensating network is $A_F = 1 + [R_F/(R_1 + R_c)]$. The gain changes because the Thévenin equivalent resistance of the compensating network, which is approximately equal to R_c, is in series with R_1. In fact, this result is based on the assumption that $R_b > R_a > R_c$. Before an external input v_{in} is applied, the op-amp in Figure 5-9 should be nulled by adjusting the wiper in pot R_a.

The closed-loop differential amplifier is somewhat more difficult to null because

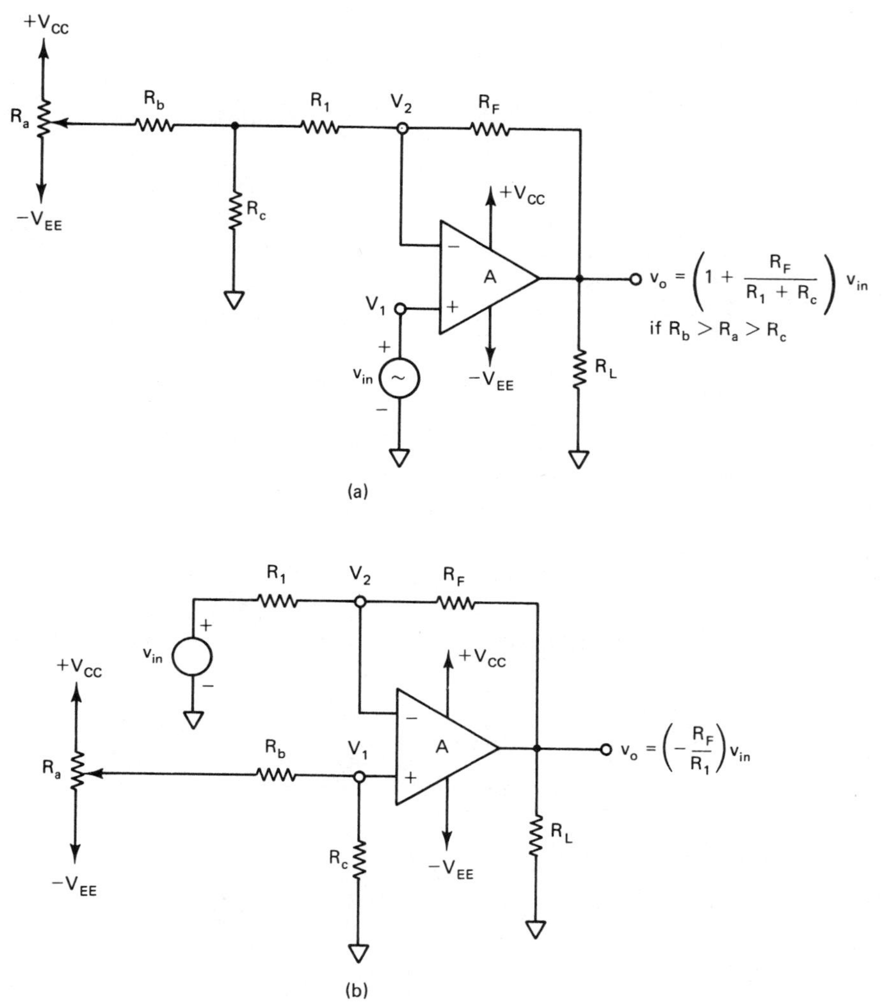

Figure 5-9 (a) Compensated noninverting amplifier with feedback. (b) Compensated inverting amplifier with feedback.

the use of the compensating network can change the common-mode rejection ratio. (See Section 5-11 for further information on the common-mode rejection ratio.) Figure 5-10 shows the differential amplifier with its input offset voltage compensating network. In order to maximum CMRR, we should use $R_1 = R_2$ and $R_F = R_3 + R_c$.

The compensated differential amplifier shown in Figure 5-11 uses the op-amp with the offset voltage null pins. Note that $R_1 = R_2$ and $R_F = R_3$. This amplifier circuit has two advantages over the one in Figure 5-10. One advantage is that it is simpler, since it uses fewer components, and the other is that the offset null circuit does not affect the CMRR. A voltage follower may be nulled using a similar compensating network, as shown in Figure 5-12.

The voltage drop across R_c is used to cancel the offset voltage. Even though $R_b > R_a$ and R_c, their finite values cause the gain of the voltage follower to increase

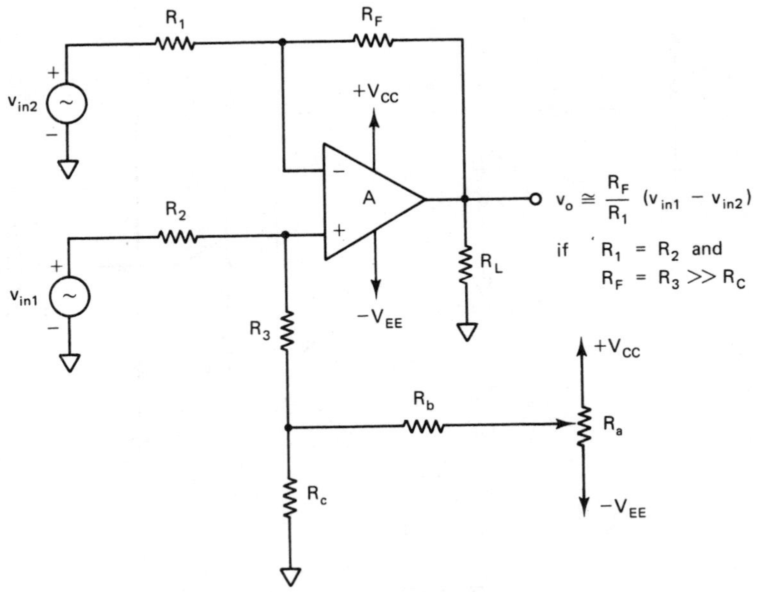

Figure 5-10 Differential amplifier with offset-voltage compensating network.

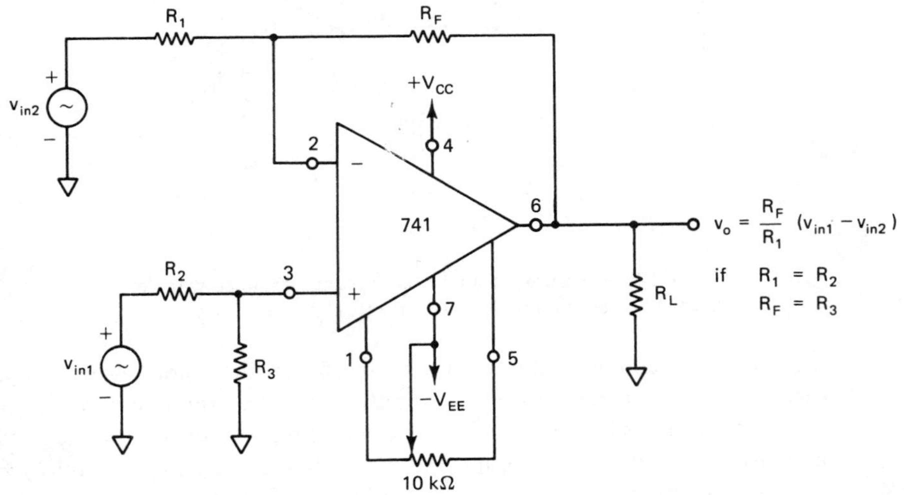

Figure 5-11 Differential amplifier with offset voltage null circuit (pins 1 and 5 are offset null pins for 741 op-amp).

slightly (see the v_o expression in Figure 5-12). This insignificant error in gain usually can be ignored, since $R_c < R_b + R_a/4$.

The offset-voltage compensating network discussed above is a versatile circuit. It is applicable to all op-amp types and can be used in different operating modes

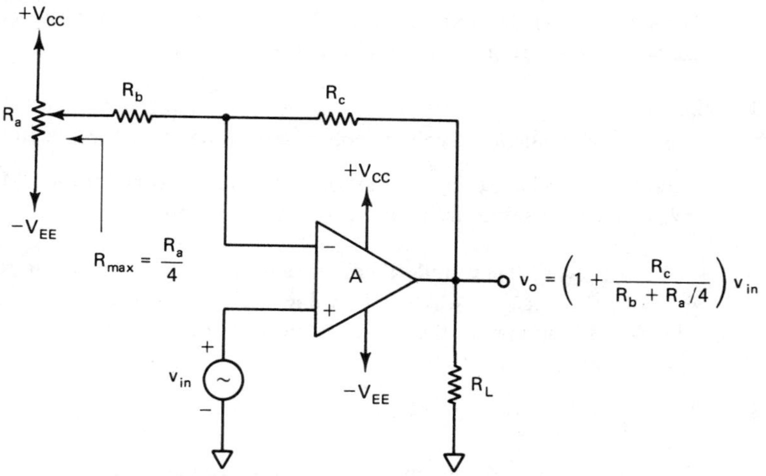

Figure 5-12 Voltage follower with offset-voltage compensating network.

without any modifications. It also allows nulling without interfering with the internal circuitry of an op-amp.

Example 5-2

The op-amp in the circuit of Figure 5-13 is the LM307 with $V_{io} = 10$ mV dc maximum. What is the maximum possible output offset voltage, V_{oo}, caused by the input offset voltage V_{io}?

Solution. To find the maximum possible value of V_{oo}, we reduce the input voltage v_{in} to zero. The closed-loop gain of the amplifier is

$$A_{oo} = 1 + \frac{R_F}{R_1} = 1 + \frac{10\text{ k}\Omega}{1\text{ k}\Omega} = 11$$

Since $V_{io} = 10$ mV dc maximum, the V_{oo} might be as large as

$$V_{oo} = A_{oo}V_{io} = (11)(10\text{ mV dc}) = 110\text{ mV dc}$$

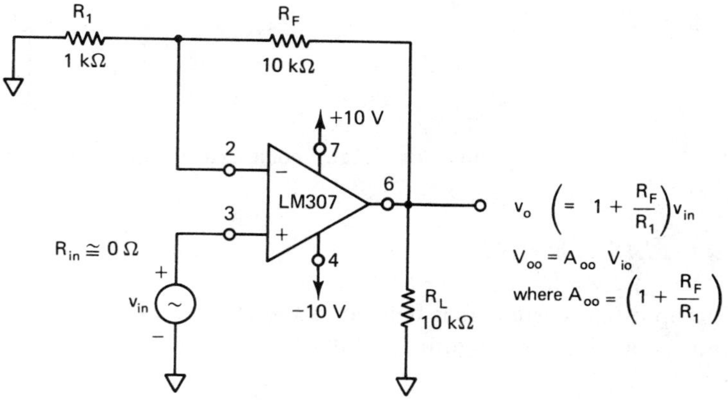

Figure 5-13 Noninverting amplifier with feedback.

This means that the output terminal can be either at a negative or a positive 110 mV dc with respect to ground even though $v_{in} = 0$ V.

Example 5-3

Design an input offset-voltage compensating network for the circuit in Figure 5-13.

Solution. We have designed the compensating network for the LM307 in Example 5-1. We can use the same circuit and connect it in the inverting terminal as shown in Figure 5-14.

Applying Thévenin's theorem across R_c, the equivalent resistance $R_{TH} = R_c$ because $R_c < R_a < R_b$. Also, the value of R_1 is much larger than R_c. This means that the closed-loop gain of the noninverting amplifier will not be affected by the use of the compensating network, that is,

$$A_F = 1 + \frac{R_F}{R_1 + R_c} = 1 + \frac{10 \text{ k}\Omega}{1 \text{ k}\Omega + 10\Omega} = 11$$

Note that we have distinguished between the voltage gains A_{oo} and A_F. The voltage gain $A_{oo} = 1 + R_F/R_1$ is used to calculate the maximum possible output offset voltage V_{oo} due to the input offset voltage V_{io} both in the inverting and noninverting amplifiers (see Figure 5-7). On the other hand, the voltage gain $A_F = 1 + R_F/R_1$ of the noninverting amplifier of Figure 5-7(a) or the voltage gain $A_F = -R_F/R_1$ of the inverting amplifier of Figure 5-7(b) is used to compute the output voltage v_o due to the input voltage v_{in}.

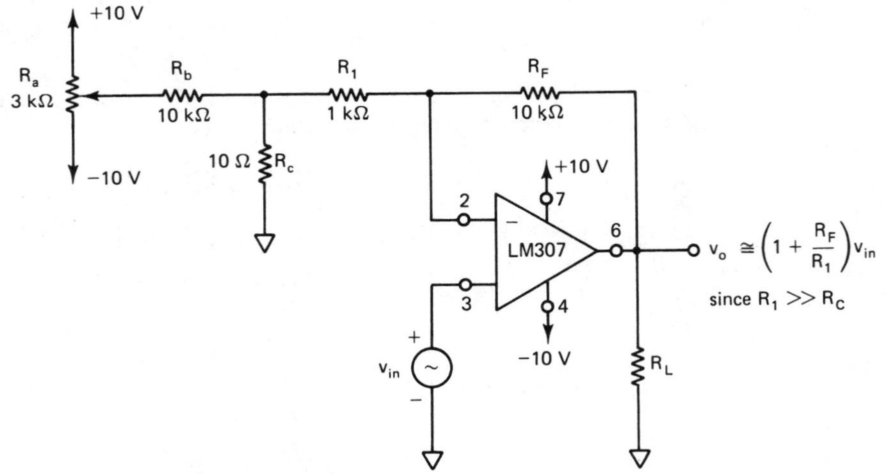

Figure 5-14 Compensated noninverting amplifier with feedback.

5-3 INPUT BIAS CURRENT

An input bias current I_B is defined as the average of the two input bias currents, I_{B1} and I_{B2}, as shown in Figure 5-15; that is,

$$I_B = \frac{I_{B1} + I_{B2}}{2} \tag{5-9}$$

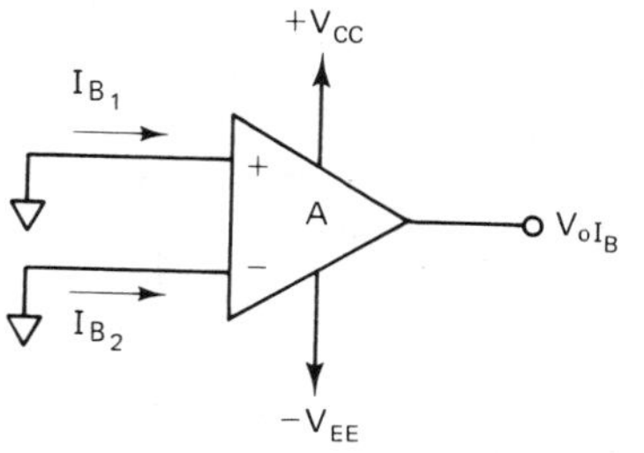

Figure 5-15 Input bias currents in an op-amp.

where I_{B1} = dc bias current flowing into the noninverting input

I_{B2} = dc bias current flowing into the inverting input

In Figure 5-15, both input terminals are grounded, so that no input voltage is applied to the op-amp. But the plus-minus supply voltages are necessary to bias the op-amp properly.

Actually, the input bias currents I_{B1} and I_{B2} are the base bias currents of the two transistors in the input (first) differential amplifier stage of the op-amp. Even though both of the input transistors are identical, it is not possible to have I_{B1} and I_{B2} exactly equal to each other because of the internal imbalance between the two inputs. In this section we assume that the specified input bias current I_B is equal to either one of the two input currents I_{B1} and I_{B2}; that is,

$$I_B = I_{B1} = I_{B2} \tag{5-10}$$

The value of input bias current I_B is very small, in the range of a few to a few hundred nanoamperes. It is a dc current and is specified on the op-amp data sheets. For example, I_B is 500 nA maximum at supply voltages = ±15 V dc, for μA741, whereas it is 75 nA maximum for MLM101A. The value of I_B has been significantly reduced in later-generation op-amps.

Even though very small, the input bias current I_B can cause a significant output offset voltage in circuits using relatively large feedback resistors. This output offset voltage may not be as large as that caused by the input offset voltage, but certain precautions must be taken to minimize it.

First we obtain the expression for the output offset voltage caused by the input bias current I_B in the inverting and noninverting amplifiers, and then devise some scheme to eliminate or minimize it. The noninverting or inverting amplifier with $v_{in} = 0$ V is redrawn in Figure 5-16. We assume for the duration of this section that the input offset voltage V_{io} is zero; that is, there is no output offset voltage due to V_{io}. Let V_{oI_B} be the output offset voltage due to input bias current I_B.

In the figure, the input bias current I_{B1} and I_{B2} are flowing into the noninverting and inverting input leads, respectively. The noninverting terminal is connected to ground; therefore, the voltage $V_1 = 0$ V. The controlled voltage source $AV_{io} = 0$ V since $V_{io} = 0$ V is assumed. With output resistance R_o negligibly small, the right end of R_F is essentially at ground potential; that is, resistors R_1 and R_F are in parallel and the bias current I_{B2} flows through them. Therefore, the voltage at the inverting terminal is

$$V_2 = (R_1 \,\|\, R_F)I_{B2} \tag{5-11}$$

$$V_2 = \frac{R_1 R_F}{R_1 + R_F} I_{B2} \tag{5-12}$$

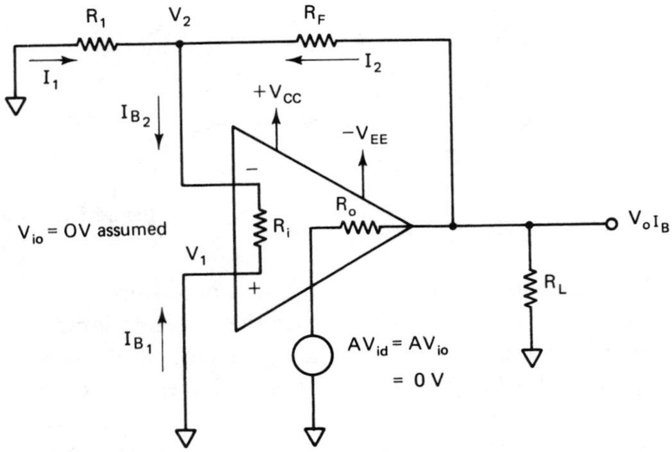

Figure 5-16 Output offset voltage due to input bias current in a noninverting or inverting amplifier.

Writing a node voltage equation for node V_2, we get

$$I_1 + I_2 = I_{B2}$$

$$\frac{0 - V_2}{R_1} + \frac{V_{oI_B} - V_2}{R_F} = \frac{V_2}{R_i} \tag{5-13}$$

where V_{oI_B} = output offset voltage due to input bias current
R_i = input resistance of the op-amp (see Figure 5-16)

Rearranging Equation (5-13), we get

$$\frac{V_{oI_B}}{R_F} = V_2\left(\frac{1}{R_1} + \frac{1}{R_F} + \frac{1}{R_i}\right)$$

Since R_i is extremely high (ideally ∞), $1/R_i \cong 0$ mhos. Therefore,

$$\frac{V_{oI_B}}{R_F} = V_2 \frac{R_1 + R_F}{R_1 R_F} \tag{5-14}$$

Substituting in Equation (5-14) the value of V_2 from Equation (5-12), we get

$$V_{oI_B} = \frac{R_1 R_F I_{B2}}{R_1 + R_F}\left(\frac{R_1 + R_F}{R_1}\right)$$

$$V_{oI_B} = R_F I_{B2} \tag{5-15}$$

From Equation (5-10),

$$V_{oI_B} = R_F I_B \tag{5-16}$$

According to Equation (5-16), the amount of output offset voltage V_{oI_B} is a function of feedback resistor R_F for a specified value of input bias current I_B. The amount of V_{oI_B} can be increased by the use of relatively large feedback resistors. Therefore, the use of small feedback resistors is recommended.

To eliminate or reduce the output offset voltage V_{oI_B} due to input bias current I_B, we have to devise some scheme at the input by which voltage V_1 can be made equal to V_2. In other words, if voltages V_1 and V_2 caused by the currents I_{B1} and I_{B2} can be made equal, there will be no output offset voltage V_{oI_B}. From Equation (5-12), we have

$$V_2 = R_p I_{B2} \tag{5-17}$$

where

$$R_p = \frac{R_1 R_F}{R_1 + R_F}$$

Equation (5-17) implies that we must express voltage V_1 at the noninverting input terminal as a fucction of I_{B1} and some specific resistor R_{OM}. This can be accomplished as follows. The input bias current I_{B1} does not produce any voltage at the noninverting input terminal because this terminal is directly connected to ground (see Figure 5-16). If we could connect the proper value of resistor R_{OM} in the noninverting terminal, the voltage V_1 would be

$$V_1 = (R_{OM})I_{B1} \tag{5-18}$$

In order to have voltage V_1 equal to V_2, the right-hand sides of Equations (5-17) and (5-18) must be equal; that is,

$$R_p I_{B2} = R_{OM} I_{B1} \tag{5-19}$$

Now if the currents I_{B1} and I_{B2} are equal, Equation (5-19) implies that

$$R_p = R_{OM} \tag{5-20}$$

or

$$\frac{R_1 R_F}{R_1 + R_F} = R_{OM} \tag{5-21}$$

Thus the proper value required of an R_{OM} resistor connected in the noninverting terminal is the parallel combination of resistors R_1 and R_F. However, the use of R_{OM} may not completely eliminate the output offset voltage V_{oI_B} because the currents I_{B1} and I_{B2} are not exactly equal. Nevertheless, the use of R_{OM} will minimize the amount of output offset voltage V_{oI_B}; therefore, the R_{OM} resistor is referred to as the "offset minimizing resistor" (see Figure 5-17).

Note that if we reduce both the inputs to zero (that is, $v_{\text{in}1} = v_{\text{in}2} = 0$ V) in the closed-loop differential amplifier, the resulting circuit becomes the same as in

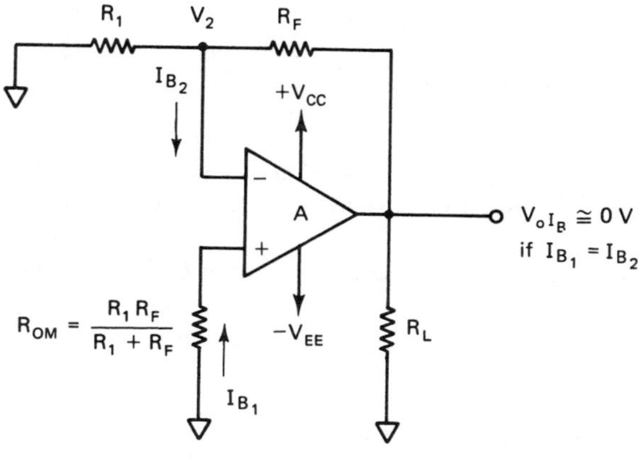

Figure 5-17 R_{OM} reduces the output offset voltage V_{oI_B} caused by the input bias current I_B.

Figure 5-17. This means that there is no need to use a separate resistor R_{OM} in the differential amplifier cirucits. Figure 5-18 shows the noninverting and inverting amplifiers with offset minimizing resistor R_{OM}.

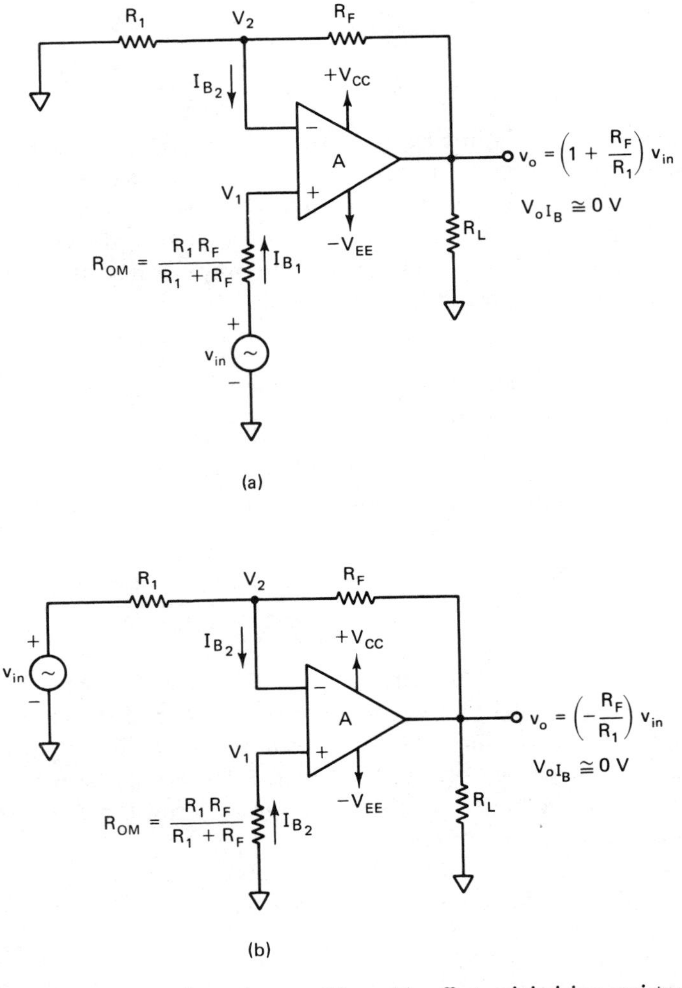

Figure 5-18 (a) Noninverting amplifier with offset minimizing resistor R_{OM}. (b) Inverting amplifier with offset minimizing resistor R_{OM}.

Example 5-4

(a) For the inverting amplifier of Figure 5-19, determine the maximum possible output offset voltage due to (1) the input offset voltage V_{io} and (2) the input bias current I_B. The op-amp is a type 741.

(b) What value of R_{OM} is needed to reduce the effect of input bias current I_B?

Solution

(a) From the 741 op-amp data sheets we have

$$V_{io} \text{ max} = 6 \text{ mV dc}$$

$$I_B \text{ max} = 500 \text{ nA dc at } T_A = 25°\text{C and } V_S = \pm 15 \text{ V}$$

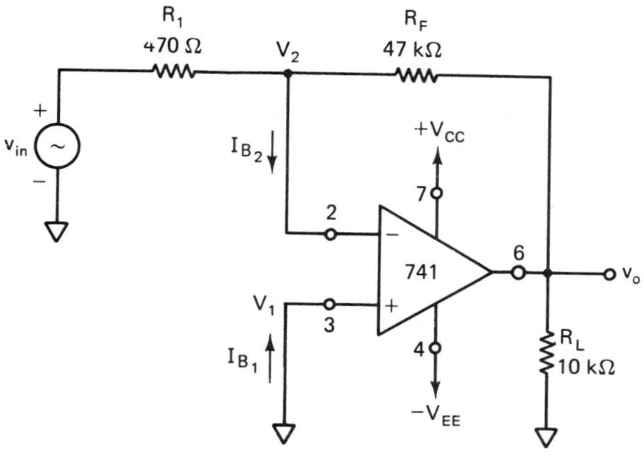

Figure 5-19 Inverting amplifier of Example 5-4.

1. Since $R_1 = 470\ \Omega$ and $R_2 = 47\ \text{k}\Omega$, the output offset due to V_{io} might be as large as

$$V_{oo} = \left(1 + \frac{R_F}{R_1}\right) V_{io}$$

$$= \left(1 + \frac{47\ \text{k}\Omega}{470\ \Omega}\right)(6\ \text{mV})$$

$$= 606\ \text{mV dc}$$

Remember that this means that the voltage at the output terminal may be below or above the ground potential by 606 mV dc without any external input signal v_{in}.

2. The value of feedback resistor $R_F = 47\ \text{k}\Omega$; therefore,

$$V_{oI_B} = R_F I_B$$

$$= (47\ \text{k}\Omega)(500\ \text{nA})$$

$$= 23.5\ \mu\text{V dc}$$

Again, this means that the output offset voltage V_{oI_B} could be positive or negative with respect to ground by as much as 23.5 μV dc.

From the results in parts 1 and 2, it is obvious that $V_{oo} \gg V_{oI_B}$; that is, input offset voltage V_{io} is potentially a greater problem than input bias current I_B.

(b) The value of R_{OM} to be used is

$$R_{OM} = \frac{R_1 R_F}{R_1 + R_F}$$

$$= \frac{(470\ \Omega)(47\ \text{k}\Omega)}{470\ \Omega + 47\ \text{k}\Omega} = 470\ \Omega$$

Example 5-5

Repeat Example 5-4 if R_1 is replaced by 1 kΩ and R_F by 100 kΩ.

Solution

(a)

1. The closed-loop voltage gain

$$A_{oo} = 1 + \frac{R_F}{R_1} = 1 + \frac{100\ \text{k}\Omega}{1\ \text{k}\Omega} = 101 \text{ is unchanged}$$

Therefore, the output offset voltage V_{oo} due to V_{io} is also unchanged; that is, $V_{oo} = 606$ mV dc as in Example 5-4.

2. With $R_F = 100\ \text{k}\Omega$, the V_{oI_B} is

$$V_{oI_B} = (100\ \text{k}\Omega)(550\ \text{nA}) = 50\ \text{mV dc}$$

This is relatively larger output offset voltage than in Example 5-4. Thus V_{oI_B} becomes larger and therefore potentially troublesome, with larger values of feedback resistor R_F.

(b) The value of R_{OM} we can use is

$$R_{OM} = 1\ \text{k}\Omega \,\|\, 100\ \text{k}\Omega = 990\ \Omega$$

Since a 990-Ω resistor is not a standard value, in practice we may use a 1-kΩ resistor as R_{OM}.

5-4 INPUT OFFSET CURRENT

We have seen in Section 5-3 that the use of R_{OM} in series with the noninverting terminal reduces the output offset voltage V_{oI_B} due to I_B. However, the value of R_{OM} was derived based on the assumption that the input bias currents I_{B1} and I_{B2} are equal. In practice, these currents are not equal because of the internal imbalances in the op-amp's circuitry. The input offset current I_{io} is used as an indicator of the degree of mismatching between these two currents. The value of I_{io} specified on the data sheets indicates the maximum amount by which the two input bias currents may differ. In fact, the smaller this value, the better.

The input offset current I_{io} is defined as the algebraic difference between two input bias currents I_{B1} and I_{B2}. In equation form,

$$I_{io} = |I_{B1} - I_{B2}| \tag{5-22}$$

For a 741-type op-amp, maximum $I_{io} = 200$ nA dc. This means that I_{B1} may be larger than I_{B2} or I_{B2} may be larger than I_{B1}, at the most by 200 nA. In other words, the maximum difference between I_{B1} and I_{B2} can be as large as 200 nA. In FET input op-amps, the value of I_{io} is extremely small. For example, the maximum I_{io} is 0.3 nA for the μA740C op-amp.

In a circuit like that in Figure 5-17, there will be an output offset voltage due to the input bias currents I_{B1} and I_{B2}. In other words, the output offset voltage in a case like this can be expressed as a function of input offset current I_{io}. Let $V_{oI_{io}}$ be the output offset voltage caused by the input offset current I_{io}. In order to separate the effect of input offset current from that of input offset voltage, let us again assume that $V_{io} = 0$ V.

Referring to Figure 5-20, we will express the voltages V_1 and V_2 as a function of I_{B1} and I_{B2}, for given values of R_1 and R_F as follows:

$$V_1 = R_{OM} I_{B1} \tag{5-18}$$

$$V_2 = R_p I_{B2} \tag{5-12}$$

where

$$R_{OM} = R_p = \frac{R_1 R_F}{R_1 + R_F}$$

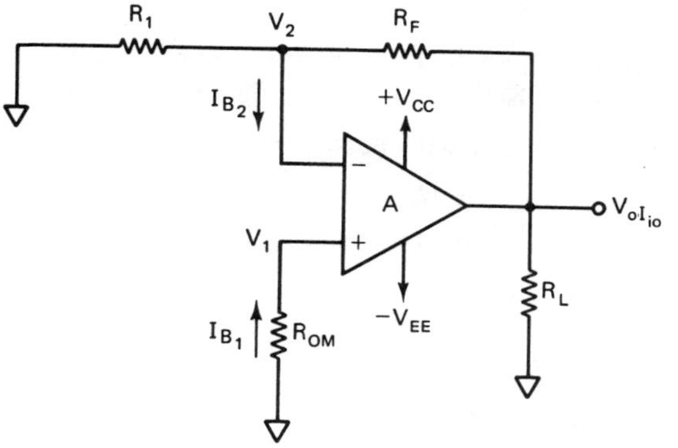

Figure 5-20 Output offset voltage $V_{oI_{io}}$, caused by the input offset current I_{io} in an inverting or noninverting amplifier.

Applying the superposition theorem, we will now find the output offset voltage due to V_1 and V_2 in terms of I_{B1}, I_{B2}, and R_F. We know, from Equation (5-15), that

$$V_{oI_{B2}} = -R_F I_{B2}$$

Here the negative sign is used because V_2 is the voltage at the inverting input terminal. This output offset voltage $V_{oI_{B2}}$ is due to voltage V_2 only in terms of I_{B2} and R_F. Similarly, the output offset voltage $V_{oI_{B1}}$ due to V_1 only in terms of I_{B1} and R_F can be obtained as follows:

$$V_{oI_{B1}} = V_1\left(1 + \frac{R_F}{R_1}\right) \tag{5-23}$$

where V_1 = voltage at the noninverting input terminal

$\left(1 + \frac{R_F}{R_1}\right)$ = gain of the noninverting amplifier

Substituting in Equation (5-23) the value of V_1 from Equation (5-18), we get

$$V_{oI_{B1}} = (R_{OM} I_{B1})\left(1 + \frac{R_F}{R_1}\right)$$

$$= \frac{R_1 R_F}{R_1 + R_F} I_{B1} \frac{R_1 + R_F}{R_1} \tag{5-24}$$

$$V_{oI_{B1}} = R_F I_{B1} \tag{5-25}$$

Therefore, the maximum magnitude of the output offset voltage due to I_{B1} and I_{B2} is

$$V_{oI_{B1}} + V_{oI_{B2}} = R_F I_{B1} - R_F I_{B2}$$

$$= R_F(I_{B1} - I_{B2})$$

$$V_{oI_{io}} = R_F(I_{io}) \tag{5-26}$$

where $V_{oI_{B1}} + V_{oI_{B2}} = V_{oI_{io}}$ is the output offset voltage due to I_{io} and $I_{io} = |I_{B1} - I_{B2}|$ the input offset current. Thus for a given value of input offset current I_{io}, the amount of output offset voltage $V_{oI_{io}}$ depends on the value of feedback resistor R_F. As in the two previous cases, the $V_{oI_{io}}$ is also a dc voltage and could be positive or negative with respect to ground. Since I_{io} is generally much smaller than I_B, the output offset caused by I_{io} is always smaller than that caused by I_B.

Example 5-6

For the inverting amplifier in Figure 5-21, determine the maximum output offset voltage $V_{oI_{io}}$ caused by the input offset current I_{io}. The op-amp is a type 741.

Solution. For the 741 op-amp, $I_{io} = 200$ nA max. Therefore,

$$\begin{aligned} V_{oI_{io}} &= (R_F)I_{io} \\ &= (100\ \text{k}\Omega)(200\ \text{nA}) \\ &= 20\ \text{mV dc} \end{aligned}$$

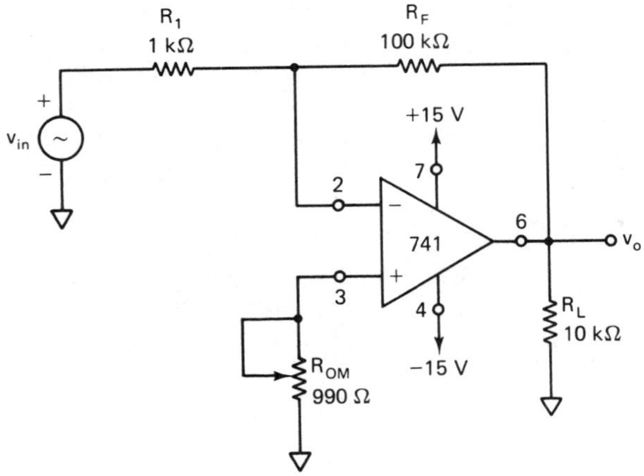

Figure 5-21 Inverting amplifier of Example 5-6.

Example 5-6 illustrates that we may not be able to eliminate completely the output offset voltage due to unequal input bias currents, even though a proper value of R_{OM} is being used. However, the output offset voltage has been reduced from 50 mV (in Example 5-5) to 20 mV (in Example 5-6) with the use of R_{OM}, which is a significant change.

5-5 TOTAL OUTPUT OFFSET VOLTAGE

We know that in a circuit like the one in Figure 5-19, the output offset voltage V_{oo} caused by V_{io} could be either positive or negative with respect to ground. Similarly, the output offset voltage V_{oI_B} caused by I_B could also be either positive or negative with respect to ground. If these output offset voltages are of different polarities, the resultant output offset will be very little. On the other hand, if both of these output offset voltages are of the same polarity, the maximum amplitude of the total output offset would be

$$V_{ooT} = V_{oo} + V_{oI_B}$$

$$V_{ooT} = \left(1 + \frac{R_F}{R_1}\right)V_{io} + (R_F)I_B \tag{5-27}$$

By the same token, in a curcuit such as that in Figure 5-21, the total output offset voltage V_{ooT} can be given by the expression

$$V_{ooT} = V_{oo} + V_{oI_{io}}$$

$$V_{ooT} = \left(1 + \frac{R_F}{R_1}\right) V_{io} + (R_F) I_{io} \tag{5-28}$$

Note the difference between Equations (5-27) and (5-28). The only difference in the circuits of Figure 5-19 and 5-21 is the R_{OM} resistor. Therefore, since $I_{io} < I_B$, the use of the R_{OM} resistor in the noninverting or inverting amplifier assures a reduction in the current generated output offset voltage.

Example 5-7

Compute the maximum possible total output offset voltages in the amplifier circuits shown in Figure 5-22. The op-amp is the MC1536 with the following specifications:

$V_{io} = 7.5$ mV maximum

$I_{io} = 50$ nA maximum

$I_B = 250$ nA maximum at $T_A = 25°C$

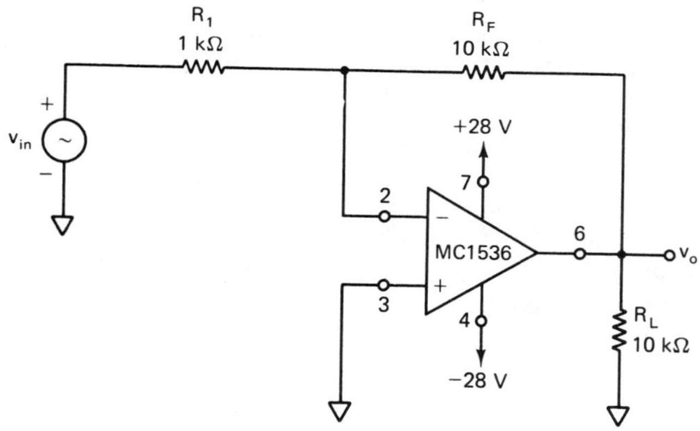

(a)

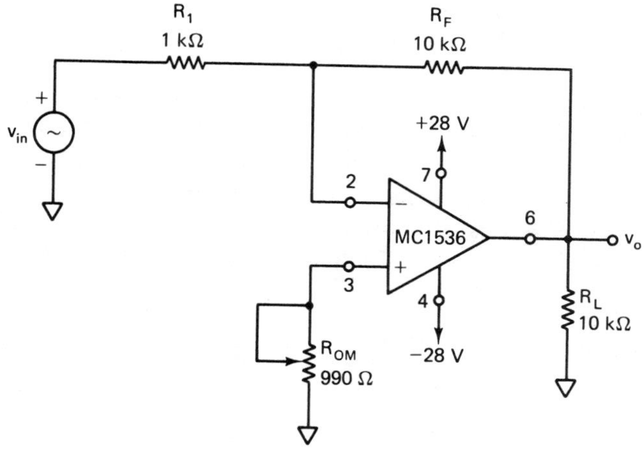

(b)

Figure 5-22 (a) MC1556 inverting amplifier of Example 5-7. (b) MC1556 inverting amplifier of Example 5-7 with R_{OM}.

Solution. In the circuit shown in Figure 5-22(a), the current generated output offset voltage is due to the input bias current I_B in the inverting lead. Therefore, to compute the total output offset voltage, we have to use Equation (5-27):

$$V_{ooT} = \left(1 + \frac{R_F}{R_1}\right) V_{io} + (R_F) I_B$$

$$= \left(1 + \frac{10\ \text{k}\Omega}{1\ \text{k}\Omega}\right)(7.5\ \text{mV}) + (10\ \text{k}\Omega)(250\ \text{nA})$$

$$= 82.5\ \text{mV} + 2.5\ \text{mV} = 85\ \text{mV}$$

The R_{OM} is used in the circuit of Figure 5-22(b), which means that the current-generated output offset is caused by the input offset current I_{io}. Therefore, using Equation (5-28), we get

$$V_{ooT} = \left(1 + \frac{R_F}{R_1}\right) V_{io} + (R_F) I_{io}$$

$$= \left(1 + \frac{10\ \text{k}\Omega}{1\ \text{k}\Omega}\right)(7.5\ \text{mV}) + (10\ \text{k}\Omega)(50\ \text{nA})$$

$$= 82.5\ \text{mV} + 0.5\ \text{mV} = 83\ \text{mV}$$

Thus it is obvious that the input offset voltage V_{io} is potentially a more troublesome problem than the input bias current I_B and the input offset current I_{io}.

5-6 THERMAL DRIFT

In previous sections we learned to compensate for the effects of input offset voltage and input bias currents. In our discussion so far, we have assumed that the parameters V_{io}, I_B, and I_{io} are constant for a given op-amp. However, in practice, the values of V_{io}, I_B, and I_{io} vary with:

1. Change in temperature
2. Change in supply voltages: $+V_{CC}$ and $-V_{EE}$
3. Time

In this section we discuss the effect of these factors on the performance of a given op-amp circuit.

The most serious variation in the values of V_{io}, I_B, and I_{io} is due to the change in temperature. Before we proceed further to find the change in output offset voltage due to the change in temperature, let us define the term "thermal drift." The average rate of change of input offset voltage per unit change in temperature is called *thermal voltage drift* and is denoted by $\Delta V_{io}/\Delta T$. It is expressed in μV/°C. By the same concept, we can also define the thermal drift in the input offset current and input bias current as follows:

$$\frac{\Delta I_{io}}{\Delta T} = \text{thermal drift in the input offset current (pA/°C)}$$

$$\frac{\Delta I_B}{\Delta T} = \text{thermal drift in the input bias current (pA/°C)}$$

It is important to note that the "drift" is not a constant value; that is, it is not uniform over a specified operating temperature range. Furthermore, the value of the input offset voltage or current may increase or decrease with increasing temperature, as shown in Figure 5-23(a) and (b). Referring to the graph in Figure 5-23(c), we can see that the value of the input bias current is not decreasing at a constant rate with the increasing temperature. The curves shown in Figure 5-23 are for the MC1741 op-amp.

Note that in Figure 5-23(a) and (b) the normalized values of the input offset voltage and input offset current are plotted versus temperature. This means that the input offset voltage and current are assumed to be zero at room temperature (25°C). Now if the change in temperature from 25°C is denoted by ΔT, the change in input offset voltage and input offset current corresponding to this change in temperature ΔT is represented by ΔV_{io} and ΔI_{io}, respectively. In fact, once we calculate these values from the graphs in Figure 5-23, for a desired operating temperature range, we can obtain

$$\frac{\Delta V_{io}}{\Delta T} = \text{thermal voltage drift}$$

$$\frac{\Delta I_{io}}{\Delta T} = \text{thermal current drift}$$

But this procedure is time consuming and also involves mathematical manipulation. Therefore, some manufacturers specify the average value of the drift in offset voltage and offset current over an entire operating temperature range instead. On the data sheet these values are listed as the average temperature coefficients of input offset voltage and input offset current. It is worth mentioning that the average temperature coefficient of input bias current is not listed on the data sheets. Instead, a graph of I_B versus temperature may be furnished by some manufacturers [see Figure 5-23 (c)]. For example, for the LM101A op-amp the average temperature coefficient of input offset voltage (that is, thermal voltage drift) is

$$\frac{\Delta V_{io}}{\Delta T} = 15\ \mu\text{V/°C maximum}$$

and the average temperature coefficient of input offset current is

$$\frac{\Delta I_{io}}{\Delta T} = 200\ \text{pA/°C maximum}$$

No information is available on the change in input bias current versus temperature. In fact, when we use an R_{OM} resistor, we are not concerned about the change in input bias current as a function of change in temperature. Note again that these values are denoted as absolute values on the data sheets because they may be positive in one temperature range and negative in another. More specifically, for the LM101A this means that the input offset voltage may change at most by $\pm 15\ \mu$V/°C change in temperature about 25°C. Similarly, the change in input offset current can be at most ± 200 pA/°C change in temperature from 25°C.

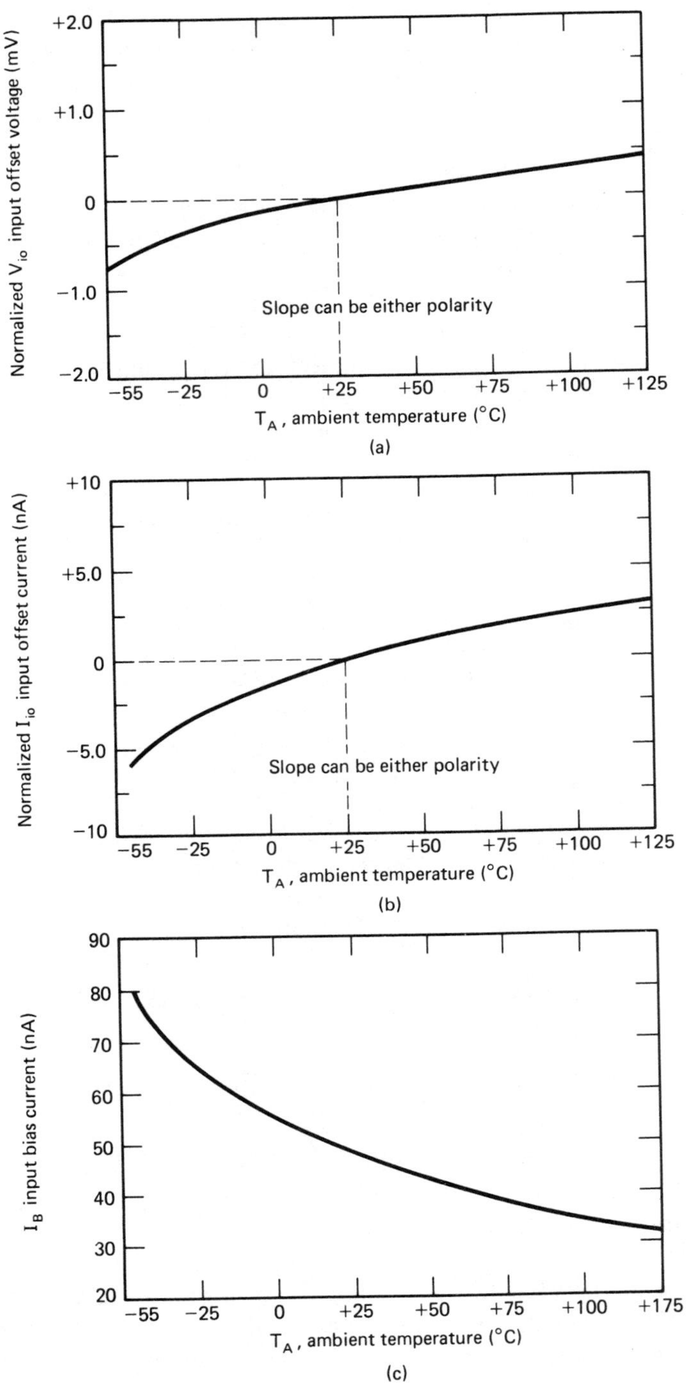

Figure 5-23 (a) Normalized V_{io}, input offset voltage versus temperature curve of MC1741. (b) Normalized I_{io}, input offset current versus temperature curve of MC1741. (c) Input bias current I_B versus temperature curve of MC1741. (Courtesy of Motorola Semiconductor Inc.)

5-6.1 Error Voltage

Next we should like to find the change in the total output offset voltage caused by the input offset voltage and input offset current drifts. This is important in evaluating the amplifier performance. To get a clear idea of the point, we should consider a specific case.

Let us consider the inverting amplifier with R_{OM} and compensating network as shown in Figure 5-24. We assume that the amplifier has been nulled at room temperature (25°C); that is, the effects of V_{io} and I_{io} have been reduced to zero by use of the potentiometer arrangement shown in Figure 5-24. We have seen that the values of V_{io} and I_{io} drift (change) with temperature. According to Equation (5-28), any change in the values of V_{io} and I_{io} results in a change in the total output offset voltage. Therefore, the total output offset voltage will not be zero at any temperature other than room temperature. This means that the desired output accuracy will be affected by the presence of the output offsets at all temperatures other than 25°C. We can conclude that thermal drifts are the major problem and not the output offsets.

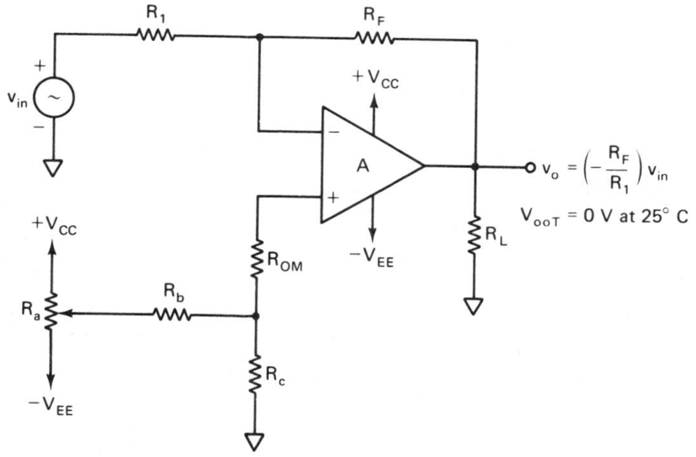

Figure 5-24 Completely compensated inverting amplifier.

To determine the effect of voltage and current drifts on the performance of an amplifier, we first need to find the average change in total output offset voltage per unit change in temperature. This value is obtained from Equation (5-28) and is given by Equation (5-29):

$$\frac{\Delta V_{ooT}}{\Delta T} = \left(1 + \frac{R_F}{R_1}\right)\frac{\Delta V_{io}}{\Delta T} + (R_F)\frac{\Delta I_{io}}{\Delta T} \tag{5-29}$$

where $\Delta V_{ooT}/\Delta T$ is the average change in total output offset voltage per unit change in temperature, in μV/°C. We know that the voltage and current drifts can be either positive or negative since they are specified as absolute values on the data sheets. Therefore, to obtain the maximum possible average change in total output offset voltage per unit change in temperature, we shall assume that the two drift effects are cumulative (additive). In other words, the $\Delta V_{ooT}/\Delta T$ value obtained from Equation

(5-29) will also be an absolute value, meaning that it could be either positive or negative. We can now obtain the maximum possible change in total output offset voltage ΔV_{ooT} resulting from a change in temperature ΔT by multiplying both sides of Equation (5-29) by ΔT. Let us define this maximum possible change in total output offset voltage ΔV_{ooT} as the "error voltage" and denote it by E_v. Therefore, from Equation (5-29), we heve

$$\Delta V_{ooT} = \left(1 + \frac{R_F}{R_1}\right)\left(\frac{\Delta V_{io}}{\Delta T}\right) \Delta T + (R_F)\left(\frac{\Delta I_{io}}{\Delta T}\right) \Delta T$$

$$E_v = \left(1 + \frac{R_F}{R_1}\right)\left(\frac{\Delta V_{io}}{\Delta T}\right) \Delta T + (R_F)\left(\frac{\Delta I_{io}}{\Delta T}\right) \Delta T \tag{5-30}$$

As we have mentioned before, E_v could be either positive or negative. Thus the expression for the output voltage for the inverting amplifier in Figure 5-24 may be written as

$$v_o = \left(-\frac{R_F}{R_1}\right) v_{in} \pm E_v \tag{5-31}$$

where E_v is the error voltage.

But v_o is expected to be directly proportional to v_{in}, which means that we would like to have E_v at zero. According to Equation (5-31), $E_v = 0$ V is possible only if either voltage and current drifts were zero or change in temperature ΔT was zero. Practically, neither of these two conditions is possible. In other words, in practice all amplifiers will exhibit some finite error E_v resulting from the change in temperature. However, if the amount of error is excessive for the application we have in mind, we can then consider a precision-type op-amp. High-performance precision op-amps are available that feature extremely low thermal voltage and current drifts. For example, the LH0044A has $\Delta V_{io}/\Delta T = 0.1\ \mu$V/°C and $\Delta I_{io}/\Delta T = 5$ pA/°C. But remember that we cannot get something for nothing; we also have to pay a higher price for these precision-type op-amps than for the general-purpose types. All in all, it is a trade-off between performance and cost.

It should be noted that the effect of voltage and current drifts is most pronounced in dc amplifiers, especially when the amplitude of the input to be amplified is relatively small. In fact, voltage and current drifts can also impose a serious problem in ac amplifiers, particularly if relatively large input voltage is to be amplified by an amplifier operating at its maximum capacity. The amplifier is said to be operating at its maximum capacity when the amplitude of the output voltage is equal to the saturation voltages. In any given amplifier circuit the saturation voltages are assumed to be slightly smaller than the supply voltages $+V_{CC}$ and $-V_{EE}$.

Example 5-8

Refer to the inverting amplifier in Figure 5-24. The op-amp is the LM307 with the following specifications:

$$\frac{\Delta V_{io}}{\Delta T} = 30\ \mu\text{V/°C maximum}$$

$$\frac{\Delta I_{io}}{\Delta T} \text{ maximum} = 300\ \text{pA/°C}$$

$$V_S = \pm 15\ \text{V}$$

$$R_1 = 1\ \text{k}\Omega,\ R_2 = 100\ \text{k}\Omega,\ \text{and } R_L = 10\ \text{k}\Omega$$

Assume that the amplifier is nulled at 25°C. Calculate the value of the error voltage E_v and the output voltage at 35°C if
(a) $V_{in} = 1$ mV dc
(b) $V_{in} = 10$ mV dc

Solution. The change in temperature $\Delta T = 35° - 25° = 10°C$. Using this change in temperature, we can calculate E_v and then V_o values as shown below.
(a) Substituting known values in Equation (5-30), we get

$$E_v = \left(1 + \frac{R_F}{R_1}\right)\left(\frac{\Delta V_{io}}{\Delta T}\right)\Delta T + (R_F)\left(\frac{\Delta I_{io}}{\Delta T}\right)\Delta T$$

$$= \left(1 + \frac{100\text{ k}\Omega}{1\text{ k}\Omega}\right)\left(\frac{30\ \mu\text{V}}{1°\text{C}}\right)(10°\text{C}) + (100\text{ k}\Omega)\left(\frac{300\text{ pA}}{1°\text{C}}\right)(10°\text{C})$$

$$= 30.3\text{ mV} + 0.3\text{ mV} = 30.6\text{ mV}$$

For $V_{in} = 1$ mV dc, the output voltage V_o using Equation (5-31) is

$$V_o = \left(-\frac{R_F}{R_1}\right)V_{in} \pm E_v$$

$$= -\left(\frac{100\text{ k}\Omega}{1\text{ k}\Omega}\right)(1\text{ mV}) \pm 30.6\text{ mV}$$

$$= -100\text{ mV} \pm 30.6\text{ mV}$$

$$= -130.6\text{ mV or } -69.4\text{ mV}$$

Thus the error voltage of ± 30.6 mV dc may cause the output voltage to equal -69.4 mV dc or -130.6 mV dc as the temperature varies from 25 to 35°C.
(b) All the other factors are unchanged except the V_{in} value; therefore, the value of error voltage is still equal to ± 30.6 mV. Substituting in Equation (5-31) the known values, we have

$$V_o = \left(-\frac{R_F}{R_1}\right)V_{in} \pm E_v$$

$$= \left(\frac{-100\text{ k}\Omega}{1\text{ k}\Omega}\right)(10\text{ mV}) \pm 30.6\text{ mV}$$

$$= -1000\text{ mV} \pm 30.6\text{ mV} = -1030.6\text{ mV or } -969.4\text{ mV}$$

This means that V_o can be equal to -969.4 mV or -1030.6 mV as the temperature changes between 25 and 35°C.

Let us now take a close look at the results in parts (a) and (b). Even though the amount of error is the same in both cases, the percentage error in V_o is larger in part (a) because the amplitude of the input is smaller. Thus for a given op-amp circuit and a given temperature change, the percentage error in the output voltage increases as the amplitude of the input signal is decreased. The percentage error in output voltage can be reduced significantly if we use a precision-type op-amp.

Example 5-9

Refer again to the amplifier circuit in Figure 5-24. Use the same circuit specifications that are given in Example 5-8. Assume that the amplifier is nulled at 25°C. If v_{in} is 10-mV peak sine wave at 1 kHz:
(a) Calculate E_v and v_o values at 55°C.
(b) Draw the output voltage waveform at 55°C.

Solution. The change in temperature ΔT is $(55° - 25°) = 30°C$.

(a) First we calculate the error voltage and then the output voltage:

$$E_v = \left(1 + \frac{R_F}{R_1}\right)\left(\frac{\Delta V_{io}}{\Delta T}\right) \Delta T + (R_F)\left(\frac{\Delta I_{io}}{\Delta T}\right) \Delta T$$

$$= \left(1 + \frac{100 \text{ k}\Omega}{1 \text{ k}\Omega}\right)\left(\frac{30 \ \mu\text{V}}{1°\text{C}}\right)(30°\text{C}) + (100 \text{ k}\Omega)\left(\frac{300 \text{ pA}}{1°\text{C}}\right)(30°\text{C})$$

$$= 90.9 \text{ mV} + 0.9 \text{ mV}$$

$$= 91.8 \text{ mV dc}$$

$$v_o = \left(-\frac{R_2}{R_1}\right) v_{\text{in}} \pm E_v$$

$$= \left(\frac{-100 \text{ k}\Omega}{1 \text{ k}\Omega}\right)(10 \text{ mV}) \pm 91.8 \text{ mV}$$

$$= -1000 \text{ mV peak} \pm 91.8 \text{ mV dc}$$

This means that the 1000-mV peak ac voltage signal rides either on a +91.8-mV or a −91.8-mV dc level. When the signal rides on a positive dc level, it is said to be shifted up, whereas if it rides on a negative dc level, it is said to be shifted down. Thus the 1000-mV peak ac voltage signal is either shifted up or down at the most by 91.8 mV.

(b) The waveform in Figure 5-25 is drawn assuming that the ac voltage signal is shifted up by 91.8 mV.

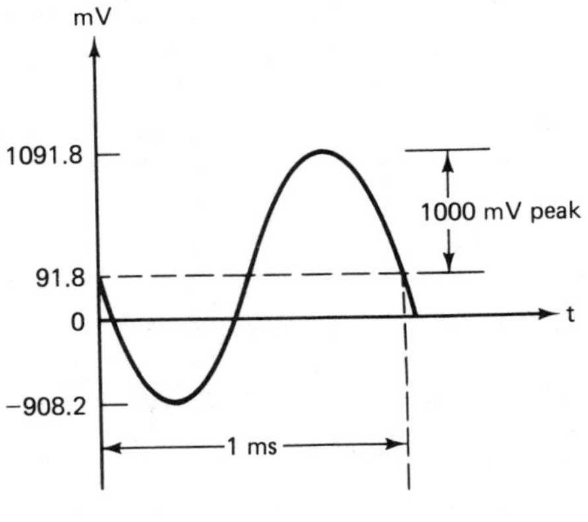

Figure 5-25 Waveform of Example 5-9.

It is clear from Example 5-9 that the error voltage causes a dc level shift in ac amplifiers. A comparatively small dc level shift may be tolerated. However, a relatively large dc level shift in an ac amplifier can cause the output waveform to distort. Namely, a dc level shift, depending on its polarity, can clip off either a positive peak or a negative peak portion of the ac output waveform. Again, to reduce the amount of dc level shift in an ac amplifier, we may use precision-type op-amps.

Thus far we have discussed the effect of drifts on the performance of an inverting amplifier. The same problems arise in a noninverting amplifier. Figure 5-26 shows the completely compensated noninverting amplifier.

Recall that the total output offset voltage Equation (5-28) is applicable to inverting as well as noninverting amplifiers. Therefore, the error voltage equation (5-30) must also be applicable to both amplifiers, since it is derived from Equation (5-28). We have seen that the gain of the noninverting amplifier with an offset-voltage compensating network is equal to $1 + [R_F/(R_1 + R_c)]$ (see Figure 5-26). Therefore, the expression for the output voltage of noninverting amplifier may be written as

$$v_o = \left(1 + \frac{R_F}{R_1 + R_c}\right) v_{in} \pm E_v \tag{5-32}$$

where

$$E_v = \left(1 + \frac{R_F}{R_1}\right)\left(\frac{\Delta V_{io}}{\Delta T}\right)\Delta T + (R_F)\left(\frac{\Delta I_{io}}{\Delta T}\right)\Delta T$$

Remember that if $R_c < R_1$, the voltage gain of the noninverting amplifier becomes equal to $(1 + R_F/R_1)$.

Example 5-10

Repeat Example 5-8 for the noninverting amplifier shown in Figure 5-26. Assume that $R_c \ll R_1$.

Solution. The error voltage E_v is the same as in Example 5-8; that is, $E_v = 30.6$ mV. (a) If $V_{in} = 1$ mV dc, then

$$\begin{aligned} V_o &= \left(1 + \frac{R_F}{R_1}\right) V_{in} \pm E_v \\ &= (101)(1 \text{ mV}) \pm 30.6 \text{ mV} \\ &= +70.4 \text{ mV or } +131.6 \text{ mV dc} \end{aligned}$$

(b) If $V_{in} = 10$ mV dc, then

$$\begin{aligned} V_o &= (101)(10 \text{ mV}) \pm 30.6 \text{ mV} \\ &= +979.4 \text{ mV or } +1040.6 \text{ mV dc} \end{aligned}$$

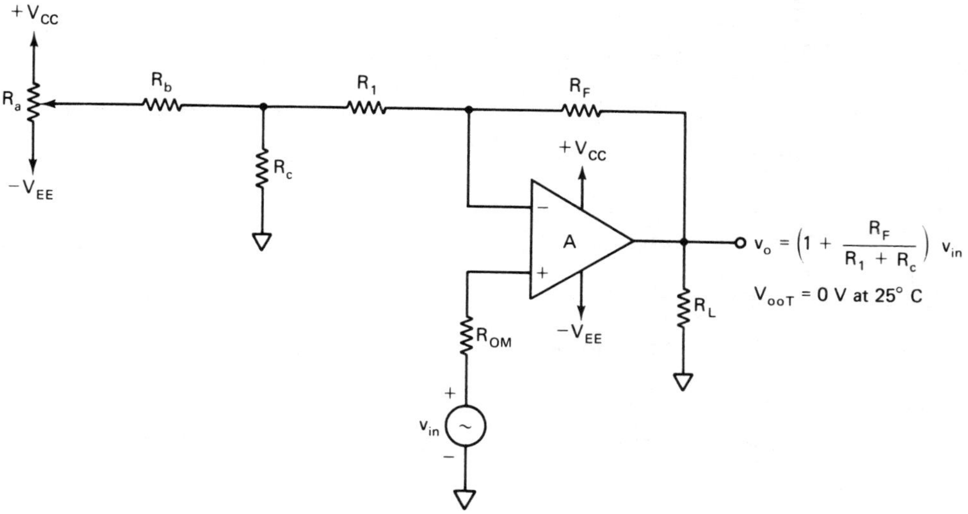

Figure 5-26 Completely compensated noninverting amplifier.

5-7 EFFECT OF VARIATION IN POWER SUPPLY VOLTAGES ON OFFSET VOLTAGE

In the preceding section we studied the effect of input offset voltage and input offset current thermal drifts on the output voltage of inverting as well as noninverting amplifiers. As we have mentioned before, the V_{io}, I_{io}, and I_B values are also susceptible to the changes in the supply voltages $+V_{CC}$ and $-V_{EE}$. Obviously, because the op-amp is capable of amplifying dc inputs, it is sensitive to changes in its supply voltages. This section is concerned with the effect of variation in supply voltages $+V_{CC}$ and $-V_{EE}$ on the values of V_{io}, I_{io}, and I_B and, in turn, the effect of changes in V_{io}, I_{io}, and I_B on output offset voltage.

Once we select the specific values for supply voltages $+V_{CC}$ and $-V_{EE}$ in a given op-amp amplifier, we do not change them deliberately. However, sometimes these voltages may change as a result of poor regulation and filtering. A poorly regulated power supply gives different values depending on the size and type of load connected to it. On the other hand, a poorly filtered power supply has a ripple voltage riding on some specific dc level.

Figure 5-27(a) shows the input bias current versus supply voltage curve for the LH0001 op-amp. A glance at output offset voltage V_{oI_B} [Equation (5-16)] reveals that for a given value of R_F any change in I_B causes a change in V_{oI_B}. Of course, you will recall that we can use the R_{OM} resistor to minimize the effect of I_B or of changes in it on the output offset voltage V_{oI_B}.

Even though input bias currents change due to the change in supply voltages, the input offset current should remain relatively constant because it is the absolute value of the difference between two input bias currents [see Figure 5-27(b)]. Thus, in practice, if we use a proper value of the R_{OM} resistor in a given amplifier circuit,

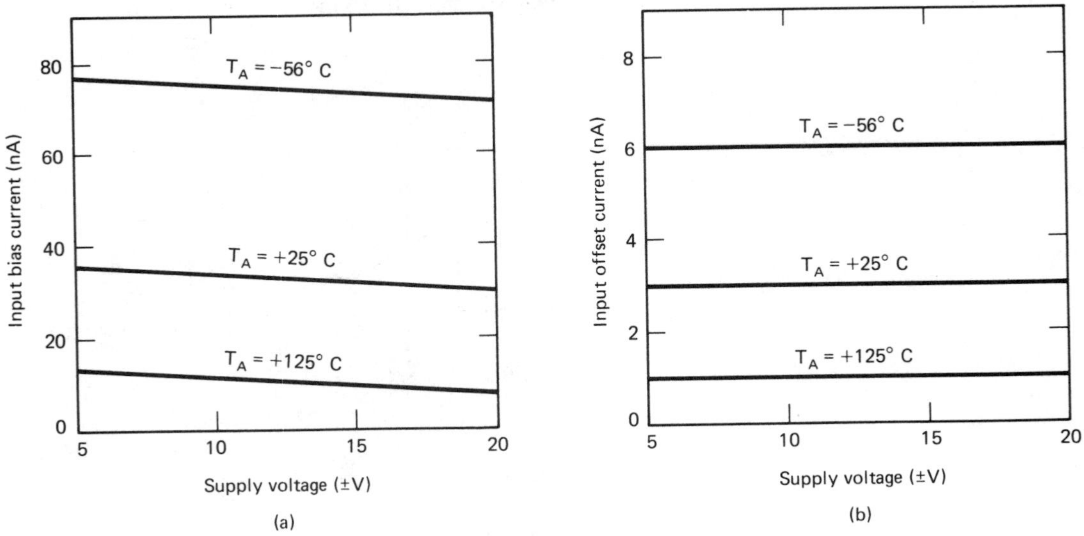

Figure 5-27 (a) Input bias current I_B, versus supply voltage for LH0001 op-amp. (b) Input offset current I_{io}, versus supply voltage for LH0001 op-amp. (Courtesy of National Semiconductor.)

there will be a negligible change in the current-generated output offset voltage due to the change in supply voltages. Therefore, manufactureres do not furnish curves like those in Figure 5-27 for all op-amps.

Recall that the supply voltages change because of poor regulation and filtering. For a given op-amp any change in the values of the supply voltages results in a change in the input offset voltage, which in turn causes a change in the output offset voltage. The change in op-amp's input offset voltage caused by variations in the supply voltages is generally specified on the data sheets by a variety of terms: The input offset voltage sensitivity, the power supply rejection ratio, the power supply sensitivity, and the supply voltage rejection ratio are some of them. All of these terms are equivalent since they convey the same information. These terms are expressed either in microvolts per volt or in decibels. For example: The supply voltage rejection ratio (SVRR) for the μA741 is $\Delta V_{io}/\Delta V = 150\ \mu\text{V/V}$ maximum and it is typically $20 \log (\Delta V/\Delta V_{io}) = 96$ dB for the LM307, where ΔV is the change in supply voltages $+V_{CC}$ and $-V_{EE}$, and ΔV_{io} is the resulting change in the input offset voltage.

Given a supply voltage rejection ratio in microvolts per volt, we can obtain an equivalent value in decibels (dB) or vice versa. For instance, SVRR ($\Delta V_{io}/\Delta V$) of 150 μV/V is equivalent to

$$20 \log \left(\frac{1}{\text{SVRR}}\right) = 20 \log \left(\frac{1}{\Delta V_{io}/\Delta V}\right) = 20 \log \left(\frac{1}{150\ \mu\text{V/V}}\right) = 20 \log \left(\frac{10^6}{150}\right) = 76.48 \text{ dB}$$

Similarly, a SVRR of 96 dB is equivalent to 15.85 μV/V as follows. Since $20 \log (1/\text{SVRR}) = 96$ dB,

$$\log \left(\frac{1}{\text{SVRR}}\right) = \frac{96}{20}$$

$$\frac{1}{\text{SVRR}} = 10^{4.8}$$

$$\text{SVRR} = \frac{1}{10^{4.8}}$$

$$= 15.85\ \mu\text{V/V}$$

Note that the higher the value of SVRR in dB, the lower is the change in input offset voltage due to the change in supply voltages, or in other words, the lower the value of SVRR in μV/V, the better for op-amp performance. In fact, ideally the value of SVRR in μV/V should be zero.

Now we should like to find for a given amplifier the change in output offset voltage due to the supply voltage rejection ratio. Refer again to the completely compensated inverting amplifier circuit shown in Figure 5-24. Let us assume that the amplifier is nulled initially. Suppose that after the circuit is in operation for a while, the supply voltages change in value due to poor regulation. We know that any change in the supply voltages results in a change in the input offset voltage. And according to Equation (5-8), any change in input offset voltage results in a change in the output offset voltage. Therefore, using Equation (5-8) we can establish a relationship between the change in output offset voltage and SVRR as follows:

$$V_{oo} = \left(1 + \frac{R_F}{R_1}\right) V_{io} \qquad (5\text{-}8)$$

where $(1 + R_F/R_1)$ is a constant for given values of R_1 and R_F. Therefore, the average change in V_{oo} per unit change in supply voltages can be

$$\frac{\Delta V_{oo}}{\Delta V} = \left(1 + \frac{R_F}{R_1}\right)\left(\frac{\Delta V_{io}}{\Delta V}\right) \tag{5-33}$$

Multiplying both sides of Equation (5-33) by ΔV, we get

$$\Delta V_{oo} = \left(1 + \frac{R_F}{R_1}\right)\left(\frac{\Delta V_{io}}{\Delta V}\right)\Delta V \tag{5-34}$$

where ΔV_{oo} = change in output offset voltage (volts)
ΔV = change in supply voltages $+V_{CC}$ and $-V_{EE}$
$\frac{\Delta V_{io}}{\Delta V}$ = supply voltage rejection ratio (μV/V)

Remember that ΔV_{oo} is a dc voltage, and it could be either positive or negative. Thus all practical op-amps are affected by changes in the supply voltages, and therefore regulated supplies are recommended.

Example 5-11

The amplifier in Figure 28(b) is nulled when the low dc supply is 20 V [see Figure 5-28(a)]. Because of poor regulation, low dc voltage varies with time from 18 to 22 V. Determine (a) the change in the output offset voltage caused by the change in supply voltages, and (b) the output voltage V_o, if V_{in} = 10 mV dc. The op-amp is the LM307 with SVRR = 96 dB.

Solution

(a) The variation in low dc voltage from 18 to 22 V, compared to its desired value of 20V ($+V_{CC} = +10$ V, and $-V_{EE} = -10$ V) implies that the change in supply voltages $\Delta V = 2$ V. The supply voltage rejection ratio in μV/V equivalent to 96 dB is

$$20 \log\left(\frac{1}{\text{SVRR}}\right) = 96 \text{ dB}$$

$$\log\left(\frac{1}{\text{SVRR}}\right) = \frac{96}{20}$$

$$\frac{1}{\text{SVRR}} = 10^{4.8}$$

$$\text{SVRR} = 15.85\ \mu\text{V/V}$$

That is,

$$\frac{\Delta V_{io}}{\Delta V} = 15.85\ \mu\text{V/V}$$

Substituting known values in Equation (5-34),

$$\Delta V_{oo} = \left(1 + \frac{100\text{ k}\Omega}{1\text{ k}\Omega}\right)(15.8\ \mu\text{V/V})(2\text{ V})$$

$$= (101)(31.6)(10^{-6})$$

$$= 3.20\text{ mV}$$

It is worth noting that the ΔV value is the same regardless of whether it is computed from the change in low dc supply or change in $+V_{CC}$ or $-V_{EE}$. In this example, as a worstcase situation, suppose that $-V_{EE}$ remains constant at -10 V; then the $+V_{CC}$ has to vary from 8 to 12 V as a result of change in low dc voltage. This means

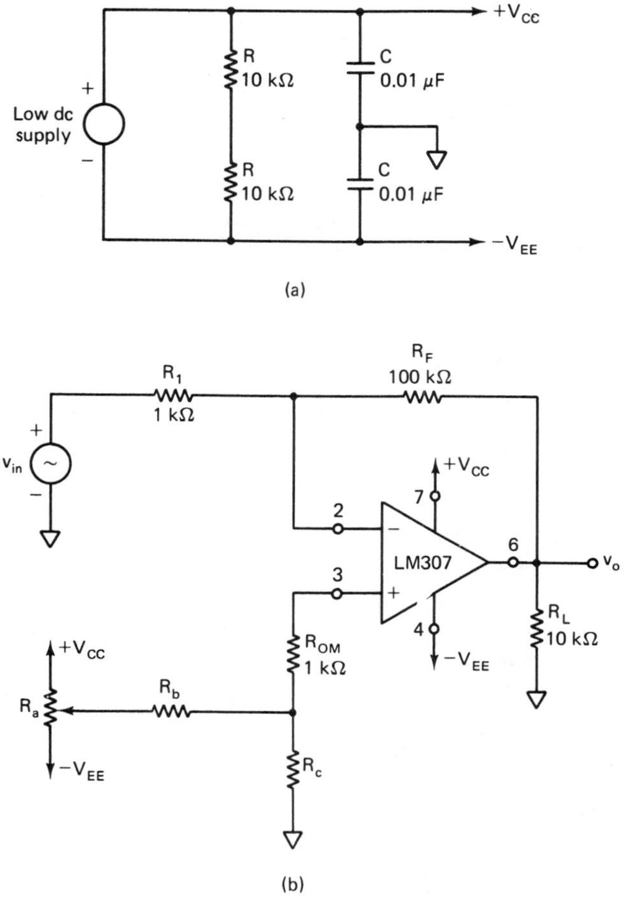

Figure 5-28 (a) Supply voltages for the inverting amplifier. (b) Inverting amplifier of Examples 5-11 and 5-12.

that the change ΔV in supply voltage $+V_{CC}$ is 2 V in either direction from 10 V. In fact, for some op-amps the positive ($-V_{EE}$ constant) and negative ($+V_{CC}$ constant) supply voltage rejection ratios are specified on the data sheets. For example, the μA1458C op-amp has positive power supply sensitivity $= 30\ \mu\text{V/V}$ and negative power supply sensitivity $= 30\ \mu\text{V/V}$.

(b) The total output voltage, including the change in output offset voltage ΔV_{oo}, is given by

$$
\begin{aligned}
V_o &= \left(-\frac{R_F}{R_1}\right) V_{\text{in}} \pm \Delta V_{oo} \\
&= \left(-\frac{100\ \text{k}\Omega}{1\ \text{k}\Omega}\right)(10\ \text{mV}) \pm 3.2\ \text{mV} \\
&= -1000\ \text{mV} \pm 3.2\ \text{mV} \\
&= -1003.2\ \text{mV} \quad \text{or} \quad -996.8\ \text{mV}
\end{aligned}
$$

It can be seen from Example 5-11 that the variation in supply voltages could be troublesome in small-signal op-amp amplifiers.

Example 5-12

Referring to Figure 5-28(b), suppose that the circuit is nulled when the voltage across terminals $+V_{CC}$ and $-V_{EE}$ measures 20 V dc. Also suppose that, because of poor filtering, 10 mV rms ac ripple is measured across the terminals $+V_{CC}$ and $-V_{EE}$. While the input signal $v_{in} = 0$ V, how much ripple voltage can we expect at the output if the op-amp is the LM307?

Solution. The SVRR = 15.85 μV/V for the LM307, and because of poor filtering, $\Delta V = 10$ mV rms. Substituting known values in Equation (5-34), we get an output ripple voltage of

$$\Delta V_{oo} = \left(1 + \frac{100\text{ k}\Omega}{1\text{ k}\Omega}\right)\left(\frac{15.85\ \mu\text{V}}{V}\right)(10\text{ mV})$$

$$= 16\ \mu\text{V}$$

5-8 CHANGE IN INPUT OFFSET VOLTAGE AND INPUT OFFSET CURRENT WITH TIME

Under operating conditions the characteristics of all semiconductor devices, such as diodes and transistors, change to some extent with time. The amount of change in semiconductor characteristics then affects the performance of a circuit, in particular its linearity and accuracy. Since an op-amp is composed of transistors and diodes among other things, the op-amp's input offset voltage and input offset current changes (drifts) with time. For long-term stability the amount of change in input offset voltage and input offset current with time is crucial. On some op-amp data sheets the input offset voltage and input offset current drifts with time are specified and are denoted by $\Delta V_{io}/\Delta t$ and $\Delta I_{io}/\Delta t$, respectively. Generally, $\Delta V_{io}/\Delta t$ is expressed in microvolts per week and $\Delta I_{io}/\Delta t$ in nanoamperes per week. For example, for the LH0041C the typical values of $\Delta V_{io}/\Delta t = 5\ \mu$V/week and $\Delta I_{io}/\Delta t = 2$ nA/week. Remember that the drifts in V_{io} and I_{io} with time are the absolute values, meaning that V_{io} may increase or decrease typically by 5 μV/week in the case of the LH0041C. Similarly, the typical change in I_{io} could be as much as either +2 nA or −2 nA per week for the LH0041C.

Thus, with the help of Equation (5-28), we should be able to determine the maximum possible change in output offset voltage as a function of input offset voltage and input offset current drifts with time over a certain time period.

$$V_{ooT} = \left(1 + \frac{R_F}{R_1}\right) V_{io} + (R_F)I_{io} \tag{5-28}$$

Therefore, the average change in V_{ooT} per unit time is

$$\frac{\Delta V_{ooT}}{\Delta t} = \left(1 + \frac{R_F}{R_1}\right)\frac{\Delta V_{io}}{\Delta t} + (R_F)\frac{\Delta I_{io}}{\Delta t} \tag{5-35}$$

Multiplying both the sides of Equation (5-35) by Δt, we get

$$\Delta V_{ooT} = \left(1 + \frac{R_F}{R_1}\right)\left(\frac{\Delta V_{io}}{\Delta t}\right)\Delta t + (R_F)\left(\frac{\Delta I_{io}}{\Delta t}\right)\Delta t \tag{5-36}$$

where Δt = time elapsed (weeks)

$\frac{\Delta V_{io}}{\Delta t}$ = input offset voltage drift with time (volts/week)

$$\frac{\Delta I_{io}}{\Delta t} = \text{input offset current drift with time (amperes/week)}$$

$$\Delta V_{ooT} = \text{change in output offset voltage (volts)}$$

Note that Equation (5-36) is applicable to noninverting as well as inverting amplifiers.

Thus, in practice, the output offset voltage in all op-amp circuits will change with time. To maintain the desired accuracy and linearity of a system it is necessary to calibrate all the op-amps in that system periodically.

Example 5-13

Suppose that the circuit in Figure 5-24 is initially nulled. Assume also that room temperature and the voltage across terminals $+V_{CC}$ and $-V_{EE}$ remain constant. Determine the maximum possible change in output offset voltage after one month if the op-amp is the LH0041C. Assume that $R_1 = 1\ \text{k}\Omega$, $R_F = 100\ \text{k}\Omega$, and $R_L = 10\ \text{k}\Omega$.

Solution. For the LH0041C, $\Delta V_{io}/\Delta t = 5\ \mu\text{V/week}$ and $\Delta I_{io}/\Delta t = 2\ \text{nA/week}$. Due to the time drift, $\Delta t = 4$ weeks,

$$\begin{aligned}\Delta V_{ooT} &= \left(1 + \frac{R_F}{R_1}\right)\left(\frac{\Delta V_{io}}{\Delta t}\right)\Delta t + (R_F)\left(\frac{\Delta I_{io}}{\Delta t}\right)\Delta t \\ &= \left(1 + \frac{100\ \text{k}\Omega}{1\ \text{k}\Omega}\right)(5 \times 10^{-6})(4) + (100\ \text{k}\Omega)(2 \times 10^{-9})(4) \\ &= 2.02\ \text{mV} + 0.8\ \text{mV} = 2.82\ \text{mV}\end{aligned}$$

5-9 OTHER TEMPERATURE- AND SUPPLY VOLTAGE-SENSITIVE PARAMETERS

The error voltage in the op-amp's output voltage is not the only possible result of temperature changes. A collection of other temperature-sensitive characteristics for the μA741C op-amp is shown in Figure 5-29. The curve in Figure 5-29(a) shows that the input resistance R_i of the 741C op-amp increases with an increasing temperature, when supply voltage equals ± 15 V. Recall that the input resistance of an op-amp is defined as the equivalent resistance which would be measured at either an inverting or noninverting input terminal with the other terminal grounded. A changing input resistance, which is the load resistance on the signal source, can cause a drifting input signal amplitude. In fact, for a given signal source, the amount of drift in input signal amplitude depends on the relative values of the external components and also on the operating mode of the op-amp amplifier (inverting or noninverting). Thus for a given signal source and op-amp circuit, we can minimize the drift in input signal amplitude without affecting the desired performance of the circuit if we select relatively small values for external components.

The power consumption as a function of ambient temperature curve for the 741C op-amp is shown in Figure 5-29(b). As can be seen from this curve, the power consumption, which is the dc power required to operate the op-amp under no-load conditions, decreases with increasing ambient temperature.

Finally, as shown in Figure 5-29(c), the output short-circuit current of the 741C op-amp decreases with an increasing temperature. In other words, the amount of

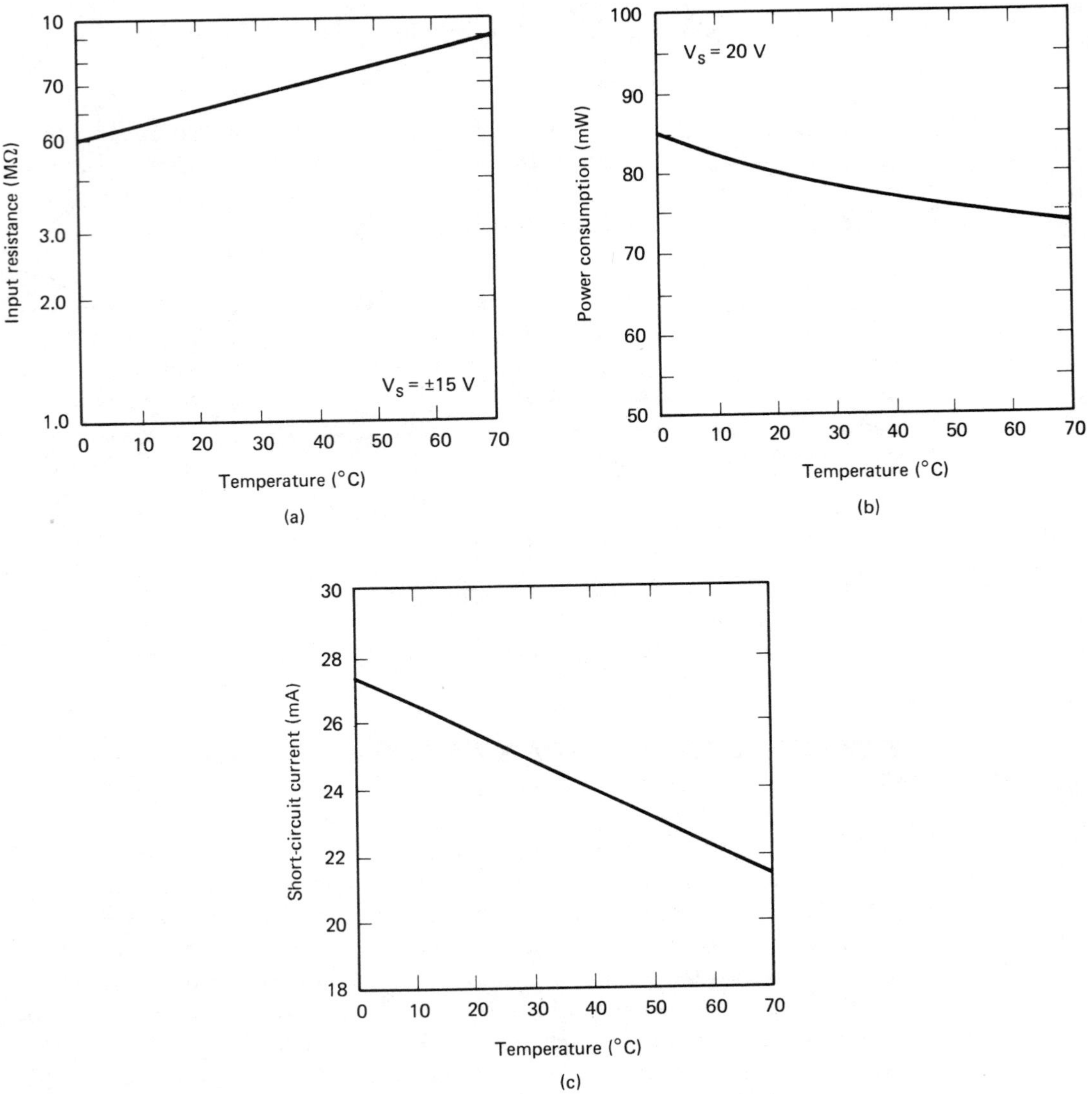

Figure 5-29 (a) Input resistance as a function of temperature. (b) Power consumption as a function of temperature. (c) Output short-circuit current as a function of temperature. (Courtesy of Fairchild Camera and Instrument Corporation.)

output current obtainable with the output terminal shorted to ground or to either $+V_{CC}$ or $-V_{EE}$ supply decreases with increasing temperature.

Figure 5-30 shows a typical collection of supply voltage-sensitive parameters of the 741C op-amp. The open-loop voltage gain in dB increases as the supply voltage values are increased, as shown in Figure 5-30(a). The open-loop voltage gain varies between 82 and 100 dB as the supply voltages are changed from ± 2 V to ± 18 V, respectively, when the ambient temperature is 25°C. Note that the voltage gain in dB is equal to 20 log (v_{out}/v_{in}).

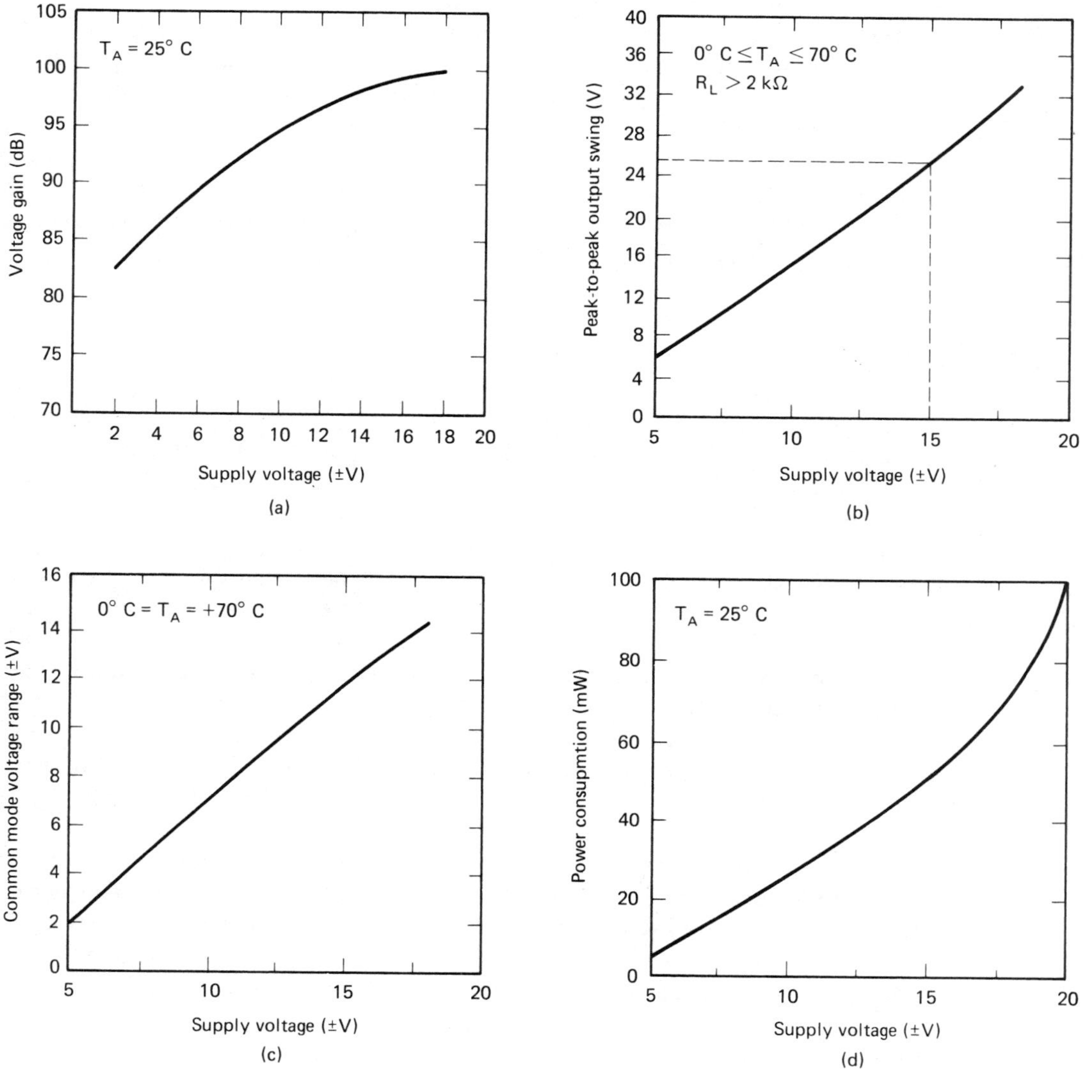

Figure 5-30 (a) Open-loop voltage gain as a function of supply voltage. (b) Output voltage swing as a function of supply voltage. (c) Input common-mode voltage range as a function of supply voltage. (d) Power consumption as a function of supply voltage. (Courtesy of Fairchild Camera and Instrument Corporation.)

The curve in Figure 5-30(b) shows the output voltage swing as a function of supply voltages $\pm V_{CC}$ and $-V_{EE}$. The peak-to-peak output swing depends on the value of the supply voltages and is always less than that value. For example, when the supply voltage is ± 15 V, the peak-to-peak output swing is 26 V, as shown in Figure 5-30(b).

Figure 5-30(c) shows that the input common-mode voltage range (CMVR) increases with increasing supply voltage values. That is, the range of common-mode input voltage over which the op-amp operates within specifications increases when the supply voltage values are increased. The CMVR varies between ± 2 V and ± 14 V as the supply voltages are changed from ± 5 V to ± 17.5 V, respectively [see Figure 5-30(c)]. For instance, when the supply voltage is ± 5 V, the common-mode voltage could be as high as either $+2$ V or as low as -2 V.

The power consumption as a function of supply voltage curve is shown in Figure 5-30(d). It can be seen here that the amount of dc power required to operate the op-amp under no-load conditions increases with an increase in supply voltages. Note that the curve is drawn at 25°C.

5-10 NOISE

A major source of interference with the desired signal in electronic systems composed of either discrete (separate) components or integrated circuits is noise. Any unwanted signal associated with the desired signal is noise. Because it is the combined effect of many sources, noise is random in nature and hard to predict or analyze. In any electronic system, noise can come from many external sources as well as be self-induced, a result of the circuitry itself. Examples of external noise sources are switching of rotating machinery, ignition systems, and various control circuits. Natural phenomena such as lightning may also be external noise sources.

The self-induced or internal noise may be caused by the ac random voltages and currents generated within conductors and semiconductors of one circuit as a result of the switching of another circuit. The rate of change of current and voltage per unit of time, the speed of operation of the circuit, and the type of coupling between two circuits are some of the factors that determine the amount of noise induced in a given circuit.

There are different types of noise phenomena associated with op-amps: Schottky noise, thermal noise, and $1/f$ noise are the most important among them. The thermal noise increases with an increase in temperature. Like thermal noise, the amount of Schottky noise is greater with wider bandwidths and larger resistances; on the other hand, $1/f$ noise increases with a decrease in frequency f. Sometimes, manufacturers provide the data on noise, as shown in Figure 5-31. Figure 5-31(a) and (b) show the input mean square noise voltage (V^2/Hz) and mean-square noise current (A^2/Hz) at various frequencies for the μA741 op-amp. These noise characteristic curves indicate that the noise level is higher at lower frequencies. Broadband noise versus source resistance curves for the 741 op-amp are shown in Figure 5-31(c). A glance at these curves reveals that the wider bandwidths for a given value of source resistance generate more noise; also, larger source resistances for a given bandwidth appreciably increase the noise level.

To reduce the effect of electrical noise on ICs, several schemes have been commonly used. Physical shielding of the ICs and associated wiring helps to prevent external electromagnetic radiation from inducing noise into the internal circuitry. Special buffering and filtering circuits can be used between the electronic circuits and signal leads. To provide a path for any radio frequency (RF), all linear IC power supply terminals should generally be bypassed to ground. The breadboard layout

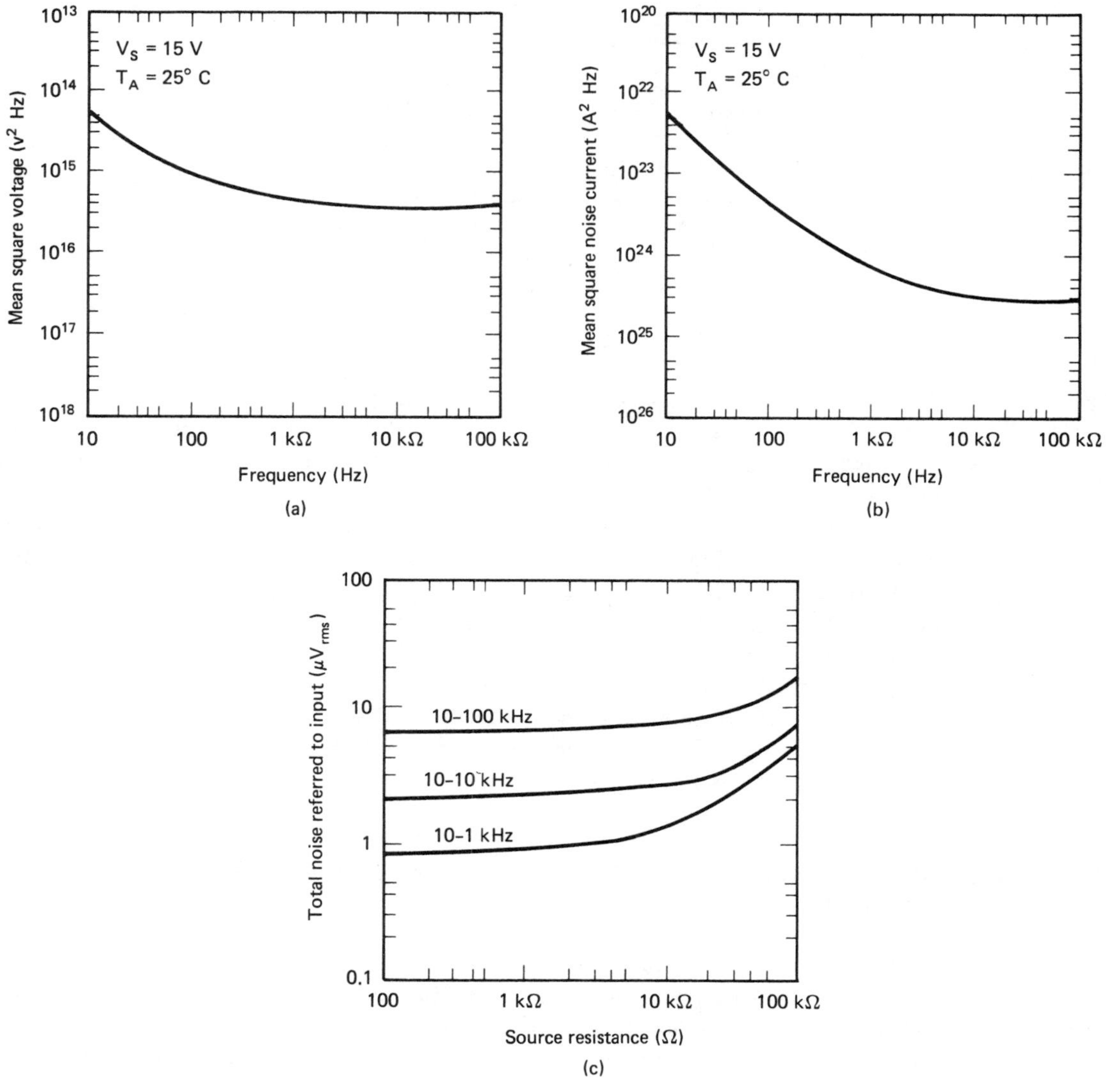

Figure 5-31 (a) Input noise voltage versus frequency. (b) Input noise current versus frequency. (c) Broadband noise for various bandwidths. (Courtesy of Fairchild Camera and Instrument Corporation.)

should be such that the bypass capacitors are as near the IC terminals as possible. Internal noise generation can be reduced by keeping input and output lead lengths as short as practically possible. Use one common tie point near the IC for all grounds. In a high electrical noise environment an IC with a high degree of noise immunity will minimize the amount of special care needed for proper circuit operation.

Among inverting, noninverting, and differential amplifiers, the latter type offers the best immunity to induced noise. The differential amplifier circuit is shown in Figure 5-32. In this circuit, v_d is the desired differential input signal to be amplified,

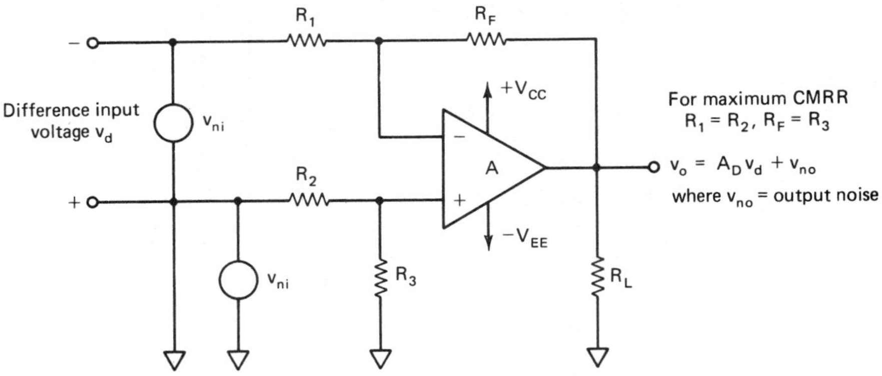

Figure 5-32 Noise in a differential amplifier.

and v_{ni} is the noise voltage induced into each input terminal with respect to ground. Since $R_1 = R_2$ and $R_F = R_3$, the voltage at the inverting and noninverting input terminals will be amplified by the same factor R_F/R_1. Thus, if the noise voltage to ground at the inverting terminals is the same as the noise voltage to ground at the noninverting terminal, the noise voltage at the output should be negligibly small if not zero. Thus the ratio of the output noise voltage to the input noise voltage in practice will be much smaller than 1.

With proper component selection such that both inverting and noninverting input impedances are equal, both of the induced noise voltages v_{ni} are equal in amplitude and phase. Therefore, in differential amplifier circuits the noise voltages at the output are greatly reduced.

5-11 COMMON-MODE CONFIGURATION AND COMMON-MODE REJECTION RATIO

When the same input voltage is applied to both input terminals of an op-amp, the op-amp is said to be operating in an common-mode configuration. Since the input voltage applied is common to both the inputs, it is referred to as a common-mode voltage v_{cm}, as shown in Figure 5-33. A common-mode voltage v_{cm} can be ac, dc, or a combination of ac and dc.

Because ideally an op-amp amplifies only differential input voltages, no common-mode output voltage v_{ocm} should appear at the output. However, due to imperfections within an actual op-amp, some common-mode voltage v_{ocm} will appear at the output. The amplitude of this v_{ocm} is very small and often insignificant compared to v_{cm}. Therefore, in practice the ratio of the output common-mode voltage v_{ocm} to the input common mode voltage v_{cm}, which is called the common-mode voltage gain A_{cm}, is generally much smaller than 1. In equation form,

$$A_{cm} = \frac{v_{ocm}}{v_{cm}} \tag{5-37}$$

Ideally, the common-mode voltage gain A_{cm} is zero.

Although A_{cm} is usually not specified on the op-amp data sheets, A_{cm} can be

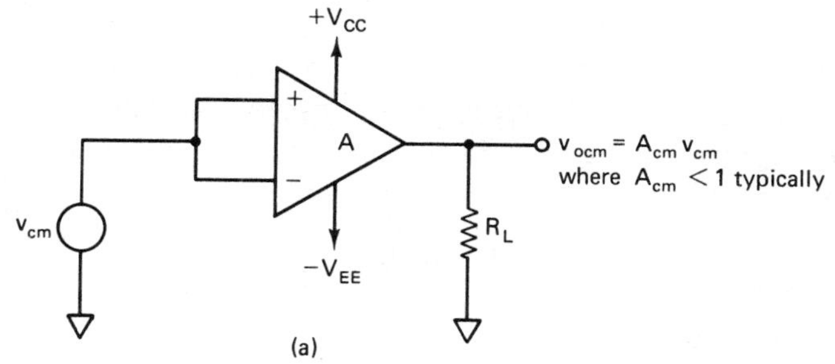

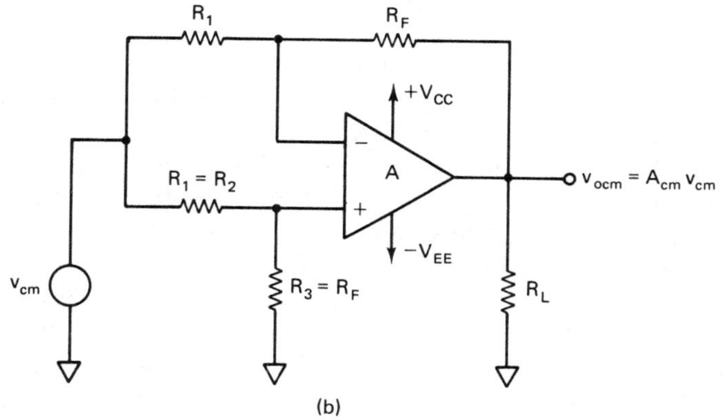

Figure 5-33 Op-amp connected in common-mode configuration. (a) Without feedback. (b) With feedback.

calculated for a given op-amp by applying a known value of common-mode input voltage v_{cm} and measuring the resultant output common-mode voltage v_{ocm}. Op-amp manufacturers usually list a common-mode rejection ratio CMRR. The CMRR is defined in several essentially equivalent ways by the various manufacturers. Generally, it can be defined as the ratio of the differential gain A_D to the common-mode gain A_{cm}, that is,

$$\text{CMRR} = \frac{A_D}{A_{cm}} \tag{5-38}$$

Note that in Figure 5-33(a), A_D is equal to internal gain A of the op-amp.

The CMRR can also be expressed as the ratio of the change in input offset voltage to the total change in common mode voltage. Thus

$$\text{CMRR} = \frac{V_{io}}{v_{cm}} \tag{5-39}$$

From Equations (5-37) and (5-38), we can then establish the relationship between the v_{ocm} and CMRR:

$$\text{CMRR} = \frac{A_D}{A_{cm}} = \frac{A_D}{v_{ocm}/v_{cm}}$$

$$= \frac{A_D v_{cm}}{v_{ocm}}$$

$$v_{ocm} = \frac{A_D v_{cm}}{\text{CMRR}} \tag{5-40}$$

Equation (5-40) indicates that the higher the value of CMRR, the smaller will be the amplitude of the output common-mode voltage v_{ocm}. Generally, the CMRR value is very large and is therefore usually specified in decibels (dB), where

$$\text{CMRR (dB)} = 20 \log \left(\frac{A_D}{A_{cm}}\right) \tag{5-41a}$$

or from Equation (5-39),

$$\text{CMRR (dB)} = 20 \log \left(\frac{1}{V_{io}/v_{cm}}\right) \tag{5-41b}$$

The value of CMRR listed on the data sheets is specified for the open-loop common-mode op-amp configuration as shown in Figure 5-33(a). For example, for μA741C op-amp CMRR (dB) = 90 dB. However, the CMRR value listed on the data sheets can be used as an approximate value for the closed-loop common-mode configuration shown in Figure 5-33(b). Being the ratio of differential gain A_D to the common-mode gain A_{cm}, CMRR should be the same for open-loop as well as closed-loop common-mode configurations [see Equation (5-38)].

Whether CMRR is defined as in Equation (5-38) or (5-39), it is a measure of the degree of matching between two input terminals; that is, the larger the value of CMRR (dB), the better is the matching between the two input terminals and the smaller is the output common-mode voltage v_{ocm}. On the other hand, a large voltage v_{ocm} for a given common-mode input voltage v_{cm} is an indication of a large degree of imbalance between the two input terminals or of poor common-mode rejection. Thus, in practice, it is advantageous to use op-amps with higher CMRRs since these op-amps have better ability to reject common-mode voltage such as 60-Hz induced noise voltages. The CMRR is a function of frequency and decreases as the frequency is increased as shown in Figure 5-34.

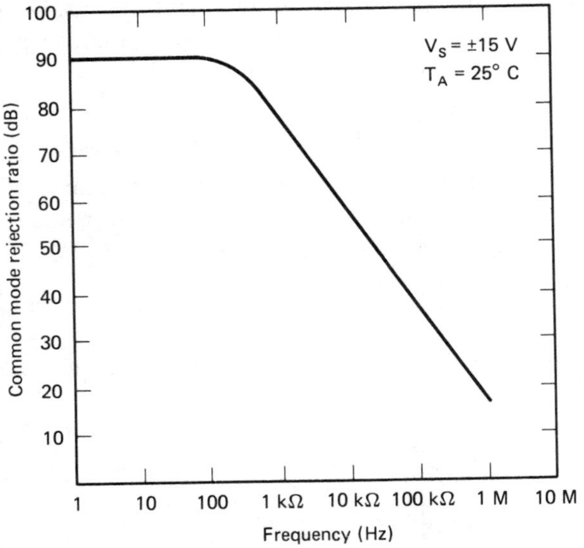

Figure 5-34 Common-mode rejection ratio as a function of frequency. (Courtesy of Fairchild Camera and Instrument Corporation.)

Example 5-14

In the circuit of Figure 5-32, if $R_1 = R_2 = 1\ \text{k}\Omega$, $R_F = R_3 = 10\ \text{k}\Omega$, $v_d = 5$ mV sine wave at 1 kHz, and $v_{ni} = 2$ mV at 60 Hz, calculate (a) the output voltage at 1 kHz and (b) the amplitude of the induced 60-Hz noise at the output. The op-amp is the μA741 with CMRR (dB) = 90 dB.

Solution

(a) The closed-loop differential gain of the circuit is

$$A_D = \frac{R_F}{R_1} = \frac{10\ \text{k}\Omega}{1\ \text{k}\Omega} = 10$$

since the input signal is applied in a differential mode, it is amplified by the differential gain A_D. Therefore, at 1 kHz the output signal is

$$v_o = A_D v_d = (10)(5\ \text{mV}) = 50\ \text{mV}$$

(b) The 60-Hz noise voltage v_{ni} appears in common mode at the input terminals of the op-amp; therefore, $v_{ni} = v_{cm}$. Thus, using Equation (5-40), we can calculate the value of output common-mode voltage v_{ocm}. In order to implement Equation (5-40), we first have to convert the CMRR (dB) value into its equivalent numerical value. The CMRR (dB) = 90 dB implies that

$$20 \log \text{CMRR} = 90\ \text{dB}$$

$$\log \text{CMRR} = \frac{90}{20}$$

$$\text{CMRR} = 10^{4.5}$$

$$\text{CMRR} = 31{,}622.78$$

Therefore,

$$v_{ocm} = \frac{A_D v_{cm}}{\text{CMRR}} = \frac{(10)(2\ \text{mV})}{31{,}622.78}$$

$$= 0.63\ \mu\text{V at 60 Hz}$$

Thus the op-amp used in the differential mode is an effective circuit in reducing noise problems since the 60-Hz output is much smaller than the input induced 60-Hz noise. The differential-mode op-amp circuit is particularly helpful in reducing induced input noise voltages when the small-amplitude differential input signal is being amplified.

SUMMARY

1. All operational amplifiers have finite values of input offset voltage, input offset current, and input bias current. These values result in a dc output offset voltage.
2. The op-amp can be nulled by using an offset-voltage compensating network. The amount of output offset voltage due to input bias currents can be significantly reduced if we use offset minimizing resistors.
3. The input offset voltage and current tend to drift with change in temperature; the drift results in a finite dc error voltage in the desired output voltage at temperatures other than 25°C.
4. The amount of error voltage affects the accuracy of the dc amplifiers and limits the amplitude of the ac signal that can be amplified without distortion.

5. The input offset voltage of an op-amp changes with the change in supply voltages; the change in input offset voltage in turn causes a change in output offset voltage. To keep the change in output offset voltage to a minimum, use a regulated power supply or use an op-amp with a relatively high supply voltage rejection ratio.
6. The op-amp's input offset voltage and input offset current changes with time. Therefore, for better long-term stability, use an op-amp with smaller drifts in input offset voltage and current per unit of time, and also calibrate the op-amp circuit periodically.
7. The input resistance, power consumption, and output short-circuit current of an op-amp are temperature-sensitive parameters. On the other hand, the open-loop voltage gain, output voltage swing, power consumption, and common-mode voltage range are all supply voltage-dependent parameters.
8. Any unwanted signal associated with the op-amp's desired output signal is noise. Schottky noise, thermal noise, and $1/f$ noise are the most important types of noise associated with op-amps.
9. The differential op-amp offers the best immunity to induced noise, since input noise voltages appear as common-mode voltage.
10. The common-mode rejection ratio of an op-amp is the ratio of differential gain to the common-mode gain. The larger the CMRR value, the smaller is the common-mode output voltage.

QUESTIONS

5-1. Define input offset voltage and explain why it exists in all op-amps.

5-2. Why is it necessary to use an external offset voltage compensating network with practical op-amp circuits?

5-3. What is the offset minimizing resistor R_{OM}?

5-4. Why is a resistor R_{OM} not needed in differential op-amp circuits?

5-5. Why is the output offset voltage generated by the input bias current always larger than that generated by the input offset current?

5-6. What are the factors that affect the input offset voltage, input bias, and input offset currents?

5-7. What is thermal drift? How does it affect the performance of an op-amp circuit?

5-8. What is error voltage? How can it be reduced?

5-9. Define supply voltage sensitivity. What is meant by a poorly regulated power supply?

5-10. What is an electrical noise? What precautions can be taken to minimize the effect of noise on an op-amp circuit?

5-11. Define the common-mode rejection ratio (CMRR) and explain the significance of a relatively large value of CMRR.

PROBLEMS

5-1. For the inverting amplifier shown in Figure 5-7(b), if $R_1 = 100\ \Omega$ and $R_F = 4.7\ \text{k}\Omega$, what is the maximum possible output offset voltage V_{oo}? The op-amp is an LM307 with $V_{io} = 10$ mV and supply voltages ± 15 V.

5-2. Repeat Problem 5-1 if the LM307 op-amp is replaced by a μA715 for which $V_{io} = 5$ mV maximum.

5-3. Design an input offset voltage compensating network for the μA715 op-amp of Problem 5-2. Draw the complete circuit diagram.

5-4. The μA715 op-amp is to be used as a voltage follower with the offset-voltage compensating network designed in Problem 5-3 (see Figure 5-12). Compute the output voltage v_o if the input is 1 V pp sine wave.

5-5. **(a)** For the noninverting amplifier of Figure 5-7(a), $R_1 = 100\ \Omega$ and $R_F = 10\ \text{k}\Omega$. Determine the maximum possible output offset voltage due to (1) the input offset voltage V_{io} and (2) the input bias current I_B. The op-amp is an LM307 with $V_{io} =$ 10 mV and $I_B = 300$ nA.

(b) What value of R_{OM} is needed to reduce the effect of input bias current I_B?

5-6. For the inverting amplifier in Figure 5-21, determine the maximum output offset voltage $V_{oI_{io}}$ if the op-amp is an LM307 with $I_{io} = 70$ nA.

5-7. Compute the maximum possible total output offset voltage V_{ooT} in the amplifier circuits shown in Figure 5-22. Assume that the op-amp is an LM307 with supply voltages ± 15 V.

5-8. Repeat Problem 5-7 if the op-amp is a μA741.

5-9. Refer to the inverting amplifier in Figure 5-24. The op-amp is a type 307 with the following specifications:

$$\frac{\Delta V_{io}}{\Delta T} = 30\ \mu\text{V}/^\circ\text{C maximum}$$

$$\frac{\Delta I_{io}}{\Delta T} = 0.3\ \text{nA}/^\circ\text{C maximum}$$

$$V_S = \pm 15\ \text{V}$$

$$R_1 = 1\ \text{k}\Omega,\ R_F = 100\ \text{k}\Omega,\ \text{and}\ R_L = 10\ \text{k}\Omega$$

Assume that the amplifier is nulled at 25°C. Calculate the value of the error voltage and the output voltage V_o at 35°C if

(a) $V_{\text{in}} = 1$ mV dc

(b) $V_{\text{in}} = 10$ mV dc

5-10. Repeat Problem 5-9 with the type 307 op-amp configured as a noninverting amplifier. Refer to Figure 5-26.

5-11. The LM312 op-amp is used as an inverting amplifier, as shown in Figure 5-24, with the following specifications:

$$\frac{\Delta V_{io}}{\Delta T} = 30\ \mu\text{V}/^\circ\text{C}, \qquad \frac{\Delta I_{io}}{\Delta T} = 10\ \text{nA}/^\circ\text{C}, \qquad V_S = \pm 15\ \text{V}$$

$$R_1 = 100\ \Omega, \qquad R_F = 8.2\ \text{k}\Omega, \qquad R_L = 10\ \text{k}\Omega$$

Assume that the amplifier is nulled at 25°C. If v_{in} is 20 mV peak sine wave at 100 Hz:

(a) Calculate E_V and v_o values at 45°C.

(b) Draw the output voltage waveform at 25°C and at 45°C.

5-12. Repeat Problem 5-11 with the same op-amp reconnected as a noninverting amplifier.

5-13. Referring to Figure 5-24, suppose that the circuit is nulled when the voltage across terminals $+V_{CC}$ and $-V_{EE}$ measures 24 V dc. Because of poor regulation, the voltage across terminals $+V_{CC}$ and $-V_{EE}$ varies with time from 20 to 28 V. Determine:

(a) The change in the output offset voltage caused by the change in dc supply.

(b) The total output voltage V_O if $V_{in} = 5$ mV dc. The op-amp is the MC1741, and $R_1 = 100\ \Omega$ and $R_1 = 4.7\ k\Omega$.

5-14. Repeat Problem 5-13 with $R_F = 10\ k\Omega$ and $v_{in} = 10$ mV peak sine wave at 100 Hz Also draw the output waveform.

5-15. Repeat Problem 5-13 if the same op-amp is reconnected as the noninverting amplifier shown in Figure 5-26. Use the same values for R_1 and R_F.

5-16. Referring to Figure 5-26, suppose that the circuit is nulled when the voltage across terminals $+V_{CC}$ and $-V_{EE}$ measures 30 V dc. Also suppose that because of poor filtering, 10 mV rms ac ripple is measured across the terminals $+V_{CC}$ and $-V_{EE}$. While the input signal $v_{in} = 0$ V, how much ripple voltage can we except at the output if the op-amp is an LM312 with SVRR = 96 dB typical? Use $R_1 = 1\ k\Omega$, and $R_F = 47\ k\Omega$.

5-17. Determine the maximum possible change in output offset voltage in the circuit of Figure 5-24 after 6 months if the op-amp is the LH0041C with the following specifications:

$$\frac{\Delta V_{io}}{\Delta t} = 5\ \mu V/\text{week} \qquad \text{and} \qquad \frac{\Delta I_{io}}{\Delta t} = 2\ nA/\text{week}$$

$$R_1 = 82\ \Omega \qquad \text{and} \qquad R_F = 8.2\ k\Omega$$

Assume that the circuit is initially nulled and that room temperature and the voltage across terminals $+V_{CC}$ and $-V_{EE}$ remain constant.

5-18. Repeat Problem 5-17 with the op-amp LH0041C configured as shown in Figure 5-26. Use the same specifications as given in Problem 5-17.

5-19. In the circuit of Figure 5-32, if $R_1 = R_2 = 1\ k\Omega$, $R_F = R_3 = 47\ k\Omega$, $v_d = 10$ mV sine wave at 1 kHz, and $v_{ni} = 1$ mV at 60 Hz, calculate:
(a) The output voltage at 1 kHz.
(b) The amplitude of the induced 60-Hz noise at the output.
The op-amp is the μA741 with CMRR (dB) = 90 dB.

5-20. Repeat Problem 5-19 if the μA741 op-amp is replaced by a LM312 for which CMRR (dB) = 100 dB typical.

5-21. In the circuit of Figure 5-33(b), $R_1 = 100\ \Omega$, $R_F = 4.7\ k\Omega$, and the op-amp is μA741 with CMRR (dB) = 90 dB. If the amplitude of the induced 60-Hz noise at the output is 5 mV (rms), calculate the amplitude of the common-mode input voltage v_{cm}.

5-22. Repeat Problem 5-21 with the op-amp LM312. Use CMRR (dB) = 100 dB.

EXPERIMENT 5-1
MEASURING AND SETTING OFFSET VOLTAGE

Objectives

In this experiment you will measure the output offset voltage and then use a "nulling" circuit to set it to zero. Also, you will investigate how an output offset voltage may vary under certain conditions. At the end of this experiment you should be able to:

1. Measure the output offset voltage and then calculate the value of input offset voltage from it.
2. Design an offset voltage compensating network.
3. Set the output offset voltage to zero using a "nulling" circuit.

4. Discuss the effect of feedback and change in supply voltages on the output offset voltage and in turn on input offset voltage.

Equipment

1. Variable low-voltage power supply
2. Multimeter

Materials

1. 741 op-amp
2. 307 op-amp
3. 100-kΩ resistor
4. 10-kΩ resistor
5. 1-kΩ resistor
6. Assortment of resistors chosen by the experimenter

Procedure

1. Connect the circuit as shown in Figure E5-1.1(a) using the 741 op-amp.
2. Turn the dc supply voltage (±15 V) on. With the multimeter set to the dc voltage mode, measure the output offset voltage V_{oo} and record it in Table E5-1.1.
3. Measure the 1-kΩ and 100-kΩ resistor values and then connect the circuit using the 741 op-amp as shown in Figure E5-1.1(b). Repeat step 2.

TABLE E5-1.1

Op-amp	V_{oo} in open-loop configuration	V_{oo} in closed-loop configuration
741		
307		

TABLE E5-1.2

Op-amp	Supply voltage, $\pm V_S$ (V)	Output offset voltage, V_{oo}	Input offset voltage, V_{io}
741	±6 ±9 ±12 ±15		
307	±6 ±9 ±12 ±15		

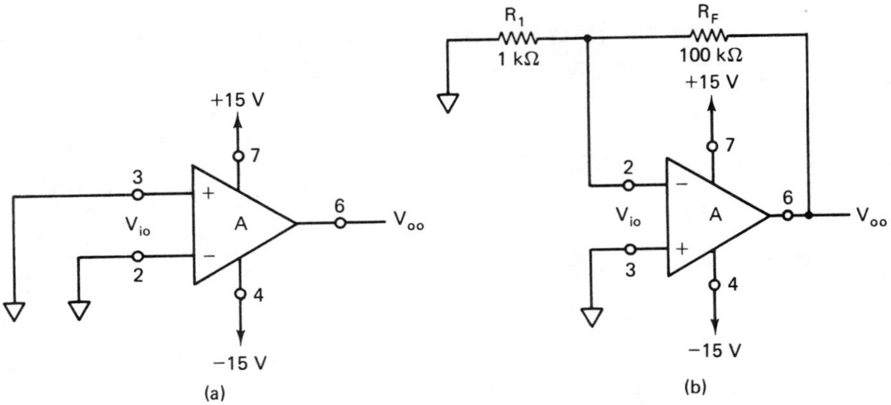

Figure E5-1.1 (a) Offset voltage in an open-loop op-amp circuit. (b) Offset voltage in a closed-loop op-amp circuit.

4. Change the supply voltage ($\pm V_S$) from ± 6 V through ± 15 V and measure the corresponding output offset voltage V_{oo}. Record these values in Table E5-1.2.
5. Calculate the magnitude of input offset voltage V_{io} from Equation (5-1.1). For maximum accuracy use the measured values of R_1 and R_F.

$$V_{io} = \frac{(R_1)V_{oo}}{(R_1 + R_F)} \qquad (5\text{-}1.1)$$

Complete Table E5-1.2.

6. Next connect the 10-kΩ potentiometer as shown in Figure E5-1.2. Note that the supply voltage is set to ± 15 V. Connect the voltmeter at the output terminal and vary the potentiometer until the meter reads "zero" volts. (*Note:* Do not expect exact zero output since it may be difficult to obtain.) However, for maximum accuracy be certain to use the appropriate meter range.
7. Repeat steps 1 through 5 using the 307 op-amp.

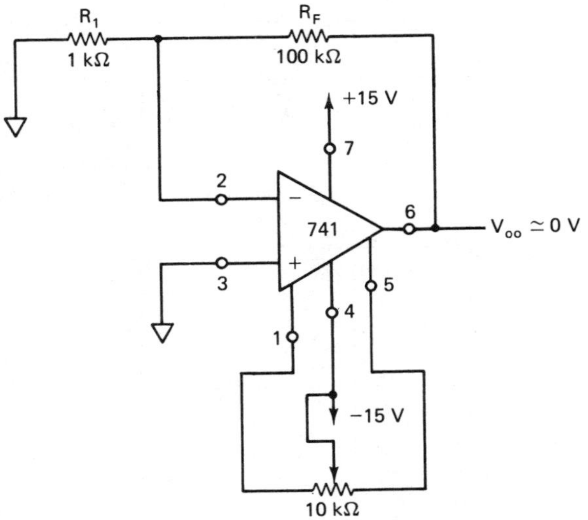

Figure E5-1.2 Offset-null arrangement of the 741 op-amp.

8. For the 307 op-amp, design an offset voltage compensating network using the supply voltage of ± 15 V and the corresponding V_{io} value.
9. Construct the compensated op-amp circuit as shown in Figure E5-1.3. Connect the voltmeter at the output terminal and adjust the R_a potentiometer for "zero" output voltage. Remember that you may not be able to force the output voltage to exactly zero.

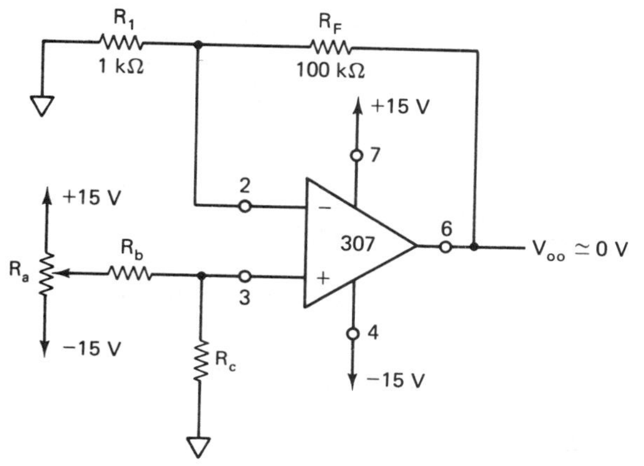

Figure E5-1.3 Op-amp with offset voltage compensating network.

Questions

1. Compare the V_{oo} value measured in open-loop configuration with that in closed-loop configuration and give the reason(s) for this difference.
2. Why is input offset voltage not directly measured between the two input terminals?
3. Based on the input offset voltages obtained (at ± 15 V) for the 741 and 307, which op-amp is better balanced? Explain.
4. Discuss briefly the effect of feedback and change in supply voltage on the input/output offset voltage.
5. Referring to the equivalent circuit of the 741 op-amp, discuss why nulling is possible with offset null circuitry (10-kΩ potentiometer).
6. Which one of the two offset null circuits is easier to use in setting the output offset to zero? Explain.

EXPERIMENT 5-2
USING THE OFFSET VOLTAGE COMPENSATING NETWORK

Objectives

In this experiment you will investigate the versatility of a compensating network. At the end of this experiment you should be able to:

1. Measure the output offset voltage in a noninverting amplifier, differential amplifier, and voltage-follower circuits.

2. Set the output voltage in the circuits above to zero using the same compensating network.

Equipment

1. Low-voltage power supply
2. Multimeter

Materials

1. 307 op-amp
2. Two 100-kΩ resistors
3. Two 1-kΩ resistors
4. Assortment of resistors chosen by the experimenter

Procedure

1. Construct the noninverting amplifier of Figure E5-2.1. Measure the output offset voltage and record it in Table E5-2.1.
2. Connect the compensating network shown in Figure E5-2.2 using the same component values for R_a, R_b, and R_c as in Experiment 5-1. Next, in Figure E5-2.1 disconnect the left-side terminal of R_1 from ground and reconnect it to point C of the compensating network of Figure E5-2.2.

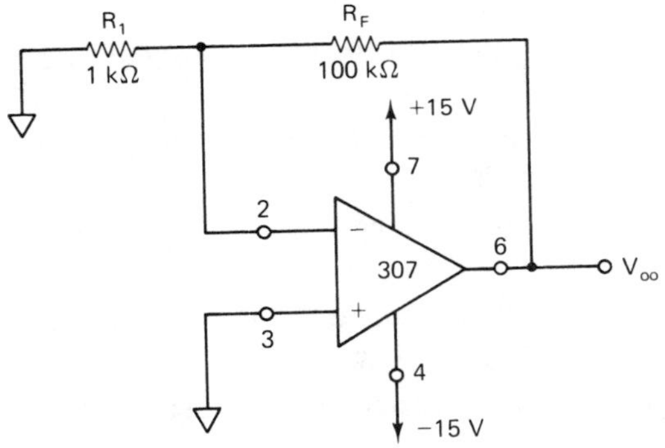

Figure E5-2.1 Offset voltage in a noninverting amplifier ($v_{in} = 0$ V).

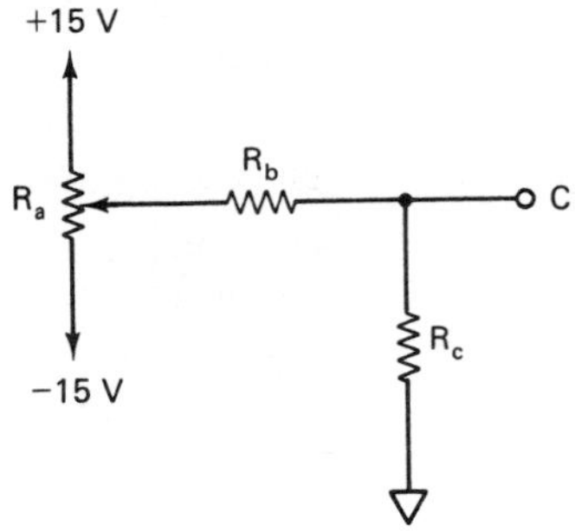

Figure E5-2.2 Offset voltage compensating network.

3. With the voltmeter connected at the output terminal of the op-amp, vary the potentiometer R_a until output voltage is zero. Remember that you may not be able to force output voltage to exactly zero.
4. Next, connect the voltage follower shown in Figure E5-2.3. Measure output offset voltage with the voltmeter and record it in Table E5-2.1.
5. Using the compensating component values obtained in Experiment 5-1, reconnect the voltage-follower circuit as shown in Figure E5-2.4.

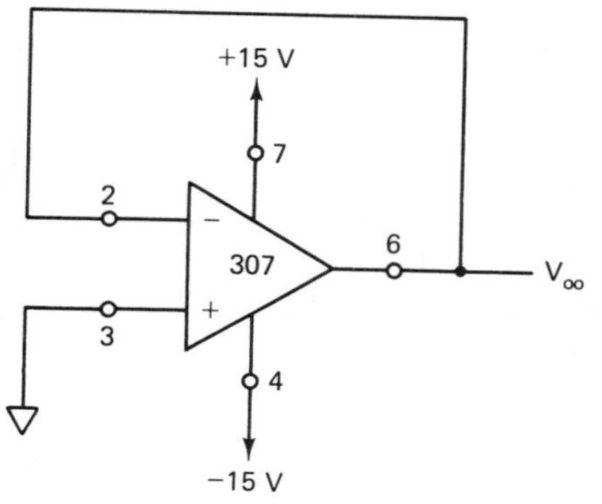

Figure E5-2.3 Offset voltage in voltage follower ($v_{in} = 0$ V).

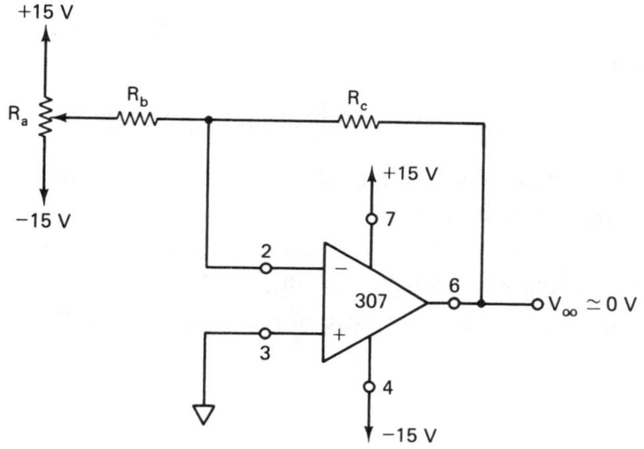

Figure E5-2.4 Compensated voltage follower ($v_{in} = 0$ V).

6. With the voltmeter connected at the output, vary the potentiometer R_a until output voltage is reduced to zero.
7. Connect the differential amplifier of Figure E5-2.5 and measure the output offset voltage. Record the measured value in Table E5-2.1.
8. Do the following modifications in Figure E5-2.5: Disconnect the lower end of R_3 and reconnect it to point C of the compensating network of Figure E5-2.2.
9. With the voltmeter connected at the output terminal, vary the potentiometer R_a until the output voltage is reduced to zero.

TABLE E5-2.1

Configuration	V_{oo}(V)
Noninverting emplifier	
Voltage follower	
Differential amplifier	

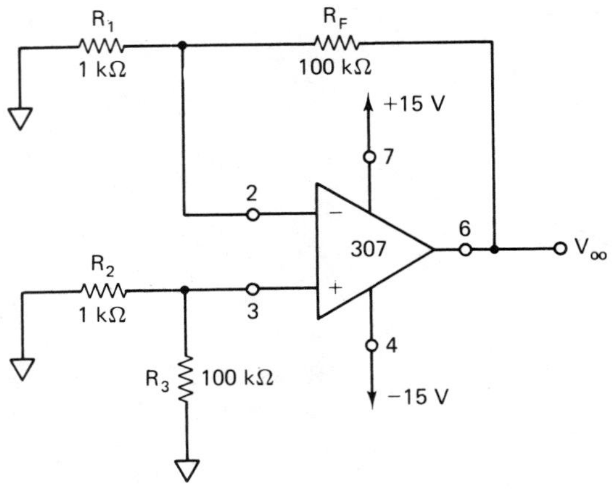

Figure E5-2.5 Offset voltage in a differential amplifier.

Questions

From the experimental data, comment on the following:

1. Is the output offset voltage the same in noninverting and differential amplifier circuits? Why or why not?
2. Does the output offset voltage in the noninverting amplifier equal that in the voltage follower? Briefly explain your answer.
3. Comment on the importance of the compensating network in op-amp circuits.

6

Frequency Response of an Op-Amp

6-1 INTRODUCTION

As we have seen in Chapter 3, some of the characteristics of the op-amp are frequency dependent. One of the most important of these characteristics is the open-loop voltage gain A. The gain A decreases as the operating frequency increases. This variation in gain as a function of frequency imposes a limitation not only on the performance of the op-amp but also on its use in ac applications. This chapter investigates the factors responsible for variations in open-loop gain as a function of frequency. It will also discuss the differences between internally and externally compensated op-amps and the effect of negative feedback on variations in open-loop gain. The chapter concludes with another important frequency-dependent parameter: slew rate and its effect on op-amp applications.

6-2 FREQUENCY RESPONSE

Up to now we have treated the gain of the op-amp as a constant. However, it is a complex number which is a function of frequency. Therefore, at a given frequency the gain will have a specific magnitude as well as a phase angle. This means that the variation in operating frequency will cause the variation in gain magnitude and its phase angle. The manner in which the gain of the op-amp *responds* to different frequencies is called the *frequency response*. A graph of the magnitude of the gain versus frequency is called a *frequency response plot*. This plot is included on the data sheets of most of the op-amps. Although gain magnitude may be expressed either in decibels (dB) or as a numerical value, the frequency is always plotted on a logarithmic scale. Remember that in order to accommodate large frequency ranges, the frequency is assigned a logarithmic scale, just as the gain magnitude is expressed in dB to accommodate very high gain, of the order of 10^5 or higher. The frequency response for the amplifiers is obtained from the experimental results by measuring its input and output voltages at different frequencies.

Another technique used in the ac analysis of networks is the *Bode plot*, composed of magnitude versus frequency and phase angle versus frequency plots. In magnitude versus frequency plots the magnitude is always taken in dB. Although both frequency response and Bode plots indicate the effect of frequency variation on gain, the Bode plot is basically an approximate method and is generally used for stability determination and network design.

6-3 COMPENSATING NETWORKS

As the operating frequency increases, the magnitude of the gain decreases, and the phase angle of the gain increases. That is, the phase shift of the output signal increases with respect to input signal, while its amplitude decreases with an increase in the operating frequency. This phase shift, which increases with frequency, has a cumulative effect on the phase shift produced by the use of the feedback circuit when the op-amp uses a feedback. Thus, for a given frequency, the total phase shift between the input and output signals is equal to the phase shift due to feedback network plus the phase shift due to the internal circuitry of the op-amp. If at any frequency total phase shift (feedback plus internal) becomes 360° and the magnitude of the loop gain is unity ($|AB| = 1$), the circuit may become unstable and break into oscillation. This condition is called positive feedback. Under this condition, even though the amplifier is configured for negative feedback, it is actually producing positive feedback. In fact, this is the most unreliable and unacceptable situation in amplifier applications.

One solution to this problem is to make sure that the total phase shift (phase of AB) never reaches 360° over the entire operating frequency range of the amplifier. This can be accomplished by using a circuit called a *compensating network*. These compensating networks control the phase shift and thus improve the stability of amplifier circuits. Compensating networks typically consist of capacitors and resistors. The *phase lag* and *phase lead* are the most commonly used compensating networks in op-amps. These two networks are indicative of their functions since phase lag contributes a negative phase angle and phase lead a positive phase angle.

In internally compensated op-amps the compensating network is designed into the circuit to control the phase shift of the op-amps. On the other hand, external (discrete) compensating components, namely resistors and/or capacitors, are added at designated terminals in noncompensated op-amps. Some of the op-amps need as many as four compensating components. For example, the μA725C requires two resistors and two capacitors. For proper operation the manufacturer recommends appropriate compensating components for the uncompensated op-amps. The 741C is an internally compensated op-amp, while the 709C is a noncompensated op-amp. The open-loop frequency response curves of these op-amps are shown in Figure 6-1.

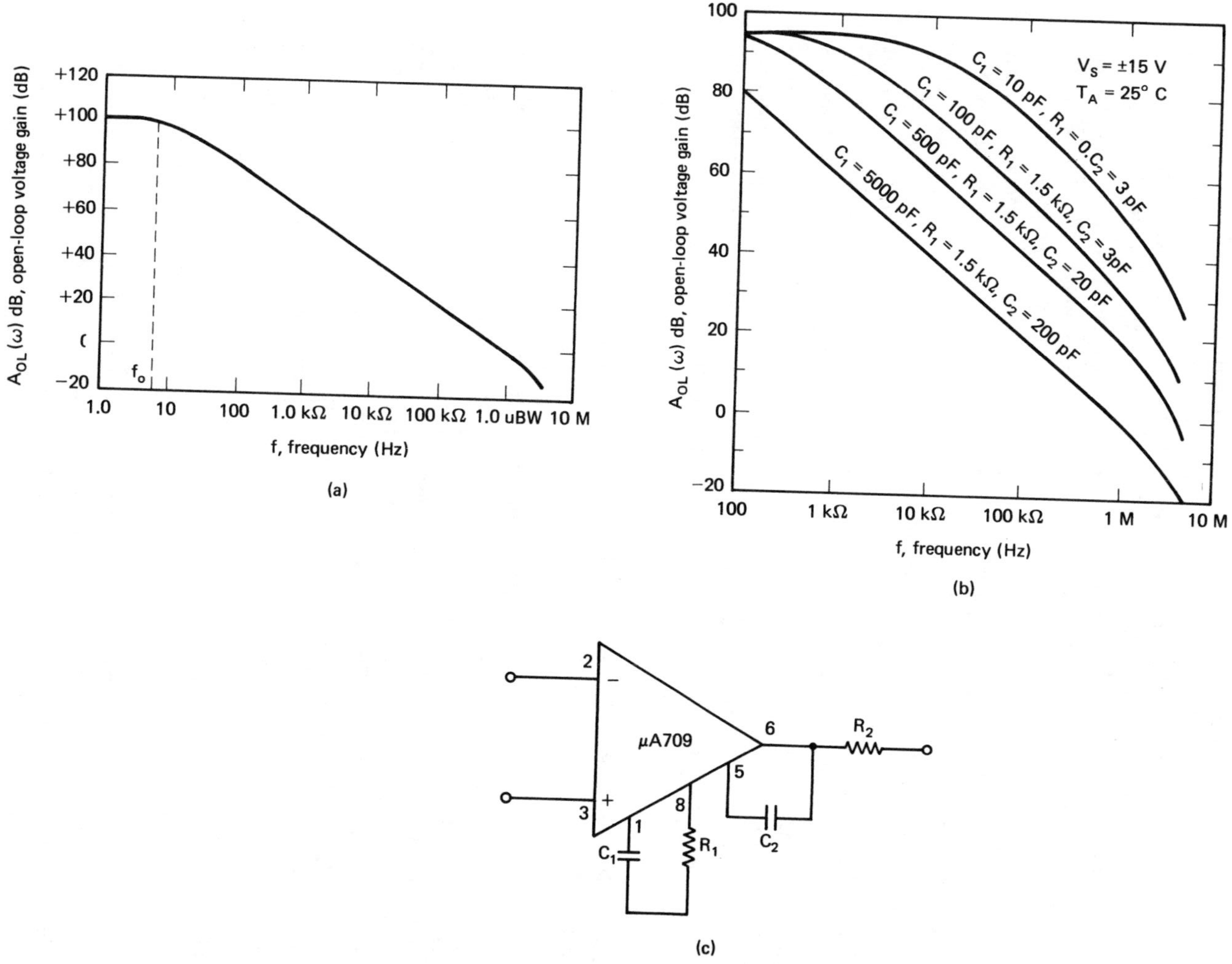

Figure 6-1 Frequency response of (a) internally compensated op-amp 741C, (b) noncompensated op-amp μA709, and (c) frequency compensation circuit of μA709. Use $R2 = 50\ \Omega$ when the amplifier is operated with capacitive loading.

6-4 FREQUENCY RESPONSE OF INTERNALLY COMPENSATED OP-AMPS

Let us examine the open-loop frequency response of the internally compensated op-amp. In the plot of Figure 6-1(a), in order to accommodate the entire operating range, the gain as well as the frequency are assigned logarithmic scales. In addition, the voltage gain is expressed in dB. The unity gain bandwidth of the 741C is approximately 1 MHz and has a single break frequency f_o before the unity gain bandwidth. Recall that at the break frequency f_o, the open-loop gain is down 3 dB or 0.707 of its value at 0 Hz (dc). The gain of the op-amp remains essentially constant from 0 Hz to the break frequency f_o, and thereafter rolls off at a constant rate, namely 20 dB per decade (tenfold increase in frequency) [see Figure 6-1(a)]. Thus the open-loop bandwidth is the frequency band extending from 0 Hz to f_o, or simply f_o. Hence the open-loop bandwidth of the 741C is approximately 5 Hz.

In the 741 op-amp a 30-pF capacitor is the internal compensating component (Figure 2-3), which helps to control the open-loop gain to allow it to roll off at a rate of 20 dB/decade. Therefore, regardless of the closed-loop gain used, for frequencies higher than the closed-loop break frequency, the gain will always roll off at a rate of 20 dB/decade. In fact, this roll-off rate assures us of stable circuit operation at any gain. The internally compensated op-amps are sometimes simply called *compensated* op-amps and generally have very small open-loop bandwidths.

6-5 FREQUENCY RESPONSE OF NONCOMPENSATED OP-AMPS

Op-amps requiring external compensating components are called *noncompensated* op-amps. They are sometimes called *tailored frequency response* op-amps because the user has to provide the compensation if it is needed to tailor the response. The open-loop gain characteristics of the noncompensated op-amp are altered by the compensating components as shown in Figure 6-1(b). The roll-off rate with various compensating components that are specified along the gain versus frequency curves is about 20 dB/decade. Note that the open-loop bandwidth of a 709C decreases from the outermost compensated curve to the innermost. That is, if $C_1 = 10$ pF, $R_1 = 0$, and $C_2 = 3$ pF, the bandwidth is approximately 5 kHz, while if $C_1 = 5000$ pF, $R_1 = 1.5$ kΩ, and $C_2 = 200$ pF, the bandwidth is 100 Hz, as shown in Figure 6-1(b). The frequency compensation circuit of the μA709 is shown in Figure 6-1(c). The uncompensated op-amps offer relatively broader open-loop bandwidths, as can be seen from Figure 6-1(a) and (b).

Next, we must obtain the general mathematical expression for the gain, primarily to confirm that it is a complex quantity and is a function of frequency. Then, to prove the validity of this equation, we need to construct the gain versus frequency plot from it. However, to derive the gain equation, a high-frequency op-amp model must first be developed.

6-6 HIGH-FREQUENCY OP-AMP EQUIVALENT CIRCUIT

What causes the gain of the op-amp to roll off after a certain frequency is reached? Obviously, there must be a capacitive component in the equivalent circuit of the op-amp since its reactance decreases as the frequency increases. This means that we have to include a capacitor in the high-frequency model of the op-amp at the output terminal. But how do we account for this capacitor? Two major sources are responsible for capacitive effects:

1. The physical characteristics of semiconductor devices. Recall that op-amps are composed of BJTs and FETs which contain junction capacitors. These junction capacitors are very small (of the order of picofarads) and act as open circuits at low frequencies but take finite values at higher frequencies. In fact, as frequency increases, the reactances of these capacitors decrease.
2. The internal construction of the op-amp is a second source of capacitive effects. In op-amps a number of transistors as well as resistors and sometimes a capacitor are integrated on the same material, called a *substrate*. In fact, the substrate acts as an insulator and helps to separate these components. The various components are connected by conducting paths, and the paths are separated by insulators. However, whenever two conducting paths are separated by an insulator, it acts as a capacitor. This means that because of its construction the op-amp may contain a number of such stray capacitors.

The cumulative effect of these capacitors due to the characteristics of semiconductor devices and the internal construction of the op-amp causes the gain to decrease as the frequency increases.

For an op-amp with only one break frequency we will represent all the capacitive effects by a single capacitor, as shown in Figure 6-2. This figure represents the high-frequency model of the op-amp with a single break frequency. Similarly, op-amps with more than one break frequency may be represented by using as many capacitors as the break frequencies they have.

Note that the high-frequency model of Figure 6-2 is a modified version of the

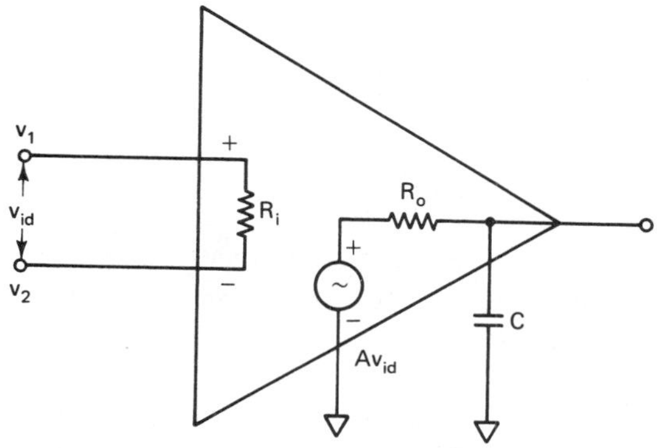

Figure 6-2 High-frequency model of an op-amp with single break frequency.

equivalent circuit of the op-amp presented in Chapter 2. Here we have simply added a capacitor C at the output.

6-7 OPEN-LOOP VOLTAGE GAIN AS A FUNCTION OF FREQUENCY

Let us now obtain an expression for the gain as a function of frequency. From Figure 6-2, using a voltage-divider rule, we get

$$v_o = \frac{-jX_C}{R_o - jX_C}(Av_{id})$$

Since $-j = 1/j$ and $X_C = 1/2\pi fC$,

$$v_o = \frac{1/j2\pi fC}{R_o + 1/j2\pi fC}(Av_{id})$$

$$= \frac{Av_{id}}{1 + j2\pi fR_oC}$$

Hence the open-loop voltage gain is

$$A_{OL} = \frac{A}{1 + j2\pi fR_oC}$$

Let $f_o = 1/2\pi R_oC$; then

$$A_{OL} = \frac{A}{1 + j(f/f_o)} \tag{6-1}$$

where A_{OL} = open-loop voltage gain
A = gain of the op-amp at 0 Hz (dc)
f = operating frequency (hertz)
f_o = break frequency of the op-amp (hertz)

Note that the break frequency f_o depends on the value of C and of output resistance R_o. Therefore, f_o is fixed for a given op-amp.

Equation (6-1) is important because it indicates that the open-loop gain of the op-amp A_{OL} is a complex quantity and is a function of operating frequency. Being complex in nature, the gain can be expressed in a polar form as follows. The open-loop gain magnitude is

$$A_{OL}(\omega) = \frac{A}{\sqrt{1 + (f/f_o)^2}} \tag{6-2a}$$

and phase angle

$$\phi(\omega) = -\tan^{-1}\left(\frac{f}{f_o}\right) \tag{6-2b}$$

where $\omega = 2\pi f$ and is used to indicate that the magnitude of gain and phase angle of gain are functions of frequency. Using Equation (6-2a), we can obtain a magnitude versus frequency plot, while the phase angle versus frequency plot can be obtained by using Equation (6-2b).

Before proceeding with the graphical representation of Equation (6-2), we will present a qualitative analysis of it. Examine first the magnitude Equation (6-2a). At 0 Hz the denominator is 1, and the open-loop gain A_{OL} is equal to A. In fact, for any frequency less than f_o the gain is approximately constant and is equal to A. However, at frequencies above f_o, the denominator value increases, causing the gain A_{OL} to decrease. Thus, as the frequency increases, the gain A_{OL} continues to drop.

As for the phase shift Equation (6-2b), at 0 Hz the phase shift between input and output voltage is zero. In fact, for any frequency below f_o, the absolute value of the phase shift is less than 45°. Above f_o the absolute value of the phase shift increases toward 90° with increases in frequency.

Now let us verify the validity of Equation (6-2) by using it to plot the Bode diagrams for the 741C op-amp. Let us begin by expressing magnitude Equation (6-2a) in dB:

$$20 \log A_{OL}(\omega) = 20 \log \left[\frac{A}{\sqrt{1 + (f/f_o)^2}} \right]$$

$$A_{OL}(\omega) \text{ dB} = 20 \log A - 20 \log \sqrt{1 + \left(\frac{f}{f_o}\right)^2} \tag{6-3}$$

Since $f_o \cong 5$ Hz and $A = 200{,}000$ for the 741C, let us substitute these values in Equation (6-3):

$$A_{OL}(\omega) \text{ dB} = 20 \log 0.2 \text{ M} - 20 \log \sqrt{1 + \left(\frac{f}{5}\right)^2}$$

We will now determine the gain in dB at different frequencies, including at f_o:

At $f = 0$ Hz, $\quad A_{OL}(\omega) \text{ dB} = 106.02 - 20 \log \sqrt{1 + \left(\frac{0}{5}\right)^2} \cong 106$

At $f = 5 \text{ Hz} = f_o$, $\quad A_{OL}(\omega) \text{ dB} = 106.02 - 3.01 \cong 103$

At $f = 50$ Hz, $\quad A_{OL}(\omega) \text{ dB} = 106.02 - 20.04 \cong 86$

At $f = 500$ Hz, $\quad A_{OL}(\omega) \text{ dB} = 106.02 - 40 \cong 66$

At $f = 5$ kHz, $\quad A_{OL}(\omega) \text{ dB} = 106.02 - 60 \cong 46$

At $f = 50$ kHz, $\quad A_{OL}(\omega) \text{ dB} = 106.02 - 80 \cong 26$

At $f = 100$ kHz, $\quad A_{OL}(\omega) \text{ dB} = 106.02 - 86.02 = 20$

At $f = 1$ MHz, $\quad A_{OL}(\omega) \text{ dB} = 106.02 - 106.02 = 0$

From the gain calculations above the following observations can be made:

1. The open-loop gain $A_{OL}(\omega)$ dB is approximately constant from 0 Hz to the break frequency f_o.
2. When the input signal frequency f is equal to the break frequency f_o, the gain $A_{OL}(\omega)$ dB is 3 dB down from its value at 0 Hz. For this reason the break frequency is sometimes called the -3 *dB frequency*. It is also known as the *corner frequency*.
3. The open-loop gain $A_{OL}(\omega)$ dB is approximately constant up to the break frequency f_o, but thereafter it decreases 20 dB each time there is a tenfold increase (one decade) in frequency. Therefore, it may be said that the gain rolls off at the rate of 20 dB/decade. Stated another way, the gain rolls off at the rate of 6 dB/octave, where *octave* represents a twofold increase in frequency. This can

be seen from the difference in gain at 50 and 100 kHz. Thus we can decsribe the roll-off rate as either −6 dB/octave or −20 dB/decade.

4. Finally, at some specific value of the input signal frequency, the open-loop gain $A_{OL}(\omega)$ dB is zero. This specific frequency is called the *unity gain bandwidth.* Other equivalent terms for it are *gain–bandwidth product, closed-loop bandwidth, small-signal bandwidth,* and *unity gain crossover frequency.* Remember that the unity gain bandwidth is equal to the gain bandwidth product only if an op-amp has a single break frequency before the unity gain bandwidth. This is precisely the case for op-amps like 741. The unity gain bandwidth of the 741C is approximately 1 MHz and is verified by the decibel gain calculation at $f = 1$ MHz. Note that a gain of 0 dB is the same as a numerical gain of 1.

The magnitude of the open-loop gain in dB versus frequency plot is shown in Figure 6-3(a). Note that this plot is almost a duplicate of the frequency response in Figure 6-1(a), which proves the validity of the gain Equation (6-2a). Thus if we know the break frequency and the gain of the op-amp, it is possible to plot the frequency response using Equation (6-2a).

Next, we will obtain the data for the phase angle versus frequency curve using Equation (6-2b). The phase angles may be computed at the same frequencies as were used to calculate the open-loop voltage gain A_{OL} in dB.

$$\phi(\omega) = -\tan^{-1}\left(\frac{f}{5}\right) \tag{6-2b}$$

At $f = 0$ Hz, $\phi(\omega) = -\tan^{-1}\left(\frac{0}{5}\right) = 0°$

At $f = f_o = 5$ Hz, $\phi(\omega) = -\tan^{-1}\left(\frac{5}{5}\right) = -45°$

At $f = 50$ Hz, $\phi(\omega) = -84.29°$

At $f = 500$ Hz, $\phi(\omega) = -89.43°$

At $f = 5$ kHz, $\phi(\omega) = -89.94°$

At $f = 50$ kHz, $\phi(\omega) = -89.99°$

At $f = 100$ kHz, $\phi(\omega) = -90°$

At $f = 1$ MHz, $\phi(\omega) = -90°$

The phase angle versus frequency plot is shown in Figure 6-3(b). From this plot it is obvious that the phase angle of the op-amp with a single break frequency varies between 0 and −90°. In other words, the phase shift between the input and output voltages is 90° maximum; that is, the output voltage lags the input voltage by 90°.

Remember that to implement gain Equation (6-1), we must know the dc gain A and break frequency f_o. However, the break frequency is obtained only from the frequency response. In other words, a frequency response plot is necessary to write the gain equation. In short, using a frequency response plot, you can obtain all the information pertinent to the gain of the op-amp as a function of frequency.

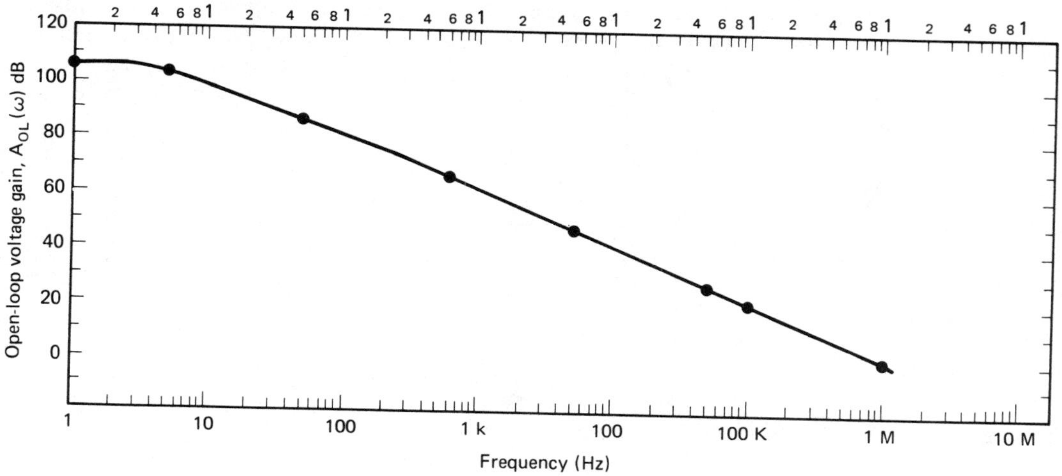

(a)

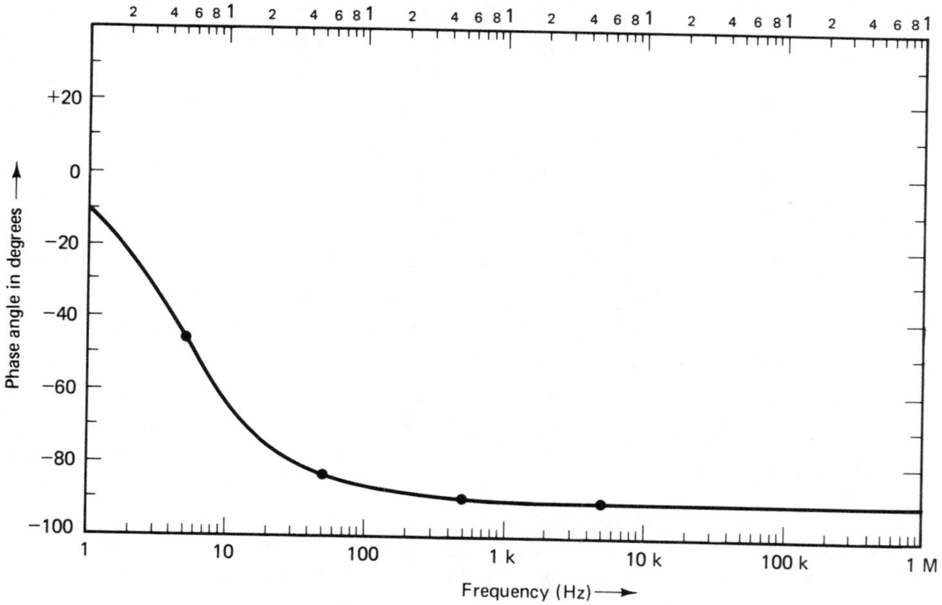

Figure 6-3 (a) Gain in dB versus frequency. (b) Phase angle versus frequency.

Example 6-1

The 741C is connected as a noninverting amplifier. What maximum gain can be used that will still keep the amplifier's response flat to 10 kHz?

Solution. The gain that will produce a flat response up to approximately 10 kHz can be obtained from the frequency response curve of the 741C. To do this, project a 10-kHz frequency vertically on the curve as shown in Figure 6-4 and then read the corresponding value of the gain on the vertical axis. This value of gain is equal to 40 dB, that is 100.

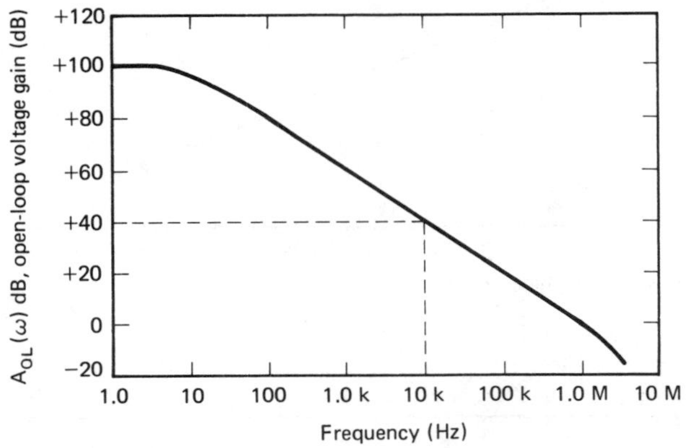

Figure 6-4 741C frequency response plot for Example 6-1.

Up to this point we have been studying op-amps with only one break frequency. Actually, the op-amp has more than one break frequency because there are quite a few capacitors present, as pointed out in Section 6-6. Often, the upper break frequencies are well above the unity gain bandwidth and are, therefore, not shown in the frequency response plot. Since each break frequency contributes at the most a 90° phase shift, the presence of multiple break frequencies should, therefore, be evident in the phase response plot. One such plot is shown in Figure 6-5(b). Our major concern is the number of break frequencies before the unity gain bandwidth because (1) we seldom use an op-amp above its unity gain bandwidth, and (2) a stable circuit always requires a phase angle less than 180°.

Using the concepts developed for a single break frequency op-amp, we can write the gain equation for the multiple break frequency op-amp, as shown in Example 6-2.

Example 6-2

Using the frequency response and phase response curves shown in Figure 6-5, obtain the gain equation for the MC1556 op-amp. Also determine the approximate values of the break frequencies.

Solution. Figure 6-5(b) shows that the phase shift is almost −180° at 10 MHz. This means that two break frequencies must be present, since each break frequency contributes at the most −90°.

Therefore, the gain equation can be written as

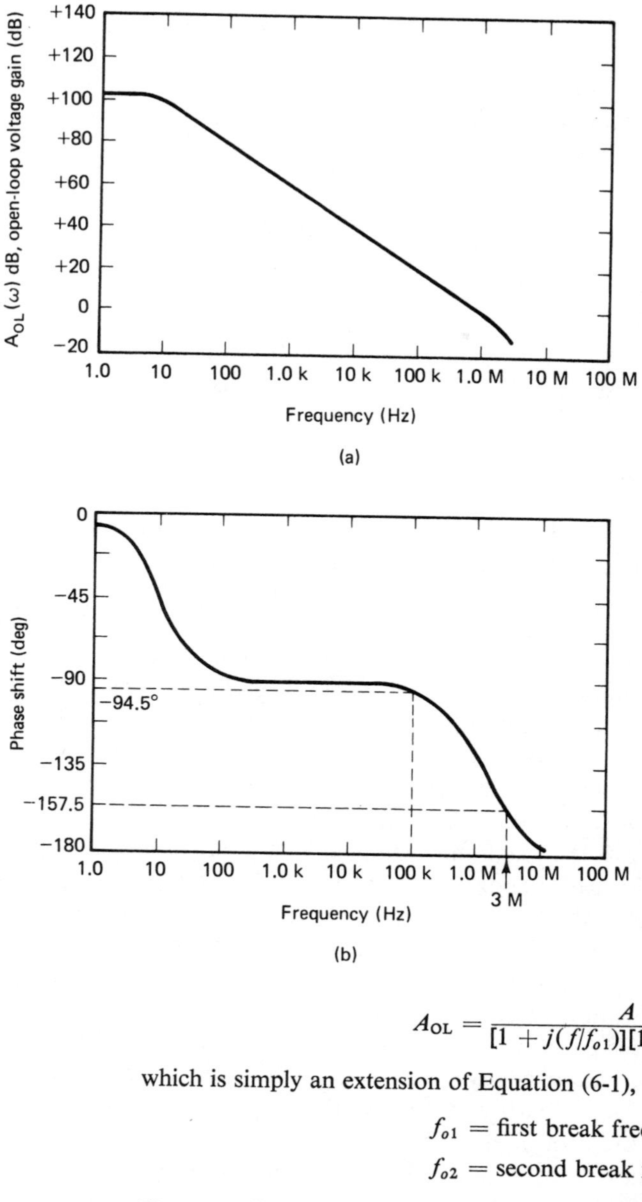

Figure 6-5 (a) Frequency response and (b) phase response curves of the MC1556. (Courtesy of Motorola Semiconductor.)

$$A_{OL} = \frac{A}{[1 + j(f/f_{o1})][1 + j(f/f_{o2})]} \tag{6-4a}$$

which is simply an extension of Equation (6-1), where

$$f_{o1} = \text{first break frequency}$$

$$f_{o2} = \text{second break frequency}$$

However, from the graph of Figure 6-5(a) the first break frequency is approximately at 6 Hz. The value of the second break frequency can be computed by substituting a specific known value of the phase shift in a phase angle equation. The equation is

$$\phi(\omega) = -\tan^{-1}\left(\frac{f}{f_{o1}}\right) - \tan^{-1}\left(\frac{f}{f_{o2}}\right) \tag{6-4b}$$

The phase shift is $-157.5°$ at about 3 MHz [see Figure 6-5(b)]. Substituting these values in the equation above, we get

$$-157.5 = -\tan^{-1}\left(\frac{3\text{ MHz}}{6\text{Hz}}\right) - \tan^{-1}\left(\frac{3\text{ MHz}}{f_{o2}}\right)$$

$$\tan^{-1}\left(\frac{3\text{ MHz}}{f_{o2}}\right) = 67.50$$

$$\frac{3\text{ MHz}}{f_{o2}} = \tan(67.50)$$

$$f_{o2} = \frac{3\text{ MHz}}{\tan(67.50)} = 1.24\text{ MHz}$$

Thus the gain equation with break frequency values is as follows:

$$A_{OL} = \frac{140{,}000}{[1 + j(f/6)][1 + j(f/1.24\text{ MHz})]}$$

where $A \cong 140{,}000$ or 103 dB, $f_{o1} = 6$ Hz, and $f_{o2} = 1.24$ MHz.

6-8 CLOSED-LOOP FREQUENCY RESPONSE

So far the discussion has been centered around the nature of the open-loop frequency response. Since the op-amp is generally used in a closed-loop configuration, a study of the closed-loop performance of the amplifier is in order.

We saw that the open-loop gain A_{OL} is constant only up to the first break frequency. Therefore, the bandwidth of the op-amp is simply the first break frequency f_{o1}; that is, f_{o1} is the maximum useful frequency of the open-loop op-amp. But this bandwidth f_{o1} is very small, and therefore the open-loop configuration is of little use practically. To increase the bandwidth of an op-amp and thus make it a versatile device, a negative feedback must be used. Ideally, of course, we expect the bandwidth to be infinite so that the gain is the same at all frequencies.

As shown in Chapter 4, the maximum bandwidth of a single break frequency op-amp is equal to its unity gain bandwidth. Recall that the 741C has a *unity gain bandwidth* of approximately 1 MHz. This means that for 741C the product of the coordinates (gain and frequency) of any point beyond the break frequency (5 Hz) on the open-loop frequency response curve is about 1 MHz. That is, given a voltage gain, the closed-loop bandwidth can be determined using a frequency response curve. For instance, if the 741C is wired for a gain of 100 or 40 dB, its bandwidth will be about 10 kHz. This bandwidth is obtained by projecting to the right on the open-loop frequency response curve from 40 dB and reading the frequency on the horizontal axis corresponding to the point of intersection (see Figure 6-4).

Often, the manufacturers' data sheets give the frequency response for various closed-loop gains. For example, Figure 6-6 shows the frequency response of 709C for various closed-loop gains using recommended compensation networks. Obviously, if high gain and relatively wide bandwidth are required, the compensating components for curve 1 should be used. Note that the closed-loop frequency response curves of Figure 6-6 can be obtained from the open-loop frequency response of Figure 6-1(b) by using the same procedure as stated above for the 741C (see Figure 6-4).

Remember that besides using the graphical method, depending on the amplifier configuration, Equation (4-10) or (4-21) can also be used to compute closed-loop bandwidth.

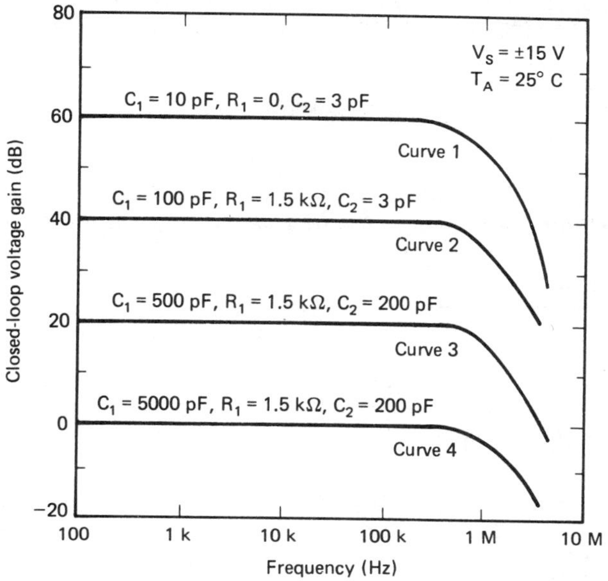

Figure 6-6 Frequency response of the μA709C for various closed-loop gains. (Courtesy of Fairchild Camera and Instrument Corporation.)

6-9 CIRCUIT STABILITY

According to feedback theory, the circuit is said to be stable if the magnitude of its loop gain $A_{OL}B$ (open-loop gain $A_{OL} \times$ gain of the feedback circuit B) is less than unity when the phase angle reaches 180°. That is,

$$[|A_{OL}B|]\,|_{\phi(\omega)=180°} < 1 \tag{6-5a}$$

where A_{OL} = open-loop voltage gain
B = gain of the feedback circuit
$\phi(\omega)$ = phase angle of the loop gain $A_{OL}B$

In other words, the phase angle of the loop gain $A_{OL}B$ must be less than 180° when its magnitude is unity. Thus

$$[\phi(\omega)]\,|_{|A_{OL}B|=1} < 180° \tag{6-5b}$$

One method of determining circuit stability is to use the open-loop frequency and phase response curves. To determine the stability of an amplifier with a resistive feedback network, the frequency corresponding to the closed-loop gain $|A_F| = 1/B$ is found on the open-loop frequency response curve. The phase shift at this frequency is then read on the phase response curve. Note that at this frequency the magnitude of the loop gain $A_{OL}B$ is 1. Therefore, if the phase shift is less than 180°, the circuit is stable. If the phase shift is more than 180°, the circuit is unstable. The same rule holds for nonresistive feedback amplifiers except that the phase shift of both the open-loop op-amp and the feedback network must be considered.

Example 6-3

The MC1556 is wired up as a noninverting amplifier with a gain of 10. Determine the stability of the amplifier using the frequency response and the phase response curves shown in Figure 6-5.

Solution. From the frequency response of Figure 6-5(a), the frequency corresponding to the closed-loop gain of 10 (20 dB) is 100 kHz. Note that at 100 kHz, $A_{OL} = A_F$; that is,

$$\left|\frac{A_{OL}}{A_F}\right| = 1 = |A_{OL}B|$$

(see Figure 6-7).

From Figure 6-5(b), the phase shift at 100 kHz is about $-94.5°$. Since $-94.5° < 180°$, the amplifier is stable. Remember that the phase shift due to the feedback network is zero because the network is purely resistive.

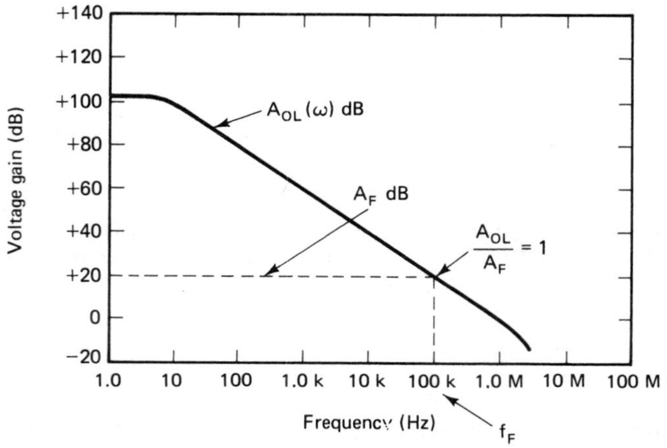

Figure 6-7 Frequency response plot showing $A_{OL} = A_F$ or $A_{OL}/A_F = 1$ at $f = f_F = 100$ kHz.

6-10 SLEW RATE

Another important frequency related parameter of an op-amp is the slew rate. The slew rate, as mentioned in Chapter 3, is the maximum rate of change of output voltage with respect to time, usually specified in V/μs. For example, a 1 V/μs slew rate means that the output rises or falls no faster than 1 V every microsecond. Ideally, we would like an infinity slew rate so that the op-amp's output voltage would change simultaneously with the input. Practical op-amps are available with slew rates from 0.1 V/μs to well above 1000 V/μs. The National Semiconductor LH0063C has a slew rate of 6000 V/μs.

Generally, the slew rate is specified for unity gain and is measured by applying a step input (dc) voltage. Slew rate is sometimes given indirectly in data sheets as *output voltage swing* as a *function of frequency* or as *voltage follower large-signal pulse response* (see Figure 6-8). Often, the slew rate improves with higher closed-loop gains and dc supply voltages, as shown in Figure 6-9(a) and (b). The slew rate is also a function of temperature and generally decreases with an increase in temperature, as shown by Figure 6-9(c). The slew rate of some op-amps is improved when they are wired with *feedforward compensation.* The 101 and 108 are such op-amps, the compensation connections for which are given in the data sheets.

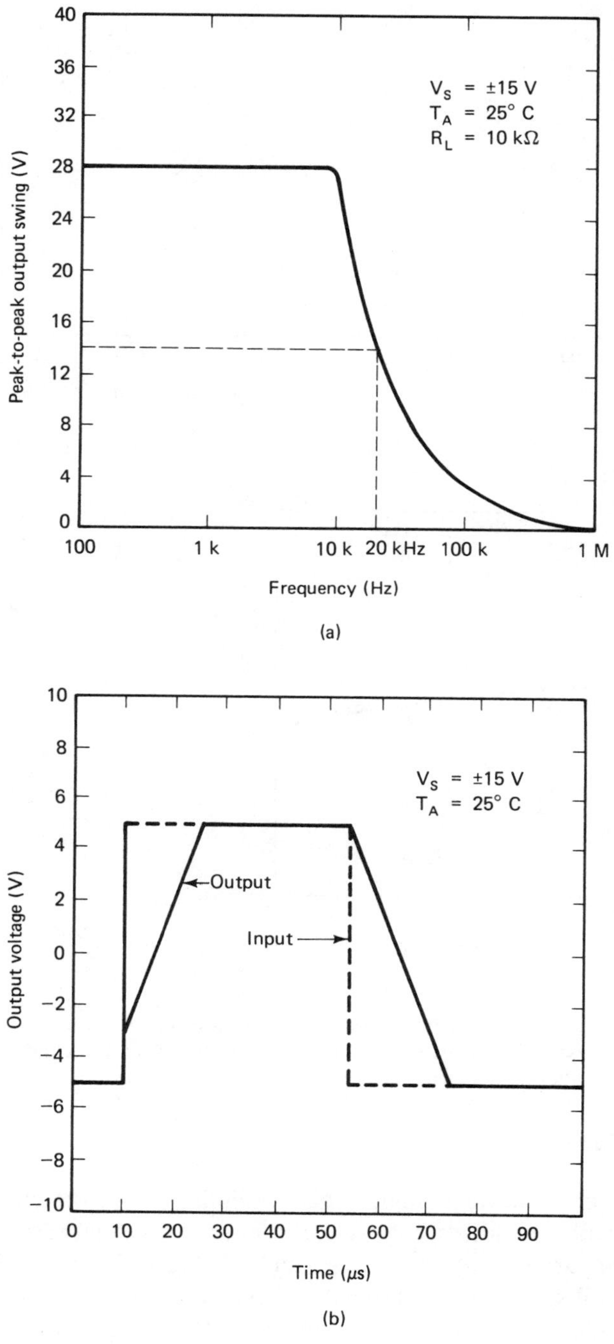

Figure 6-8 Specifying slew rate indirectly for the 741C as (a) output voltage swing versus frequency and (b) voltage follower large signal pulse response. (Courtesy of Fairchild Camera and Instrument Corporation.)

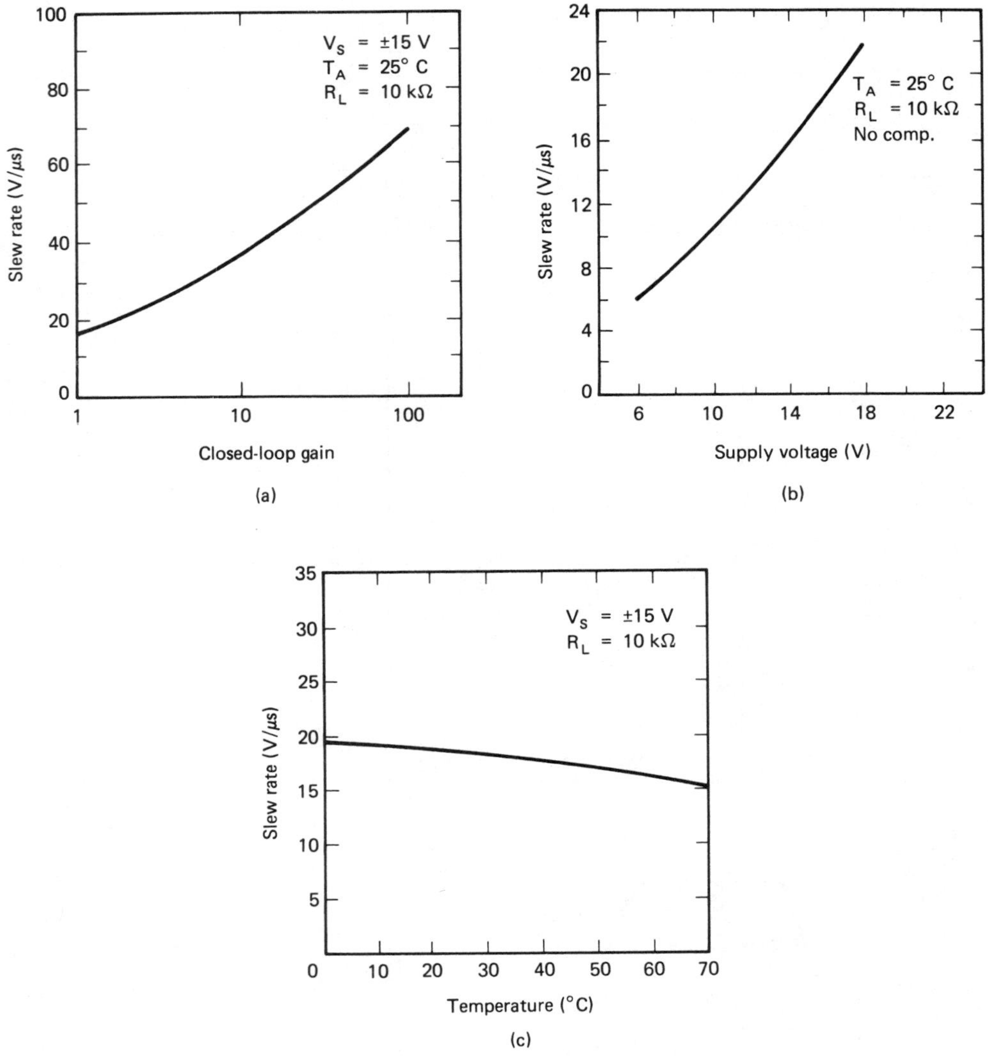

Figure 6-9 Slew rate of the μA715C. (a) Increases with closed loop gain. (b) Increases with supply voltage. (c) Decreases with increase in temperature. (Courtesy of Fairchild Camera and Instrument Corporation.)

6-10.1 Causes of Slew Rate

The unity gain bandwidth is the *small-signal* high-frequency limitation on the use of the op-amp. Slew rate, on the other hand, is a *large-signal* phenomenon. A large signal is one whose amplitude is comparable with the power supply voltages. Typically, the large signal is of the order of volts, while a small signal is in the range of milli- or even microvolts. Slew rate is caused by current limiting and the saturation of internal stages of an op-amp when a high-frequency, large-amplitude signal is applied. The resulting current is the maximum current available to charge the compensation capacitance network. We know that the capacitor requires a finite amount of time to charge and discharge. This means that internal capacitors prevent the output voltage from responding immediately to a fast-changing input. The rate at which the voltage across the capacitor rises is

$$\frac{dV_C}{dt} = \frac{I}{C} \tag{6-6}$$

Thus slew rate limiting is caused by this capacitor charging rate, in which the voltage across the capacitor is the output voltage.

6-10.2 Slew Rate Equation

Since the slew rate on a data sheet is generally listed for a unity gain, let us consider the voltage follower shown in Figure 6-10. Furthermore, let us assume that the input is a large-amplitude and high-frequency sine wave. The equation for the sine wave is

$$v_{in} = V_P \sin \omega t$$

or

$$v_o = V_P \sin \omega t \tag{6-7}$$

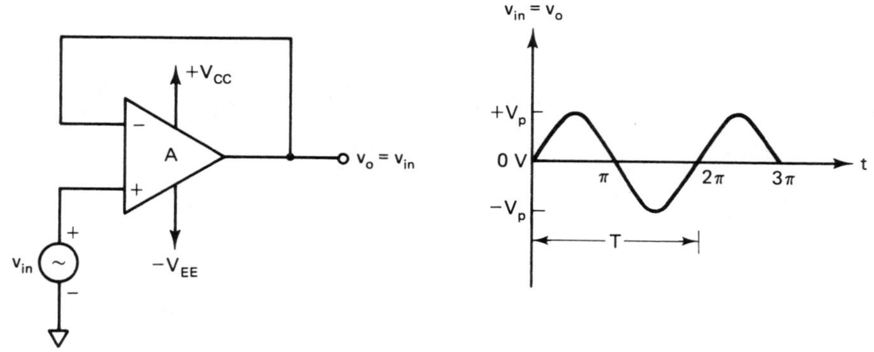

Figure 6-10 Deriving the slew rate equation.

The rate of change of the output is

$$\frac{dv_o}{dt} = V_p \, \omega \cos \omega t$$

and the maximum rate of change of the output occurs when $\cos \omega t = 1$. That is,

$$\left.\frac{dv_o}{dt}\right|_{max} = V_p \, \omega$$

or

$$SR = 2\pi f V_P \qquad \text{V/s}$$

$$= \frac{2\pi f V_P}{10^6} \qquad \text{V}/\mu\text{s} \tag{6-8}$$

where SR = slew rate in (V/μs)
f = input frequency (hertz)
V_P = peak value of the output sine wave (volts)

For output free of distortion, the slew rate determines the maximum frequency of operation for a desired output swing. Stated another way, with a lower frequency the slew rate determines the maximum undistorted output voltage swing. Thus, as long as the value of the right-hand side of Equation (6-8) is less than the slew rate of the op-amp, the output waveform will always be sinusoidal. If either the frequency or the amplitude of the input signal is increased to exceed the slew rate of the op-amp, the output will distort.

6-10.3 Effect of Slew Rate in Applications

The slew rate has important effects on both open-loop and closed-loop op-amp circuits. Figure 6-11(a) shows the open-loop configuration using the 741C. Since the open-loop voltage gain is very large, the output will go to about +14 V and then to −14 V each time the input sine wave crosses zero volts, as shown in Figure 6-11(b). The time taken by the output to go from +14 V to −14 V can be determined by using the slew rate of the 741C listed in the data sheet. The slew rate can also be calculated from the slope of the voltage-follower large-signal pulse response curve of Figure 6-8(b).

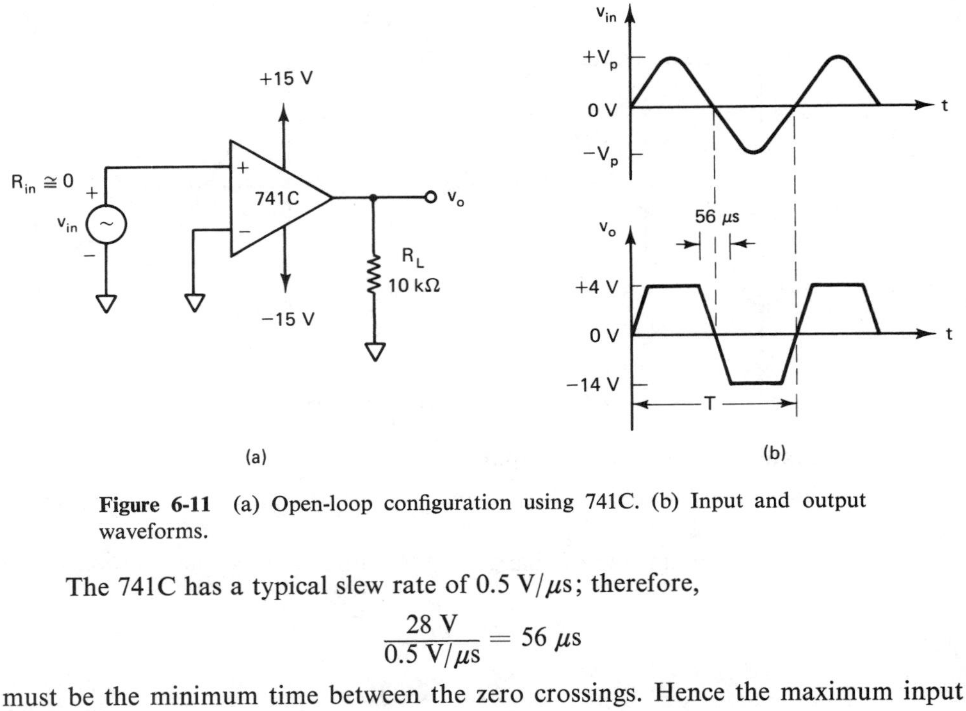

Figure 6-11 (a) Open-loop configuration using 741C. (b) Input and output waveforms.

The 741C has a typical slew rate of 0.5 V/μs; therefore,

$$\frac{28\text{ V}}{0.5\text{ V}/\mu\text{s}} = 56\ \mu\text{s}$$

must be the minimum time between the zero crossings. Hence the maximum input frequency should be

$$f_{max} = \frac{1}{(2)(56\ \mu\text{s})} = 8.93\text{ kHz}$$

However, at this frequency the output may be a triangular instead of square wave. Therefore, to have a more square-wave output either we keep the input frequency below f_{max} or choose an op-amp with a faster slew rate. However, faster slew rate op-amps are sometimes characterized by overshoot and ringing, which cause the output to take longer to reach a steady state than with slower slew rate op-amps. For this reason, settling time is another important parameter that should be considered, especially in applications such as digital-to-analog (D/A) or analog-to-digital (A/D) converters.

Next, consider the effect of slew rate on the inverting amplifier with feedback. Figure 6-12 shows the 741C op-amp used as an inverting amplifier with a gain of 50. From the voltage gain versus frequency curve of the 741C in Figure 6-1(a), obviously the amplifier will operate with a gain of 50 up to about 20 kHz (50 × 20 kΩ = 1 MHz). Using Equation (6-8), the maximum output voltage at 20 kHz is

$$0.5 = \frac{(2\pi)(20\text{ k})(V_P)}{10^6}$$

$$V_P = \frac{(0.5)(10^6)}{(2\pi)(20\text{ k})} = 3.98\text{ V peak}$$

or

$$v_o = 7.96\text{ V peak-to-peak sine wave}$$

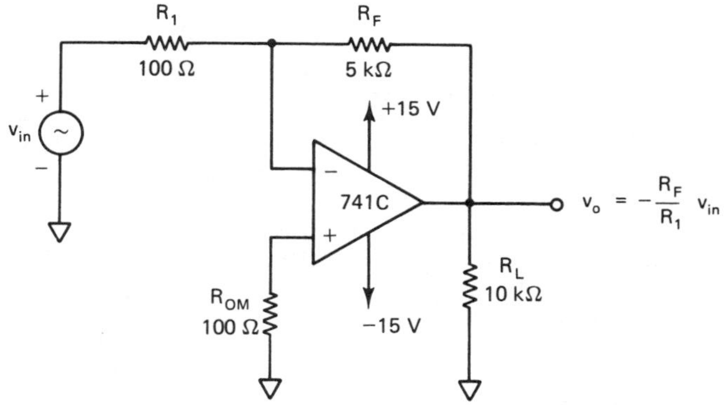

Figure 6-12 Inverting amplifier with a gain of 50.

Hence, for the output to be a sine wave, the maximum input signal should be less than

$$\frac{7.96}{50} = 159\text{ mV peak to peak}$$

We can also easily read the maximum output voltage swing from the curve showing output voltage swing as a function of frequency in Figure 6-8(a). This curve enables us to estimate the maximum output swing without distortion for different frequencies. From the curve the maximum output voltage at 20 kHz is approximately 14 V pp. This value, however, is much higher than the value calculated above for the following reasons:

1. The graph in Figure 6-8(a) is approximate because the same graph is to be used for different grades of op-amps (741, 741A, 741E, and 741C), even though the slew rate of the 741C is 0.5 V/μs and that of the 741A is 0.7 V/μs.
2. The slew rate computed from the graph is much higher than the slew rate

(0.5 V/μs) specified in the data sheets of the 741C. For instance, referring to Figure 6-8(a), the output voltage swing is 28 V pp at 10 kHz, which in turn implies a slew rate of

$$SR = (2\pi)(14)(10\text{ k})(10^{-6}) = 0.88\text{ V}/\mu\text{s}$$

In short, the graph of Figure 6-8(a) should be used as a rough estimate of output voltage swing or slew rate and not as an accurate measure.

6-10.4 Difference between Bandwidth, Transient Response, and Slew Rate

Table 6-1 is a summary of three important ac parameters of the op-amp: bandwidth, transient response, and slew rate. A clear idea of the differences between these parameters is important in ac applications.

TABLE 6-1 SUMMARY OF AC PARAMETERS

Bandwidth	Transient response	Slew rate
1. A small-signal phenomenon	A small-signal phenomenon	A large-signal phenomenon
2. Band of frequencies for which the gain remains constant	That part of the total response before the response reaches a steady state	The maximum time rate of change of the output voltage
3. Depends on compensating components and closed-loop gain	Composed of overshoot and rise time; rise time is related to bandwidth and overshoot is a measure of stability	Slew rate limiting depends on both frequency and amplitude; often increases with closed-loop gain and power supply voltages
4. If exceeded, results in a reduction of output voltage but causes no distortion	Affects settling time	If exceeded, results in distortion

SUMMARY

1. Frequency response is the manner in which the gain magnitude and the phase angle between the input and output respond to different frequencies.
2. Compensating networks are used to control the phase shift and thus improve the stability of the op-amps. These networks are typically composed of resistors and/or capacitors. The compensating network is either designed into the circuit or is added at designated terminals.
3. In internally compensated op-amps the compensating network is designed into the circuit. On the other hand, a compensating network is added externally in noncompensated op-amps. Generally, open-loop noncompensated op-amps have wider bandwidths than those of compensated op-amps.

4. Because of capacitances within the op-amp, the gain decreases and the phase shift between input and output voltages increases as frequency increases.
5. The open-loop gain of the op-amp is relatively constant at frequencies below the break frequency but decreases at a rate of -20 dB/decade after each break frequency.
6. The very small bandwidth of an open-loop op-amp limits its use. The closed-loop op-amp has the advantage of a much larger bandwidth.
7. To achieve a stable circuit, the phase angle of the loop gain must be less than 180° when its magnitude reaches unity.
8. Slew-rate limiting occurs with all large, fast-changing signals. If slew rate is exceeded, distortion of the output waveform results.

QUESTIONS

6-1. What is a frequency response?

6-2. Briefly explain the need for compensating networks in op-amps.

6-3. What is the difference between compensated and noncompensated op-amps?

6-4. How does the high-frequency model differ from the equivalent circuit of an op-amp? Explain.

6-5. Define the *break frequency* and *bandwidth*.

6-6. Explain the effect of negative feedback on the frequency response.

6-7. What is a slew rate? List causes of the slew rate and explain its significance in applications.

6-8. Explain the difference between the slew rate and transient response.

PROBLEMS

6-1. The 741C is connected as an inverting amplifier and is required to have a closed-loop gain of 20. What is its bandwidth? Is the circuit stable?

6-2. The 709C is used as a noninverting amplifier with a gain of 50 dB. What is its bandwidth if $C_1 = 500$ pF, $R_1 = 1.5$ kΩ, and $C_2 = 20$ pF?

6-3. Repeat Problem 6-2 with a gain of 70 dB.

6-4. Repeat Problem 6-2 with the compensating components as follows: $C_1 = 5{,}000$ pF, $R_1 = 1.5$ kΩ, and $C_2 = 200$ pF.

6-5. What maximum gain can be used while keeping the amplifier's response flat to 1 kHz? Assume that the 741C is wired as an inverting amplifier.

6-6. The frequency response of a certain op-amp is shown in Figure 6-13. Write the open-loop gain equation for the op-amp as a function of break frequencies and dc gain A.

6-7. Determine the magnitude of the open-loop gain and the phase shift between the input and output for the op-amp in Problem 6-6 if the frequency of operation is 10 MHz. Is the circuit stable? Explain.

6-8. Compute the unity gain bandwidth of the op-amp whose frequency response is given in Figure 6-13.

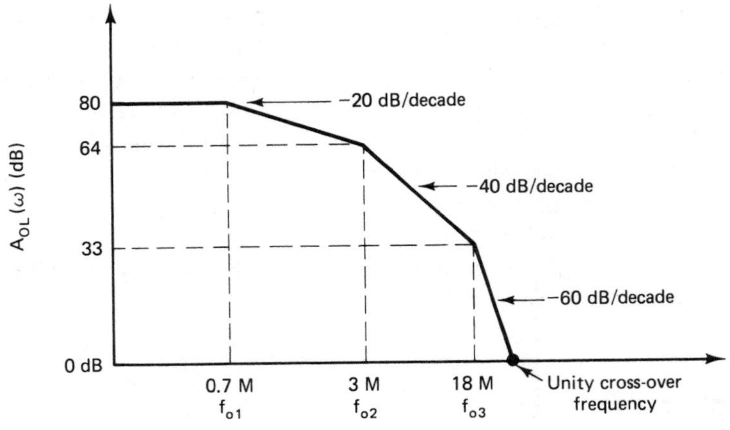

Figure 6-13 Frequency response for Problem 6-6.

6-9. Noncompensated op-amp MC1539 has a dc gain $A = 120{,}000$ and the following break frequencies: $f_{o1} = 5$ kHz, $f_{o2} \cong 320$ kHz, $f_{o3} = 1$ MHz, and $f_{o4} = 2$ MHz. Write the open-loop gain equation for the op-amp as a function of break frequencies and dc gain A. Then plot the Bode diagrams.

6-10. For the op-amp of Problem 6-9, compute the phase shift between input and output signals at 50 kHz and 1.5 MHz.

6-11. The μA709C is connected as shown in Figure 6-14. What is its bandwidth when the 20 kΩ pot is adjusted to (a) 1 kΩ and (b) 10 kΩ? [*Hint:* Use characteristic curves of Figure 6-1(b).]

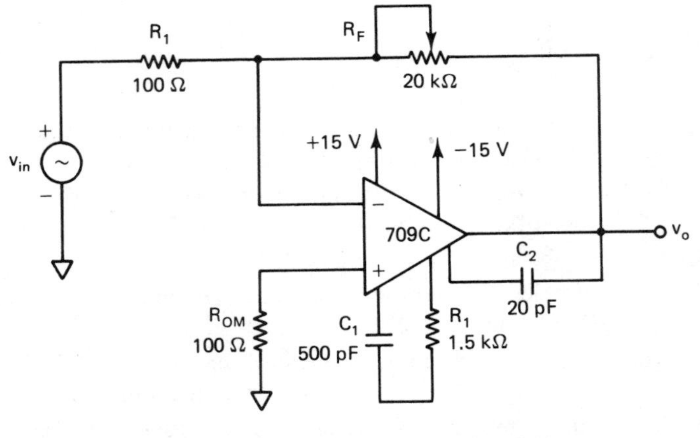

Figure 6-14 Amplifier of Problem 6-11.

6-12. The μA709C is wired up as a noninverting amplifier for a gain of 1000 and is required to give a flat response to about 300 kHz. What compensating components would you recommend for the op-amp? (*Hint:* Refer to the characteristic curves shown in Figure 6-6.)

6-13. The 741C is connected as an inverting amplifier for a gain of 100. Determine the stability of the amplifier at this gain.

6-14. The 741C is used as an inverting amplifier with a gain of 50. The sinusoidal input signal has a variable frequency and maximum amplitude of 20 mV peak. What is the maximum frequency of the input at which the output will be undistorted? Assume that the amplifier is initially nulled.

6-15. Determine the slew rate of the 741C using the voltage-follower large-signal pulse response of Figure 6-8(b).

6-16. An inverting amplifier using the 741C must have a flat response up to 40 kHz. The gain of the amplifier is 10. What maximum peak-to-peak input signal can be applied without distorting the output? [Refer to the op-amp characteristic in Figure 6-8(a).]

6-17. Determine the slew rate of the 715C if it is used as a noninverting amplifier with a gain of (a) 5 and (b) 50. [*Note:* Refer to Figure 6-9(a).]

6-18. When an 8 V peak-to-peak square wave of 3.6 MHz frequency is the input to a voltage follower, the output is a triangular wave, as shown in Figure 6-15. What is the slew rate of the op-amp?

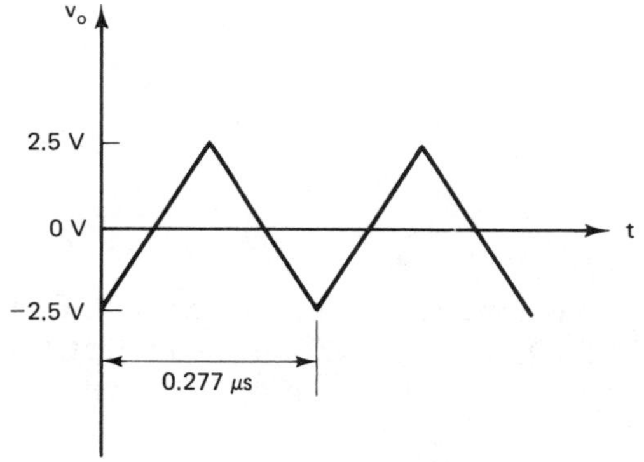

Figure 6-15 Waveform of Problem 6-18.

6-19. In Problem 6-18, what must the minimum slew rate of the op-amp be in order to get the square-wave output?

6-20. The 715C is connected as a noninverting amplifier with a gain of 10. If the input is a sine wave at a frequency of 2 MHz, what is the largest undistorted output voltage? [Refer to the op-amp characteristic in Figure 6-9(a).]

EXPERIMENT 6-1
FREQUENCY RESPONSE OF NONCOMPENSATED OP-AMP 709C

Objectives

In this experiment you will plot the frequency response of the noncompensated op-amp 709C with two different sets of compensating networks. After you complete this experiment, you should be able to:

1. Explain the effect of compensating networks on the bandwidth of the amplifier.

2. Select a suitable compensating network for a desired bandwidth and closed-loop gain.

Equipment

1. Oscilloscope
2. Audio-frequency generator: $f_{max} = 1$ MHz
3. $\pm$15-V power supply

Materials

1. 709C op-amp
2. Two 100-Ω resistors
3. 1.5-kΩ resistor
4. 10.0-kΩ resistor
5. 20-pF capacitor
6. 200-pF capacitor
7. 500-pF capacitor
8. 5000-pF capacitor
9. Semilog paper

Procedure

1. Connect the op-amp as an inverting amplifier, as shown in Figure E6-1.1 with $C_1 = 5000$ pF, $R_1 = 1.5$ kΩ, and $C_2 = 200$ pF. Note that the pin numbers refer to the 14-pin DIP. If you have the 8-pin mini DIP, refer to the data sheet of the 709C.

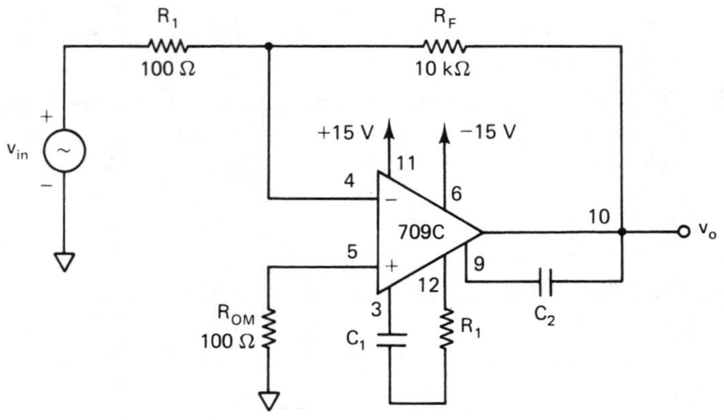

Figure E6-1.1 709C configured as an inverting amplifier with a gain of 100.

2. Turn on the dc supply voltages and signal generator. Connect the scope at the output terminal of the amplifier. Then set the generator frequency at 100 Hz sine wave and adjust the amplitude of the input so that the output is a measur-

able, undistorted sine wave. Measure the input and output voltages and enter the values in Table E6-1.1.

TABLE E6-1.1

Compensating components	Input frequency, (Hz)	Input amplitude, v_{in} pp	Output amplitude, v_o pp	Voltage gain, v_o/v_{in}	Voltage gain (dB) = 20 log (v_o/v_{in})
$C_1 = 5000$ pF,	100				
$R_1 = 1.5$ kΩ	300				
$C_2 = 200$ pF	700				
	1,000				
	3,000				
	7,000				
	10,000				
	30,000				
	70,000				
	100,000				
$C_1 = 500$ pF,	100				
$R_1 = 1.5$ kΩ	300				
$C_2 = 20$ pF	700				
	1,000				
	3,000				
	7,000				
	10,000				
	30,000				
	70,000				
	100,000				
	500,000				
	700,000				
	1,000,000				

3. Change the frequency of the input to 300 Hz and again measure the input and output voltages with the scope. Enter the measured values in Table E6-1.1. Repeat the procedure for the remaining frequencies.
4. Turn off the supply voltages and the signal generator. Replace the compensating capacitors with the following values: $C_1 = 500$ pF and $C_2 = 20$ pF.
5. Repeat steps 2 and 3 to complete Table E6-1.1.
6. Draw the frequency response plots for the 709C op-amp using both the compensating networks, and determine the bandwidths.

Questions

1. Which set of compensating components used with 709C results in a larger bandwidth?
2. How do these bandwidths compare with those obtained from the *open-loop frequency response of the 709C for various values of the compensation*? (The open-loop frequency response is given in the data sheets of the 709C.)

EXPERIMENT 6-2
MEASURING SLEW RATE

Objectives

In this experiment you will measure the slew rate of the 741C op-amp using sine and square waves. When you complete this experiment, you should have the following knowledge and capabilities:

1. Given a particular op-amp, you should be able to measure its slew rate using either a sine- or square-wave input.
2. You should be able to verify that slew rate limiting depends on both amplitude and frequency.
3. You should also be able to verify that the output distorts if slew rate is exceeded.
4. For a specified frequency and amplitude of the output, you should be able to select an op-amp with a suitable slew rate.

Equipment

1. Oscilloscope
2. Sine- and square-wave audio-frequency generator: $f_{\max} = 1$ MHz
3. ± 15-V power supply

Materials

1. 741C op-amp
2. 100-Ω resistor
3. 10-kΩ resistor
4. 1-kΩ potentiometer

Procedure

1. Connect the 741C as shown in Figure E6-2.1(a). Turn on the supply voltages and signal generator, making sure that the generator is set on the sine wave.
2. Connect the scope at the output terminal and adjust the signal generator so that the output is 1 V peak sine wave at 1 kHz.
3. Slowly increase the frequency of the signal generator until the output is just barely distorted. Enter this value of frequency and the amplitude of the output in Table E6-2.1.
4. Calculate the slew rate using the following equation:

$$\text{SR} = \frac{2\pi f V_P}{10^6} \qquad \text{V}/\mu\text{s}$$

where V_P = the peak output amplitude (volts)
f = frequency (hertz)

Enter the calculated value of the slew rate in Table E6-2.1.

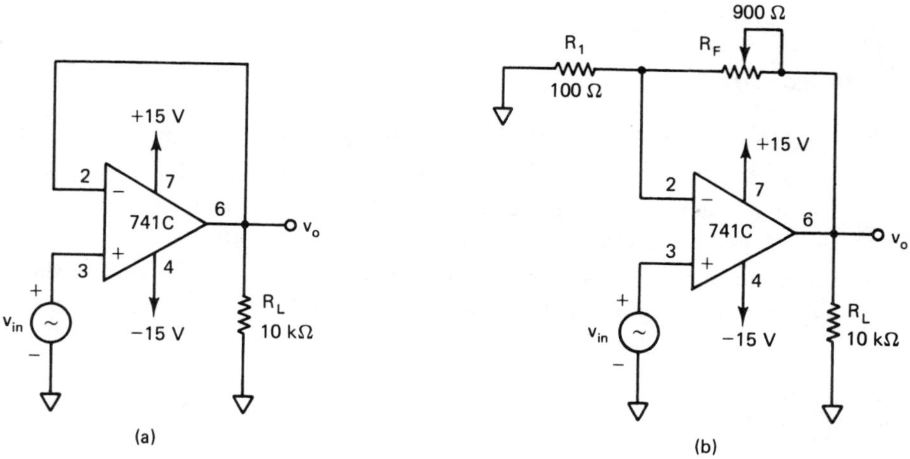

Figure E6-2.1 Measuring slew rate with (a) voltage follower and (b) noninverting amplifier.

TABLE E6-2.1

Network	Input waveform	Maximum input frequency applied (Hz)	Output amplitude (V)	Slew rate (V/μs)
1. Voltage follower	Sinusoidal			
2. Voltage follower	Square wave			
3. Noninverting amplifier	Sinusoidal			

5. Switch the signal generator to a square-wave input and repeat step 2.
6. Slowly increase the frequency of the input until the output is just barely a triangular wave. Enter this value of frequency as well as the output amplitude in Table E6-2.1.
7. Calculate the slew rate using the following equation:

$$\text{SR} = \frac{\Delta V_o}{\Delta t} \quad \text{V}/\mu\text{s}$$

where ΔV_o = change in the output voltage amplitude (volts)
Δt = time required for ΔV_o (μs).

8. Connect the 741C as shown in Figure E6-2.1(b). Turn on the supply voltages and switch the signal generator back to sine wave. Repeat steps 2, 3, and 4.

You may have to increase an amplitude of the input signal in order to cause an output to distort.

Questions

1. How does the slew rate of the voltage follower using sine wave compare to that with square wave? If the values are more than 20% apart, give the reasons.
2. Is the slew rate of the noninverting amplifier the same as the voltage follower? What conclusion can be drawn from this comparison?
3. Compare the slew rate you measured at unity gain to that given in the data sheet of the 741C. If the difference is more than 20%, give the reasons.

7

General Linear Applications

7-1 INTRODUCTION

All op-amps are basically alike in being direct-coupled high-gain differential amplifiers. However, many different specific op-amps are available because no one circuit design can possibly optimize all the dc and ac parameters. For particular applications, op-amps are designed to optimize a parameter such as high slew rate, wide bandwidth, high gain, or low power consumption. Therefore, to achieve the best performance in a desired application, it is essential to select a suitable op-amp. Often, a choice has to be made between performance and cost since an op-amp optimized for a particular parameter is more expensive than a general-purpose op-amp.

The op-amp parameters were defined and discussed in Chapter 3. These parameters are very useful in circuit design as well as in selecting an op-amp for a specific application. Also, Chapter 6 showed how data sheet information can be used to determine circuit stability. The data books contain a wealth of information, and we should consult them to select the most suitable device for a desired application.

Having considered the op-amp's basic construction, characteristics, parameter limitations, and various configurations, our next logical step is to discuss op-amp applications. Since there are literally countless applications of op-amps, they will be arranged into the following categories: general linear applications, filter and ocsillator applications, comparator and detector applications, special integrated circuit applications, and selected system applications. Each category will include the most important applications in that area. Thus the remainder of the book is devoted entirely to op-amp applications. We begin with *general linear applications* in this chapter. Recall that in linear circuits the output signal is of the same nature as the input and varies in accordance with the input within the limits set by the saturation levels and slew rate. The first part of the chapter presents the important types of amplifiers and the last part discusses simple modifications of basic amplifier configurations.

In all these applications a *general-purpose op-amp* such as the 741 will be used. Remember that no one op-amp has all the dc and ac parameters optimized, and the 741 is no exception. However, using the concepts developed in this chapter, the reader may select a more specific op-amp that will improve the performance and accuracy of the circuit.

7-2 DC AND AC AMPLIFIERS

As pointed out in Chapter 4, the properties of closed-loop amplifiers depend primarily on characteristics of the feedback network rather than those of the op-amp itself. Since a typical feedback network is composed of resistors and capacitors, the accuracy and stability of circuits using them can be improved significantly because these components are available in conditions of high precision and low drift.

7-2.1 DC Amplifier

Basically, an op-amp can amplify two types of signals: dc and ac. In a *dc amplifier* the output signal changes in response to changes in its dc input levels. A dc amplifier can be inverting, noninverting, or differential, as shown in Figure 7-1. To reduce the output offset voltage to zero, that is, to improve the accuracy of the dc amplifier, the offset null circuitry of the op-amp should be used. For op-amps without offset null capability, the external offset voltage compensating network should be used, as shown in Figure 7-2. Otherwise, a high-precision op-amp such as the μA714, which has smaller offsets and drifts, must be used. Recall that all the circuits in Figures 7-1 and 7-2 have been discussed previously and their output voltage equations derived.

7-2.2 AC Amplifier

The circuits of Figures 7-1 and 7-2 respond to ac input signals as well. However, if the designer needs the ac response characteristics of the op-amp, that is, low- and high-frequency limits, or if the ac input is riding on some dc level, it is necessary to use an *ac amplifier* with a coupling capacitor. For example, in an *audio receiver system* which consists of a number of stages, because of thermal drift, component tolerances, and variations, the dc level is produced. To prevent the amplification of such dc levels the coupling capacitors must be used between the stages. Figure 7-3 shows the ac inverting and noninverting amplifiers with coupling capacitors.

The coupling capacitor not only blocks the dc voltage but also sets the low-frequency cutoff limit, which is given by

$$f_L = \frac{1}{2\pi C_i(R_{iF} + R_o)} \tag{7-1}$$

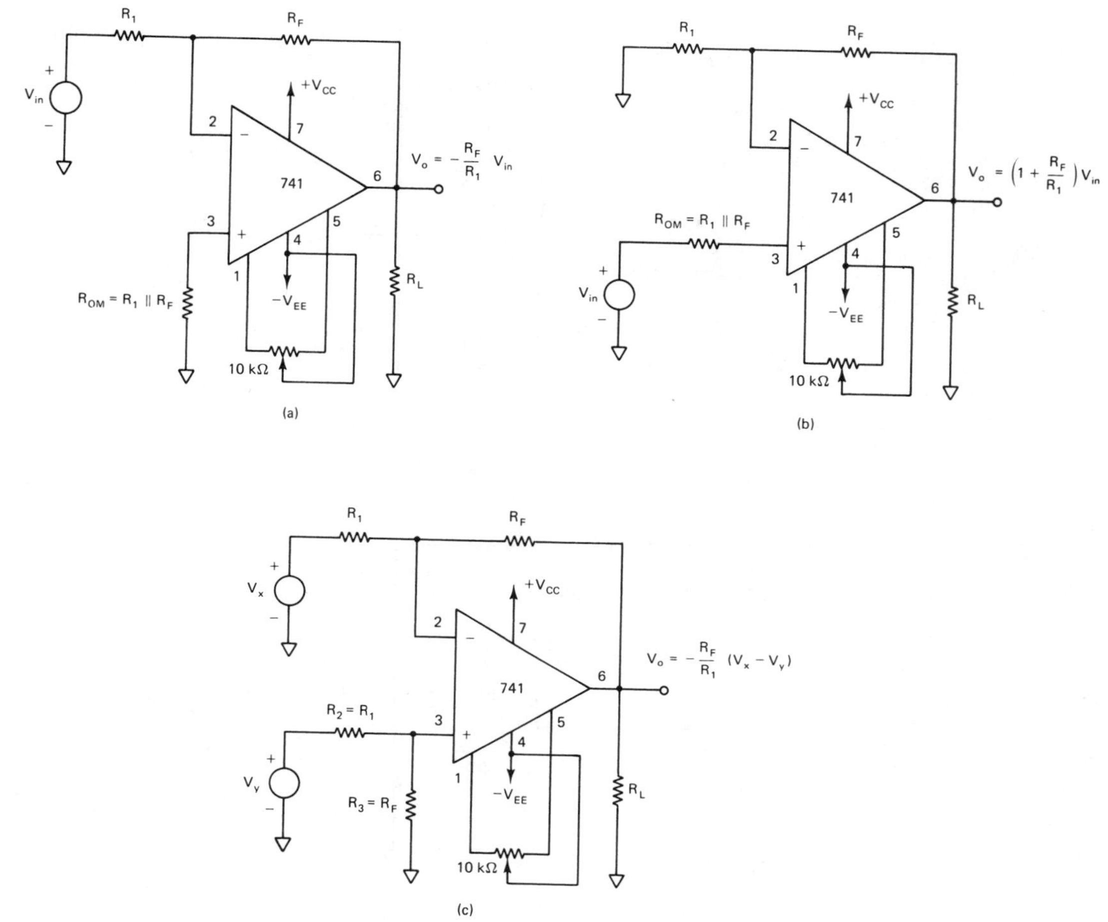

Figure 7-1 (a) Inverting amplifier, (b) noninverting amplifier, and (c) differential amplifier with offset null circuitry.

where f_L = low-frequency cutoff or low end of the bandwidth
C_i = capacitance between two stages being coupled or dc blocking capacitor
R_{iF} = ac input resistance of the second stage
R_o = ac output resistance of the first stage or the source resistance

Remember that the high-frequency cutoff f_H or high end of the bandwidth depends on the closed-loop gain of the amplifier.

Thus the bandwidth of the amplifier,

$$BW = (f_H - f_L) \tag{7-2}$$

depends on the desired value of f_L and the closed-loop gain of the amplifier. The required value of capacitor C_i can be calculated using Equation (7-1). The coupling capacitor C_i, besides providing a low-frequency cutoff limit, also helps to eliminate dc level amplification from stage to stage. In Figure 7-3, for example, the input voltage

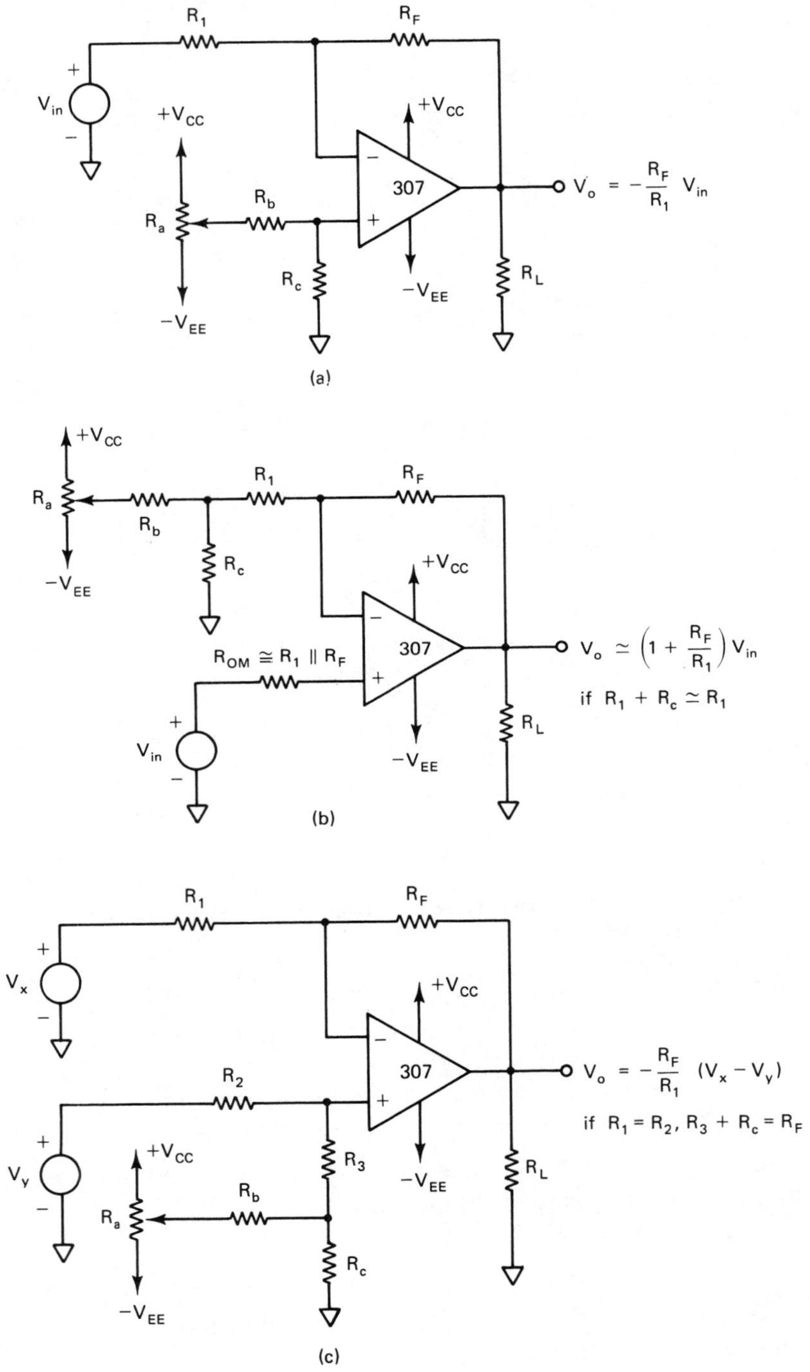

Figure 7-2 (a) Inverting amplifier, (b) noninverting amplifier, and (c) differential amplifier with external offset-voltage compensating network.

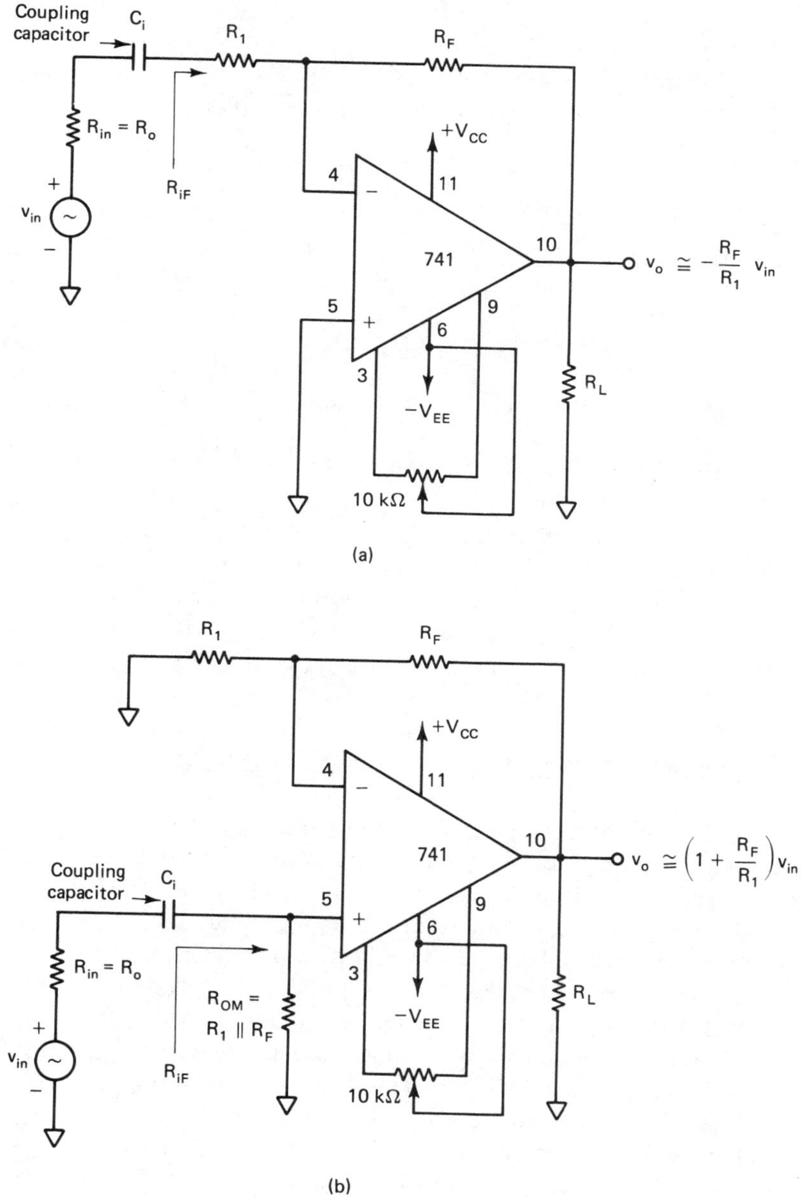

Figure 7-3 (a) Ac inverting amplifier. (b) Ac noninverting amplifier. (Pin numbers are for the 14-pin DIP.)

v_{in} could be a preceding amplifier output with a dc level; the coupling capacitor C_i will block this dc level. Thus the coupling capacitor prevents the amplification of the dc level.

Figure 7-3(a) shows an ac inverting amplifier, the closed-loop gain of which is still

$$A_F \cong -\frac{R_F}{R_1} \tag{7-3}$$

since the reactance of C_i is negligible within the bandwidth. Similarly, the closed-loop gain of the noninverting amplifier of Figure 7-3(b) is

$$A_F \cong 1 + \frac{R_F}{R_1} \tag{7-4}$$

Example 7-1

In the circuit of Figure 7-3(a), $R_{in} = 50\ \Omega$, $C_i = 0.1\ \mu F$, $R_1 = 100\ \Omega$, $R_F = 1\ k\Omega$, $R_L = 10\ k\Omega$, and supply voltages $= \pm 15$ V. Determine the bandwidth of the amplifier.

Solution. Recall that the input resistance of the inverting amplifier with feedback is

$$R_{iF} = R_1 + \frac{R_F}{A} \cong 100\ \Omega$$

The source resistance $R_{in} = R_o = 50\ \Omega$. Therefore, from Equation (7-1),

$$f_L = \frac{1}{(2\pi)(10^{-7})(150)} = 10.6\text{ kHz}$$

Since $A_F = -R_F/R_1 = -10$, the high-frequency cutoff f_H using Equation (4-21b) is

$$f_H = \frac{\text{unity gain bandwidth}}{A_F} = \frac{1\text{ MHz}}{10} = 100\text{ kHz}$$

Thus, from Equation (7-2), BW $= 100$ kHz $- 10.6$ kHz $= 89.4$ kHz.

7-3 AC AMPLIFIERS WITH A SINGLE SUPPLY VOLTAGE

An ac amplifier can be powered by a single supply voltage if an additional coupling capacitor is used on the output of an amplifier, as shown in Figure 7-4. This capacitor blocks dc; therefore, an output offset voltage as well as a dc level in the output signal has little effect on the operation of the amplifier. Besides that, in Figure 7-4(a) positive dc level is intentionally inserted using a voltage-divider network at the noninverting input terminal so that output can swing both positively as well as negatively. Furthermore, the dc level inserted is generally $(+V_{CC}/2)$, so that the positive output swing equals the negative. This is accomplished by selecting $R_2 = R_3$. In the inverting amplifier of Figure 7-4(a), the voltage at the noninverting terminal is approximately $(+V_{CC}/2)$ V dc. Since the reactance of input capacitor C_i at 0 Hz (dc) is infinity, the resistor R_1 is an open component; hence the circuit effectively operates as a voltage follower, producing an output of $(+V_{CC}/2)$ V dc. However, with ac input single v_{in}, the reactance of the input coupling capacitor C_i is negligible, and the ac output voltage v_o at frequencies within the bandwidth is $v_o = (-R_F/R_1)v_{in}$. Thus, as shown in Figure 7-4(b) the ac output is riding on a dc level of $(+V_{CC}/2)$ volts. The output coupling capacitor C_o blocks this dc voltage, and the resultant waveform is therefore ac. Note that the maximum peak-to-peak output voltage swing is approximately equal to $+V_{CC}$ volts. The noninverting ac amplifier using a single supply is shown in Figure 7-4(c). Again, at 0 Hz or dc the input capacitors C_i and C_1 act as open components; therefore, the circuit effectively operates as a voltage follower, producing

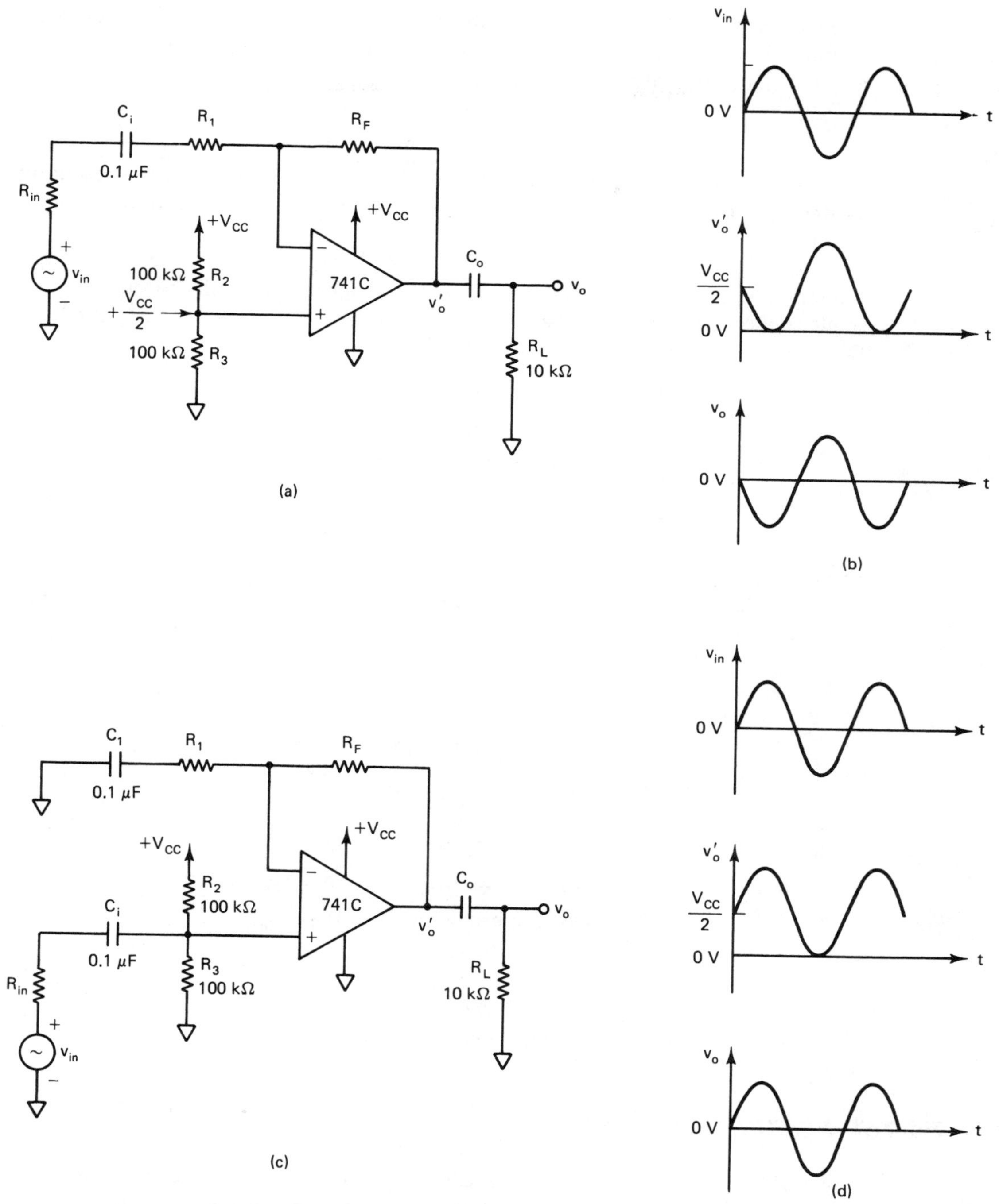

Figure 7-4 (a) Ac inverting amplifier with single supply. (b) Its input and output waveforms. (c) Ac noninverting amplifier with single supply. (d) Its input and output waveforms.

the dc output voltage of $(+V_{CC}/2)$ volts. Note that the noninverting amplifier requires an extra capacitor C_1 to produce an output dc level of $(+V_{CC}/2)$ volts. At frequencies within the bandwidth the input capacitors C_i and C_1 have negligible reactances; as a result, the ac output voltage v_o of the amplifier is

$$v_o = \left(1 + \frac{R_F}{R_1}\right) v_{\text{in}}$$

The total output voltage (dc plus ac) v_o' is as shown in Figure 7-4(d). The output capacitor C_o blocks dc voltage of $(+V_{CC}/2)$ volts, and hence the ac output voltage v_o results.

Although the 741C was used as an ac amplifier with a single supply voltage, some op-amps, such as the LM124 series, are designed specifically to operate on a single supply. These op-amps have a wide power supply range and low input offsets. For example, for the LM324:

$$\text{Power supply range} = 3 \text{ to } 30 \text{ V dc}$$
$$I_B = 45 \text{ nA}$$
$$I_{io} = 5 \text{ nA}$$
$$V_{io} = 2 \text{ mV}$$
$$\text{Supply current drain} = 800\ \mu\text{A independent of supply voltage}$$

Example 7-2

For the noninverting amplifier of Figure 7-4(c), $R_{\text{in}} = 50\ \Omega$, $C_i = C_1 = 0.1\ \mu\text{F}$, $R_1 = R_2 = R_3 = 100\ \text{k}\Omega$, $R_F = 1\ \text{M}\Omega$, and $+V_{CC} = +15$ V. Determine (a) the bandwidth of the amplifier and (b) maximum output voltage swing.

Solution

(a) The ac input resistance of the amplifier is

$$R_{iF} \cong 100\ \text{k}\Omega \,||\, 100\ \text{k}\Omega = 50\ \text{k}\Omega$$

Therefore, from Equation (7-1),

$$f_L = \frac{1}{(2\pi)(10^{-7})(50\ \text{k}\Omega + 50\Omega)} = 31.8\ \text{Hz}$$

The gain of the amplifier is $(1 + R_F/R_1) = 11$. Hence from Equation (4-10b)

$$f_H = \frac{\text{unity gain bandwidth}}{A_F} = \frac{1\ \text{MHz}}{11} = 90.91\ \text{kHz}$$

and

$$\text{BW} = f_H - f_L = 90.91\ \text{k}\Omega - 0.0318\ \text{k}\Omega = 90.88\ \text{kHz}$$

(b) The ideal maximum output voltage swing $= +V_{CC} = +15$ V pp.

7-4 THE PEAKING AMPLIFIER

The *peaking response*, that is, the frequency response that peaks at a certain frequency can be obtained by using a parallel LC network with the op-amp. Figure 7-5(a) shows a peaking amplifier that uses a parallel LC network in the feedback path. The frequency response of the amplifier is shown in Figure 7-5(b). The resonant

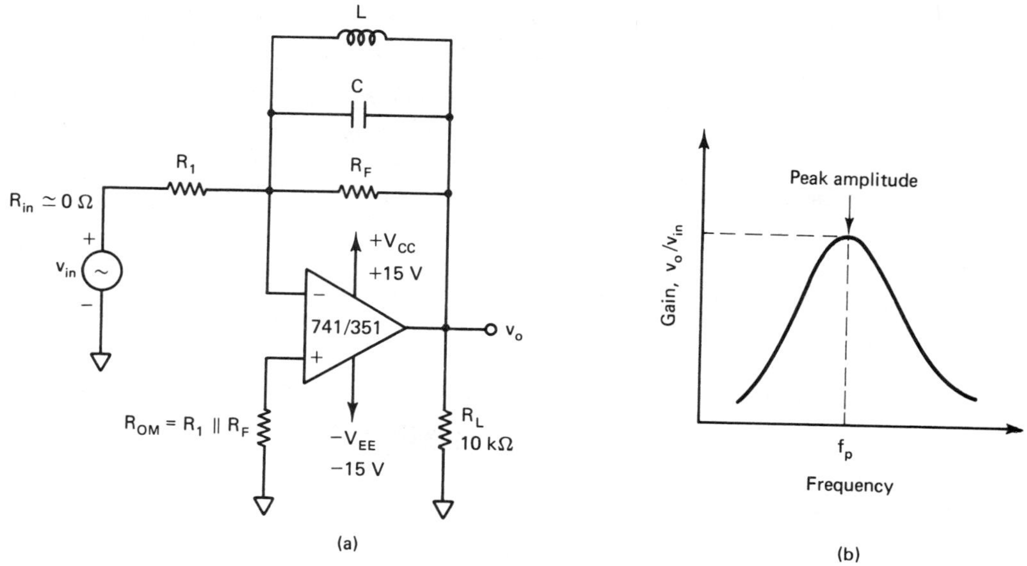

Figure 7-5 The peaking amplifier. (a) Schematic diagram. (b) Frequency response.

frequency or peak frequency at which the peaking occurs is determined by the combination of L and C.

$$f_p = \frac{1}{2\pi\sqrt{LC}} \quad \text{if} \quad Q_{\text{coil}} \geq 10 \tag{7-5a}$$

where Q_{coil} = figure of merit of the coil. The impedance of the parallel-LC network is very large at the resonant frequency. Therefore, the gain of the amplifier at resonance is maximum and given by

$$A_F = -\frac{R_F \parallel R_P}{R_1} \tag{7-5b}$$

where R_p = equivalent parallel resistance of the tank circuit = $(Q^2_{\text{coil}} R)$ and,

R = internal resistance of the coil.

The impedance of the parallel LC network below and above the peak frequency is less than R_p; therefore the gain of the amplifier is less than $(R_F \parallel R_P/R_1)$ at any other frequency than f_p. Note that using Equation (7-5b) the impedance of the tank circuit R_p at resonance can easily be computed from the measured value of gain and known values of R_1 and R_F. The bandwidth of the peaking amplifier of Figure (7-5a) can be determined by using the following equation:

$$BW = \frac{f_p}{Q_p} \tag{7-5c}$$

where f_p = peaking frequency

and, Q_p = figure of merit of the parallel resonant circuit = $(R_F \parallel R_P/X_L)$

Example 7-3

The circuit of Figure 7-5(a) is to provide a gain of 10 at a peak frequency of 16 kHz. Determine the values of all components.

Solution. The gain times peak frequency = (10)(16 kHz) = 160 kHz. Therefore, the 741C can be used in the circuit of Figure 7-5(a) since its 1 MHz unity gain bandwidth is larger than 160 kHz.

Although any combination of L and C that provides the peak frequency of 16 kHz can be used, the value of the capacitor should be less than 1 μF to avoid leakage problems. Similarly to reduce the large size of the inductors, it should be less than 1 H in value. Let $C = 0.01\ \mu\text{F}$. Then, using Equation (7-5a), the value of the inductor is:

$$L = \frac{1}{(2\pi)^2(f_p)^2(C)}$$

$$L = \frac{1}{(4\pi^2)(16\text{ k})^2(10^{-8})} = 9.9\text{ mH}$$

Let L be 10 mH. Also, let the internal resistance R of the inductor be 30-Ω. Then,

$$Q_{\text{coil}} = \frac{X_L}{R} = \frac{(2\pi)(16\text{k})(10)(10^{-3})}{30} = 33.5$$

and,

$$R_p = (Q_{\text{coil}})^2(R) = (33.5)^2(30) = 33.67\text{ k}\Omega$$

Since a gain of 10 is needed at 16 kHz, let R_1 be equal to 100 Ω; therefore using Equation (7-5b) the value of R_F is

$$10 = \frac{(R_F)(33.67\text{k})}{100(R_F + 33.67\text{k})}$$

that is,

$$R_F = 1.03\text{ k}\Omega. \quad (\text{Use } R_F = 1\text{ k}\Omega)$$

To minimize the effect of input bias currents, use R_{OM} approximately equal to $R_1 \| R_F = 100\ \Omega$. Thus, the component values are as follows:

Op-amp: 741C

$R_1 = 100\ \Omega$

$R_F = 1\text{ k}\Omega$

$L = 10$ mH having $R_1 = 30\ \Omega$

$C = 0.01\ \mu\text{F}$

7-5 SUMMING, SCALING, AND AVERAGING AMPLIFIERS

This section shows how the inverting, noninverting, and differential configurations are useful in such applications as summing, scaling, and averaging amplifiers.

7-5.1 Inverting Configuration

Figure 7-6 shows the inverting configuration with three inputs V_a, V_b, and V_c. Depending on the relationship between the feedback resistor R_F and the input resistors R_a, R_b, and R_c, the circuit can be used as either a summing amplifier, scaling amplifier, or averaging amplifier. The circuit's function can be verified by examining the expression for the output voltage V_o, which is obtained from Kirchhoff's current equation

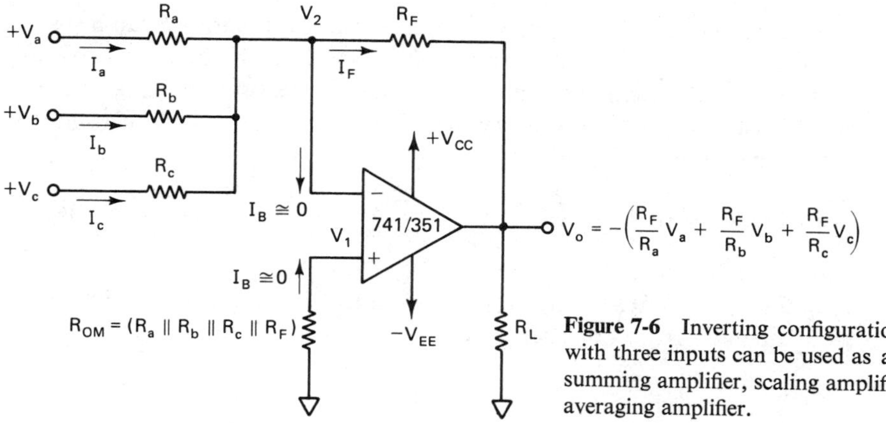

Figure 7-6 Inverting configuration with three inputs can be used as a summing amplifier, scaling amplifier, or averaging amplifier.

written at node V_2. Referring to Figure 7-6,

$$I_a + I_b + I_c = I_B + I_F \tag{7-6}$$

Since R_i and A of the op-amp are ideally infinity, $I_B = 0$ A and $V_1 = V_2 \cong 0$ V. Therefore,

$$\frac{V_a}{R_a} + \frac{V_b}{R_b} + \frac{V_c}{R_c} = -\frac{V_o}{R_F}$$

or

$$V_o = -\left(\frac{R_F}{R_a}V_a + \frac{R_F}{R_b}V_b + \frac{R_F}{R_c}V_c\right) \tag{7-7}$$

7-5.1(a) Summing amplifier. If in the circuit of Figure 7-6, $R_a = R_b = R_c = R$, for example, then Equation (7-7) can be rewritten as

$$V_o = -\frac{R_F}{R}(V_a + V_b + V_c) \tag{7-8a}$$

This means that the output voltage is equal to the *negative* sum of all the inputs times the gain of the circuit R_F/R; hence the circuit is called a *summing amplifier*. Obviously, when the gain of the circuit is 1, that is, $R_a = R_b = R_c = R_F$, the output voltage is equal to the *negative* sum of all input voltages. Thus

$$V_o = -(V_a + V_b + V_c) \tag{7-8b}$$

7-5.1(b) Scaling or weighted amplifier. If each input voltage is amplified by a different factor, in other words, weighted differently at the output, the circuit in Figure 7-6 is then called a *scaling* or *weighted amplifier*. This condition can be accomplished if R_a, R_b, and R_c are different in value. Thus the output voltage of the scaling amplifier is

$$V_o = -\left(\frac{R_F}{R_a}V_a + \frac{R_F}{R_b}V_b + \frac{R_F}{R_c}V_c\right) \tag{7-8c}$$

where

$$\frac{R_F}{R_a} \neq \frac{R_F}{R_b} \neq \frac{R_F}{R_c}$$

7-5.1(c) Averaging circuit. The circuit of Figure 7-6 can be used as an *averaging circuit*, in which the output voltage is equal to the average of all the input voltages. This is accomplished by using all input resistors of equal value, $R_a = R_b = R_c = R$ say. In addition, the gain by which each input is amplified must be equal to 1 over the number of inputs; that is,

$$\frac{R_F}{R} = \frac{1}{n}$$

where n is the number of inputs.

Thus, if there are three inputs (as shown in Figure 7-6), we want $R_F/R = 1/3$. Consequently, from Equation (7-7).

$$V_o = -\left(\frac{V_a + V_b + V_c}{3}\right) \tag{7-8d}$$

Remember that in the applications above the inputs V_a, V_b, and V_c could be either ac or dc. These circuits are commonly used in analog computers and audio mixers, in which a number of inputs is added up (mixed) to produce a desired output.

In Figure 7-6 the offset minimizing resistor R_{OM} is used to minimize the effect of input bias currents on the output offset voltage. However, to reduce the output offset voltage to zero, the offset voltage compensating network must be used, especially when the inputs are dc voltages.

Example 7-4

In the circuit of Figure 7-6, $V_a = +1$ V, $V_b = +2$ V, $V_c = +3$ V, $R_a = R_b = R_c = 3$ kΩ, $R_F = 1$ kΩ, $R_{OM} = 270$ Ω, and supply voltages $= \pm 15$ V. Assuming that the op-amp is initially nulled, determine the output voltage V_o.

Solution. Using Equation (7-8a), we obtain

$$V_o = -\frac{1\text{ k}\Omega}{3\text{ k}\Omega}(1 + 2 + 3) = -2\text{ V}$$

This value is equal to the average of three inputs with a negative sign.

7-5.2 Noninverting Configuration

If input voltage sources and resistors are connected to the noninverting terminal as shown in Figure 7-7, the circuit can be used either as a summing or averaging amplifier through selection of appropriate values of resistors, namely R_1 and R_F.

Again, to verify the functions of the circuit, the expression for tne output voltage must be obtained. Recall that the input resistance R_{iF} of the noninverting amplifier is very large (see Figure 7-7). Therefore, using the superposition theorem, the voltage V_1 at the noninverting terminal is

$$V_1 = \frac{R/2}{R + R/2}V_a + \frac{R/2}{R + R/2}V_b + \frac{R/2}{R + R/2}V_c$$

or

$$V_1 = \frac{V_a}{3} + \frac{V_b}{3} + \frac{V_c}{3} = \frac{V_a + V_b + V_c}{3} \tag{7-9}$$

Hence the output voltage V_o is

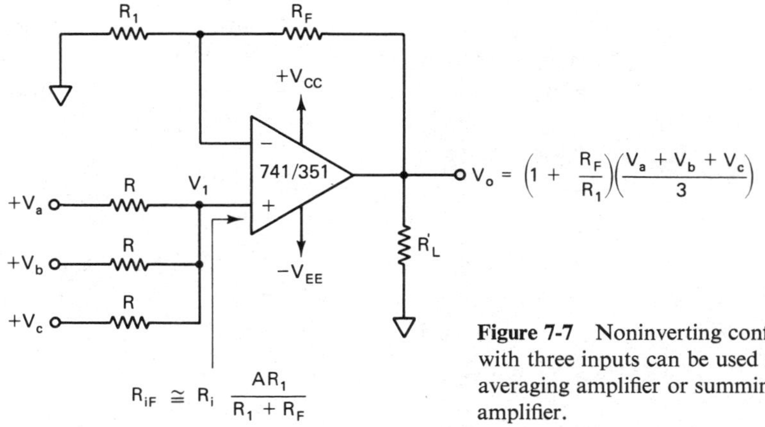

Figure 7-7 Noninverting configuration with three inputs can be used as averaging amplifier or summing amplifier.

$$V_o = \left(1 + \frac{R_F}{R_1}\right) V_1$$

$$= \left(1 + \frac{R_F}{R_1}\right) \frac{V_a + V_b + V_c}{3} \quad (7\text{-}10a)$$

7-5.2(a) Averaging amplifier. Equation (7-10a) shows that the output voltage is equal to the average of all input voltages times the gain of the circuit $(1 + R_F/R_1)$, hence the name *averaging amplifier*. Depending on the application requirement, the gain $(1 + R_F/R_1)$ can be set to a specific value. Obviously, if the voltage follower is used, the output voltage will be equal to the average of all input voltages.

Note that there are two basic differences between this averaging amplifier and that using the inverting configuration [see Equations (7-8d) and (7-10a)]:

1. No sign change or phase reversal occurs between the average of the inputs and output.
2. The noninverting input voltage V_1 is the average of all inputs, where as in the inverting averaging amplifier, the output is the average of all inputs, with a negative sign.

7-5.2(b) Summing amplifier. A close examination of Equation (7-10a) reveals that if the gain $(1 + R_F/R_1)$ is equal to the number of inputs, the output voltage becomes equal to the *sum* of all input voltages. That is, if $(1 + R_F/R_1) = 3$ [from Equation (7-10a)],

$$V_o = (V_a + V_b + V_c) \quad (7\text{-}10b)$$

Hence the circuit is called a *noninverting summing amplifier*.

Again, when the circuit of Figure 7-7 is used as either an averaging or summing amplifier, offset null circuitry or an offset null compensating network must be used to improve its accuracy.

Example 7-5

In the circuit of Figure 7-7, supply voltages $= \pm 15$ V, $V_a = +2$ V, $V_b = -3$ V, $V_c = +4$ V, $R = R_1 = 1$ kΩ, and $R_F = 2$ kΩ. Determine the voltage V_1 at the noninverting terminal and the output voltage V_o. Assume that the op-amp is initially nulled.

Solution. Using Equation (7-9),

$$V_1 = \frac{2 - 3 + 4}{3} = 1 \text{ V}$$

which is the average of three inputs: +2 V, −3 V, and +4 V. From Equation (7-10a),

$$V_o = \left(1 + \frac{2 \text{ k}\Omega}{1 \text{ k}\Omega}\right)(1) = 3 \text{ V}$$

which is the sum of the three inputs.

7-5.3 Differential Configuration

Using a basic differential op-amp configuration, a subtractor and a summing amplifier may be constructed as described below.

7-5.3(a) A subtractor. A basic differential amplifier can be used as a *subtractor* as shown in Figure 7-8. In this figure input signals can be scaled to the desired values by selecting appropriate values for the external resistors; when done so the circuit is referred to as *scaling amplifier*. However, in Figure 7-8, all external resistors are equal in value, so that the gain of the amplifier is equal to 1.

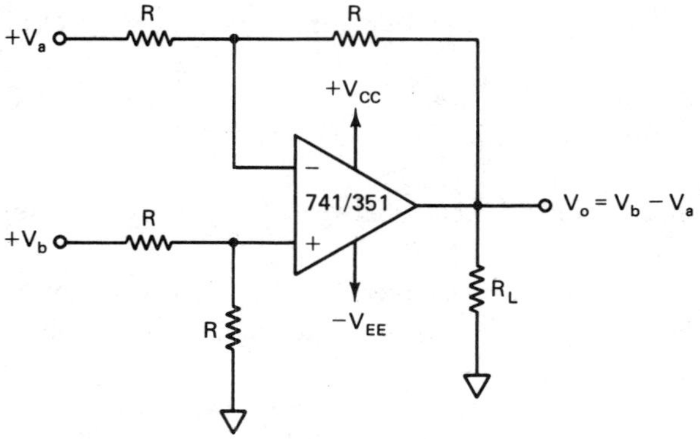

Figure 7-8 Basic differential amplifier used as a subtractor.

From this figure, the output voltage of the differential amplifier with a gain of 1 is

$$V_o = -\frac{R}{R}(V_a - V_b) \qquad \text{[see Equation (4-25)]}$$

That is,

$$V_o = V_b - V_a \tag{7-11}$$

Thus the output voltage V_o is equal to the voltage V_b applied to the noninverting terminal *minus* the voltage V_a applied to the inverting terminal; hence the circuit is called a *subtractor*.

7-5.3(b) Summing amplifier. A four-input summing amplifier may be constructed using the basic differential amplifier of Figure 7-8 if two additional input sources are connected, one each to the inverting and noninverting input terminals through resistor R (see Figure 7-9).

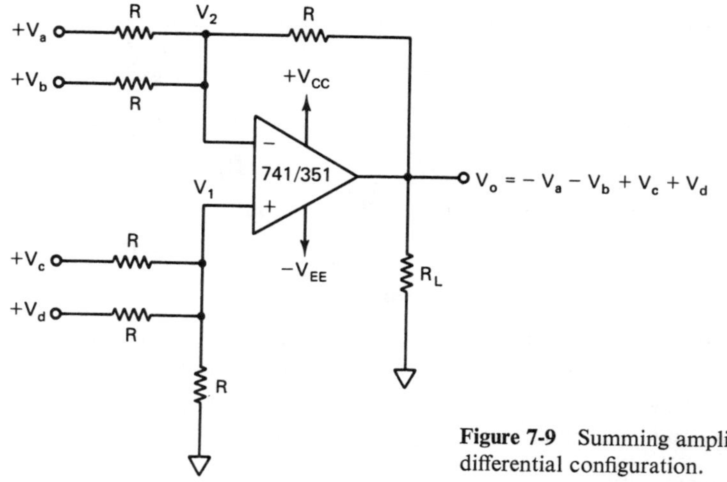

Figure 7-9 Summing amplifier using differential configuration.

The output voltage equation for this circuit can be obtained by using the superposition and Thévenin's theorems. For instance, to find the output voltage due to V_a alone, reduce all other input voltages V_b, V_c, and V_d to zero as shown in Figure 7-10(a). Then Thévenize the circuit to the left of node V_2. The resulting circuit is shown in Figure 7-10(b). In fact, this circuit is an inverting amplifier whose output voltage is

$$V_{oa} = -\frac{R}{R/2}\frac{V_a}{2} = -V_a$$

Similarly, the output voltage due to V_b alone is

$$V_{ob} = -V_b$$

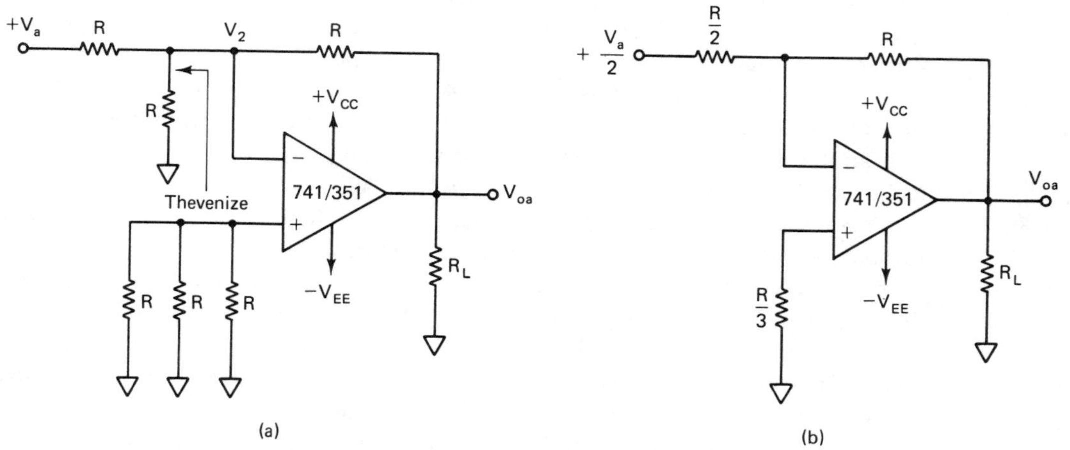

Figure 7-10 Deriving the output voltage equation for the summing amplifier of Figure 7-9.

Now if input voltages V_a, V_b, and V_d are set to zero, the circuit in Figure 7-9 becomes a noninverting amplifier in which the voltage V_1 at the noninverting input is

$$V_1 = \frac{R/2}{R + R/2} V_c = \frac{V_c}{3}$$

This means that the output voltage due to V_c alone is

$$V_{oc} = \left(1 + \frac{R}{R/2}\right) V_1 = (3)\left(\frac{V_c}{3}\right) = V_c$$

Similarly, the output voltage due to input voltage V_d alone is

$$V_{od} = V_d$$

Thus the output voltage due to all four input voltages is given by

$$\begin{aligned} V_o &= V_{oa} + V_{ob} + V_{oc} + V_{od} \\ &= -V_a - V_b + V_c + V_d \end{aligned} \tag{7-12}$$

Notice that the output voltage is equal to the *sum* of the input voltages applied to the noninverting terminal plus the negative sum of the input voltages applied to the inverting terminal. Even though in Figure 7-9 the gain of the summing amplifier is 1, any scale factor can be used for the inputs by selecting proper external resistors.

Example 7-6

In the circuit of Figure 7-9, $R = 1\ \text{k}\Omega$, $V_a = +2\ \text{V}$, $V_b = +3\ \text{V}$, $V_c = +4\ \text{V}$, $V_d = +5\ \text{V}$, and supply voltages $= \pm 15\ \text{V}$. Determine the output voltage V_o. Assume that the op-amp is initially nulled.

Solution. From Equation (7-12),

$$V_o = -2 - 3 + 4 + 5 = +4\ \text{V}$$

7-6 INSTRUMENTATION AMPLIFIER

In many industrial and consumer applications the measurement and control of physical conditions are very important. For example, measurements of temperature and humidity inside a dairy or meat plant permit the operator to make necessary adjustments to maintain product quality. Similarly, precise temperature control of plastic furnace is needed to produce a particular type of plastic.

Generally, a *transducer* is used at the measuring site to obtain required information easily and safely. The *transducer* is a device that converts one form of energy into another. For example, a strain gage when subjected to *pressure* or force (physical energy) undergoes a change in its *resistance* (electrical energy). An instrumentation system is used to measure the output signal produced by a transducer and often to control the physical signal producing it. Figure 7-11 shows a simplified form of such a system. The input stage is composed of a preamplifier and some sort of transducer, depending on the physical quantity to be measured. The output stage may use devices such as meters, oscilloscopes, charts, or magnetic recoders. In automatic instrumentation systems the output stage also includes an automatic process controller that compensates for changes in operating conditions.

In Figure 7-11 the connecting lines between the blocks represent *transmission lines*, especially when the transducer is at a remote test site monitoring hazardous

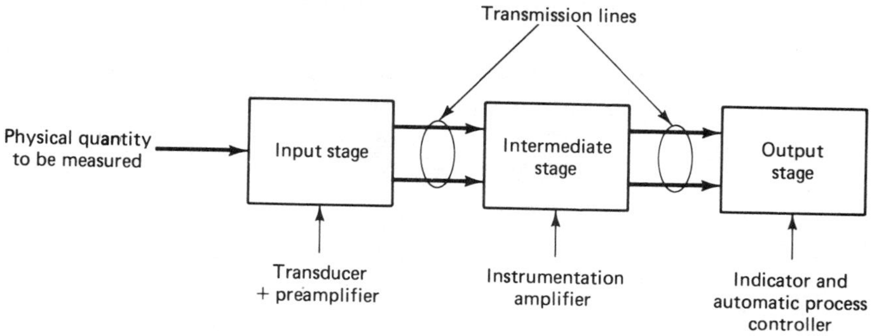

Figure 7-11 Block diagram of an instrumentation system.

conditions such as high temperatures or the liquid levels of flammable chemicals. These transmission lines permit signal transfer from unit to unit. The length of transmission lines depends primarily on the physical quantities to be monitored and on system requirements.

The signal source of the instrumentation amplifier is the output of the transducer. Although some transducers produce outputs with sufficient strength to permit their use directly, many do not. To amplify the low-level output signal of the transducer so that it can drive the indicator or display is the major function of the *instrumentation amplifier*. In short, the instrumentation amplifier is intended for precise, low-level signal amplification where low noise, low thermal and time drifts, high input resistance, and accurate closed-loop gain are required. Besides that, low power consumption, high common-made rejection ratio, and high slew rate are desirable for superior performance.

There are many instrumentation operational amplifiers, such as the μA725, ICL7605, and LH0036, which make a circuit extremely stable and accurate. These ICs are, however, relatively expensive; they are very precise special-purpose circuits in which most of the electrical parameters, such as offsets, drifts, and power consumption, are minimized, whereas input resistance, CMRR, and supply range are optimized. Some instrumentation amplifiers are even available in modular form to suit special installation requirements.

Obviously, the requirements for instrumentation op-amp are more rigid than those for general-purpose applications. However, where the requirements are not too strict, the general-purpose op-amp can be employed in the differential mode. The differential amplifiers discussed in Chapter 4 can be employed as instrumentation amplifiers. We will call such amplifiers *differential instrumentation amplifiers*. Since most of the instrumentation systems use a transducer in a bridge circuit, we will consider a simplified differential instrumentation system arrangement using a transducer bridge circuit.

7-6.1 Instrumentation Amplifier Using Transducer Bridge

Figure 7-12 shows a simplified differential instrumentation amplifier using a transducer bridge. A *resistive transducer* whose resistance changes as a function of some physical energy is connected in one arm of the bridge with a small circle around it and is denoted by $(R_T \pm \Delta R)$, where R_T is the resistance of the transducer and ΔR the change in resistance R_T.

The bridge in the circuit of Figure 7-12 is dc excited but could be ac excited as well. For the balanced bridge at some reference condition:

$$V_b = V_a$$

or

$$\frac{R_B(V_{dc})}{R_B + R_C} = \frac{R_A(V_{dc})}{R_A + R_T}$$

That is,

$$\frac{R_C}{R_B} = \frac{R_T}{R_A} \tag{7-13}$$

Generally, resistors R_A, R_B, and R_C are selected so that they are equal in value to the transducer resistance R_T at some *reference condition*. The reference condition is the specific value of the physical quantity under measurement at which the bridge is balanced. This value is normally established by the designer and depends on the transducer's characteristics, the type of physical quantity to be measured, and the desired application.

The bridge is balanced initially at a desired reference condition. However, as the physical quantity to be measured changes, the resistance of the transducer also changes, which causes the bridge to unbalance ($V_a \neq V_b$). The output voltage of the

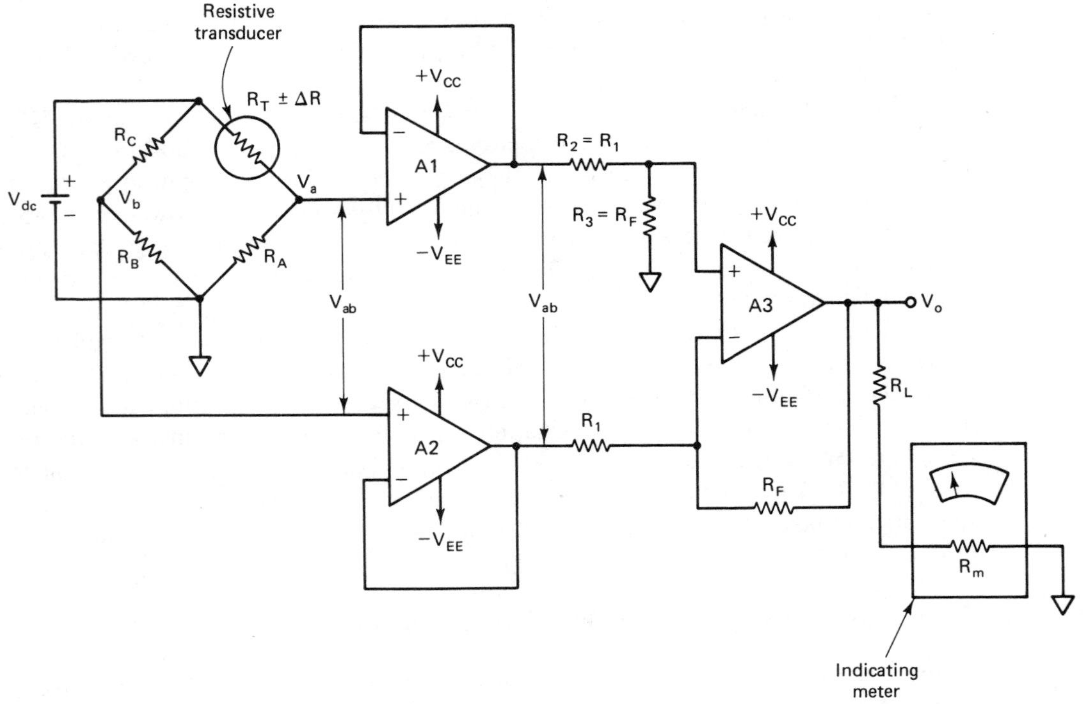

Figure 7-12 Differential instrumentation amplifier using transducer bridge.

bridge can be expressed as a function of the change in resistance of the transducer, as described below.

Let the change in resistance of the transducer be ΔR. Since R_B and R_C are fixed resistors, the voltage V_b is constant. However, voltage V_a varies as a function of the change in transducer resistance. Therefore, according to the voltage divider rule,

$$V_a = \frac{R_A(V_{dc})}{R_A + (R_T + \Delta R)}$$

$$V_b = \frac{R_B(V_{dc})}{R_B + R_C}$$

Consequently, the voltage V_{ab} across the output terminals of the bridge is

$$V_{ab} = V_a - V_b$$

$$= \frac{R_A V_{dc}}{R_A + R_T + \Delta R} - \frac{R_B V_{dc}}{R_B + R_C}$$

However, if $R_A = R_B = R_C = R_T = R$, say then

$$V_{ab} = -\frac{\Delta R(V_{dc})}{2(2R + \Delta R)} \tag{7-14}$$

The output voltage V_{ab} of the bridge is then applied to the differential instrumentation amplifier composed of three op-amps (see Figure 7-12). The voltage followers preceding the basic differential amplifier help to eliminate loading of the bridge circuit. The gain of the basic differential amplifier is (R_F/R_1); therefore, the output V_o of the circuit in Figure 7-12 is

$$V_o = V_{ab}\left(\frac{R_F}{R_1}\right) = \frac{-(\Delta R)V_{dc}}{2(2R + \Delta R)}\frac{R_F}{R_1} \tag{7-15}$$

This means that V_o is a function of the change in resistance ΔR of the transducer. Since the change in resistance is caused by change in physical energy, a meter connected at the output can be calibrated in terms of the units of that physical energy. If a transducer operates within its specified limits, the change in resistance ΔR can be assumed to be linear.

Before proceeding with specific bridge applications, let us briefly consider the important characteristics of some resistive types of transducers. In these resistive types of transducers the resistance of the transducer changes as a function of some physical quantity. The thermistors, photoconductive cells, and strain gages are some of the most commonly used resistive transducers; hence they will be further discussed here.

Thermistors are essentially semiconductors which behave as resistors, usually with a *negative temperature coefficient of resistance*. That is, as the temperature of a thermistor increases, its resistance decreases. The temperature coefficient of resistance is expressed in ohms per unit change in degrees Celsius (°C). Thermistors with a high temperature coefficient of resistance are more sensitive to temperature change and are therefore well suited to temperature measurement and control. Thermistors are available in a wide variety of shapes and sizes. However, thermistor beads sealed in the tips of glass rods are most commonly used because they are relatively easy to mount.

The photoconductive cell belongs to the family of photodetectors (photosensitive devices) whose resistance varies with an incident radiant energy or with light. As the intensity of incident light increases, the resistance of the cell decreases. The resistance of the photoconductive cell in darkness is typically of the order of 100 kΩ. Generally, the resistance of the cell in darkness and at particular light intensities is listed on the data sheet. The intensity of light is expressed in meter candles (lux).

Materials such as cadmium sulfide and silicon, whose conductivity is a function of incident radiant energy, are used for photoconductive cells. Some cells are extremely sensitive to light and hence can be used into the ultraviolet and infrared regions. The photoconductive cell is typically composed of a ceramic base, a layer of photoconductive material, a moisture-proof enclosure, and metallic leads. Photoconductive cells are also known as photocells or light-dependent resistors (LDRs).

Another important resistive transducer is the strain gage, whose resistance changes due to elongation or compression when an external stress is applied. The stress is defined as force per unit area [Newtons/(meter)^2] and can be related to pressure, torque, and displacement. Therefore, a strain gage may be used to monitor change in applied pressure, torque, and displacement by measuring the corresponding change in the gage's resistance.

Two basic types of strain gages are the *wire* and *semiconductor*. Semiconductor strain gages are much more sensitive than the wire type, and therefore provide better accuracy and resolution. The *sensitivity* of a strain gage is defined as unit change in resistance per unit change in length and is a dimensionless quantity.

The thermistor, photocell, and strain gage are all passive transducers, meaning that they require external voltage (ac or dc) for their operation.

7-6.1(a) Temperature indicator. The circuit of Figure 7-12 can be used as a temperature indicator if the transducer in the bridge circuit is a thermistor and the output meter is calibrated in degrees centigrade or Fahrenheit. The bridge can be balanced at a desired reference condition, say 25°C. As the temperature varies from its reference value, the resistance of the thermistor changes and the bridge becomes unbalanced. This unbalanced bridge in turn produces the meter movement. The meter can be calibrated to read a desired temperature range by selecting an appropriate gain for the differential instrumentation amplifier. In Figure 7-12 the meter movement is dependent on the amount of imbalance in the bridge, that is, the change ΔR in the value of the thermistor resistance. The ΔR for the thermistor, however, can be determined as follows: ΔR = (temperature coefficient of resistance) (final temperature − reference temperature.) See Example 7-7.

Example 7-7

In the circuit of Figure 7-12, $R_1 = 1$ kΩ, $R_F = 4.7$ kΩ, $R_A = R_B = R_C = 100$ kΩ, $V_{dc} = +5$ V, and op-amp supply voltages = ±15 V. The transducer is a thermistor with the following specifications: $R_T = 100$ kΩ at a reference temperature of 25°C; temperature coefficient of resistance = −1 kΩ/°C or 1%/°C. Determine the output voltage at 0°C and at 100°C.

Solution. At 25°C, $R_A = R_B = R_C = R_T = 100$ kΩ. Therefore, the bridge is balanced ($V_a = V_b$) and $V_o = 0$ V. However, at 0°C the change ΔR in the resistance of the thermistor is $\Delta R = \frac{(-1\text{ k}\Omega)}{°\text{C}}(0°\text{C} - 25°\text{C}) = 25\text{ k}\Omega$

Therefore; using Equation 7-15, at 0°C,

$$V_o = \frac{-(25\text{ K})(5)}{2(200\text{ K} + 25\text{ K})} \cdot \frac{4.7\text{ K}}{1\text{ K}} = -1.31\text{ V}$$

Similarly, at 100°C, $\Delta R = \frac{-1\text{ K}\Omega}{°\text{C}}(100°\text{C} - 25°\text{C}) = -75\text{ K}\Omega$ and,

$$V_o = \frac{-(-75\text{ K})(5)}{2(200\text{ K} - 75\text{ K})} \cdot \frac{4.7\text{ K}}{1\text{ K}} = 7.05\text{ V}$$

Thus, when $V_o = -1.31$ V, the meter face can be marked as 0°C and when $V_o =$ 7.05 V, it can be marked as 100°C. Note that at 25°C, $V_o = 0$ V; therefore, a center-zero meter is required. Assuming that the resistance-temperature characteristic of the thermistor is linear, the meter may be interpolated linearly between 25° to 0°C and 25° to 100°C.

Thermistors with relatively higher resistance ($R_T \geq 1$ MΩ) and sensitivity (temperature coefficient of resistance $\geq 3\%/°$C) are best suited for *remote measurements* because the effect of transmission line resistance is negligible.

7-6.1(b) Temperature controller. A simple and inexpensive temperature control circuit may be constructed by using a thermistor in the bridge circuit and by replacing a meter with a relay in the circuit of Figure 7-12. The output of the differential instrumentation amplifier drives a relay that controls the current in the heat-generating circuit. A properly designed circuit should energize a relay when the temperature of the thermistor drops below a desired value, causing the heat unit to turn on.

7-6.1(c) Light-intensity meter. The circuit in Figure 7-12 can be used as a light-intensity meter if a transducer is a photocell. The bridge can be balanced for darkness conditions. Therefore, when exposed to light, the bridge will be unbalanced and cause the meter to deflect. The meter can be calibrated in terms of *lux* to measure the change in light intensity.

The light-intensity meter using an instrumentation bridge amplifier is more accurate and stable than single input inverting or noninverting configuration because the common-mode (noise) voltages are effectively rejected by the differential configuration.

7-6.1(d) Measurement of flow and thermal conductivity. A *flow meter* or a *thermal conductivity meter* may be constructed using the circuit of Figure 7-12, provided that two thermistors are used adjacent to each other in the bridge. For instance, assume that R_C and R_T represent two identical thermistors. For the flow measurement one thermistor is sealed in a small-cavity copper cylinder and the other installed in a small copper pipe. When no air flows through the pipe, the bridge can be balanced so that the output voltage is zero. When air flows over the thermistor, its temperature decreases and in turn the resistance increases, causing the bridge to be unbalanced. This unbalanced voltage is then amplified by the differential instrumentation amplifier and applied to the meter. Thus the amount of meter deflection is proporational to the flow rate of the air in the pipe. The meter can be calibrated in $(\text{meter})^3/\text{s}$ to accommodate a desired flow rate range.

For the thermal conductivity measurement, two thermistors are mounted in separate small copper cylinders. With air in both cylinders, the bridge can be balanced, and hence the output will be zero. When air in one cylinder is replaced by a medium being tested, which has a different thermal conductivity than air, the bridge becomes unbalanced. This happens because the temperature of the thermistor changes, thus changing its resistance. Suppose that the medium being tested is carbon dioxide $(CO)_2$. Because of its lower thermal conductivity, CO_2 will increase the temperature of the thermistor, which will cause a decrease in the thermistor's resistance. This results in unbalancing the bridge, which in turn produces a meter deflection. The meter can be calibrated in terms of relative thermal conductivity (cal/s-cm-°C).

7-6.1(e) Analog weight scale. By connecting a strain gage in the bridge, the circuit of Figure 7-12 can be converted into a simple and inexpensive *analog weight scale*.

In the analog weight scale, strain gage elements are connected in all four arms of the bridge. The elements are mounted on the base of the weight platform so that when an external force or weight is applied to the platform, one pair of elements in the opposite arms elongates, whereas the other pair of elements in the opposite arms compresses. In other words, when the weight is placed on the platform, R_{T1} and R_{T3} both decrease in resistance, and R_{T2} and R_{T4} both increase in resistance, or vice versa (see Figure 7-13).

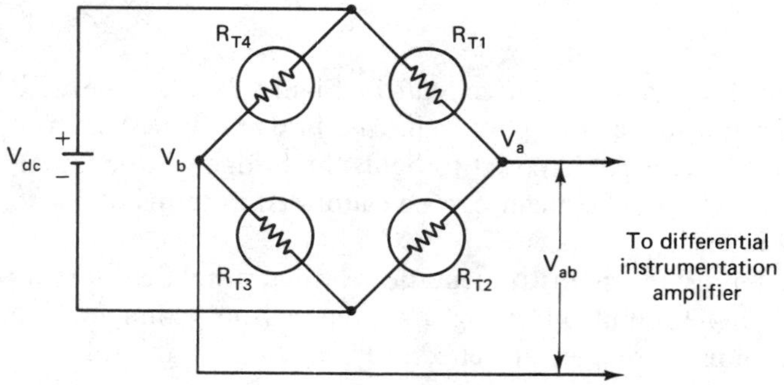

Figure 7-13 Strain-gage bridge circuit for analog weight scale $R_{T1} = R_{T2} = R_{T3} = R_{T4}$.

When no weight is placed on the platform, the bridge is balanced $R_{T1} = R_{T2} = R_{T3} = R_{T4} = R$, say, and the output voltage of the weight scale can be zero. When a weight is placed on the scale platform, the bridge becomes unbalanced. Assuming that R_{T1} and R_{T3} decrease in resistance and R_{T2} and R_{T4} increase in resistance by the same number of ohms ΔR, the unbalanced voltage V_{ab} is given by

$$V_{ab} = +V_{dc}\left(\frac{\Delta R}{R}\right) \tag{7-16}$$

where V_{dc} = dc excitation voltage of the bridge

$R = R_{T1} = R_{T2} = R_{T3} = R_{T4}$ = unstrained gage resistance

ΔR = change in gage resistance

Remember that if the decrease in gage resistance R_{T1} and R_{T3} is ΔR, the increase in resistance of R_{T2} and R_{T4} is also ΔR.

The voltage V_{ab} is then amplified by the differential instrumentation amplifier, which drives the meter. Since the gain of the amplifier is $(+R_F/R_1)$, the output voltage V_o is

$$V_o = V_{dc}\left(\frac{\Delta R}{R}\right)\frac{R_F}{R_1} \tag{7-17}$$

The gain of the amplifier can be selected according to the sensitivity of the strain gage and the full-scale deflection requirements of the meter. The meter can be calibrated in terms of kilograms.

For better accuracy and resolution, a microprocessor-based digital weight scale may be constructed. However, such a scale is much more complex and expensive than the analog scale.

Example 7-8

The circuit of Figure 7-12 is used as an analog weight scale with the following specifications. The gain of the differential instrumentation amplifier = 100. Assume that $V_{dc} = +10\,\text{V}$ and that the op-amp supply voltages = $\pm 10\,\text{V}$. The unstrained resistance of each of the four elements of the strain gage is 100 Ω. When a certain weight is placed on the scale platform, the output voltage $V_o = 1$ V. Assuming that the output is initially zero, determine the change in the resistance of each strain-gage element.

Solution. Using Equation (7-17),

$$1 = (10)\frac{\Delta R}{100}(100)$$

Therefore, $\Delta R = 0.1\,\Omega$. This means that R_{T1} and R_{T3} will decrease by 0.1 Ω if R_{T2} and R_{T4} increase by 0.1 Ω when a certain weight is placed on the scale platform.

7-7 DIFFERENTIAL INPUT AND DIFFERENTIAL OUTPUT AMPLIFIER

In all the applications discussed so far, the op-amp is used with a single-ended or unbalanced output. However, in certain applications a differential output is required. The differential input and differential output amplifier is most commonly used as preamplifier and in driving push-pull arrangements.

Figure 7-14(a) shows one possible arrangement of a differential input and differential output amplifier using two identical op-amps, that is, a dual op-amp. For example, the μAF772 dual op-amp incorporates two well-matched op-amps. The connection diagrams of the 8-pin mini DIP and 14-pin DIP are shown in Figure 7-14(b). The analysis of the circuit in Figure 7-14(a) can be accomplished by determining the output of each op-amp due to the differential input. Using the superposition theorem, the output V_{ox} due to inputs V_x and V_y is

$$V_{ox} = \left(1 + \frac{R_F}{R_1}\right)V_x - \left(\frac{R_F}{R_1}\right)V_y \tag{7-18a}$$

Similarly, the output V_{oy} is

$$V_{oy} = \left(1 + \frac{R_F}{R_1}\right)V_y - \left(\frac{R_F}{R_1}\right)V_x \tag{7-18b}$$

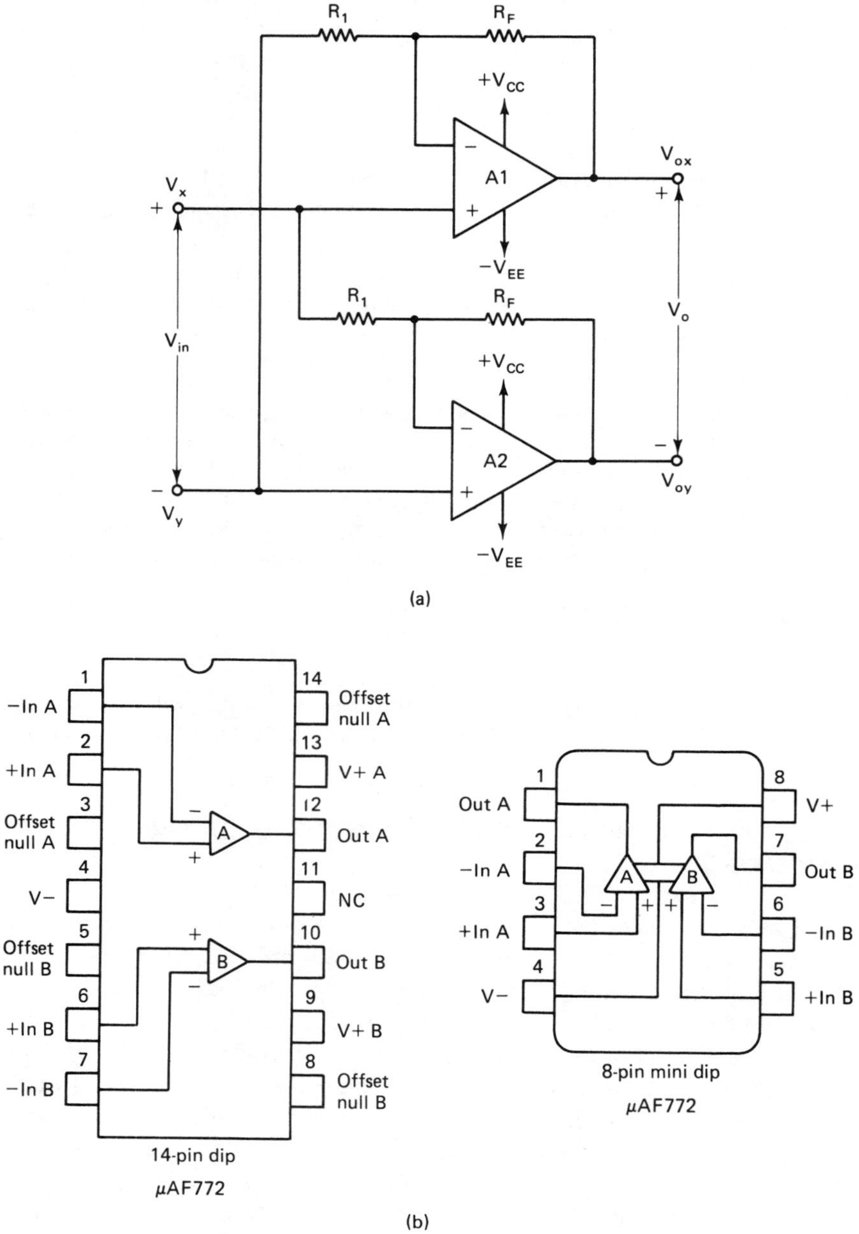

Figure 7-14 (a) Differential input and differential output amplifier using a dual op-amp. (b) The connection diagram of a typical dual op-amp. (Note especially the difference between the supply voltage and offset-null connections.) (Courtesy of Fairchild Camera and Instrument Corporation.)

However, the differential output V_o is

$$V_o = V_{ox} - V_{oy}$$

Therefore, from Equations (7-18a) and (7-18b),

$$V_o = \left(1 + \frac{R_F}{R_1}\right)V_x - \left(\frac{R_F}{R_1}\right)V_y - \left(1 + \frac{R_F}{R_1}\right)V_y + \left(\frac{R_F}{R_1}\right)V_x$$

$$= \left(1 + \frac{2R_F}{R_1}\right)(V_x - V_y)$$

or

$$V_o = \left(1 + \frac{2R_F}{R_1}\right)V_{in} \tag{7-18c}$$

This means that the differential input and output are in phase or of the same polarity provided that $V_{in} = V_x - V_y$ and $V_o = V_{ox} - V_{oy}$.

The differential input and output amplifier of Figure 7-14(a) is very useful in noisy environments, especially if the input signal is relatively smaller because it rejects the common-mode noise voltages. Recall also that the output voltage swing with differential output is twice as large as that with a single-ended output.

Another important advantage of the differential input and output amplifier is that it produces a bipolar output with a single supply.

Example 7-9

The differential input and output amplifier of Figure 7-14(a) is used as a preamplifier and requires a differential output of at least 3.7 V. Determine the gain of the circuit if the differential input $v_{in} = 100$ mV.

Solution. Using Equation (7-18c), we get

$$3.7 = \left(1 + \frac{2R_F}{R_1}\right)(100)(10^{-3})$$

$$1 + \frac{2R_F}{R_1} = 37 \quad \text{or} \quad R_F = 18R_1$$

If $R_1 = 100\ \Omega$, then $R_F = 1.8\ \text{k}\Omega$.

7-8 VOLTAGE-TO-CURRENT CONVERTER WITH FLOATING LOAD

Figure 7-15 shows a voltage-to-current converter in which load resistor R_L is *floating* (not connected to ground). The input voltage is applied to the noninverting input terminal, and the feedback voltage across R_1 drives the inverting input terminal. This circuit is also called a *current-series negative feedback* amplifier because the feedback voltage across R_1 (applied to the inverting terminal) depends on the output current i_o and is in series with the input difference voltage v_{id}.

Writing Kirchhoff's voltage equation for the input loop,

$$v_{in} = v_{id} + v_f$$

But $v_{id} \cong 0$ V, since A is very large; therefore,

$$v_{in} = v_f$$

$$v_{in} = R_1 i_o$$

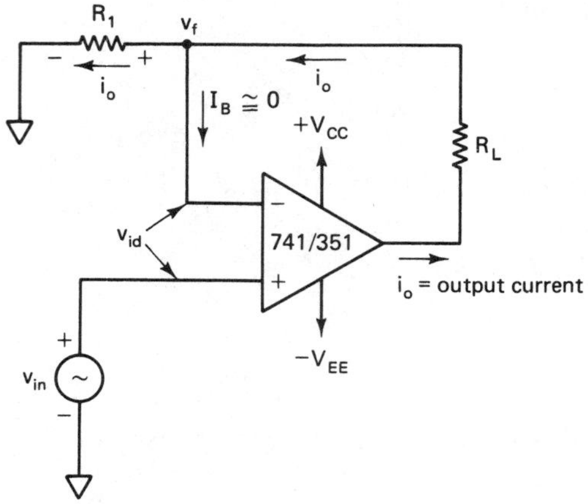

Figure 7-15 Voltage-to-current converter with floating load.

or

$$i_o = \frac{v_{in}}{R_1} \tag{7-19}$$

This means that in the circuit of Figure 7-15, an input voltage v_{in} is converted into an output current of v_{in}/R_1. In other words, input voltage v_{in} appears across R_1. If R_1 is a precision resistor, the output current ($i_o = v_{in}/R_1$) will be precisely fixed.

The voltage-to-current converter can be used in such applications as low-voltage dc and ac voltmeters, diode match finders, light-emitting diodes (LEDs), and zener diode testers.

7-8.1 Low-Voltage DC Voltmeter

If in the converter circuit of Figure 7-15, we replace the load resistor R_L by an ammeter (D'Arsonval meter movement) with a full-scale deflection of 1 mA, the resulting circuit is the dc voltmeter of Figure 7-16. This figure shows an external offset voltage compensating network which is used because it makes the nulling of the op-amp relatively easy. Using the *offset null circuitry* of the 741 (10-kΩ potentiometer), the op-amp sometimes cannot be nulled because the output is very sensitive to even slight variations in wiper position.

The Thévenin equivalent resistance of the compensating network is approximately equal to 926 Ω (12.5 kΩ || 1 kΩ). Therefore, when the switch is in ×1 position,

$$R_1 = 926 + 100 = 1.026 \text{ k}\Omega \cong 1 \text{ k}\Omega$$

If $V_{in} = 1$ V, then

$$I_o = \frac{V_{in}}{R_1} \cong \frac{1 \text{ V}}{1 \text{ k}\Omega} = 1 \text{ mA}$$

This means that 1 V causes the full-scale deflection of the ammeter.

If the range switch is changed to ×10 position (926 Ω + 9 kΩ ≅ 10 kΩ), it will require a 10-V input to get full-scale deflection. Thus successively higher resistance values are required to measure relatively higher input voltages. However, the input voltage range for the 741C op-amp is ±14 V. Therefore, with ±15 V supply voltages,

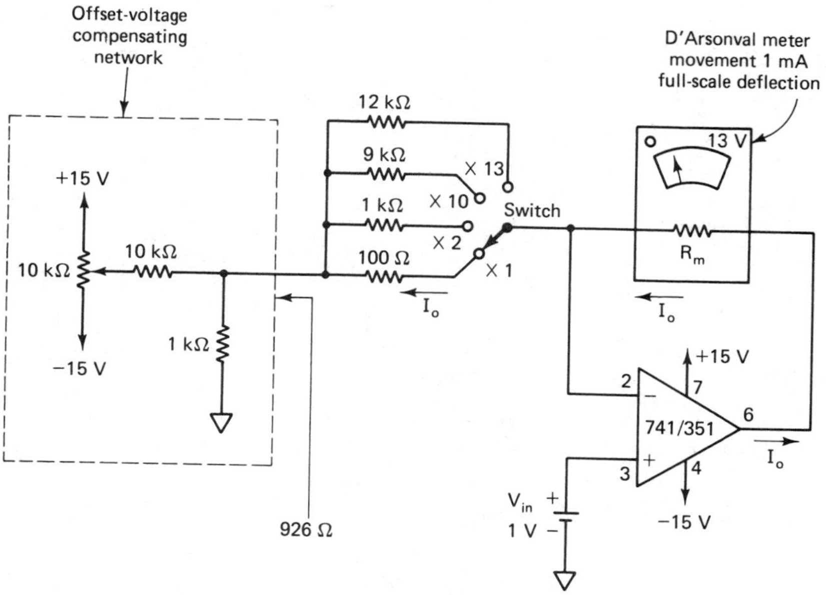

Figure 7-16 Dc voltmeter with 1 to 13 V full-scale range.

the maximum input voltage has to be $\leq \pm 14$ V. In Figure 7-16 the maximum full-scale input voltage of 13 V can be applied when the range switch is in $\times$ 13 position.

Thus by calibrating the face of the ammeter in volts, a dc voltmeter with a full-scale voltage range of 1 to 13 V can be constructed. Note that meter resistance R_m does not affect I_o. Only V_{in} and R_1 determine the I_o value.

A centered-zero ammeter can be used to measure positive as well as negative input voltages. To improve the accuracy of voltage measurements, the ammeter should be nulled each time before input voltage is applied.

7-8.2 Low-Voltage AC Voltmeter

It is also possible to modify the meter movement so that it will indicate values of ac voltage or current. A combination of an ammeter and a full-wave rectifier can be employed in the feedback loop to form an ac voltmeter, as shown in Figure 7-17. In this circuit an alternating current is converted into a direct current. During the positive half-cycle of the ac input diodes D_1 and D_3 conduct, whereas diodes D_2 and D_4 conduct during the negative half-cycle of v_{in}. Thus the current through the ammeter flows only in one direction (A to B) for the entire cycle of the ac input. In other words, the ammeter registers the average (dc) value of the rectified current. The amount of deflection of the pointer must be considered for proper scale calibration. For full-wave rectification, meter current can be expressed as

$$i_o = \frac{0.9 v_{in}}{R_1} \tag{7-20a}$$

or

$$v_{in} = (1.1 R_1) i_o \tag{7-20b}$$

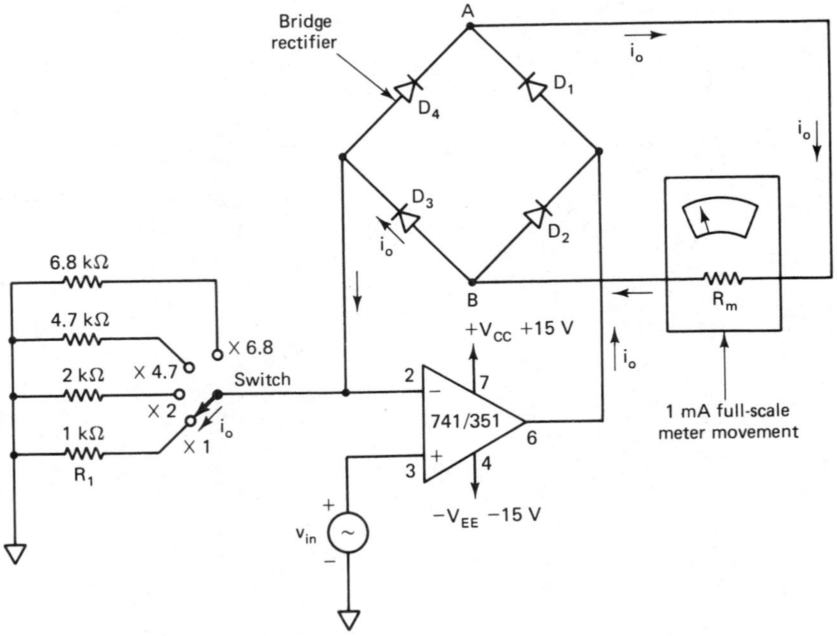

Figure 7-17 Ac voltmeter.

where i_o = dc meter current (amperes)
v_{in} = ac rms input voltage (volts)
R_1 = resistance in series with the meter (ohms)

Using Equation (7-20b), the ammeter face can be calibrated to read an rms value of the input voltage.

Rectifier-type meters are designed primarily for ac frequencies in the audio range. However, because of the low slew rate of the 741C op-amp, the frequency of the input signal should be less than 4 kHz for proper operation of the voltmeter. Also, the *peak inverse voltage* (PIV) of the diodes is recommended to be larger than the saturation voltages of the op-amp.

Example 7-10

In the circuit of Figure 7-17, for the indicated values of resistors, determine the full-scale range for the input voltage.

Solution. The full-scale meter movement is 1 mA. Therefore, substituting minimum and maximum values of resistor R_1 in Equation (7-20b), we can determine the range for the input voltage.

$$v_{in}(\text{rms})|_{\text{minimum}} = (1.1)(1\text{ k}\Omega)(1\text{ mA}) = 1.1\text{ V}$$
$$v_{in}(\text{rms})|_{\text{maximum}} = (1.1)(6.8\text{ k}\Omega)(1\text{ mA}) = 7.48\text{ V}$$

Thus the full-scale input voltage ranges from 1.1 to 7.48 V rms or 3.1 to 21.16 V peak to peak. Note that since the input voltage range for the 741C is ±14 V, the input voltage swing must be less than 28 V peak to peak for safe operation of the circuit.

7-8.3 Diode Match Finder

In some circuits, such as the *ring modulator* and *Foster–Sealy discriminator*, it is necessary to have matched diodes with equal voltage drops at a particular value of diode current. The circuit shown in Figure 7-18 can be used in finding matched diodes and is obtained from Figure 7-15 by replacing R_L with a diode.

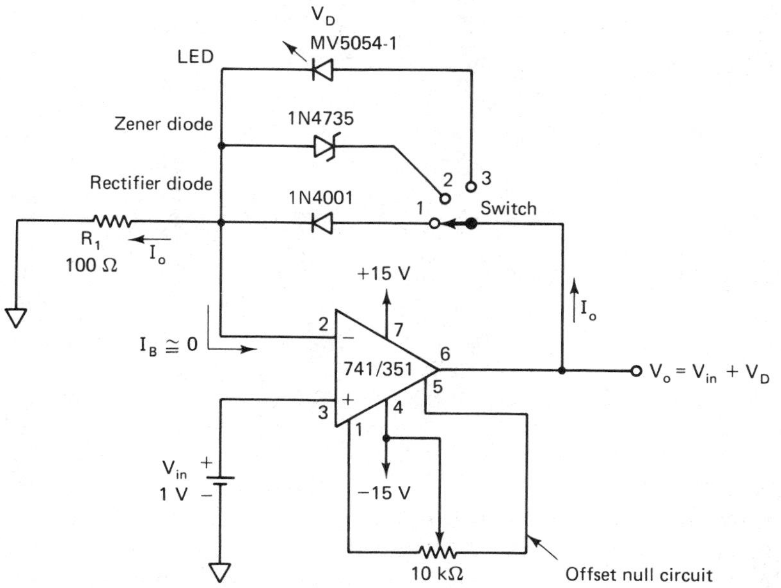

Figure 7-18 Diode match finder.

When the switch is in position 1, the rectifier diode 1N4001 is placed in the feedback loop; the current through this loop is set by input voltage V_{in} and resistor R_1. For $V_{in} = 1$ V and $R_1 = 100\ \Omega$, the current through the diode is

$$I_o = \frac{V_{in}}{R_1} = \frac{1}{100} = 10 \text{ mA}$$

As long as V_{in} and R_1 are constant, the current I_o will be constant. The voltage drop across the diode can be found either by measuring the voltage across it or the output voltage. The output voltage is equal to $(V_{in} + V_D)$. To avoid an error in output voltage, the op-amp should be *initially* nulled.

Thus the matched diodes can be found by connecting diodes one after another in the feedback path and measuring the voltage across them. A desired testing current I_o can be established by selecting an appropriate combination of V_{in} and R_1.

Example 7-11

In the circuit of Figure 7-18 when the switch is in position 1, $V_{in} = 0.5$ V and $V_o = 1.2$ V. Determine the current through the diode and the voltage drop across it. Assume that the op-amp is initially nulled.

Solution

$$I_o = \frac{V_{\text{in}}}{R_1} = \frac{0.5}{100} = 5 \text{ mA}$$

$$V_D = V_o - V_{\text{in}} = 1.2 - 0.5 = 0.7 \text{ V}$$

7-8.4 Zener Diode Tester

The circuit of Figure 7-18 becomes a zener diode tester when the switch is placed in position 2. This circuit can be used to find the breakdown voltage of zener diodes. V_{in} and R_1 set the zener current at a constant value. If this current is larger than the knee current (I_{zk}) of the zener, the zener blocks V_z volts. For example, the 1N4735 zener in the circuit of Figure 7-18 has $I_{zk} = 1$ mA and $V_z = 6.2$ V. Since the current through the zener is

$$I_o = \frac{V_{\text{in}}}{R_1} = \frac{1}{100} = 10 \text{ mA} > I_{zk}$$

the voltage across the zener will be approximately equal to 6.2 V.

Matched zeners are useful in obtaining split ($\pm$) supply voltages from a single supply.

7-8.5 Light-Emitting Diode Tester

The circuit of Figure 7-18 can be converted into a light-emitting diode (LED) tester when the switch is in position 3. Again the LED current is set at a constant value by V_{in} and R_1. The LEDs can be tested for brightness one after another at this current. Matched LEDs with equal brightness at a specific value of current are useful as indicators and display devices in digital applications.

In all the applications described above it should be remembered that the maximum current through the load (ammeter, rectifier and zener diodes, or LED) cannot exceed the *short-circuit curent* of the op-amp. For the 741C this current is 25 mA.

7-9 VOLTAGE-TO-CURRENT CONVERTER WITH GROUNDED LOAD

Another version of the voltage-to-current converter is shown in Figure 7-19. In this circuit one terminal of the load is grounded, and load current is controlled by an input voltage. The analysis of the circuit is accomplished by first determining the voltage V_1 at the noninverting input terminal and then establishing the relationship between V_1 and the load current.

Writing Kirchhoff's current equation at node V_1,

$$I_1 + I_2 = I_L$$

$$\frac{V_{\text{in}} - V_1}{R} + \frac{V_0 - V_1}{R} = I_L$$

$$V_{\text{in}} + V_0 - 2V_1 = I_L R$$

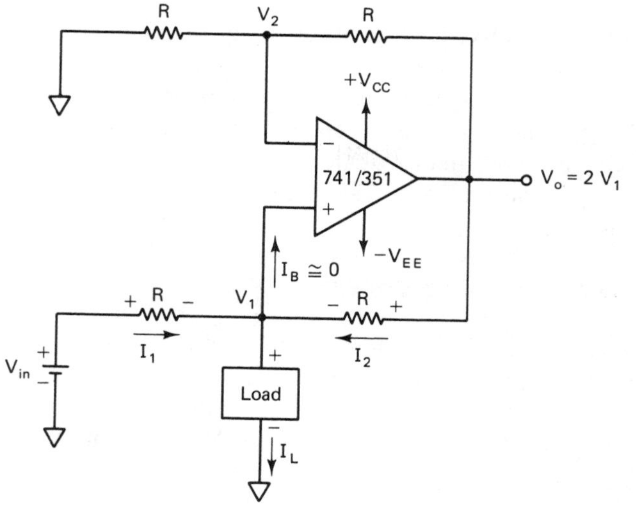

Figure 7-19 Voltage-to-current converter with grounded load.

Therefore,

$$V_1 = \frac{V_{in} + V_0 - I_L R}{2} \tag{7-21a}$$

Since the op-amp is connected in the noninverting mode, the gain of the circuit in Figure 7-19 is $1 + R/R = 2$. Then the output voltage is

$$V_0 = 2V_1 = V_{in} + V_0 - I_L R$$

That is,

$$V_{in} = I_L R$$

or

$$I_L = \frac{V_{in}}{R} \tag{7-21b}$$

This means that the load current depends on the input voltage V_{in} and resistor R. Notice that all resistors must be equal in value.

The voltage-to-current converter of Figure 7-19 may also be used in testing such devices as zeners and LEDs forming a ground load. However, the circuit will perform satisfactorily provided that load size $\leq R$ value.

Example 7-12

In the circuit of Figure 7-19, $V_{in} = 5$ V, $R = 10$ kΩ, and $V_1 = 1$ V. Find (a) the load current and (b) the output voltage V_o. Assume that the op-amp is initially nulled.

Solution

(a) Using Equation (7-21b),

$$I_L = \frac{V_{in}}{R} = \frac{5}{10\text{ k}\Omega} = 0.50\text{ mA}$$

(b) Since $V_{in} = I_L R$, from Equation (7-21a), $V_o = 2V_1 = 2$ V.

7-10 CURRENT-TO-VOLTAGE CONVERTER

In Chapter 4 the current-to-voltage converter was presented as a special case of the inverting amplifier in which an input current is converted into a proportional output voltage. One of the most common uses of the current-to-voltage converter is in digital-to-analog circuits (DACs) and in sensing current through photodetectors such as photocells, photodiodes, and photovoltaic cells. Photosensitive devices produce a current that is proportional to an incident radiant energy or light and therefore can be used to detect the light.

7-10.1 DAC Using Current-to-Voltage Converter

Figure 8-20 shows a combination of a DAC and current-to-voltage converter. The eight-digit binary signal is the input to the MC1408 DAC and V_o is the corresponding analog output of the current-to-voltage converter. The output of the MC1408 is current I_o, the value of which depends on the logic state (0 ro 1) of the binary inputs as indicated by the following equation:

$$I_o = \frac{V_{\text{ref}}}{R_1}\left(\frac{D_7}{2} + \frac{D_6}{4} + \frac{D_5}{8} + \frac{D_4}{16} + \frac{D_3}{32} + \frac{D_2}{64} + \frac{D_1}{128} + \frac{D_0}{256}\right) \tag{7-22a}$$

where I_o = output current of the DAC (mA)
R_1 = resistance (kΩ)
V_{ref} = reference voltage (volts)
D_0 through D_7 = eight binary inputs

This means that I_o is zero when all the inputs are logic "0" and I_o is maximum when all the inputs are logic "1." Thus the value of I_o is a function of the state of eight binary inputs. The variations in I_o can be converted into a desired output voltage range by selecting a proper value for R_F since

$$V_o = I_o R_F \tag{7-22b}$$

where I_o is given by Equation (7-22a). It is common to parallel R_F with a capacitance C to minimize overshoot and ringing. Note that in Figure 7-20 the output voltage V_o of the current-to-voltage converter is positive because the direction of input current I_o is opposite to that in Figure 4-12.

Example 7-13

In the circuit of Figure 7-20, $V_{\text{ref}} = 2$ V, $R_1 = 1$ kΩ, and $R_F = 2.7$ kΩ. Assuming that the op-amp is initially nulled, determine the range for the output voltage V_o.

Solution. When all the binary inputs D_0 through D_7 are logic 0, the current $I_o = 0$; therefore, the minimum value of $V_o = 0$ V. However, when all the inputs are logic 1, the I_o [using Equation (7-22a)] is

$$I_o = \frac{2}{1\text{ k}\Omega}\left(\frac{1}{2} + \frac{1}{4} + \frac{1}{8} + \frac{1}{16} + \frac{1}{32} + \frac{1}{64} + \frac{1}{128} + \frac{1}{256}\right)$$
$$= 1.992 \text{ mA}$$

Hence the maximum value of output voltage is

$$V_o = I_o R_F = (1.992 \text{ mA})(2.7 \text{ k}\Omega) = 5.38 \text{ V}$$

Thus the output voltage range is 0 to 5.38 V.

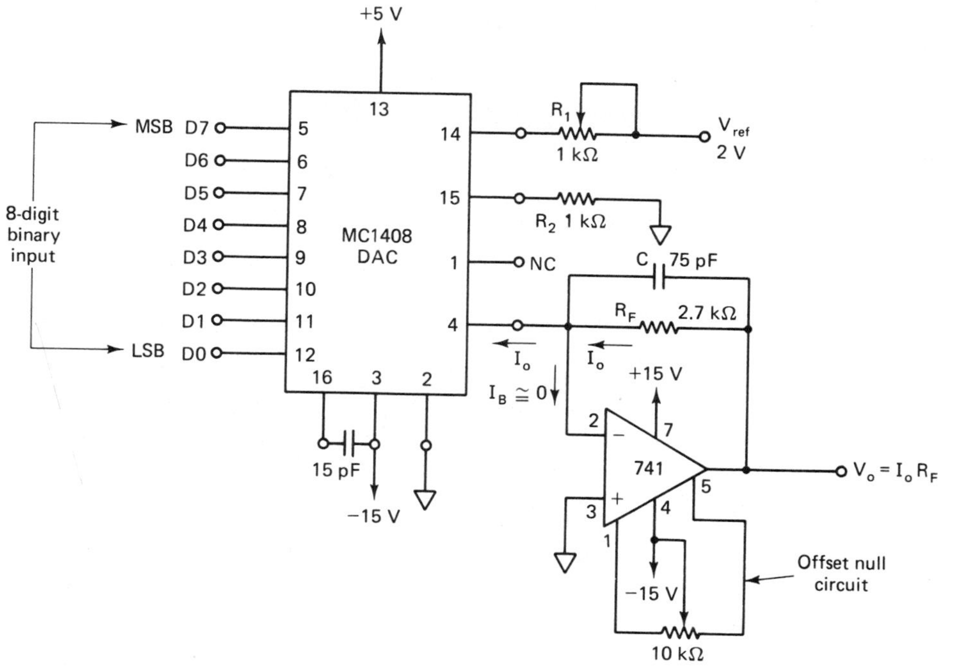

$$I_o = \frac{V_{ref}}{R_1}\left(\frac{D7}{2} + \frac{D6}{4} + \frac{D5}{8} + \frac{D4}{16} + \frac{D3}{32} + \frac{D2}{64} + \frac{D1}{128} + \frac{D0}{256}\right)$$

Figure 7-20 DAC using current-to-voltage converter.

7-10.2 Detecting Current through Photosensitive Devices

Photocells, photodiodes, and photovoltaic cells give an output current that depends on the intensity of an incident radiant energy or light and is independent of the load. The current through these devices can be converted to voltage by using an *I*-to-*V* converter and can be used as a measure of the amount of light or radiant energy incident in the device.

Figure 7-21 shows a photocell, the CL505L, connected to the *I*-to-*V* converter. Note that since the photocell is a *passive* transducer, it requires an external voltage V_{dc}. The Clairex CL505L photocell has the following specifications:

Resistance when illuminated (at 0.61 lux) = 1.5 kΩ
Minimum dark resistance = 100 kΩ
Measurement voltage = 10 V
Temperature range = −50° to 75°C

Example 7-14

In the circuit of Figure 7-21, $V_{dc} = 5$ V and $R_F = 3$ kΩ. Determine the change in the output voltage if the photocell is exposed to a light of 0.61 lux from a dark condition. Assume that the op-amp is initially nulled.

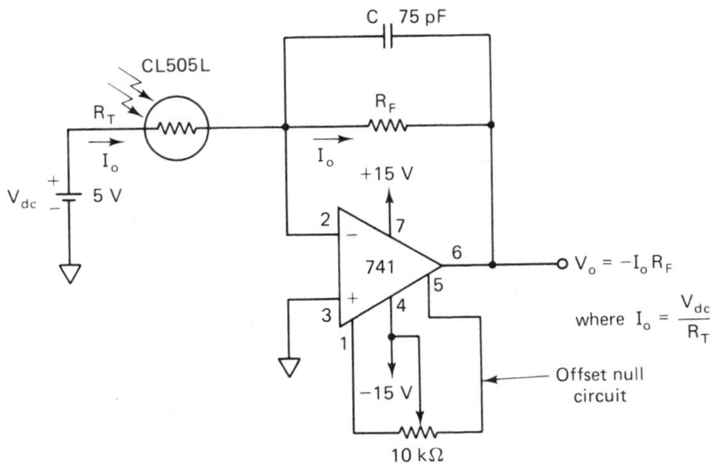

Figure 7-21 *I*-to-*V* converter used to measure the current through the photocell.

Solution. The resistance R_T of the CL505L in darkness is 100 kΩ. The minimum output voltage in darkness is

$$V_{omin} = -\frac{V_{dc}}{R_T}R_F = -\frac{5}{100\ \text{k}\Omega}(3\ \text{k}\Omega) = -0.15\ \text{V}$$

When the CL505L is illuminated, its resistance $R_T = 1.5$ kΩ. Therefore, the maximum output voltage is

$$V_{omax} = -\frac{V_{dc}}{R_T}R_F = -\frac{5}{1.5\ \text{k}\Omega}(3\ \text{k}\Omega) = -10\ \text{V}$$

Thus V_o varies from −0.15 V to −10 V as the photocell is exposed to light from a dark condition. The capacitor C is used to reduce high-frequency noise.

The circuit in Figure 7-21 can be used as a *light-intensity meter* by connecting at the output a meter that is calibrated for light intensity. The dc voltage V_{dc} in Figure 7-21 can be eliminated if a photovoltaic cell is used instead of a photocell. The photovoltaic cell is a semiconductor jucntion device that converts radiation energy into electrical power. It is a *self-generating* transducer because it does not require external voltage. The most common example of a photovoltaic cell is the *soler cell* used in space applications and watches.

The lower limit on current measurement with an *I*-to-*V* converter is set by the bias current I_B of the op-amp. This means that op-amps with smaller I_B, values such as the μA714 ($I_B = 3$ nA), can be used to detect lower currents.

7-11 VERY HIGH INPUT IMPEDANCE CIRCUIT

Recall that the voltage follower has the highest input resistance of any op-amp circuit [see section (4-3.8)]. For this reason it is used to reduce voltage error caused by source loading and to isolate high-impedance sources from following circuity. Figure

7-22(a) shows a direct-coupled dc voltage follower in which the input voltage is directly applied to the noninverting input terminal. Bias current I_B for the op-amp is supplied through the input source resistances R_{in}.

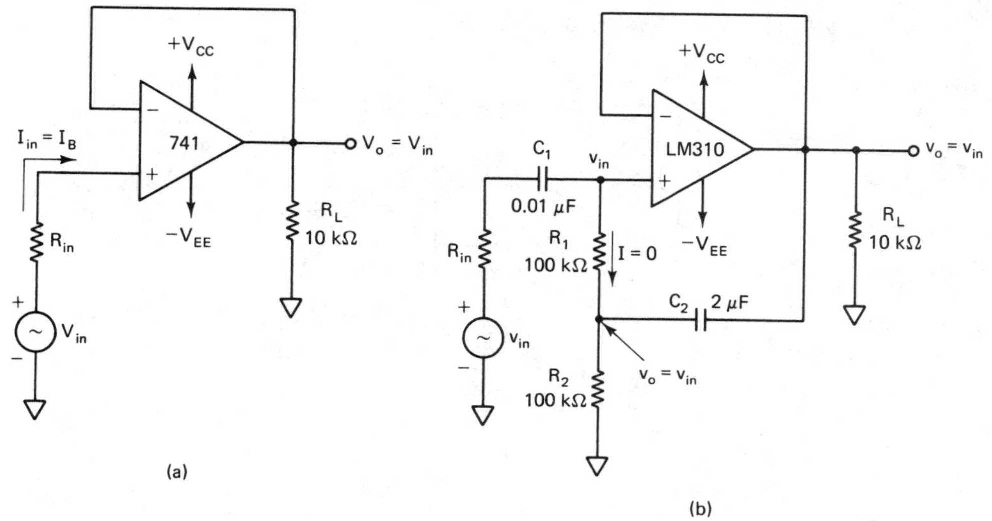

Figure 7-22 (a) Dc-coupled voltage follower. (b) Ac-coupled voltage follower with input resistance bootstrapped.

If the input source resistance is high (or if the voltage follower is driven from high-input source resistance) because of the voltage drop across it, the voltage at the noninverting input will be in error. In other words, the output and input voltages will not be eqaul. To remedy this problem an op-amp with low input bias current should be chosen as a voltage follower when working from high-input source resistances. Obviously, the best choices are op-amps specially designed as buffers or voltage followers. These op-amps have low bias currents and generally higher slew rates. For example, the LM310 has $I_B = 10$ nA maximum and SR $= 30$ V/μs.

When an ac input voltage rides on a dc level, in order to block the dc level a coupling capacitor must be used in series with the input, as shown in Figure 7-22(b). But whenever a voltage follower is ac-coupled, it is necessary to connect a bias resistor to provide a ground path for I_B. However, this bias resistor drastically reduces the input resistance of the follower circuit. In fact, the input resistance is equal to the bias resistance. Therefore, to get higher input resistance the bias resistance is *bootstrapped* as shown in Figure 7-22(b). In this circuit the input is applied through a coupling capacitor C_1 to the top of R_1, and simultaneously the output voltage is coupled through capacitor C_2 to the bottom of R_1. Since the gain of the circuit is 1, the voltage drop across R_1 is $(v_{in} - v_o)$, which is almost zero. Therefore, the current through R_1 is almost zero, and the input resistance, in turn, is incredibly high.

Voltage followers are useful in such applications as active filters, sample-and-hold circuits, and bridge circuits using transducers.

7-12 THE INTEGRATOR

A circuit in which the output voltage waveform is the integral of the input voltage waveform is the *integrator* or the *integration amplifier*. Such a circuit is obtained by using a basic inverting amplifier configuration if the feedback resistor R_F is replaced by a capacitor C_F [see Figure 7-23(a)].

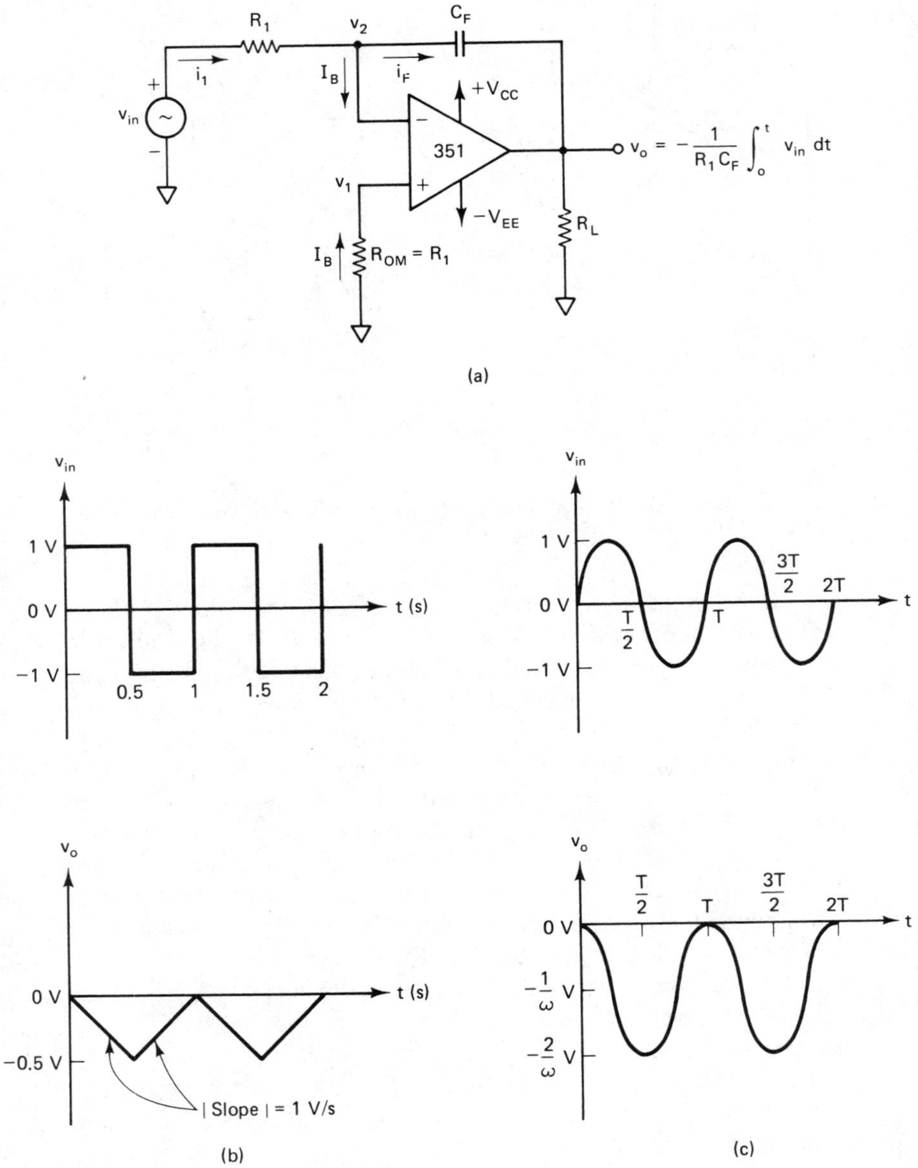

Figure 7-23 (a) The integrator circuit. (b, c) Input and ideal output waveforms using sine wave and square wave, respectively. $R_1C_F = 1$ second and $V_{OO} = 0$ V assumed.

The expression for the output voltage v_o can be obtained by writing Kirchhoff's current equation at node v_2:

$$i_1 = I_B + i_F$$

Since I_B is negligibly small,

$$i_1 \cong i_F$$

Recall that the relationship between current through and voltage across the capacitor is

$$i_c = C\frac{dv_c}{dt}$$

Therefore,

$$\frac{v_{in} - v_2}{R_1} = C_F\left(\frac{d}{dt}\right)(v_2 - v_o)$$

However, $v_1 = v_2 \cong 0$ because A is very large. Therefore,

$$\frac{v_{in}}{R_1} = C_F\frac{d}{dt}(-v_o)$$

The output voltage can be obtained by integrating both sides with respect to time:

$$\int_0^t \frac{v_{in}}{R_1}\,dt = \int_0^t C_F\frac{d}{dt}(-v_o)\,dt$$

$$= C_F(-v_o) + v_o\big|_{t=0}$$

Therefore,

$$v_o = -\frac{1}{R_1C_F}\int_0^t v_{in}\,dt \qquad (7\text{-}23)$$

Equation (7-23) says that the output voltage is directly proporational to the negative integral of the input voltage and inversely proportional to the time constant R_1C_F. For example, if the input is a sine wave, the output will be a cosine wave; or if the input is a square wave, the output will be a triangular wave, as shown in Figure 7-23(b) and (c), respectively. Note that these waveforms are drawn with the assumption that $R_1C_F = 1$ second and $V_{oo} = 0$ V.

For accurate integration of the input waveform, the time period of the input signal T must be longer than or equal to R_1C_F. Again, the R_{OM} is used to minimize the effect of input bias current on the output offset voltage. The input offset voltage V_{io} and the part of the input current charging capacitor C_F produce the error voltage at the output of the integrator. Therefore, in the practical integrator shown in Figure 7-24, a resistor R_F is connected across the feedback capacitor C_F to provide dc stabilization. In other words, R_F limits the low-frequency gain and hence minimizes the variations in the output voltage. Generally, $R_F \cong 10R_1$. The integrator is most commonly used in analog computers and analog-to-digital (ADC) and signal-wave-shaping circuits.

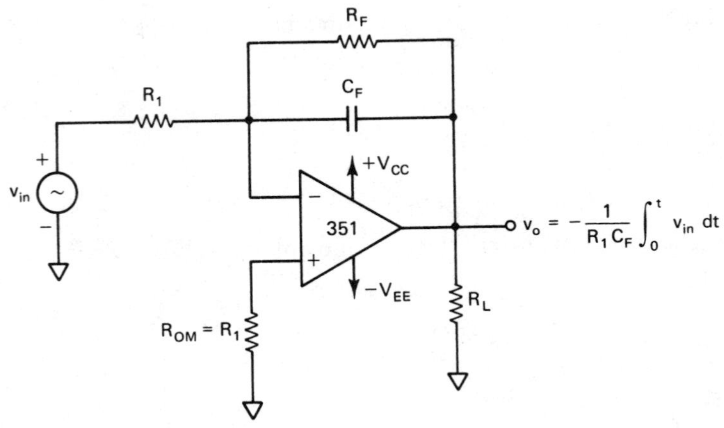

Figure 7-24 Practical integrator.

Example 7-15

In the circuit of Figure 7-23, $R_1C_F = 1\text{s}$, and the input is a step (dc) voltage, as shown in Figure 7-25(a). Determine the output voltage and sketch it. Assume that the op-amp is initially nulled.

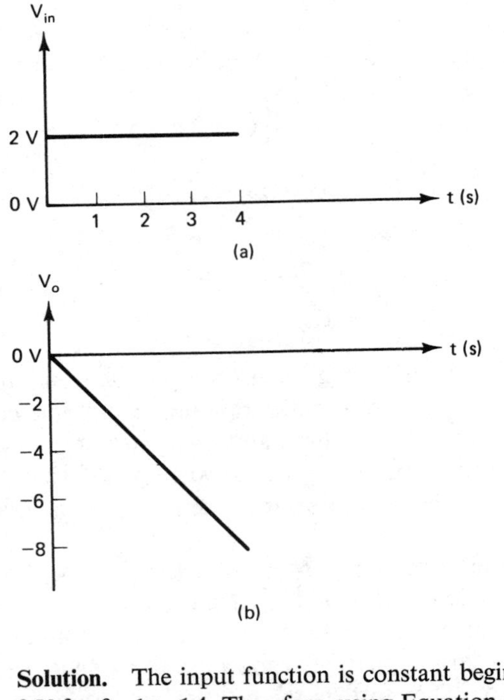

Figure 7-25 (a) Input and (b) output waveforms for Example 7-15.

Solution. The input function is constant beginning at $t = 0$ seconds. That is, $V_{in} = 2$ V for $0 \leq t \leq 4$. Therefore, using Equation (7-23),

$$\begin{aligned} V_o &= -\int_0^{t=4} 2\,dt \\ &= -\left[\int_0^1 2\,dt + \int_1^2 2\,dt + \int_2^3 2\,dt + \int_3^4 2\,dt\right] \\ &= -(2 + 2 + 2 + 2) = -8\text{ V} \end{aligned}$$

The output voltage waveform is drawn in Figure 7-25(b); the waveform is called a *ramp function*. The slope of the ramp is −2 volts/sec. Thus, with a constant voltage applied at the input the integrator gives a ramp at the output.

7-13 THE DIFFERENTIATOR

Figure 7-26(a) shows the *differentiator* or *differentiation amplifier*. As its name implies, the circuit performs the mathematical operation of differentiation; that is, the output waveform is the derivative of the input waveform. The differentiator may be constructed from a basic inverting amplifier if an input resistor R_1 is replaced by a capacitor C_1.

The expression for the output voltage can be obtained from Kirchhoff's current equation written at node v_2 as follows:

$$i_C = I_B + i_F$$

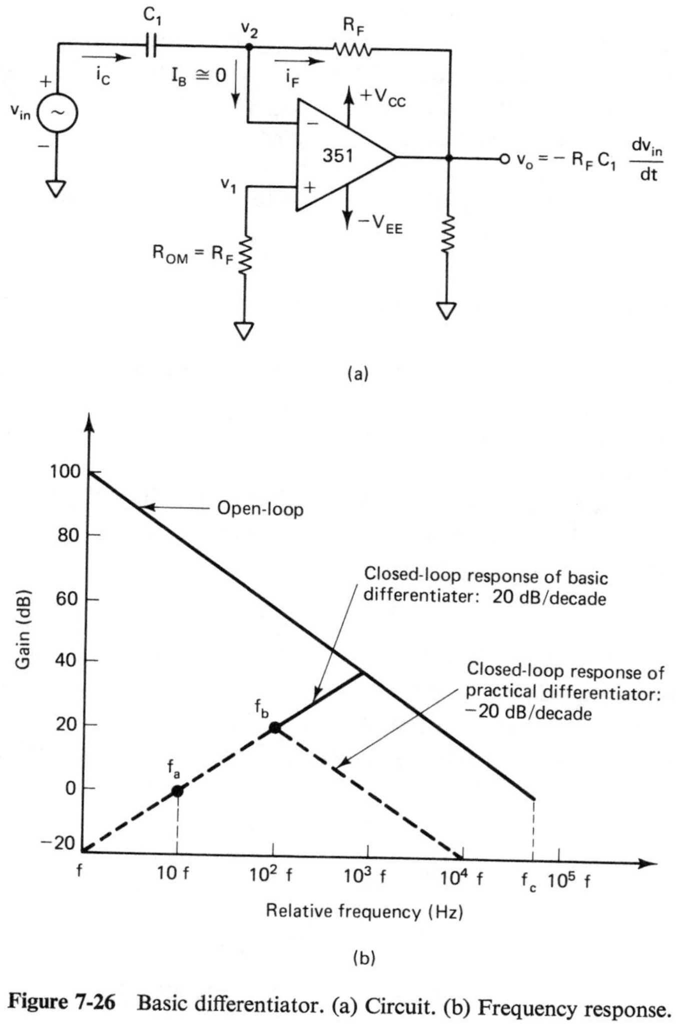

Figure 7-26 Basic differentiator. (a) Circuit. (b) Frequency response.

Since $I_B \cong 0$,

$$i_C = i_F$$

$$C_1 \frac{d}{dt}(v_{\text{in}} - v_2) = \frac{v_2 - v_o}{R_F}$$

But $v_1 = v_2 \cong 0$ V, because A is very large. Therefore,

$$C_1 \frac{dv_{\text{in}}}{dt} = -\frac{v_o}{R_F}$$

or

$$v_o = -R_F C_1 \frac{dv_{\text{in}}}{dt} \tag{7-24}$$

Thus the output v_o is equal to the $R_F C_1$ times the negative instantaneous rate of change of the input voltage v_{in} with time. Since the differentiator performs the reverse of the integrator's function, a cosine-wave input will produce a sine-wave output, or a triangular input will produce a square-wave output. However, the differentiator of Figure 7-26(a) will not do this because it has some practical problems. The gain of the circuit (R_F/X_{C1}) increases with increase in frequency at a rate of 20 dB/decade. This makes the circuit unstable. Also, the input impedance X_{C1} decreases with increase in frequency, which makes the circuit very susceptible to high-frequency noise. When amplified, this noise can completely override the differentiated output signal. The frequency response of the basic differentiator is shown in Figure 7-26(b). In this figure, f_a is the frequency at which the gain is 0 dB and is given by

$$f_a = \frac{1}{2\pi R_F C_1} \tag{7-25}$$

Also, f_c is the unity gain bandwidth of the op-amp, and f is some relative operating frequency.

Both the stability and the high-frequency noise problems can be corrected by the addition of two components: R_1 and C_F, as shown in Figure 7-27(a). This circuit is a *practical differentiator*, the frequency response of which is shown in Figure 7-26(b) by a dashed line. From frequency f to f_b, the gain increases at 20 dB/decade. However, after f_b the gain decreases at 20 dB/decade. This 40-dB/decade change in gain is caused by the R_1C_1 and R_FC_F combinations. The gain-limiting frequency f_b is given by

$$f_b = \frac{1}{2\pi R_1 C_1} \tag{7-26}$$

where $R_1C_1 = R_FC_F$. Thus R_1C_1 and R_FC_F help to reduce significantly the effect of high-frequency input, amplifier noise, and offsets. Above all, it makes the circuit more stable by preventing the increase in gain with frequency. Generally, the value of f_b and in turn R_1C_1 and R_FC_F values should be selected such that

$$f_a < f_b < f_c \tag{7-27}$$

where

$$f_a = \frac{1}{2\pi R_F C_1}$$

$$f_b = \frac{1}{2\pi R_1 C_1} = \frac{1}{2\pi R_F C_F}$$

$$f_c = \text{unity gain bandwidth}$$

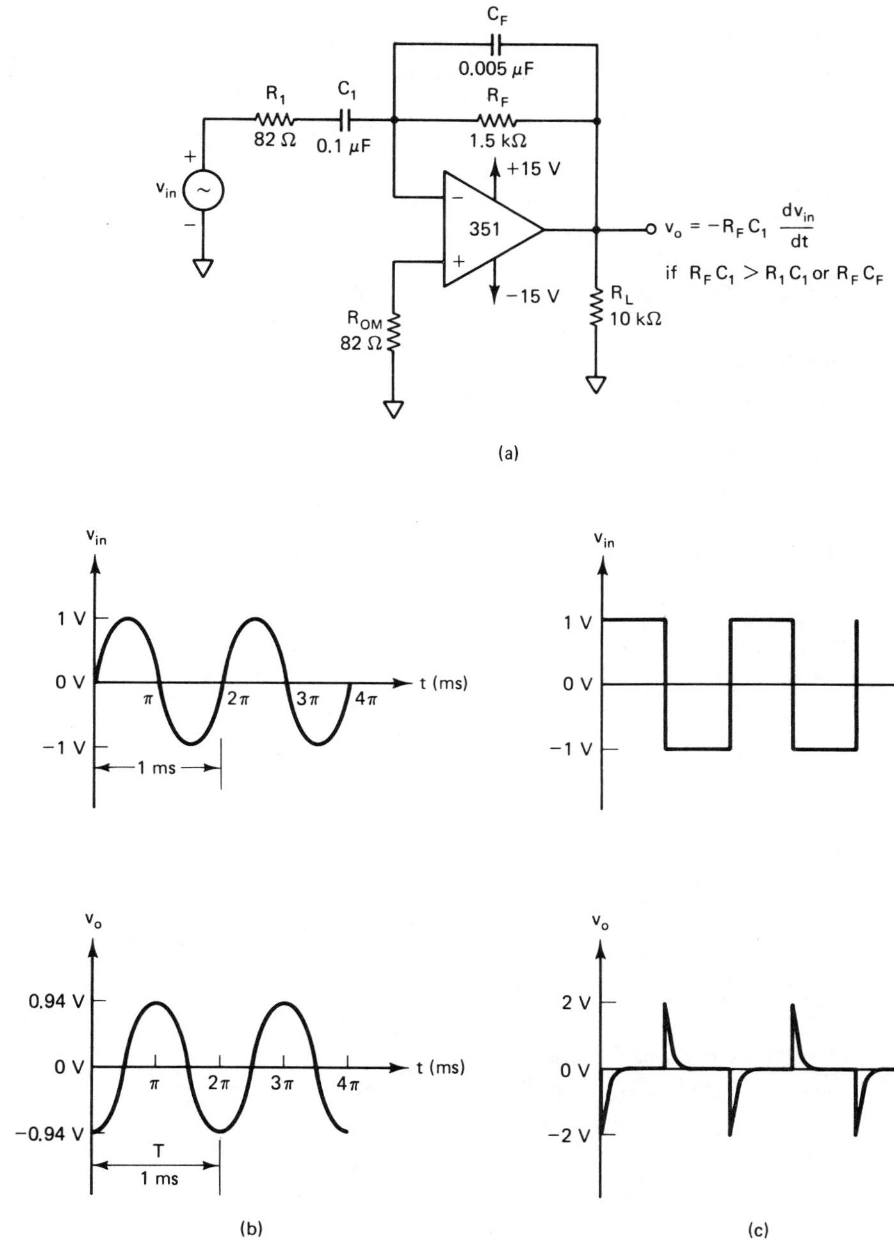

Figure 7-27 Practical differentiator. (a) Circuit. (b) Sine-wave input and resulting cosine-wave output. (c) Square-wave input and resulting spike output.

The input signal will be differentiated properly if the time period T of the input signal is larger than or equal to $R_F C_1$. That is,

$$T \geq R_F C_1$$

Figure 7-27(b) and (c) show the sine-wave and square-wave inputs and resulting differentiated outputs, respectively, for the practical differentiator. A workable differentiator can be designed by implementing the following steps (see Example 7-16):

1. Select f_a equal to the highest frequency of the input signal to be differentiated. Then, assuming a value of $C_1 < 1\ \mu F$, calculate the value of R_F.
2. Choose $f_b = 20 f_a$ and calculate the values of R_1 and C_F so that $R_1 C_1 = R_F C_F$.

The differentiator is most commonly used in waveshaping circuits to detect high-frequency components in an input signal and also as a rate-of-change detector in FM modulators.

Example 7-16

(a) Design a differentiator to differentiate an input signal that varies in frequency from 10 Hz to about 1 kHz.
(b) If a sine wave of 1 V peak at 1000 Hz is applied to the differentiator of part (a), draw its output waveform.

Solution

(a) To design a differentiator we simply follow the steps outlined above:

1. $f_a = 1$ kHz

$$= \frac{1}{2\pi R_F C_1} \quad \text{[from Equation (7-25)]}$$

Let $C_1 = 0.1\ \mu F$; then

$$R_F = \frac{1}{(2\pi)(10^3)(10^{-7})} = 1.59\ k\Omega$$

Let R_F be 1.5 kΩ.

2. $f_b = 20$ kHz

$$= \frac{1}{2\pi R_1 C_1} \quad \text{[from Equation (7-26)]}$$

Hence

$$R_1 = \frac{1}{(2\pi)(2)(10^4)(10^{-7})} = 79.5\ \Omega$$

Let R_1 be 82 Ω.
Since $R_1 C_1 = R_F C_F$,

$$C_F = \frac{(82)(10^{-7})}{1.5\ k\Omega} \cong 0.0055\ \mu F$$

Let C_F be 0.005 μF
Finally, $R_{OM} = R_1 \| R_F \cong 78\ \Omega$ (use 82 Ω).

The complete circuit with component values is shown in Figure 7-27(a).
(b) Since $V_p = 1$ V and $f = 1000$ Hz, the input voltage is

$$v_{in} = V_p \sin \omega t$$
$$= \sin (2\pi)(10^3)t$$

Hence, from Equation (7-24),

$$v_o = -R_F C_1 \frac{dv_{in}}{dt}$$
$$= -(1.5\ k\Omega)(0.1\ \mu F)\frac{d}{dt}[\sin (2\pi)(10^3)t]$$
$$= -(1.5\ k\Omega)(0.1\ \mu F)(2\pi)(10^3) \cos [(2\pi)(10^3)t]$$
$$= -0.94 \cos [(2\pi)(10^3)t]$$

The input and differentiated output waveforms are as shown in Figure 7-27(b).

SUMMARY

1. So many different op-amps are available because no one circuit design can possibly optimize all the dc and ac parameters. To select the most suitable device for a desired application, it is therefore necessary to refer to the op-amp data book.
2. Op-amps are designed to optimize a parameter such as high slew rate, wide bandwidth, or low power consumption. However, the use of a general-purpose op-amp such as the 741 will give satisfactory results in most applications.
3. Although a direct-coupled op-amp can amplify both ac and dc inputs, a capacitive coupling is used when ac response characteristics are needed or when it is essential to block the dc associated with the input signal. An ac amplifier can be powered by a single supply, provided that a dc level is injected at the input and removed at the output by using a voltage divider and coupling capacitor, respectively. Besides that, the output voltage swing is limited to a single supply voltage.
4. A peaking or narrow bandpass response can be obtained by using a parallel *LC* circuit in the feedback path of an inverting amplifier.
5. Summing, scaling, and averaging amplifiers can be constructed by using inverting, noninverting, and differential configurations.
6. The instrumentation amplifier is intended for precise, low-level signal amplification where low noise, low thermal and time drifts, and high input resistance are required. Various differential amplifier configurations can be used as instrumentation amplifiers, depending on the application requirements. When employed in conjunction with the transducer bridge and indicating meter, the instrumentation amplifier can be used for a variety of applications, such as temperature and light indicators, temperature controller, and analog weight scale.
7. The differential input and output amplifier can be used as a preamplifier and also in driving push-pull arrangements.
8. The *V*-to-*I* converter can be used in such applications as dc and ac voltmeters, diode, zener, and LED testers.
9. The *I*-to-*V* converter, on the other hand, is used with digital-to-analog circuits as well as in testing photosensitive devices.
10. Using a bootstrap technique, the input impedance of an ac voltage follower can be increased significantly.
11. The integrator and differentiator are most commonly used in signal wave-shaping applications. In addition, the integrator is used in analog computers and the differentiator is used as a rate-of-change detector in FM modulators.

QUESTIONS

7-1. Briefly explain the difference between the dc and ac amplifiers.

7-2. What is the major advantage and disadvantage of a single-supply ac amplifier?

7-3. What determines the peak frequency f_p in the peaking amplifier?

7-4. Explain the difference between (a) inverting and differential summing amplifiers and (b) inverting and noninverting averaging amplifiers.

7-5. What is an instrumentation amplifier? List three applications of the instrumentation amplifier besides those discussed in this chapter.

7-6. Explain briefly the characteristics of a thermistor, photocell, and strain gage.

7-7. What is the major difference between the photocell and the photovoltaic cell?

7-8. Explain briefly the advantages of the differential input and output amplifier.

7-9. What modifications are needed in the dc voltmeter circuit of Figure 7-16 so that it may be used to measure higher dc voltages ($> \pm 13$ V)?

7-10. The maximum ideal value of current I_0 in the circuit of Figure 7-18 is equal to the output short circuit current of the op-amp. True or false?

7-11. Is there any limitation on the size of the load in the *V*-to-*I* converter with grounded load of Figure 7-19? Explain.

7-12. If a 741C is configured as an *I*-to-*V* converter, what is the lowest value of current that may be measured?

7-13. Explain the difference between the dc and ac voltage followers.

7-14. Explain the difference between the integrator and differentiator and give one application of each.

PROBLEMS

(In all of the following problems, assume that the op-amp is initially nulled and the supply voltages $= \pm 15$ V unless otherwise specified.)

7-1. In the circuit of Figure 7-1(a), $R_F = 4.7$ kΩ, $R_1 = 100$ Ω, $R_L = 10$ kΩ, and $R_{OM} = 100$ Ω. Determine the output voltage V_o if $V_{in} = 100$ mV.

7-2. Repeat Problem 7-1 for the circuit in Figure 7-1(b).

7-3. For the noninverting amplifier of Figure 7-3(b), $R_{in} = 50$ Ω, $C_i = 0.01$ μF, $R_1 = 1$ kΩ, $R_{OM} = 820$ Ω, $R_F = 5.6$ kΩ, $R_L = 10$ kΩ. Determine:
(a) The gain
(b) The bandwidth of the amplifier.

7-4. For the inverting amplifier with a single supply shown in Figure 7-4(a), $R_{in} = 50$ Ω, $R_1 = 10$ kΩ, $R_2 = R_3 = R_F = 100$ kΩ, $C_i = C_o = 0.1$ μF.
(a) Determine the bandwidth of the amplifier.
(b) Determine the maximum ideal output voltage swing.
(c) Sketch the output voltage waveform v_o' and v_o if $v_{in} = 200$ mVP sinewave at 1 KHz.

7-5. The peaking amplifier of Figure 7-5(a) has the following values: $R_1 = 1$ kΩ, $L = 100$ μH with 3 Ω internal resistance, $C = 0.01$ μF, $R_F = 6.8$ kΩ, $R_{OM} = 820$ Ω, and $R_L = 10$ kΩ. Determine:
(a) The peak frequency f_p,
(b) The gain of the amplifier at f_p,
(c) The bandwidth of the amplifier

7-6. In the circuit of Figure 7-6, $V_a = 100$ mV, $V_b = -200$ mV, $V_c = 300$ mV, $R_a = 3$ kΩ, $R_b = 2.2$ kΩ, $R_c = 1$ kΩ, $R_F = 4.7$ kΩ, $R_{OM} = 470$ Ω, and $R_L = 10$ kΩ.
(a) Determine the output voltage V_o.
(b) Identify the circuit from its operation.

7-7. Repeat Problem 7-6 with $R_a = R_b = R_c = 1$ kΩ.

7-8. The circuit of Figure 7-7 is to be used as an averaging amplifier with the following specifications: $V_a = V_b = 1.5$ V, $V_c = 3$ V, $R_1 = R = 1.5$ kΩ, and $V_o = 5.2$ V. Determine the required value of R_F.

7-9. In the circuit of Figure 7-9, $V_a = 1$ V, $V_b = 2$ V, $V_c = -3$ V, and $V_d = -4$ V; $R = 1$ kΩ, $R_L = 10$ kΩ. Determine the output voltage V_o.

7-10. In the circuit of Figure 7-12, $R_1 = 1.8$ kΩ, $R_F = 4.7$ kΩ, $R_A = R_B = R_C = 500$ kΩ, $V_{dc} = 10$ V, and supply voltages $= \pm 15$ V. The transducer is a thermistor with the following specifications. $R_T = 500$ kΩ at 25°C. The temperature coefficient of resistance $= -10$ kΩ/°C. If the temperature changes from 0°C to 70°C, find the variations in (a) the input signal V_{ab} and (b) the output signal V_o.

7-11. The transducer in the circuit of Figure 7-12 is a CL605L photocell with the following specifications: minimum dark resistance = 500 kΩ; resistance at 0.61 meter-candle (lux) = 7.5 kΩ. Using the circuit specifications of Problem 7-10, determine the input voltage V_{ab} and output voltage V_o (a) at a dark condition and (b) at 0.61 lux.

7-12. For the strain gage bridge circuit shown in Figure 7-13, show that

$$V_{ab} = +V_{dc}\left(\frac{\Delta R}{R}\right)$$

Assume that under the strained condition the resistance of R_{T1} and R_{T3} decreases and that of R_{T2} and R_{T4} increases by the same amount ΔR Ω. Also, $R_{T1} = R_{T2} = R_{T3} = R_{T4} = R$ under the unstrained condition.

7-13. The circuit of Figure 7-12 uses a strain gage and has the following specifications. The unstrained resistance of each of the four elements of the strain gage is 120 Ω, $V_{dc} =$ 5 V, and the op-amp supply voltages $= \pm 10$ V. $R_1 = 100$ Ω and $R_F = 47$ kΩ. Determine the output voltage V_o if the change in the resistance of each strain gage element is 0.1 Ω.

7-14. In the differential amplifier of Figure 7-14(a), the desired output voltage $V_o = 5$ V. Determine the required value of differential input voltage V_{in} if $R_1 = 100$ Ω and $R_F =$ 22 kΩ. The op-amp supply voltages $= \pm 10$ V.

7-15. In the circuit of Figure 7-16, $V_{in} = 2$ V and the switch is in position 1. Determine the current through the meter.

7-16. A 100-μA full-scale meter movement is used in the ac voltmeter of Figure 7-17. Determine the full-scale input voltage v_{in} if $R_1 = 4.7$ kΩ.

7-17. In the circuit of Figure 7-18, if $R_1 = 1.5$ kΩ and $V_{in} = 5$ V, determine (a) the current through the diode and (b) the output voltage V_o.

7-18. Repeat Problem 7-17 if a diode is replaced by a zener with $V_z = 5.1$ V and $I_{zk} = 1$ mA.

7-19. Determine the load current I_L in the circuit of Figure 7-19 if $R = 10$ kΩ and $V_{in} = 2$ V. For the proper operation of this circuit the value of the load has to be less than R Why?

7-20. Referring to the circuit of Figure 7-20, determine the output voltage V_o if D_0 through D_3 are connected to ground and D_4 through D_7 are connected to +5 V (logic 1.)

7-21. In the circuit of Figure 7-21, $V_{dc} = 10$ V, the dark resistance of the CL505L is 100 kΩ, and $R_F = 10$ kΩ. Determine the output voltage for the dark condition. What is the function of capacitor C?

7-22. In the integrator circuit of Figure 7-24, the input is a sine wave with a peak-to-peak amplitude of 5 V at 1 kHz. Draw the output voltage waveform if $R_1C_F = 0.1$ ms and $R_F = 10R_1$. Assume that the voltage across C_F is initially zero.

7-23. In the differentiator circuit of Figure 7-27(a), the input is a sine wave with a peak-to-peak amplitude of 3 V at 200 Hz. Sketch the output voltage waveform.

7-24. Design a differentiator that will differentiate an input signal with $f_{max} = 100$ Hz.

EXPERIMENT 7-1
AC INVERTING AMPLIFIER

Objectives

In this experiment you will study the characteristics of an ac amplifier and an amplifier that uses a single power supply. At the end of this experiment you should be able to:

1. Verify that the low-cutoff frequency depends on the value of the input capacitor–resistor combination and the high-cutoff frequency depends on the gain of the amplifier.
2. Select the proper values of external components for a desired bandwidth.
3. Discuss the differences between an ac amplifier with two supply voltages and one with a single supply voltage.

Equipment

1. Oscilloscope
2. Audio-signal generator
3. $\pm$15-V power supply

Materials

1. 741 or 351 op-amp
2. Three 10-kΩ resistors
3. Two 100-kΩ resistors
4. 470-kΩ resistor
5. 0.01-μF capacitor
6. 0.1-μF capacitor
7. Semilog paper

Procedure

1. Connect the circuit as shown in Figure E7-1.1(a).
2. Set $v_{in} = 100$ mV pp at 100 Hz sine wave. Using the scope, measure the input and output voltages and enter the results in Table E7-1.1.
3. Repeat step 2 for the frequencies listed in Table E7-1.1.
4. Next, connect the circuit of Figure E7-1.1(b) and repeat steps 2 and 3.
5. Complete Table E7-1.1.
6. Using the data of Table E7-1.1, plot the frequency versus gain in dB curves for the inverting amplifier with dual and single supply on the semilog graph paper. Use the log scale for frequency and linear scale for gain.
7. From the graphs first find the low-cutoff and high-cutoff frequencies and then determine the bandwidths.
8. Calculate the theoretical bandwidths for the amplifiers shown in Figure E7-1.1.

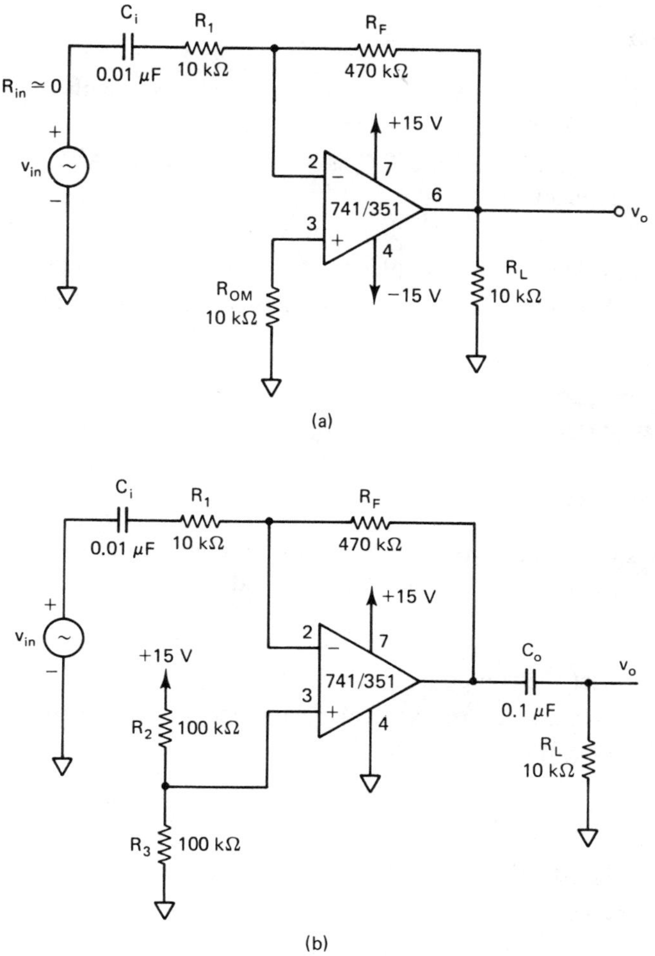

Figure E7-1.1 Ac inverting amplifier with (a) dual supply and (b) single supply. (Pin numbers refer to the 8-pin mini DIP.)

TABLE E7-1.1

Input voltage, v_{in} pp	Input frequency, f_{in} (Hz)	Output voltage, v_o, of an ac inverting amplifier with:		Voltage gain, $\frac{v_o}{v_{in}}$	Voltage gain (dB) $= 20 \log (v_o/v_{in})$	
		Dual supply	Single supply		Dual supply	Single supply
	100					
	1,000					
	1,300					
	4,000					
100 mV	7,000					
	10,000					
	13,000					
	20,000					
	50,000					
	100,000					

Questions

1. How do the calculated bandwidths compare with the measured bandwidths? If the difference is more than 20%, give the reasons.
2. In Figure E7-1.1(a), if C_i is change to 0.1 μF, what would the new low-cutoff frequency be?
3. What determines the high-cutoff frequency of an amplifier?
4. For a high-cutoff frequency of 50 kHz, what should be the gain of an amplifier?
5. What are the advantages and disadvantages of a single-supply amplifier?

EXPERIMENT 7-2
AN INTEGRATOR

Objectives

In this experiment you will learn what integration is and how the integrator works. At the end of this experiment you should be able to:

1. Determine the integral of certain waveforms.
2. Select the components for proper integration.
3. Learn how to use the dual-trace scope to measure the phase shift between input and output waveforms.

Equipment

1. Dual-trace oscilloscope
2. Audio-function generator
3. ±15-V power supply

Materials

1. 351 op-amp
2. Two 10-kΩ resistors
3. 100-kΩ resistor
4. 0.01-μF capacitor
5. 0.05-μF capacitor
6. 0.1-μF capacitor

Procedure

1. Connect the circuit as shown in Figure E7-2.1. (The pin numbers indicated in this figure refer to the 8-pin mini DIP).
2. Set the function generator for a square-wave output and adjust its amplitude to 1 V pp. Set the frequency of the square wave at 1 kHz initially.

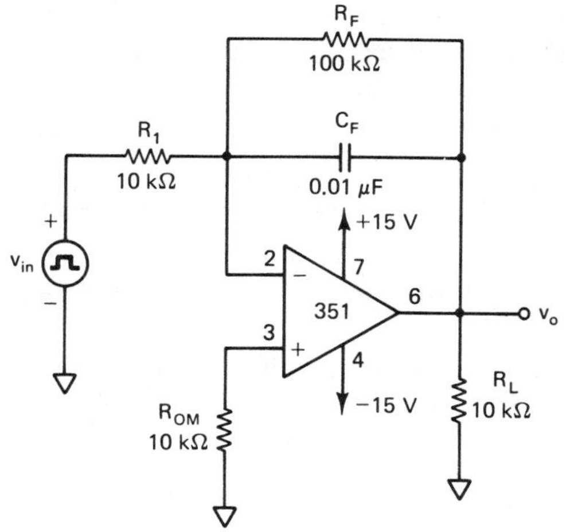

Figure E7-2.1 Integrator. (Pin numbers refer to the 8-pin mini DIP.)

3. Connect one channel of the scope at the input and the other at the output terminal of the op-amp and slowly adjust the input frequency until the output is as good (linear) a triangle wave as possible. Measure the amplitude and frequency of the input and output waveforms. Enter the measured values in Table E7-2.1.
4. Repeat step 3 for $C_F = 0.05\ \mu F$ and $0.1\ \mu F$.
5. Set the function generator for a sine-wave output. Adjust the amplitude of the sinewave to 1 V pp and set the frequency at 1 kHz initially. Remove the 0.1-μF capacitor and reconnect the 0.01-μF capacitor in its place.
6. Connect one channel of the scope at the input and the other at the output terminal of the op-amp. Adjust the frequency of the input until the output is a negative-going cosine wave. Make sure that the scope is triggered properly. Measure the frequency and amplitude of the input and output waveforms. Enter the measured values in Table E7-2.1.

TABLE E7-2.1

R_1 (kΩ)	C_F (μF)	Input waveform			Output waveform		
		Type	Amplitude	Frequency	Type	Amplitude	Frequency
10	0.01	Square	1 V pp				
10	0.05	Square	1 V pp				
10	0.1	Square	1 V pp				
10	0.01	Sine	1 V pp				

Questions

1. For proper integration, what must be the relationship between R_1C_F and input frequency?

2. What characteristic is common to both the integrator and amplifier?
3. From the results of Table E7-2.1, discuss the relationship between the amplitude of the triangle wave and the amplitude and frequency of the square wave.
4. When the input of an integrator is a sine wave, what is the phase shift between the input and output?
5. What parameter must be considered in selecting an op-amp for integrator application?
6. If the input to an integrator is a pulse waveform (20% duty cycle), what is its output waveform?
7. In the integrator of Figure E7-2.1, what is the function of R_F?

8

Active Filters and Oscillators

8-1 INTRODUCTION

In Chapter 7 you saw how op-amp circuits are used to provide ac/dc amplification, perform such mathematical operations as summing, averaging, differentiation, and integration, convert I-to-V and V-to-I signals, and provide very high input impedance. This chapter presents another important field of application using op-amps: filters and oscillators. The chapter begins with the analysis and design of basic and inexpensive filter types and then discusses the various oscillator circuits. At the end of the chapter, a voltage-controlled oscillator (VCO) using the NE/SE566 integrated circuit is presented.

8-2 ACTIVE FILTERS

An electric filter is often a *frequency-selective* circuit that passes a specified band of frequencies and blocks or attenuates signals of frequencies outside this band. Filters may be classified in a number of ways:

1. Analog or digital
2. Passive or active
3. Audio (AF) or radio frequency (RF)

Analog filters are designed to process analog signals, while *digital* filters process analog signals using digital techniques. Depending on the type of elements used in their construction, filters may be classified as passive or active. Elements used in *passive* filters are resistors, capacitors, and inductors. *Active* filters, on the other hand, employ transistors or op-amps in addition to the resistors and capacitors. The type of element used dictates the operating frequency range of the filter. For example, *RC* filters are commonly used for audio or low-frequency operation, whereas *LC* or crystal filters are employed at RF or high frequencies. Especially because of their high *Q* value (figure of merit), the crystals provide more stable operation at higher frequencies.

First, this chapter presents the analysis and design of analog active-*RC* (audio-frequency) filters using op-amps. In the audio frequencies, inductors are often not used because they are very large, costly, and may dissipate more power. Inductors also eminate magnetic fields.

An active filter offers the following advantages over a passive filter:

1. *Gain and frequency adjustment flexibility.* Since the op-amp is capable of providing a gain, the input signal is not attenuated as it is in a passive filter. Besides that, the active filter is easier to tune or adjust.
2. *No loading problem.* Because of the high input resistance and low output resistance of the op-amp, the active filter does not cause loading of the source or load.
3. *Cost.* Typically, active filters are more economical than passive filters. This is because of the variety of cheaper op-amps and the absence of inductors.

Although active filters are most extensively used in the field of communications and signal processing, they are employed in one form or another in almost all sophisticated electronic systems. Radio, television, telephone, radar, space satellites, and biomedical equipment are but a few systems that employ active filters.

The most commonly used filters are these:

1. Low-pass filter
2. High-pass filter
3. Band-pass filter
4. Band-reject filter
5. All-pass filter

Each of these filters uses an op-amp as the active element and resistors and capacitors as the passive elements. Although the 741 type op-amp works satisfactorily in these filter circuits, high-speed op-amps such as the LM318 or the ICL8017 improve the filter's performance through their increased slew rate and higher unity gain bandwidth.

Figure 8-1 shows the frequency response characteristics of the five types of filters. The ideal response is shown by dashed curves, while the solid lines indicate the practical filter response. A low-pass filter has a constant gain from 0 Hz to a high cutoff frequency f_H. Therefore, the bandwidth is also f_H. At f_H the gain is down by 3 dB; after that ($f > f_H$) it decreases with the increase in input frequency. The frequencies between 0 Hz and f_H are known as the *passband* frequencies, whereas the range of frequencies, those beyond f_H, that are attenuated includes the *stopband* frequencies.

Figure 8-1(a) shows the frequency response of the low-pass filter. As indicated by the dashed line, an *ideal* filter has a zero loss in its passband and infinite loss in

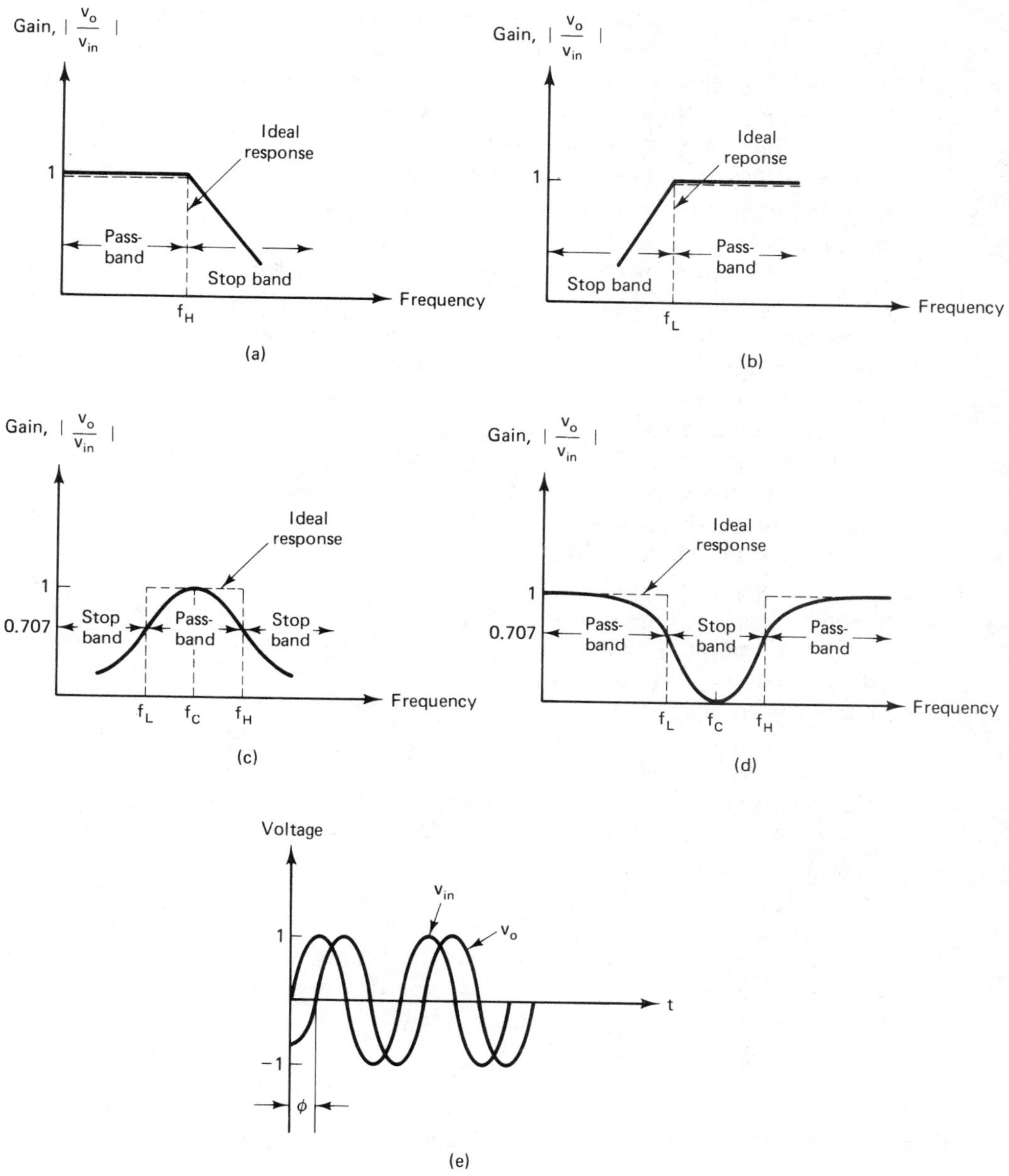

Figure 8-1 Frequency response of the major active filters. (a) Low-pass. (b) High-pass. (c) Band-pass. (d) Band-reject. (e) Phase shift between input and output voltages of an all-pass filter.

its stopband. Unfortunately, ideal filter response is not practical because linear networks cannot produce the discontinuities. However, it is possible to obtain a practical response that approximates the ideal response by using special design techniques as well as precision component values and high-speed op-amps.

Butterworth, Chebyshev, and Cauer filters are some of the most commonly used practical filters that approximate the ideal response. The key characteristic of the Butterworth filter is that it has a flat passband as well as stopband. For this reason, it is sometimes called a *flat-flat* filter. The Chebyshev filter has a ripple passband and flat stopband, while the Cauer filter has a ripple passband and a ripple stopband. Generally, the Cauer filter gives the best stopband response among the three. Because of their simplicity of design, the low-pass and high-pass Butterworth filters are discussed here.

Figure 8-1(b) shows a high-pass filter with a stopband $0 < f < f_L$ and a passband $f > f_L$. f_L is the low cutoff frequency, and f is the operating frequency. A band-pass filter has a passband between two cutoff frequencies f_H and f_L, where $f_H > f_L$, and two stopbands: $0 < f < f_L$ and $f > f_H$. The bandwidth of the band-pass filter, therefore, is equal to $(f_H - f_L)$. The band-reject filter performs exactly opposite to the band-pass; that is, it has a bandstop between two cutoff frequencies f_H and f_L and two passbands: $0 < f < f_L$ and $f > f_H$. The band-reject is also called a *band-stop* or *band-elimination* filter. The frequency response of band-pass and band-reject filters is shown in Figure 8-1(c) and (d). In these figures, f_C is called the center frequency since it is approximately at the center of the passband or stopband.

Figure 8-1(e) shows the phase shift between input and output voltages of an all-pass filter. This filter passes all frequencies equally well; that is, output and input voltages are equal in amplitude for all frequencies, with the phase shift between the two a function of frequency. The highest frequency up to which the input and output amplitudes remain equal is dependent on the unity gain bandwidth of the op-amp. At this frequency, however, the phase shift between the input and output is maximum.

Before proceeding with specific filter types, let us reexamine the filter characteristics, especially in the stopband region. As shown in Figure 8-1(a)–(d), the actual response curves of the filters in the stopband either steadily decrease or increase or both with increase in frequency. The rate at which the gain of the filter changes in the stopband is determined by the order of the filter. For example, for the first-order low-pass filter the gain rolls off at the rate of 20 dB/decade in the stopband that is, for $f > f_H$; on the other hand, for the second-order low-pass filter the roll-off rate is 40 dB/decade; and so on. By contrast, for the first-order high-pass filter the gain increases at the rate of 20 dB/decade in the stopband, that is, until $f = f_L$; the increase is 40 dB/decade for the second-order high-pass filter; and so on.

8-3 FIRST-ORDER LOW-PASS BUTTERWORTH FILTER

Figure 8-2 shows a first-order low-pass Butterworth filter that uses an *RC* network for filtering. Note that the op-amp is used in the noninverting configuration; hence it does not load down the *RC* network. Resistors R_1 and R_F determine the gain of the filter.

According to the voltage-divider rule, the voltage at the noninverting terminal (across capacitor C) is

$$v_1 = \frac{-jX_C}{R - jX_C} v_{\text{in}} \tag{8-1a}$$

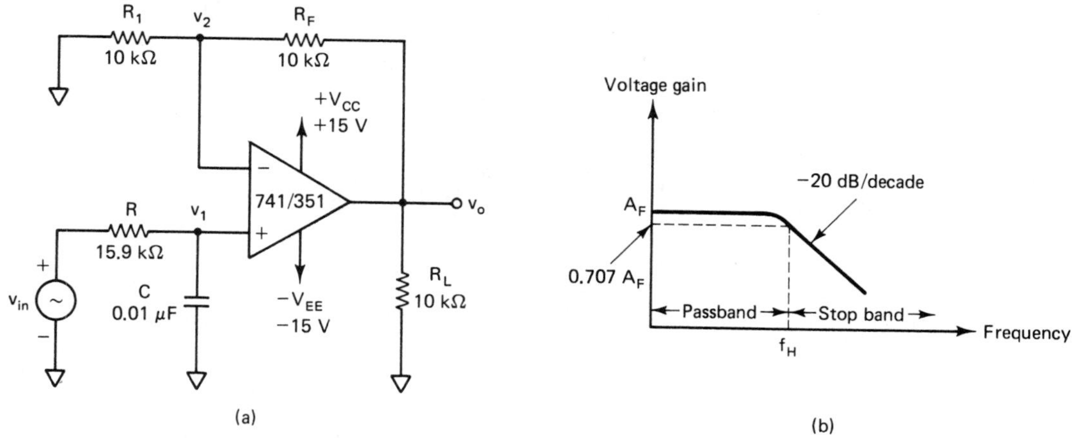

Figure 8-2 First-order low-pass Butterworth filter. (a) Circuit. (b) Frequency response.

where

$$j = \sqrt{-1} \qquad \text{and} \qquad -jX_C = \frac{1}{j2\pi fC}$$

Simplifying Equation (8-1a), we get

$$v_1 = \frac{v_{\text{in}}}{1 + j2\pi fRC}$$

and the output voltage

$$v_o = \left(1 + \frac{R_F}{R_1}\right) v_1$$

That is,

$$v_o = \left(1 + \frac{R_F}{R_1}\right) \frac{v_{\text{in}}}{1 + j2\pi fRC}$$

or

$$\frac{v_o}{v_{\text{in}}} = \frac{A_F}{1 + j(f/f_H)} \tag{8-1b}$$

where $\dfrac{v_o}{v_{\text{in}}}$ = gain of the filter as a function of frequency

$A_F = 1 + \dfrac{R_F}{R_1}$ = passband gain of the filter

f = frequency of the input signal

$f_H = \dfrac{1}{2\pi RC}$ = high cutoff frequency of the filter

The gain magnitude and phase angle equations of the low-pass filter can be obtained by converting Equation (8-1b) into its equivalent polar form, as follows:

$$\left|\frac{v_o}{v_{\text{in}}}\right| = \frac{A_F}{\sqrt{1 + (f/f_H)^2}} \tag{8-2a}$$

$$\phi = -\tan^{-1}\left(\frac{f}{f_H}\right) \tag{8-2b}$$

where ϕ is the phase angle in degrees.

The operation of the low-pass filter can be verified from the gain magnitude Equation (8-2a):

1. At very low frequencies, that is, $f < f_H$,

$$\left|\frac{v_o}{v_{\text{in}}}\right| \cong A_F$$

2. At $f = f_H$,

$$\left|\frac{v_o}{v_{\text{in}}}\right| = \frac{A_F}{\sqrt{2}} = 0.707A_F$$

3. At $f > f_H$,

$$\left|\frac{v_o}{v_{\text{in}}}\right| < A_F$$

Thus the low-pass filter has a constant gain A_F from 0 Hz to the high cutoff frequency f_H. At f_H the gain is $0.707A_F$, and after f_H it decreases at a constant rate with an increase in frequency [see Figure 8-2(b)]. That is, when the frequency is increased tenfold (one decade), the voltage gain is divided by 10. In other words, the gain decreases 20 dB (= 20 log 10) each time the frequency is increased by 10. Hence the rate at which the gain rolls off after f_H is 20 dB/decade or 6 dB/octave, where octave signifies a twofold increase in frequency. The frequency $f = f_H$ is called the *cutoff frequency* because the gain of the filter at this frequency is down by 3 dB (= 20 log 0.707) from 0 Hz. Other equivalent terms for cutoff frequency are *−3 dB frequency*, *break frequency*, or *corner frequency*.

8-3.1 Filter Design

A low-pass filter can be designed by implementing the following steps:

1. Choose a value of high cutoff frequency f_H.
2. Select a value of C less than or equal to 1 μF. Mylar or tantalum capacitors are recommended for active filters.
3. Calculate the value of R using

$$R = \frac{1}{2\pi f_H C}$$

4. Finally, select values of R_1 and R_F dependent on the desired passband gain A_F using

$$A_F = 1 + \frac{R_F}{R_1}$$

8-3.2 Frequency Scaling

The procedure that is used to convert an original cutoff frequency f_H to a new cutoff frequency f'_H is called *frequency scaling*. Frequency scaling is accomplished as follows. To change a high cutoff frequency, multiply R or C but not both by the ratio of the original cutoff frequency to the new cutoff frequency. In filter design the needed values of R and C are often not standard. Besides that, a variable capacitor C is not commonly used. Therefore, choose a standard value of capacitor, and then calculate the value of resistor for a desired cutoff frequency. This is because for a nonstandard value of resistor a potentiometer can be used (see Examples 8-1 and 8-2).

Example 8-1

Design a low-pass filter at a cutoff frequency of 1 kHz with a passband gain of 2.

Solution. Following the design steps described above,

1. $f_H = 1$ kHz.
2. Let $C = 0.01\ \mu F$.
3. Then $R = 1/(2\pi)(10^3)(10^{-8}) = 15.9\ k\Omega$. (Use a 20-kΩ potentiometer.)
4. Since the passband gain is 2, R_1 and R_F must be equal. Therefore, let $R_1 = R_F = 10\ k\Omega$. The complete circuit with component values is shown in Figure 8-2.

Example 8-2

Using the frequency scaling technique, convert the 1-kHz cutoff frequency of the low-pass filter of Example 8-1 to a cutoff frequency of 1.6 kHz.

Solution. To change a cutoff frequency from 1 kHz to 1.6 kHz, we multiply the 15.9-kΩ resistor by

$$\frac{\text{original cutoff frequency}}{\text{new cutoff frequency}} = \frac{1\text{ kHz}}{1.6\text{ kHz}} = 0.625$$

Therefore, new resistor $R = (15.9\ k\Omega)(0.625) = 9.94\ k\Omega$. However, 9.94 kΩ is not a standard value. Therefore, use $R = 10\ k\Omega$ potentiometer. Thus the new cutoff frequency is

$$f_H = \frac{1}{(2\pi)(0.01\ \mu F)(9.94\ k\Omega)}$$
$$= 1.6\text{ kHz}$$

Example 8-3

Plot the frequency response of the low-pass filter of Example 8-1.

Solution. To plot the frequency response, we have to use Equation (8-2a). The data of Table 8-1 are, therefore, obtained by substituting various values for f in this equation.

TABLE 8-1 FREQUENCY RESPONSE DATA FOR EXAMPLE 8-3

Input frequency, f (Hz)	Gain magnitude, $\lvert v_o/v_{in} \rvert$	Magnitude (dB) = $20 \log \lvert v_o/v_{in} \rvert$
1. 10	2	6.02
2. 100	1.99	5.98
3. 200	1.96	5.85
4. 700	1.64	4.29
5. 1,000	1.41	3.01
6. 3,000	0.63	−3.98
7. 7,000	0.28	−10.97
8. 10,000	0.20	−14.02
9. 30,000	0.07	−23.53
10. 100,000	0.02	−33.98

Equation (8-2a) will be repeated here for convenience:

$$\left|\frac{v_o}{v_{in}}\right| = \frac{A_F}{\sqrt{1 + (f/f_H)^2}}$$

where $A_F = 2$ and $f_H = 1$ kHz. The data of Table 8-1 are plotted as shown in Figure 8-3.

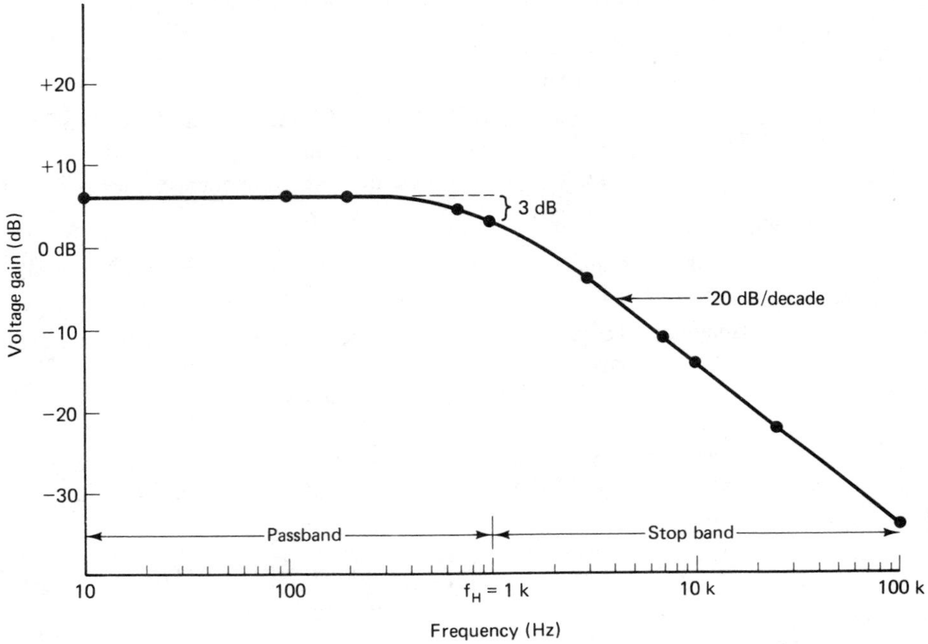

Figure 8-3 Frequency response for Example 8-3.

8-4 SECOND-ORDER LOW-PASS BUTTERWORTH FILTER

A stop-band response having a 40-dB/decade roll-off is obtained with the second-order low-pass filter. A first-order low-pass filter can be converted into a second-order type simply by using an additional *RC* network, as shown in Figure 8-4.

Second-order filters are important because higher-order filters can be designed using them. The gain of the second-order filter is set by R_1 and R_F, while the high cutoff frequency f_H is determined by R_2, C_2, R_3, and C_3, as follows:

$$f_H = \frac{1}{2\pi\sqrt{R_2 R_3 C_2 C_3}} \tag{8-3}$$

Furthermore, for a second-order low-pass Butterworth response, the voltage gain magnitude equation is

$$\left|\frac{v_o}{v_{in}}\right| = \frac{A_F}{\sqrt{1 + (f/f_H)^4}} \tag{8-4}$$

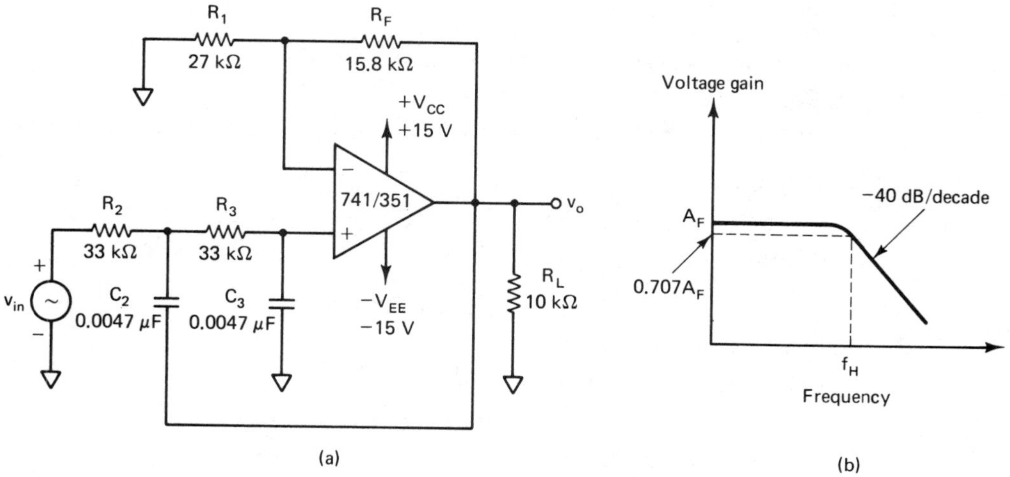

Figure 8-4 Second-order low-pass Butterworth filter. (a) Circuit. (b) Frequency response.

where $A_F = 1 + \dfrac{R_F}{R_1}$ = passband gain of the filter

f = frequency of the input signal (hertz)

$f_H = \dfrac{1}{2\pi\sqrt{R_2 R_3 C_2 C_3}}$ = high cutoff frequency (hertz)

8-4.1 Filter Design

Except for having twice the roll-off rate in the stopband, the frequency response of the second-order low-pass filter is identical to that of the first-order type. Therefore, the design steps of the second-order filter are identical to those of the first-order filter, as shown below:

1. Choose a value for the high cutoff frequency f_H.
2. To simplify the design calculations, set $R_2 = R_3 = R$ and $C_2 = C_3 = C$. Then choose a value of $C \leq 1\ \mu F$.
3. Calculate the value of R using Equation (8-3):

$$R = \frac{1}{2\pi f_H C}$$

4. Finally, because of the equal resistor ($R_2 = R_3$) and capacitor ($C_2 = C_3$) values, the passband voltage gain $A_F = (1 + R_F/R_1)$ of the second-order low-pass filter has to be equal to 1.586. That is, $R_F = 0.586R_1$. This gain is necessary to guarantee Butterworth response. Hence choose a value of $R_1 \leq 100\ k\Omega$ and calculate the value of R_F.

As outlined in Section 8-3.2, the frequency scaling method of the first-order filter is also applicable to the second-order low-pass filter.

Example 8-4

(a) Design a second-order low-pass filter at a high cutoff frequency of 1 kHz.
(b) Draw the frequency response of the network in part (a).

Solution

(a) To design the second-order low-pass filter, simply follow the steps presented above:

1. $f_H = 1$ kHz.
2. Let $C_2 = C_3 = 0.0047\ \mu$F.
3. Then

$$R_2 = R_3 = \frac{1}{(2\pi)(10^3)(47)(10^{-10})} = 33.86\ \text{k}\Omega$$

(Use $R_2 = R_3 = 33\ \text{k}\Omega$.)

4. Since R_F must be equal to $0.586R_1$, let R_1 equal 27 kΩ. Therefore,

$$R_F = (0.586)(27\ \text{k}\Omega) = 15.82\ \text{k}\Omega$$

(Use $R_F = 15.8\ \text{k}\Omega$.) Thus the required components are:

$$R_2 = R_3 = 33\ \text{k}\Omega$$
$$C_2 = C_3 = 0.0047\ \mu\text{F}$$
$$R_1 = 27\ \text{k}\Omega \quad \text{and} \quad R_F = 15.8\ \text{k}\Omega$$

Recall that the first-order low-pass filter at a cutoff frequency of 1 kHz was designed in Example 8-1. Since the cutoff frequency of the second-order low-pass filter is also 1 kHz, we could use the resistor and capacitor values obtained in Example 8-1, That is, $R_2 = R_3 = 15.9\ \text{k}\Omega$, and $C_2 = C_3 = 0.01\ \mu$F. However, the values of R_1 and R_F must be chosen such that $R_F = 0.586R_1$. Therefore, let $R_1 = 27\ \text{k}\Omega$ and $R_F = 15.8\ \text{k}\Omega$.

(b) The frequency response data shown in Table 8-2 are obtained from the magnitude

TABLE 8-2 FREQUENCY RESPONSE DATA FOR EXAMPLE 8-4

Frequency, f (Hz)	Gain magnitude, $\lvert v_o/v_{in} \rvert$	Magnitude (dB) = $20 \log \lvert v_o/v_{in} \rvert$
1. 10	1.59	4.01
2. 100	1.59	4.01
3. 200	1.58	4.00
4. 700	1.42	3.07
5. 1,000	1.12	1.00
6. 3,000	0.18	−15.13
7. 7,000	0.03	−29.80
8. 10,000	0.02	−35.99
9. 30,000	1.76×10^{-3}	−55.08
10. 100,000	1.59×10^{-4}	−75.99

Equation (8-4) by substituting various values from 10 Hz to 100 kHz for f. Equation (8-4) is repeated here for convenience:

$$\left|\frac{v_o}{v_{in}}\right| = \frac{A_F}{\sqrt{1 + (f/f_H)^4}}$$

where $A_F = 1.586$ and $f_H = 1$ kHz. The frequency response of the second-order low-pass filter of Example 8-4 is shown in Figure 8-5.

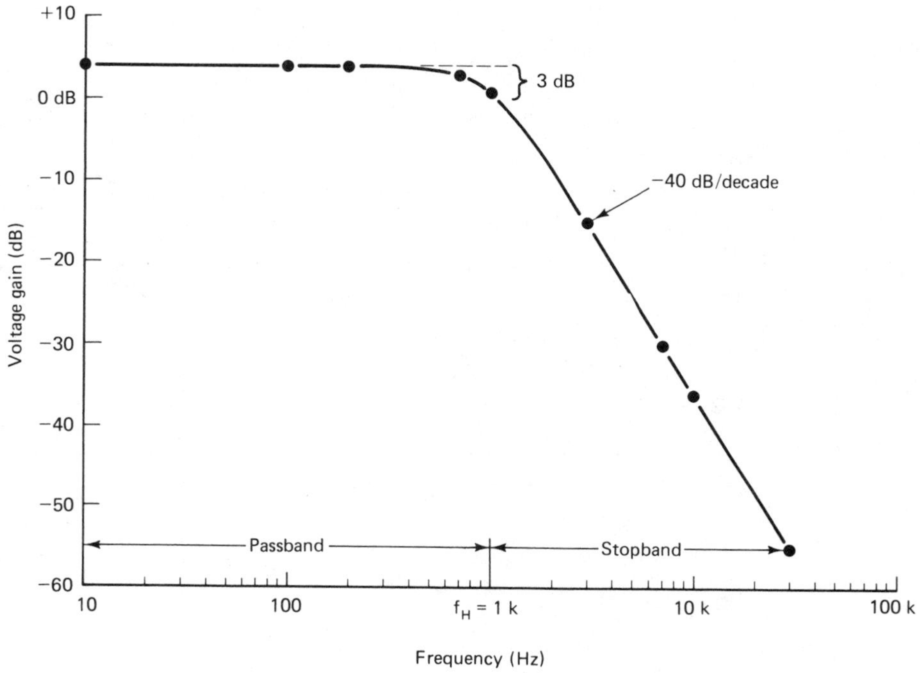

Figure 8-5 Frequency response for Example 8-4.

8-5 FIRST-ORDER HIGH-PASS BUTTERWORTH FILTER

High-pass filters are often formed simply by interchanging frequency determining resistors and capacitors in low-pass filters. That is, a first-order high-pass filter is formed from a first-order low-pass type by interchanging components R and C. Similarly, a second-order high-pass filter is obtained from a second-order low-pass filter if R and C are interchanged, and so on. Figure 8-6 shows a first-order high-pass Butterworth filter with a low cutoff frequency of f_L. This is the frequency at which the magnitude of the gain is 0.707 times its passband value. Obviously, all frequencies higher than f_L are passband frequencies with the highest frequency determined by the closed-loop bandwidth of the op-amp.

Note that the high-pass filter of Figure 8-6(a) and the low-pass filter of Figure 8-2(a) are the same circuits, except that the frequency determining components (R and C) are interchanged.

For the first-order high-pass filter of Figure 8-6(a), the output voltage is

$$v_o = \left(1 + \frac{R_F}{R_1}\right) \frac{j2\pi fRC}{1 + j2\pi fRC} v_{\text{in}}$$

or

$$\frac{v_o}{v_{\text{in}}} = A_F \left[\frac{j(f/f_L)}{1 + j(f/f_L)}\right] \tag{8-5}$$

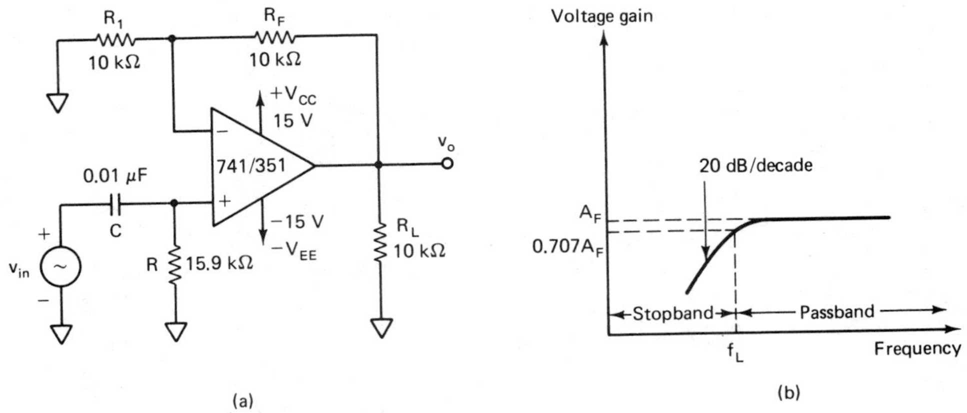

Figure 8-6 (a) First-order high-pass Butterworth filter. (b) Its frequency response.

where $A_F = 1 + \dfrac{R_F}{R_1}$ = passband gain of the filter

f = frequency of the input signal

$f_L = \dfrac{1}{2\pi RC}$ = low cutoff frequency

Hence the magnitude of the voltage gain is

$$\left|\frac{v_o}{v_{\text{in}}}\right| = \frac{A_F(f/f_L)}{\sqrt{1+(f/f_L)^2}} \tag{8-6}$$

Since high-pass filters are formed from low-pass filters simply by interchanging R's and C's, the design and frequency scaling procedures of the low-pass filters are also applicable to the high-pass filters (see Sections 8-3.1 and 8-3.2).

Example 8-5

(a) Design a high-pass filter at a cutoff frequency of 1 kHz with a passband gain of 2.
(b) Plot the frequency response of the filter in part (a).

Solution

(a) Use the same values of R and C that were determined for the low-pass filter of Example 8-1, since $f_L = f_H = 1$ kHz. That is, $C = 0.01\ \mu$F and $R = 15.9$ kΩ. Similarly, use $R_1 = R_F = 10$ kΩ, since $A_F = 2$.

(b) The data for the frequency response plot can be obtained by substituting for the input frequency f values from 100 Hz to 100 kHz in Equation (8-6). These data are included in Table 8-3. Equation (8-6) is repeated here for convenience:

$$\left|\frac{v_o}{v_{\text{in}}}\right| = \frac{A_F(f/f_L)}{\sqrt{1+(f/f_L)^2}}$$

where $A_F = 2$ and $f_L = 1$ kHz. The frequency response data of Table 8-3 are plotted in Figure 8-7. In the stopband (from 100 Hz to 1 kHz) the gain increases at the rate of 20 dB/decade. However, in the passband (after $f = f_L = 1$ kHz) the gain remains constant at 6.02 dB. The upper-frequency limit of the passband is set by the closed-loop bandwidth of the op-amp.

TABLE 8-3 FREQUENCY RESPONSE DATA FOR THE FIRST-ORDER HIGH-PASS FILTER OF EXAMPLE 8-5

Frequency, f (Hz)	Gain magnitude, $\lvert v_o/v_{in} \rvert$	Magnitude (dB) = $20 \log \lvert v_o/v_{in} \rvert$
1. 100	0.20	−14.02
2. 200	0.39	−8.13
3. 400	0.74	−2.58
4. 700	1.15	1.19
5. 1,000	1.41	3.01
6. 3,000	1.90	5.56
7. 7,000	1.98	5.93
8. 10,000	1.99	5.98
9. 30,000	2	6.02
10. 100,000	2	6.02

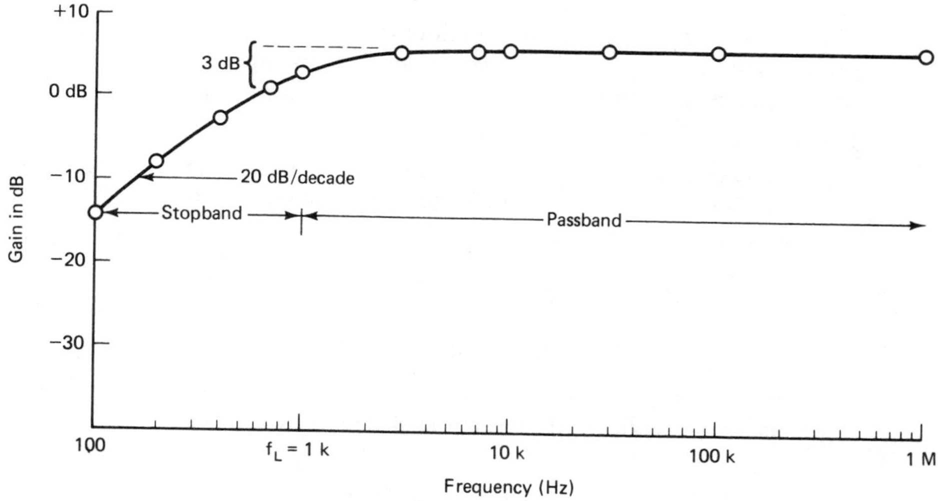

Figure 8-7 Frequency response for Example 8-5.

8-6 SECOND-ORDER HIGH-PASS BUTTERWORTH FILTER

As in the case of the first-order filter, a second-order high-pass filter can be formed from a second-order low-pass filter simply by interchanging the frequency determining resistors and capacitors. Figure 8-8(a) shows the second-order high-pass filter.

The voltage gain magnitude equation of the second-order high-pass filter is as follows:

$$\left| \frac{v_o}{v_{in}} \right| = \frac{A_F}{\sqrt{1 + (f_L/f)^4}} \tag{8-7}$$

where $A_F = 1.586$ = passband gain for the second-order Butterworth response
f = frequency of the input signal (hertz)
f_L = low cutoff frequency (hertz)

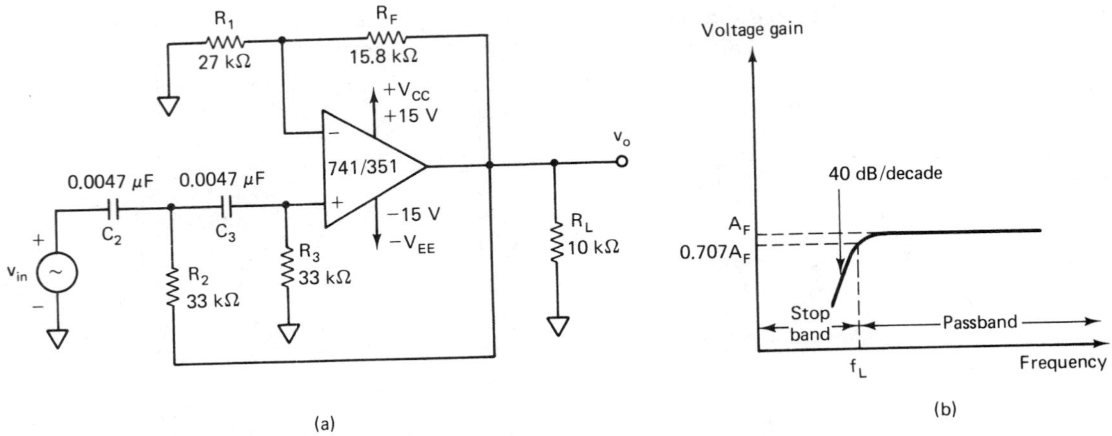

Figure 8-8 (a) Second-order high-pass Butterworth filter. (b) Its frequency response.

Since second-order low-pass and high-pass filters are the same circuits except that the positions of resistors and capacitors are interchanged, the design and frequency scaling procedures for the high-pass filter are the same as those for the low-pass filter.

Example 8-6

(a) Determine the low cutoff frequency f_L of the filter shown in Figure 8-8(a).
(b) Draw the frequency response plot of the filter.

Solution

(a) $$f_L = \frac{1}{2\pi\sqrt{R_2C_2R_3C_3}}$$

$$= \frac{1}{2\pi\sqrt{(33\text{ k}\Omega)^2(0.0047\ \mu\text{F})^2}} \cong 1\text{ kHz}$$

(b) The frequency response data in Table 8-4 are obtained from voltage-gain-magnitude Equation (8-7), which is repeated here for convenience:

$$\left|\frac{v_o}{v_{\text{in}}}\right| = \frac{A_F}{\sqrt{1 + (f_L/f)^4}}$$

TABLE 8-4 FREQUENCY RESPONSE DATA FOR SECOND-ORDER HIGH-PASS FILTER OF EXAMPLE 8-6

Input frequency, f (Hz)	Gain magnitude, $\lvert v_o/v_{\text{in}} \rvert$	Magnitude (dB) = 20 log $\lvert v_o/v_{\text{in}} \rvert$
1. 100	0.01586	−35.99
2. 200	0.0634	−23.96
3. 700	0.6979	−3.124
4. 1,000	1.1215	0.9960
5. 3,000	1.5763	3.953
6. 7,000	1.5857	4.004
7. 10,000	1.5859	4.006
8. 30,000	1.5860	4.006
9. 100,000	1.5860	4.006

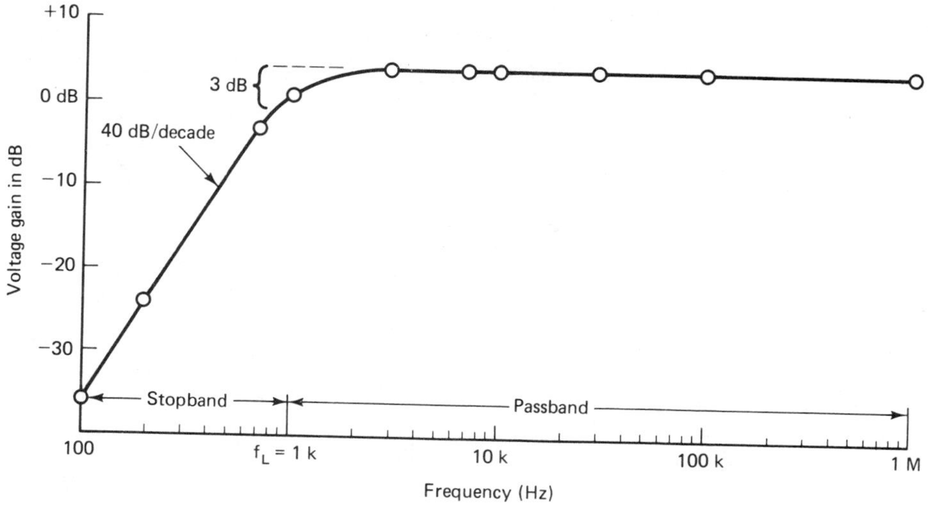

Figure 8-9 Frequency response for Example 8-6.

where $A_F = 1.586$ and $f_L = 1$ kHz. The resulting frequency response plot is shown in Figure 8-9.

8-7 HIGHER-ORDER FILTERS

From the preceding discussions of filters we can conclude that in the stopband the gain of the filter changes at the rate of 20 dB/decade for first-order filters and at 40 dB/decade for second-order filters. This means that as the order of the filter is increased, the actual stopband response of the filter approaches its ideal stopband characteristic.

Higher-order filters, such as third, fourth, fifth, and so on, are formed simply by using the first- and second-order filters. For example, a third-order low-pass filter is formed by connecting in series or cascading first- and second-order low-pass filters; a fourth-order low-pass filter is composed of two cascaded second-order low-pass sections, and so on. Although there is no limit to the order of the filter that can be formed, as the order of the filter increases, so does its size. Also, its accuracy declines, in that the difference between the actual stopband response and the theoretical stopband response increases with an increase in the order of the filter. Figure 8-10 shows third- and fourth-order low-pass Butterworth filters. Note that in the third-order filter the voltage gain of the first-order section is *one*, and that of the second-order section is *two*. On the other hand, in the fourth-order filter the gain of the first section is *1.152*, while that of the second section is *2.235*. These gain values are necessary to guarantee Butterworth response and have to remain the same regardless of the filter's cutoff frequency. Furthermore, the overall gain of the filter is equal to the product of the individual voltage gains of the filter sections. Thus the overall gain of the third-order filter is 2.0, and that of the fourth order is $(1.152)(2.235) = 2.57$.

Since the frequency-determining resistors are equal, and the frequency-determining capacitors are also equal, the high cutoff frequencies of the third- and fourth-order low-pass filters in Figure 8-10(a) and (b) must also be equal. That is,

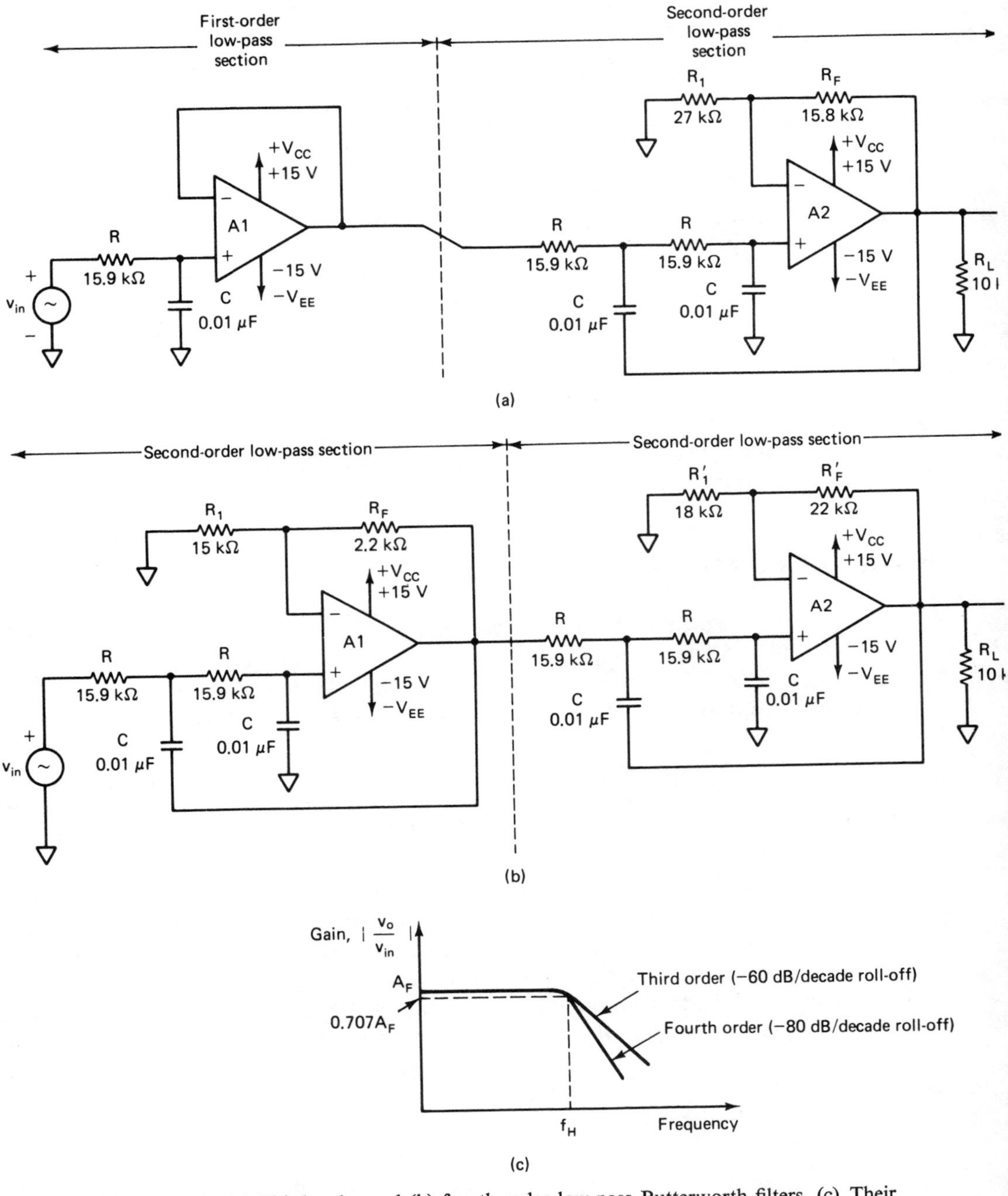

Figure 8-10 (a) Third-order and (b) fourth-order low-pass Butterworth filters. (c) Their frequency responses. A_1 and A_2 dual op-amp: 1458/353.

$$f_H = \frac{1}{2\pi RC} \tag{8-8}$$

As with the first- and second-order filters, the third- and fourth-order high-pass filters are formed by simply interchanging the positions of the frequency-determining resistors and capacitors in the corresponding low-pass filters. The high-order filters can be designed by following the procedures outlined for the first- and second-order filters. However, note that the overall gain of the higher-order filters is *fixed* because all the frequency-determining resistors and capacitors are equal.

Generally, the minimum-order filter required depends on the application specifications. Although a higher-order filter than necessary gives a better stopband response, the higher-order type is more complex, occupies more space, and is more expensive.

8-8 BAND-PASS FILTERS

Basically, there are two types of band-pass filters:

1. Wide band pass
2. Narrow band pass

Unfortunately, there is no set dividing line between the two. However, we will define a filter as wide band pass if its *figure of merit* or *quality factor* $Q < 10$. On the other hand, if $Q > 10$, we will call the filter a narrow band-pass filter. Thus Q is a measure of selectivity, meaning the higher the value of Q, the more selective is the filter or the narrower its bandwidth. The relationship between Q, the 3-dB bandwidth, and the center frequency f_C is given by

$$Q = \frac{f_C}{\text{BW}} = \frac{f_C}{f_H - f_L} \tag{8-9a}$$

For the wide band-pass filter the center frequency f_C can be defined as

$$f_C = \sqrt{f_H f_L} \tag{8-9b}$$

where f_H = high cutoff frequency
f_L = low cutoff frequency of the wide band-pass filter

In a narrow band-pass filter, the output voltage peaks at the center frequency.

8-8.1 Wide Band-Pass Filter

A number of possible circuits will realize the band-pass responses. A wide band-pass filter can be formed by simply cascading high-pass and low-pass sections and is generally the choice for simplicity of design and performance. To obtain a ± 20 dB/decade band-pass, first-order high-pass and first-order low-pass sections are cascaded; for a ± 40-dB/decade band-pass filter, second-order high-pass and second-order low-pass sections are connected in series, and so on. In other words, the order of the band-pass filter depends on the order of the high-pass and low-pass filter sections.

Figure 8-11 shows the $\pm$20-dB/decade wide band-pass filter, which is composed of a first-order high-pass filter and a first-order low-pass filter. To realize a band-pass response, however, f_H must be larger than f_L, as illustrated in Example 8-7.

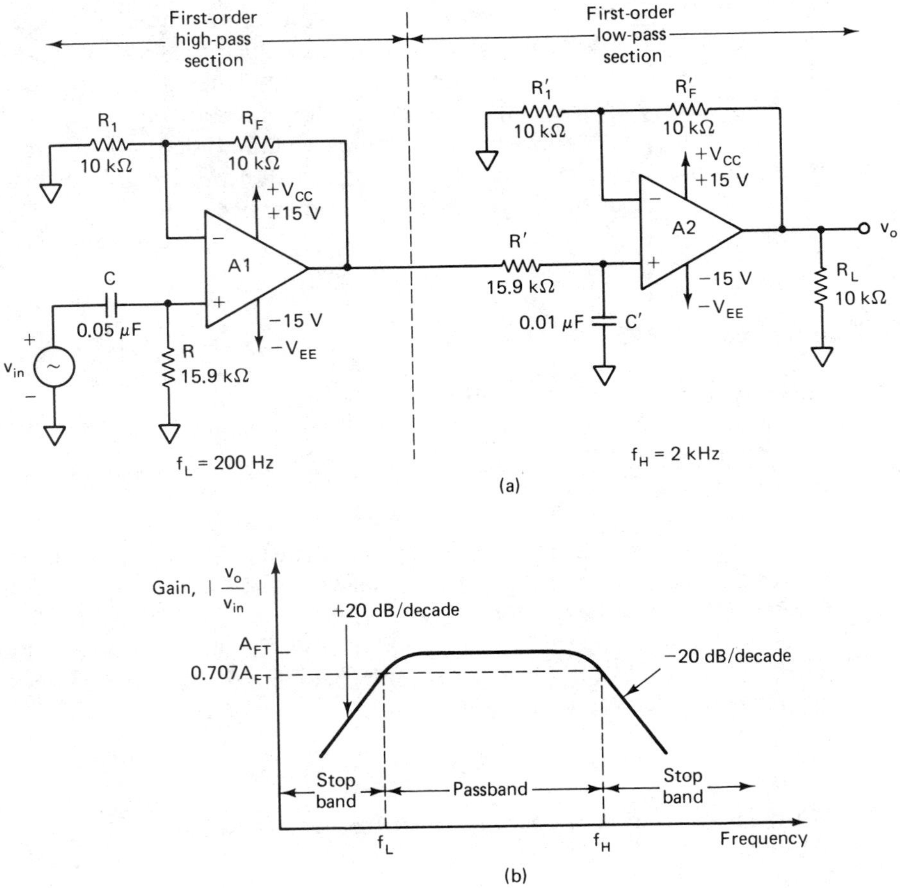

Figure 8-11 (a) $\pm$20-dB/decade-wide band-pass filter. (b) Its frequency response. A_1 and A_2 dual op-amp: 1458/353.

Example 8-7

(a) Design a wide band-pass filter with $f_L = 200$ Hz, $f_H = 1$ kHz, and a passband gain $= 4$.
(b) Draw the frequency response plot of this filter.
(c) Calculate the value of Q for the filter.

Solution

(a) A low-pass filter with $f_H = 1$ kHz was designed in Example 8-1; therefore, the same values of resistors and capacitors can be used here, that is, $R' = 15.9$ kΩ and $C' = 0.01$ μF. As in the case of the high-pass filter, it can be designed by following the steps of Section 8-3.1:

1. $f_L = 200$ Hz.
2. Let $C = 0.05\ \mu$F.
3. Then

$$R = \frac{1}{2\pi f_L C} = \frac{1}{(2\pi)(200)(5)(10^{-8})}$$
$$= 15.9\ \text{k}\Omega$$

Since the band-pass gain is 4, the gain of the high-pass as well as low-pass sections could be set equal to 2. That is, input and feedback resistors must be equal in value, say 10 kΩ each. The complete band-pass filter is shown in Figure 8-11(a).
(b) The voltage gain magnitude of the band-pass filter is equal to the product of the voltage gain magnitudes of the high-pass and low-pass filters. Therefore, from Equations (8-2a) and (8-6),

$$\left|\frac{v_o}{v_{in}}\right| = \frac{A_{FT}(f/f_L)}{\sqrt{[1 + (f/f_L)^2][1 + (f/f_H)^2]}} \tag{8-10}$$

where A_{FT} = total passband gain
f = frequency of the input signal (hertz)
f_L = low cutoff frequency (hertz)
f_H = high cutoff frequency (hertz)

Here $A_{FT} = 4$, $f_L = 200$ Hz, and $f_H = 1$ kHz. The frequency response data in Table 8-5 are obtained by substituting into Equation (8-10) the values of f from 10 Hz to

TABLE 8-5 FREQUENCY RESPONSE DATA FOR THE BAND-PASS FILTER OF EXAMPLE 8-7

Input frequency, f (Hz)	Gain magnitude, $\lvert v_o/v_{in} \rvert$	Magnitude (dB) = 20 log $\lvert v_o/v_{in} \rvert$
1. 10	0.1997	−13.99
2. 30	0.5931	−4.54
3. 100	1.780	5.01
4. 200	2.774	8.861
5. 447.2	3.33	10.46
6. 700	3.151	9.969
7. 1,000	2.774	8.861
8. 2,000	1.780	5.001
9. 7,000	0.5655	−4.95
10. 10,000	0.3979	−8.004

10 kHz. The frequency response plot is shown in Figure 8-12.
(c) From Equation (8-9b),

$$f_C = \sqrt{(1000)(200)} = 447.2 \text{ Hz}$$

Substituting this value in Equation (8-9a),

$$Q = \frac{447.2}{1000 - 200} = 0.56$$

Thus Q is less than 10, as expected for the wide band-pass filter.

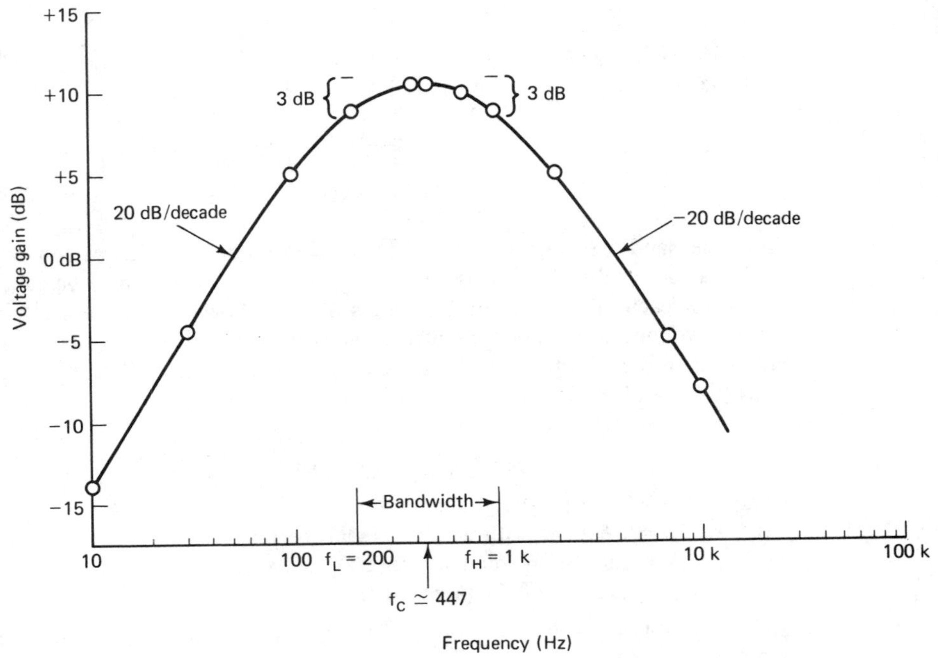

Figure 8-12 Frequency response for Example 8-7.

8-8.2 Narrow Band-Pass Filter

The narrow band-pass filter using multiple feedback is shown in Figure 8-13. As shown in this figure, the filter uses only one op-amp. Compared to all the filters discussed so far, this filter is unique in the following respects:

1. It has two feedback paths, hence the name *multiple-feedback filter*.
2. The op-amp is used in the *inverting* mode.

Generally, the narrow band-pass filter is designed for specific values of center frequency f_C and Q, or f_C and bandwidth [see Equation (8-9a)]. The circuit components are determined from the following relationships:

To simplify the design calculations, choose $C_1 = C_2 = C$.

$$R_1 = \frac{Q}{2\pi f_C C A_F} \tag{8-11}$$

$$R_2 = \frac{Q}{2\pi f_C C(2Q^2 - A_F)} \tag{8-12}$$

$$R_3 = \frac{Q}{\pi f_C C} \tag{8-13}$$

where A_F is the gain at f_C, given by

$$A_F = \frac{R_3}{2R_1} \tag{8-14a}$$

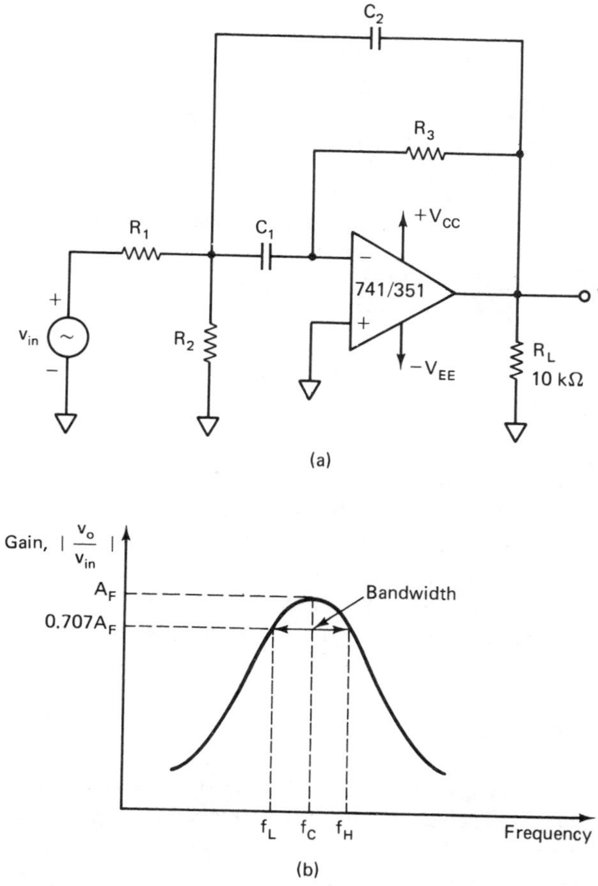

Figure 8-13 (a) Multiple-feedback narrow band-pass filter. (b) Its frequency response.

The gain A_F, however, must satisfy the condition

$$A_F < 2Q^2 \tag{8-14b}$$

Another advantage of the multiple feedback filter of Figure 8-13 is that its center frequency f_C can be changed to a new frequency f'_C without changing the gain or bandwidth. This is accomplished simply by changing R_2 to R'_2, so that

$$R'_2 = R_2 \left(\frac{f_C}{f'_C}\right)^2 \tag{8-15}$$

(see Example 8-8).

Example 8-8

(a) Design the bandpass filter shown in Figure 8-13(a) so that $f_C = 1$ kHz, $Q = 3$, and $A_F = 10$.

(b) Change the center frequency to 1.5 kHz, keeping A_F and the bandwidth constant.

Solution

(a) Choose the values of C_1 and C_2 first and then calculate the values of R_1, R_2, and R_3 from Equations (8-11) through (8-13). Let $C_1 = C_2 = C = 0.01\ \mu$F.

$$R_1 = \frac{3}{(2\pi)(10^3)(10^{-8})(10)} = 4.77\ \text{k}\Omega$$

$$R_2 = \frac{3}{(2\pi)(10^3)(10^{-8})[2(3)^2 - 10]} = 5.97\ \text{k}\Omega$$

$$R_3 = \frac{3}{(\pi)(10^3)(10^{-8})} = 95.5\ \text{k}\Omega$$

Use $R_1 = 4.7\ \text{k}\Omega$, $R_2 = 6.2\ \text{k}\Omega$, and $R_3 = 100\ \text{k}\Omega$.

(b) Using Equation (8-15), the value of R_2' required to change the center frequency from 1 kHz to 1.5 kHz is

$$R_2' = (5.97\ \text{k}\Omega)\left(\frac{1\ \text{k}\Omega}{1.5\ \text{k}\Omega}\right) = 2.65\ \text{k}\Omega$$

(Use $R_2' = 2.7\ \Omega$.)

8-9 BAND-REJECT FILTERS

The band-reject filter is also called a *band-stop* or *band-elimination* filter. In this filter, frequencies are attenuated in the stopband while they are passed outside this band, as shown in Figure 8-1(d). As with band-pass filters, the band-reject filters can also be classified as (1) wide band-reject or (2) narrow band-reject. The narrow band-reject filter is commonly called the *notch filter*. Because of its higher Q (> 10), the bandwidth of the narrow band-reject filter is much smaller than that of the wide band-reject filter.

8-9.1 Wide Band-Reject Filter

Figure 8-14(a) shows a wide band-reject filter using a low-pass filter, a high-pass filter, and a summing amplifier. To realize a band-reject response, the low cutoff frequency f_L of the high-pass filter must be larger than the high cutoff frequency f_H of the low-pass filter. In addition, the passband gain of both the high-pass and low-pass sections must be equal (see Example 8-9). The frequency response of the wide band-reject filter is shown in Figure 8-14(b).

Example 8-9

Design a wide band-reject filter having $f_H = 200$ Hz, $f_L = 1$ kHz.

Solution. In Example 8-7 a wide band-pass filter was designed with $f_L = 200$ Hz and $f_H = 1$ kHz. In this example these band frequencies are interchanged, that is, $f_L = 1$ kHz and $f_H = 200$ Hz. This means that we can use the same components as in Example 8-7, but interchanged between high-pass and low-pass sections. Therefore, for the low-pass section, $R' = 15.9\ \text{k}\Omega$ and $C' = 0.05\ \mu\text{F}$, while for the high-pass section

$$R = 15.9\ \text{k}\Omega \qquad \text{and} \qquad C = 0.01\ \mu\text{F}$$

Since there is no restriction on the passband gain, use a gain of 2 for each section. Hence let

$$R_1 = R_F = R_1' = R_F' = 10\ \text{k}\Omega$$

Further, the gain of the summing amplifier is set at 1; therefore,

$$R_2 = R_3 = R_4 = 10\ \text{k}\Omega$$

Finally, the value of $R_{OM} = R_2 \,||\, R_3 \,||\, R_4 \cong 3.3\ \text{k}\Omega$.

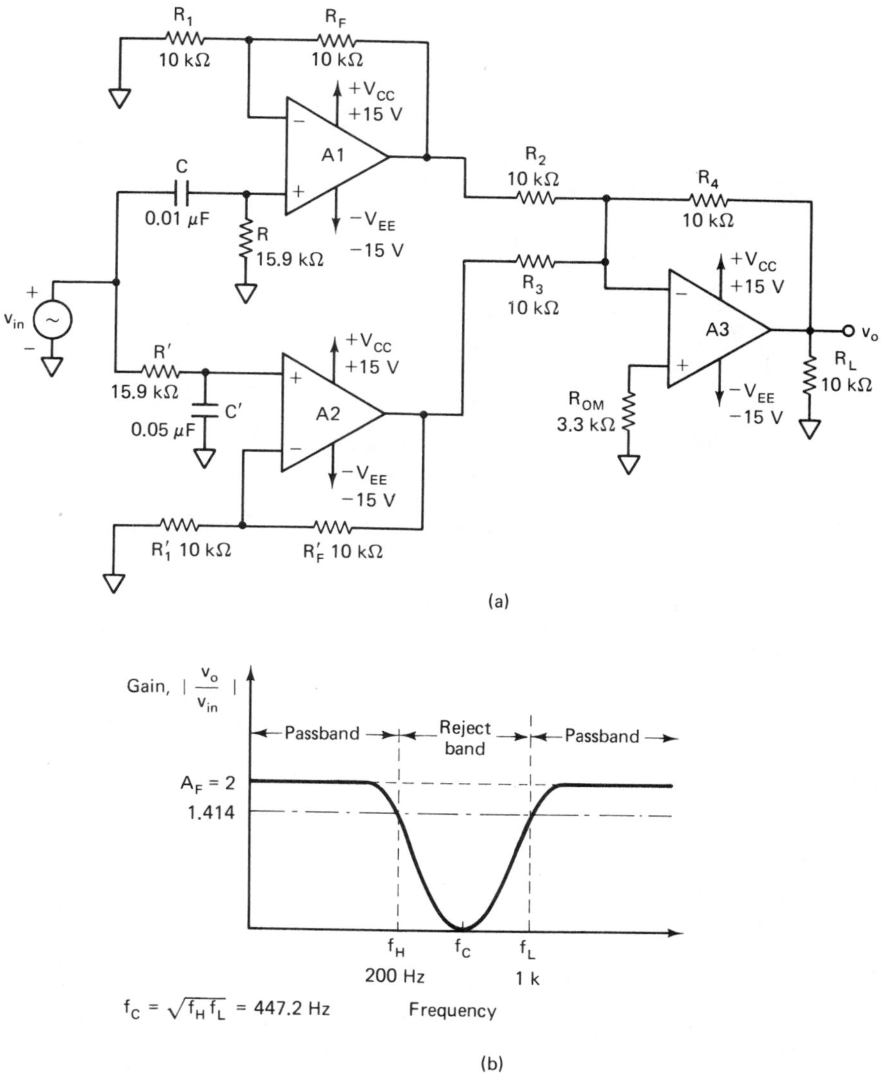

Figure 8-14 Wide band-reject filter. (a) Circuit. (b) Frequency response. For all A_1, A_2, and A_3 use quad op-amp μAF774/MC34004.

The complete circuit is shown in Figure 8-14(a), and its response is shown in Figure 8-14(b). The voltage gain changes at the rate of 20 dB/decade above f_H and below f_L, with a maximum attenuation occurring at f_C.

8-9.2 Narrow Band-Reject Filter

The narrow band-reject filter, often called the *notch filter*, is commonly used for the rejection of a single frequency, such as the 60-Hz power line frequency hum. The most commonly used notch filter is the *twin-T* network shown in Figure 8-15(a). This is a *passive filter* composed of two "T"-shaped networks. One T network is made up

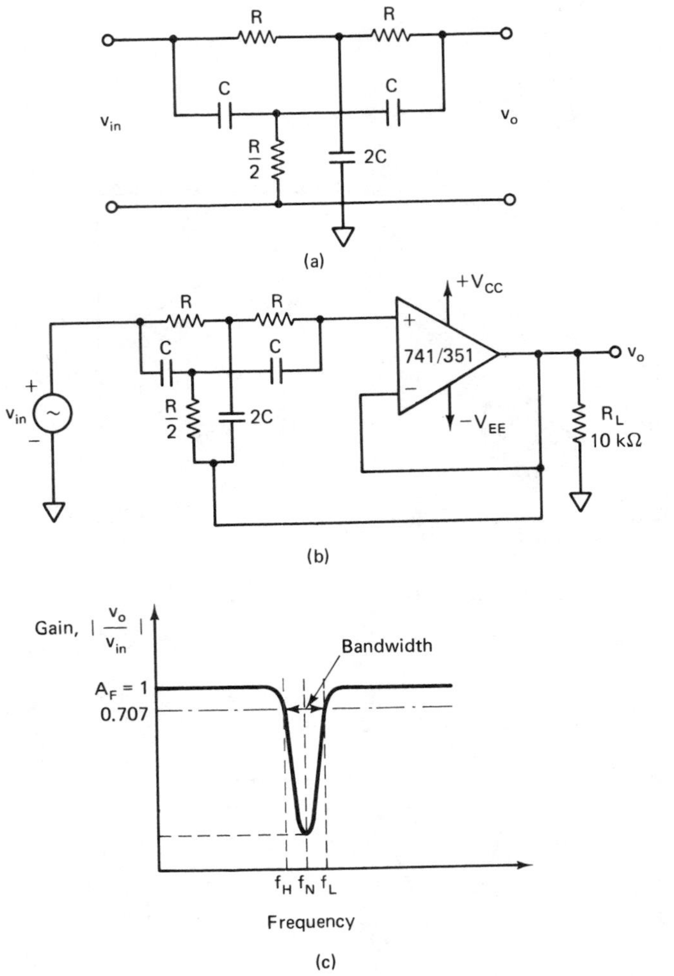

Figure 8-15 (a) Twin-T notch filter. (b) Active notch filter. (c) Frequency response of the active notch filter.

of two resistors and a capacitor, while the other uses two capacitors and a resistor. The "notch-out" frequency is the frequency at which maximum attenuation occurs; it is given by

$$f_N = \frac{1}{2\pi RC} \tag{8-16}$$

Unfortunately, the passive twin-T network has a relatively low figure of merit Q. However, the Q of the network can be increased significantly if it is used with the voltage follower as shown in Figure 8-15(b). Here the output of the voltage follower is fed back to the junction of $R/2$ and $2C$. The frequency response of the active notch filter of Figure 8-15(b) is shown in Figure 8-15(c). The most common use of notch filters is in communications and biomedical instruments for eliminating undesired frequencies. In order to design an active notch filter for a specific notch-out frequency f_N, choose the value of $C \leq 1\ \mu$F and then calculate the required value of R from Equation (8-16). For the best response, the circuit components should be very close to their indicated values.

Example 8-10

Design a 60-Hz active notch filter.

Solution. Let $C = 0.068\ \mu\text{F}$. Then, from Equation (8-16), the value of R is

$$R = \frac{1}{2\pi f_N C} = \frac{1}{(2\pi)(60)(68)(10^{-9})} = 39.01\ \text{k}\Omega$$

(Use 39 kΩ.) For $R/2$, parallel two 39-kΩ resistors; for the $2C$ component, parallel two 0.068-μF capacitors.

8-10 ALL-PASS FILTER

As the name suggests, an all-pass filter passes all frequency components of the input signal without attenuation, while providing predictable phase shifts for different frequencies of the input signal. When signals are transmitted over transmission lines, such as telephone wires, they undergo change in phase. To compensate for these phase changes, all-pass filters are required. The all-pass filters are also called *delay equalizers* or *phase correctors*. Figure 8-16(a) shows an all-pass filter wherein $R_F = R_1$. The output voltage v_o of the filter can be obtained by using the superposition theorem:

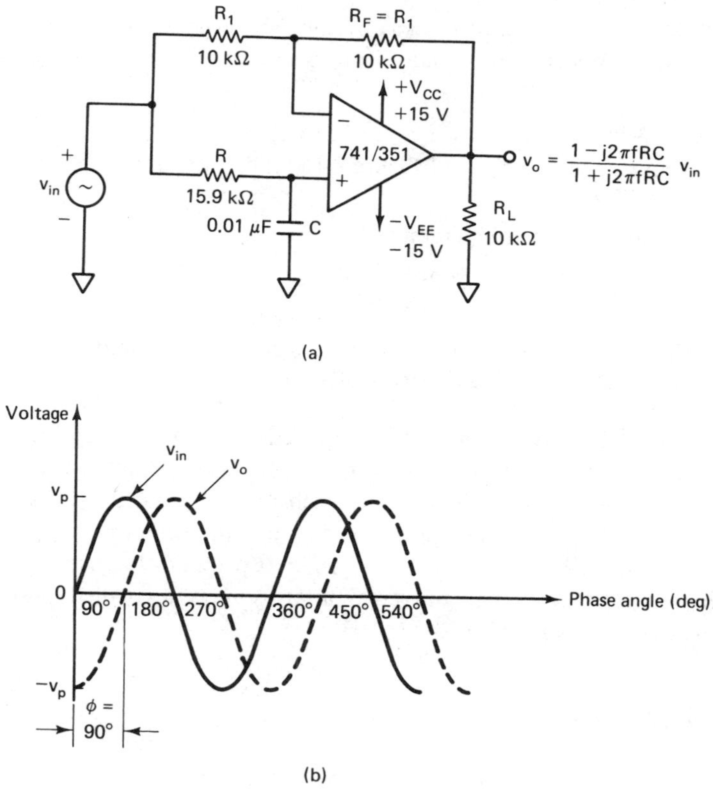

Figure 8-16 All-pass filter. (a) Circuit. (b) Phase shift between input and output voltages.

$$v_o = -v_{in} + \frac{-jX_C}{R - jX_C} v_{in}(2) \tag{8-17}$$

But $-j = 1/j$ and $X_C = 1/2\pi fC$. Therefore, substituting for X_C and simplifying, we get

$$v_o = v_{in}\left(-1 + \frac{2}{j2\pi fRC + 1}\right)$$

or

$$\frac{v_o}{v_{in}} = \frac{1 - j2\pi fRC}{1 + j2\pi fRC} \tag{8-18}$$

where f is the frequency of the input signal in hertz.

Equation (8-18) indicates that the amplitude of v_o/v_{in} is unity; that is, $|v_o| = |v_{in}|$ throughout the useful frequency range, and the phase shift between v_o and v_{in} is a function of input frequency f. The phase angle ϕ is given by

$$\phi = -2\tan^{-1}\left(\frac{2\pi fRC}{1}\right) \tag{8-19}$$

where ϕ is in degrees, f in hertz, R in ohms, and C in farads. Equation (8-19) is used to find the phase angle ϕ if f, R, and C are known. Figure 8-16(b) shows a phase shift of 90° between the input v_{in} and output v_o. That is, v_o lags v_{in} by 90°. For fixed values of R and C, the phase angle ϕ changes from 0 to −180° as the frequency f is varied from 0 to ∞. In Figure 8-16(a), if the positions of R and C are interchanged, the phase shift between input and output becomes positive. That is, output v_o leads input v_{in}.

Example 8-11

For the all-pass filter of Figure 8-16(a), find the phase angle ϕ if the frequency of v_{in} is 1 kHz.

Solution. From Equation (8-19),

$$\phi = -2\tan^{-1}\left[\frac{(2\pi)(10^3)(15.9)(10^3)(10^{-8})}{1}\right]$$
$$= -90°$$

This means that the output voltage v_o has the same frequency and amplitude but lags v_{in} by 90°, as shown in Figure 8-16(b).

With the advancement of integrated-circuit technology, a number of manufacturers now offer ready-to-use *universal filters* having simultaneous low-pass, high-pass, and band-pass output responses. Notch and all-pass functions are also available by combining these output responses in the uncommitted op-amp. Because of its versatility, this filter is called the *universal filter*. It provides the user with easy control of the gain and Q factor. The universal filter, sometimes called a *state-variable filter*, is presented in Chapter 10.

8-11 OSCILLATORS

Thus far we have examined op-amps wired as amplifiers or filters. This section will introduce the use of op-amps as oscillators capable of generating a variety of output waveforms. Basically, the function of an oscillator is to generate alternating current or voltage waveforms. More precisely, an oscillator is a circuit that generates a repetitive waveform of fixed amplitude and frequency without any external input signal. Oscillators are used in radio, television, computers, and communications. Although there are different types of oscillators, all of them work on the same basic principle.

8-11.1 Oscillator Principles

An oscillator as such is a type of feedback amplifier in which part of the output is fed back to the input via a feedback circuit. If the signal fed back is of proper magnitude and phase, the circuit produces alternating currents or voltages. To visualize the requirements of an oscillator, consider the block diagram of Figure 8-17. This diagram looks identical to that of the feedback amplifiers of Chapter 4 (see Figures 4-3 and 4-9). However, here the input voltage is zero ($v_{in} = 0$). Besides that, the feedback is positive because most oscillators use positive feedback. Finally, the closed-loop gain of the amplifier is denoted by A_V rather than A_F.

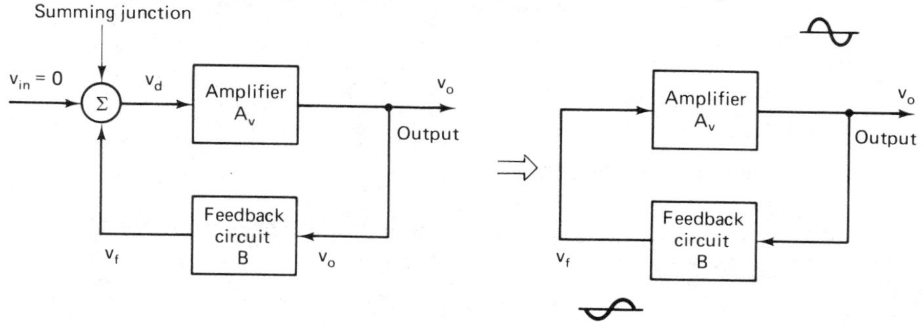

Figure 8-17 Oscillator block diagram.

In the block diagram of Figure 8-17,

$$v_d = v_f + v_{in}$$

$$v_o = A_v v_d$$

$$v_f = Bv_o$$

Using these relationships, the following equation is obtained:

$$\frac{v_o}{v_{in}} = \frac{A_v}{1 - A_v B}$$

However, $v_{in} = 0$ and $v_o \neq 0$ implies that

$$A_v B = 1 \tag{8-20}$$

Expressed in polar form,

$$A_v B = 1\underline{/0^\circ \text{ or } 360^\circ} \tag{8-21}$$

Equation (8-21) gives the two requirements for oscillation: (1) the magnitude of the loop gain A_vB must be at least 1, and (2) the total phase shift of the loop gain A_vB must be equal to 0° or 360°. For instance, as indicated in Figure 8-17, if the amplifier causes a phase shift of 180°, the feedback circuit must provide an additional phase shift of 180° so that the total phase shift around the loop is 360°. The waveforms shown in Figure 8-17 are sinusoidal and are used to illustrate the circuit's action. The type of waveform generated by an oscillator depends on the components in the circuit and hence may be sinusoidal, square, or triangular. In addition, the frequency of oscillation is determined by the components in the feedback circuit.

8-11.2 Oscillator Types

Because of their widespread use, many different types of oscillators are available. These oscillator types are summarized in Table 8-6.

TABLE 8-6 OSCILLATOR TYPES

Types of components used	Frequency of oscillation	Types of waveform generated
1. *RC* oscillator 2. *LC* oscillator 3. Crystal oscillator	1. Audio frequency (AF) 2. Radio frequency (RF)	1. Sinusoidal 2. Square wave 3. Triangular wave 4. Sawtooth wave etc.

8-11.3 Frequency Stability

The ability of the oscillator circuit to oscillate at one exact frequency is called *frequency stability*. Although there may be a number of factors that cause changes in oscillator frequency, the primary factors are temperature changes and changes in the dc power supply. Temperature and power supply changes cause variations in the op-amp's gain, in junction capacitances and resistances of the transistors in an op-amp, as well as in external circuit components. In most cases these variations can be kept small by careful design, by using regulated power supplies, and by temperature control.

Another important factor that determines frequency stability is the *figure of merit* Q of the circuit. The higher the Q, the greater the stability. For this reason, crystal oscillators are far more stable than *RC* or *LC* oscillators, especially at higher frequencies. *LC* circuits and crystals are generally used for the generation of high-frequency signals, while *RC* components are most suitable for audio-frequency applications. Here we discuss audio-frequency *RC* oscillators only. We begin with the sinusoidal oscillators.

8-12 PHASE SHIFT OSCILLATOR

Figure 8-18 shows a phase shift oscillator which consists of an op-amp as the amplifying stage and three *RC* cascaded networks as the feedback circuit. The feedback circuit provides feedback voltage from the output back to the input of the amplifier. The op-amp is used in the inverting mode; therefore, any signal that appears at the inverting terminal is shifted by 180° at the output. An additional 180° phase shift required for oscillation is provided by the cascaded *RC* networks. Thus the total phase shift around

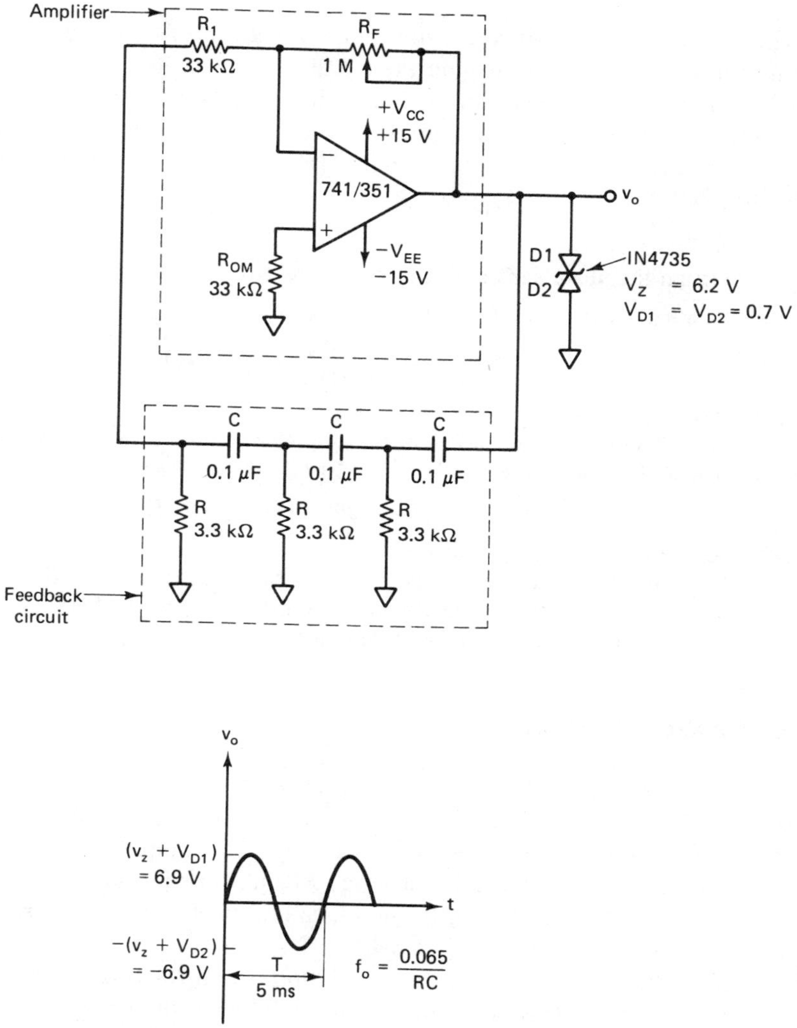

Figure 8-18 Phase shift oscillator and its output waveform.

the loop is 360° (or 0°). At some specific frequency when the phase shift of the cascaded RC networks is exactly 180° and the gain of the amplifier is sufficiently large, the circuit will oscillate at that frequency. This frequency is called the frequency of oscillation f_o and is given by

$$f_o = \frac{1}{2\pi\sqrt{6}\,RC} = \frac{0.065}{RC} \tag{8-22a}$$

At this frequency, the gain A_v must be *at least 29*. That is,

$$\left|\frac{R_F}{R_1}\right| = 29$$

or

$$R_F = 29\,R_1 \tag{8-22b}$$

Thus the circuit will produce a sinusoidal waveform of frequency f_o if the gain is 29 and the total phase shift around the circuit is exactly 360°. For a desired frequency of oscillation, choose a capacitor C, and then calculate the value of R from Equation (8-22a). A desired output amplitude, however, can be obtained with back-to-back zeners connected at the output terminal. For more information on zeners, refer to Section 9-7.

Example 8-12

Design the phase shift oscillator of Figure 8-18 so that $f_o = 200$ Hz.

Solution. Let $C = 0.1\ \mu$F. Then from Equation (8-22a),

$$R = \frac{0.065}{(200)(10^{-7})} = 3.25\text{ k}\Omega$$

(Use $R = 3.3$ kΩ.)

To prevent the loading of the amplifier because of RC networks, it is necessary that $R_1 \geq 10R$. Therefore, let $R_1 = 10R = 33$ kΩ. Then, from Equation (8-22b),

$$R_F = 29(33\text{ k}\Omega) = 957\text{ k}\Omega$$

(Use $R_F = 1$ MΩ potentiometer.)

When choosing an op-amp, type 741 can be used at lower frequencies (<1 kHz); however, at higher frequencies, an op-amp such as the LM318 or LF351 is recommended because of its increased slew rate.

8-13 WIEN BRIDGE OSCILLATOR

Because of its simplicity and stability, one of the most commonly used audio-frequency oscillators is the Wien bridge. Figure 8-19 shows the Wien bridge oscillator in which the Wien bridge circuit is connected between the amplifier input terminals and the output terminal. The bridge has a series RC network in one arm and a parallel RC network in the adjoining arm. In the remaining two arms of the bridge, resistors R_1 and R_F are connected (see Figure 8-19).

The phase angle criterion for oscillation is that the total phase shift around the circuit must be 0°. This condition occurs only when the bridge is balanced, that is, *at resonance*. The frequency of oscillation f_o is exactly the resonant frequency of the balanced Wien bridge and is given by

$$f_o = \frac{1}{2\pi RC} = \frac{0.159}{RC} \tag{8-23a}$$

assuming that the resistors are equal in value, and capacitors are equal in value in the reactive leg of the Wien bridge. At this frequency the gain required for sustained oscillation is given by

$$A_v = \frac{1}{B} = 3$$

That is,

$$1 + \frac{R_F}{R_1} = 3$$

or

$$R_F = 2R_1 \tag{8-23b}$$

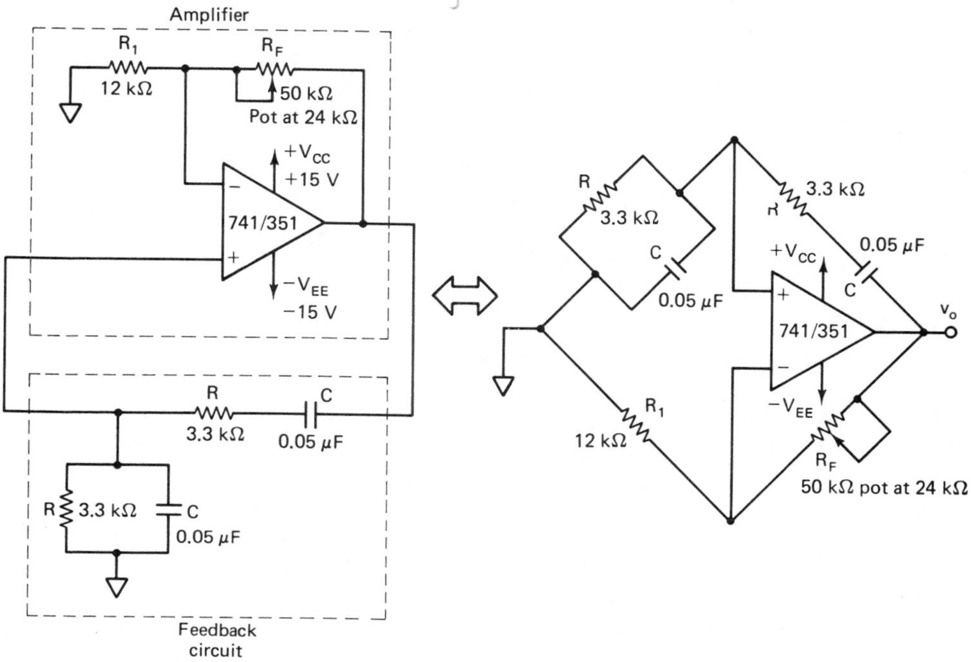

Figure 8-19 Wien bridge oscillator.

The Wien bridge oscillator is designed using Equations (8-23a) and (8-23b), as illustrated in Example 8-13.

Example 8-13

Design the Wien bridge oscillator of Figure 8-19 so that $f_o = 965$ Hz.

Solution. Let $C = 0.05\ \mu$F. Therefore, from Equation (8-23a),

$$R = \frac{0.159}{(5)(10^{-8})(965)} = 3.3\ \text{k}\Omega$$

Now, let $R_1 = 12\ \text{k}\Omega$. Then, from Equation (8-23b),

$$R_F = (2)(12\ \text{k}\Omega) = 24\ \text{k}\Omega$$

(Use $R_F = 50\ \text{k}\Omega$ potentiometer.)

8-14 QUADRATURE OSCILLATOR

As its name implies, the quadrature oscillator generates two signals (sine and cosine) that are in quadrature, that is, out of phase by 90°. Although the actual location of the sine and cosine is arbitrary, in the quadrature oscillator of Figure 8-20 the output of A_1 is labeled a sine and the output of A_2 is a cosine. This oscillator requires a dual op-amp and three RC combinations. The first op-amp A_1 is operating in the noninverting mode and appears as a noninverting integrator. The second op-amp A_2 is working as a pure integrator. Further, A_2 is followed by a voltage divider consisting

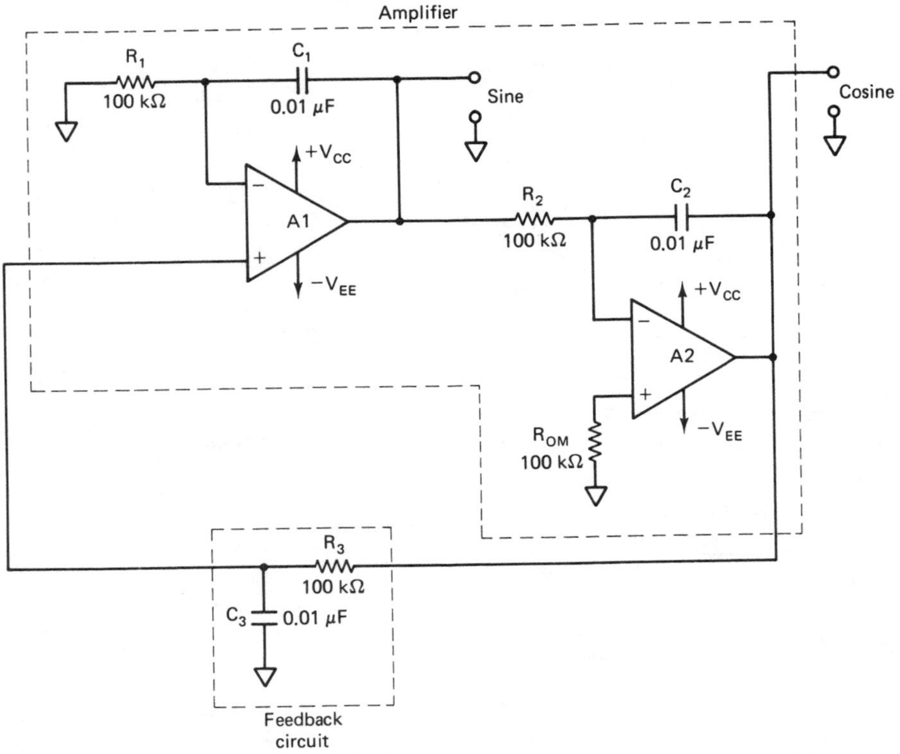

Figure 8-20 Quadrature oscillator. A_1 and A_2: dual op-amp 1458/353.

of R_3 and C_3. The divider network forms a feedback circuit, whereas A_1 and A_2 form the amplifier stage.

The total phase shift of 360° around the loop required for oscillation is obtained in the following way. The op-amp A_2 is a pure integrator and inverter. Hence it contributes −270° or (90°) of phase shift. The remaining −90° (or 270°) of phase shift needed are obtained at the voltage divider R_3C_3 and the op-amp A_1. The total phase shift of 360°, however, is obtained at only one frequency f_o, called the *frequency of oscillation*. This frequency is given by

$$f_o = \frac{1}{2\pi RC} \tag{8-24a}$$

where $R_1C_1 = R_2C_2 = R_3C_3 = RC$. At this frequency,

$$A_v = \frac{1}{B} = 1.414 \tag{8-24b}$$

which is the second condition for oscillation.

Thus to design a quadrature oscillator for a desired frequency f_o, choose a value of C; then from Equation (8-24a), calculate the value of R. To simplify design calculations, choose $C_1 = C_2 = C_3$ and $R_1 = R_2 = R_3$. In addition, R_1 may be a potentiometer in order to eliminate any possible distortion in the output waveforms.

Example 8-14

Design the quadrature oscillator of Figure 8-20 so that $f_o = 159$ Hz. The op-amp is the 1458/772.

Solution. Let $C = 0.01\ \mu$F. Then, from Equation (8-24a),

$$R = \frac{0.159}{(159)(10^{-8})} = 100\ \text{k}\Omega$$

Thus $C_1 = C_2 = C_3 = 0.01\ \mu$F and $R_1 = R_2 = R_3 = 100\ \text{k}\Omega$. R_1 may be a 200-kΩ potentiometer, which can be adjusted for undistorted output waveforms.

Using the oscillator principles of Section 8-11.1, a number of other sine-wave *RC* oscillators, such as the twin-T and biquad (biquadratic), can also be designed.

Before proceeding with the other types of oscillators, consider the difference between the signal generator and the function generator. The function generator has more than one function, such as sine wave, square wave, and triangular wave; the signal generator often generates only one type of waveform, such as a sine wave. However, both types of generators have amplitude and frequency modulation capabilities, unlike the oscillator. Since the generator and the oscillator are equivalent terms, they can be used interchangeably. Usually, circuits producing sine waves are called oscillators, while those generating a square wave, triangular wave, or sawtooth wave are called generators.

8-15 SQUARE-WAVE GENERATOR

In contrast to sine-wave oscillators, square-wave outputs are generated when the op-amp is forced to operate in the saturated region. That is, the output of the op-amp is forced to swing repetitively between positive saturation $+V_{sat}$ ($\cong +V_{CC}$) and negative saturation $-V_{sat}$ ($\cong -V_{EE}$), resulting in the square-wave output. One such circuit is shown in Figure 8-21(a). This square-wave generator is also called a *free-running* or *astable* multivibrator. The output of the op-amp in this circuit will be in positive or negative saturation, depending on whether the differential voltage v_{id} is negative or positive, respectively.

Assume that the voltage across capacitor C is zero volts at the instant the dc supply voltages $+V_{CC}$ and $-V_{EE}$ are applied, see Figure 8-21(b)]. This means that the voltage at the inverting terminal is zero initially. At the same instant, however, the voltage v_1 at the noninverting terminal is a very small finite value which is a function of the output offset voltage V_{ooT} and the values of R_1 and R_2 resistors. Thus the differential input voltage v_{id} is equal to the voltage v_1 at the noninverting terminal. Although very small, voltage v_1 will start to drive the op-amp into saturation. For example, suppose that the output offset voltage V_{ooT} is positive and that, therefore, voltage v_1 is also positive. Since initially the capacitor C acts as a short circuit, the gain of the op-amp is very large (A), hence v_1 drives the output of the op-amp to its positive saturation $+V_{sat}$. With the output voltage of the op-amp at $+V_{sat}$, the capacitor C starts charging toward $+V_{sat}$ through resistor R. However, as soon as the voltage v_2 across capacitor C is slightly more positive than v_1, the output of the op-amp is forced to switch to a negative saturation $-V_{sat}$. With the op-amps' output

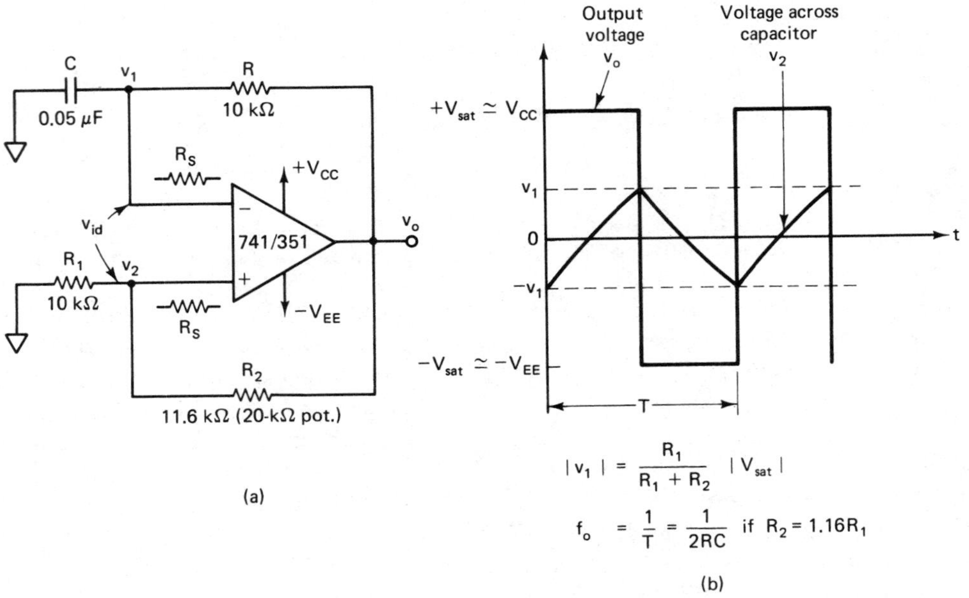

Figure 8-21 (a) Square-wave generator. (b) Waveforms of output voltage v_o and capacitor voltage v_2 of the square-wave generator.

voltage at negative saturation $-V_{sat}$, the voltage v_1 across R_1 is also negative, since

$$v_1 = \frac{R_1}{R_1 + R_2}(-V_{sat}) \tag{8-25a}$$

Thus the net differential voltage $v_{id} = v_1 - v_2$ is negative, which holds the output of the op-amp in negative saturation. The output remains in negative saturation until the capacitor C discharges and then recharges to a negative voltage slightly higher than $-v_1$ [see Figure 8-21(b)]. Now, as soon as the capacitor's voltage v_2 becomes more negative than $-v_1$, the net differential voltage v_{id} becomes positive and hence drives the output of the op-amp back to its positive saturation $+V_{sat}$. This completes one cycle. With output at $+V_{sat}$, voltage v_1 at the noninverting input is

$$v_1 = \frac{R_1}{R_1 + R_2}(+V_{sat}) \tag{8-25b}$$

The time period T of the output waveform is given by

$$T = 2RC \ln\left(\frac{2R_1 + R_2}{R_2}\right) \tag{8-26a}$$

or

$$f_o = \frac{1}{2RC \ln[(2R_1 + R_2)/R_2]} \tag{8-26b}$$

Equation (8-26b) indicates that the frequency of the output f_o is not only a function of the RC time constant but also of the relationship between R_1 and R_2. For example, if $R_2 = 1.16R_1$, Equation (8-26b) becomes

$$f_o = \frac{1}{2RC} \tag{8-27}$$

Equation (8-27) shows that the smaller the RC time constant, the higher the output frequency f_o, and vice versa. As with the sine-wave oscillators, the highest frequency generated by the square-wave generator is also set by the slew rate of the op-amp. An attempt to operate the circuit at relatively higher frequencies causes the oscillator's output to become triangular. In practice, each inverting and noninverting terminal needs a series resistance R_S to prevent excessive differential current flow because the inputs of the op-amp are subjected to large differential voltages. The resistance R_S used should be 100 kΩ or higher. A reduced peak-to-peak output voltage swing can be obtained in the square-wave generator of Figure 8-21(a) by using back-to-back zeners at the output terminal.

Example 8-15

Design the square-wave oscillator of Figure 8-21(a) so that $f_o = 1$ kHz. The op-amp is a 741 with dc supply voltages $= \pm 15$ V.

Solution. Use $R_2 = 1.16R_1$, so that the simplified frequency Equation (8-27) can be applied. Let $R_1 = 10$ kΩ. Then

$$R_2 = (1.16)(10\ \text{k}\Omega) = 11.6\ \text{k}\Omega$$

(Use $R_2 = 20$ kΩ potentiometer.)

Next, choose a value of C and calculate the value of R from Equation (8-27). Hence let $C = 0.05\ \mu$F. By Equation (8-27),

$$R = \frac{1}{(10)(10^{-8})(10^3)} = 10\ \text{k}\Omega$$

Thus

$$R_1 = 10\ \text{k}\Omega$$

$$R_2 = 11.6\ \text{k}\Omega\ (20\ \text{k}\Omega\ \text{potentiometer})$$

$$R = 10\ \text{k}\Omega$$

$$C = 0.05\ \mu\text{F}$$

8-16 TRIANGULAR-WAVE GENERATOR

Recall that the output waveform of the integrator is triangular if its input is a square wave (refer to Section 7-12). This means that a triangular-wave generator can be formed by simply connecting an integrator to the square-wave generator of Figure 8-21. The resultant circuit is shown in Figure 8-22(a). This circuit requires a dual op-amp, two capacitors, and at least five resistors. The frequency of the square wave and triangular wave is the same. For fixed R_1, R_2, and C values, the frequency of the square-wave as well as the triangular wave depends on the resistance R [see Equation (8-26b)]. As R is increased or decreased, the frequency of the triangular wave will decrease or increase, respectively. Although the amplitude of the square wave is constant ($\pm V_{sat}$), the amplitude of the triangular wave decreases with an increase in its frequency, and vice versa. [see Figure 8-22(b)].

The input of integrator A_2 is a square wave, while its output is a triangular wave. However, the output of A_2 to be a triangular wave requires that $5R_3C_2 > T/2$, where T is the period of the square-wave input. As a general rule, R_3C_2 should be equal to

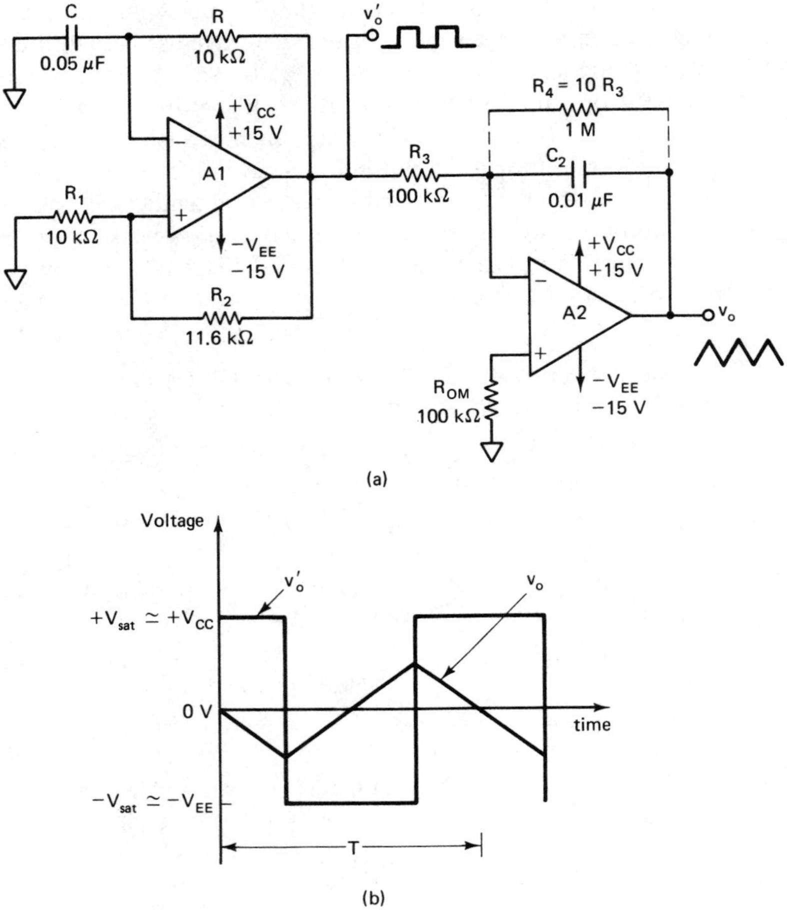

Figure 8-22 Triangular-wave generator. (a) Circuit. (b) Its output waveform. A_1 and A_2: dual op-amp 1458/353.

T. To obtain a stable triangular wave, it may also be necessary to shunt the capacitor C_2 with resistance $R_4 = 10R_3$ and connect an offset voltage compensating network at the noninverting terminal of A_2. As with any other oscillator, the frequency of the triangular-wave generator is limited by the slew rate of the op-amp. Therefore, high slew rate op-amps should be used for the generation of relatively higher frequencies.

Another triangular-wave generator which requires fewer components is shown in Figure 8-23(a). The generator consists of a *comparator* A_1 and an *integrator* A_2. The comparator A_1 compares the voltage at point P continuously with the inverting input that is at 0 V. When the voltage at P goes slightly below or above 0 V, the output of A_1 is at the negative or positive saturation level, respectively.

To illustrate the circuit's operation, let us set the output of A_1 at positive saturation $+V_{sat}$ ($\cong +V_{CC}$). This $+V_{sat}$ is an input of the integrator A_2. The output of A_2, therefore, will be a negative-going ramp. Thus one end of the voltage divider R_2–R_3

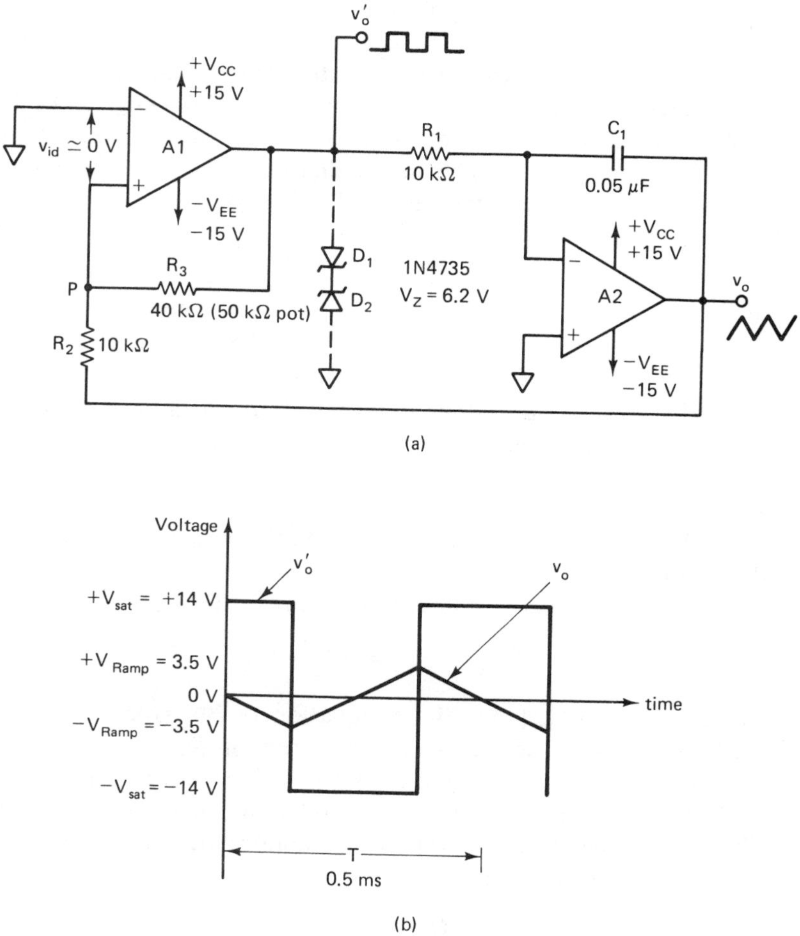

Figure 8-23 Triangular wave generator. (a) Circuit. (b) Its waveforms. A_1 and A_2 dual op-amp: 1458/353.

is the positive saturation voltage $+V_{sat}$ of A_1 and the other is the negative-going ramp of A_2. When the negative-going ramp attains a certain value $-V_{Ramp}$, point P is slightly below 0 V; hence the output of A_1 will switch from positive saturation to negative saturation $-V_{sat}$ ($\cong -V_{EE}$). This means that the output of A_2 will now stop going negatively and will begin to go positively. The output of A_2 will continue to increase until it reaches $+V_{Ramp}$. At this time the point P is slightly above 0 V; therefore, the output of A_1 is switched back to the positive saturation level $+V_{sat}$. The sequence then repeats. The output waveform is as shown in Figure 8-23(b).

The frequency of the square wave and the triangular wave is the same. The amplitude of the square wave is a function of the dc supply voltages. However, a desired amplitude can be obtained by using appropriate zeners at the output of A_1 [see Figure 8-23(a)].

The amplitude and the frequency of the triangular wave can be determined as follows:

From Figure 8-23(b), when the output of the comparator A_1 is $+V_{sat}$, the output of the integrator A_2 steadily decreases until it reaches $-V_{Ramp}$. At this time the output of A1 switches from $+V_{sat}$ to $-V_{sat}$. Just before this switching occurs, the voltage at point P (+input) is 0 V. This means that the $-V_{Ramp}$ must be developed across R_2, and $+V_{sat}$ must be developed across R_3. That is,

$$\frac{-V_{Ramp}}{R_2} = -\frac{+V_{sat}}{R_3}$$

or

$$-V_{Ramp} = -\frac{R_2}{R_3}(+V_{sat}) \tag{8-28a}$$

Similarly, $+V_{Ramp}$, the output voltage of A_2 at which the output of A1 switches from $-V_{sat}$ to $+V_{sat}$, is given by

$$+V_{Ramp} = -\frac{R_2}{R_3}(-V_{sat}) \tag{8-28b}$$

Thus, from Equations (8-28a) and (8-28b), the peak-to-peak (pp) output amplitude of the triangular wave is

$$v_o(\text{pp}) = +V_{Ramp} - (-V_{Ramp})$$
$$= (2)\frac{R_2}{R_3}(V_{sat}) \tag{8-29}$$

where $V_{sat} = |+V_{sat}| = |-V_{sat}|$. Equation (8-29) indicates that the amplitude of the triangular wave decreases with an increase in R_3.

The time it takes for the output waveform to swing from $-V_{Ramp}$ to $+V_{Ramp}$ (or from $+V_{Ramp}$ to $-V_{Ramp}$) is equal to half the time period $T/2$. [see Figure 8-23(b)]. This time can be calculated from the integrator output Equation (7-23) by substituting $v_i = -V_{sat}$ and $v_o = v_o(\text{pp})$:

$$v_o(\text{pp}) = -\frac{1}{R_1C_1}\int_o^{T/2}(-V_{sat})\,dt \tag{7-23}$$
$$= \left(\frac{V_{sat}}{R_1C_1}\right)\left(\frac{T}{2}\right)$$

Hence

$$\frac{T}{2} = \frac{v_o(\text{pp})}{V_{sat}}(R_1C_1)$$

or

$$T = (2R_1C_1)\frac{v_o(\text{pp})}{V_{sat}} \tag{8-30a}$$

where $V_{sat} = |+V_{sat}| = |-V_{sat}|$. Substituting the value of $v_o(\text{pp})$ from Equation (8-29), the time period of the triangular wave is

$$T = \frac{4R_1C_1R_2}{R_3} \tag{8-30b}$$

The frequency of oscillation then is

$$f_o = \frac{R_3}{4R_1C_1R_2} \tag{8-30c}$$

Equation (8-30c) shows that the frequency of oscillation f_o increases with an increase in R_3.

The triangular-wave generator is designed for a desired amplitude and frequency f_o by using Equations (8-29) and (8-30c).

Example 8-16

Design the triangular-wave generator of Figure 8-23 so that $f_o = 2$ kHz and $v_o(\text{pp}) = 7$ V. The op-amp is a 1458/772 and supply voltages $= \pm 15$ V.

Solution. For the 1458, $V_{\text{sat}} = 14$ V. Therefore, from Equation (8-29),

$$\frac{R_2}{R_3} = \frac{7}{(2)(14)}$$

$$R_2 = \frac{R_3}{4}$$

Let $R_2 = 10\ \text{k}\Omega$; then $R_3 = 40\ \text{k}\Omega$. (Use a 50-kΩ potentiometer.)

Now, from Equation (8-30c),

$$2\ \text{kHz} = \frac{40\ \text{k}\Omega}{(4)(R_1C_1)(10\ \text{k}\Omega)}$$

Therefore, $R_1C_1 = 0.5$ ms. Let $C_1 = 0.05\ \mu\text{F}$; then $R_1 = 10\ \text{k}\Omega$. Thus $R_1 = R_2 = 10\ \text{k}\Omega$, $C_1 = 0.05\ \mu\text{F}$, and $R_3 = 40\ \text{k}\Omega$ (50 kΩ potentiometer) [see Figure 8-23(a)].

8-17 SAWTOOTH-WAVE GENERATOR

The difference between the triangular and sawtooth waveforms is that the rise time of the triangular wave is always equal to its fall time. That is, the same amount of time is required for the triangular wave to swing from $-V_{\text{Ramp}}$ to $+V_{\text{Ramp}}$ as from $+V_{\text{Ramp}}$ to $-V_{\text{Ramp}}$ [see Figure 8-23(b)]. On the other hand, the sawtooth waveform has unequal rise and fall times. That is, it may rise positively many times faster than it falls negatively, or vice versa. The triangular-wave generator of Figure 8-23(a) can be converted into a sawtooth-wave generator by simply connecting a potentiometer to the noninverting input terminal of the integrator A_2. The resultant circuit is shown in Figure 8-24(a). Depending on the R_4 setting, a certain dc level is inserted in the output of A_2. Now, suppose that the output of A_1 is a square wave and the potentiometer R_4 is adjusted for a certain dc level. This means that the output of A_2 will be a triangular wave, riding on some dc level that is a function of R_4 setting. The duty cycle of the square wave will be determined by the polarity and amplitude of this dc level. A duty cycle less than 50% will then cause the output of A_2 to be a sawtooth [see Figure 8-24(b)]. With the wiper at the center of R_4, the output of A_2 is a triangular wave. For any other position of R_4 wiper, the output is a sawtooth waveform. Specifically, as the R_4 wiper is moved toward $-V_{EE}$, the rise time of the sawtooth wave becomes longer than the fall time, as shown in Figure 8-24(b). On the other hand, as the wiper is moved toward $+V_{CC}$, the fall time becomes longer than the rise time. Also the frequency of the sawtooth wave decreases as R_4 is adjusted toward $+V_{CC}$ or $-V_{EE}$. However, the amplitude of the sawtooth wave is independent of the R_4 setting.

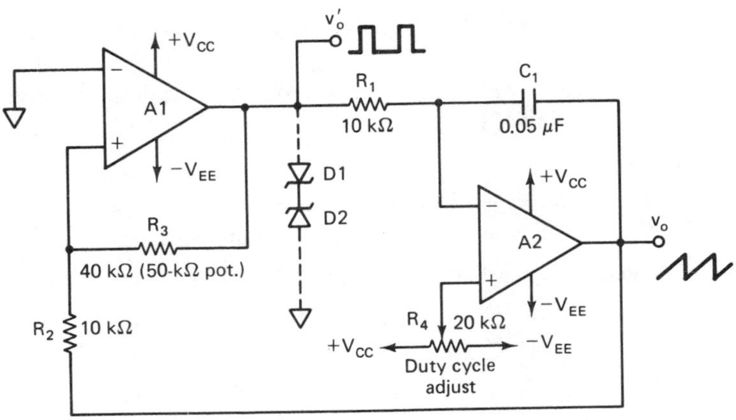

(a)

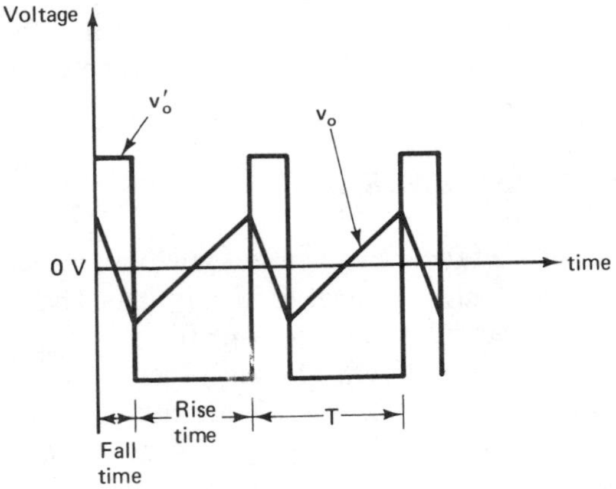

(b)

Figure 8-24 Sawtooth-wave generator. (a) Circuit. A_1 and A_2 dual op-amp: 1458/353. D_1 and D_2: 1N4735 with $V_z = 6.2$ V. (b) Output waveform when noninverting input of A_2 is at some negative dc level.

8-18 VOLTAGE-CONTROLLED OSCILLATOR

In all the preceding oscillators, the frequency is determined by the *RC* time constant. However, there are applications, such as *frequency modulation* (FM), *tone generators*, and *frequency shift keying* (FSK), where the frequency needs to be controlled by means of an input voltage called *control voltage*. This function is achieved in the *voltage-controlled oscillator* (VCO), also called a *voltage-to-frequency converter*. A typical example is the Signetics NE/SE 566 VCO, which provides simultaneous square-wave and triangular-wave outputs as a function of input voltage. Figure 8-25(b) is a block

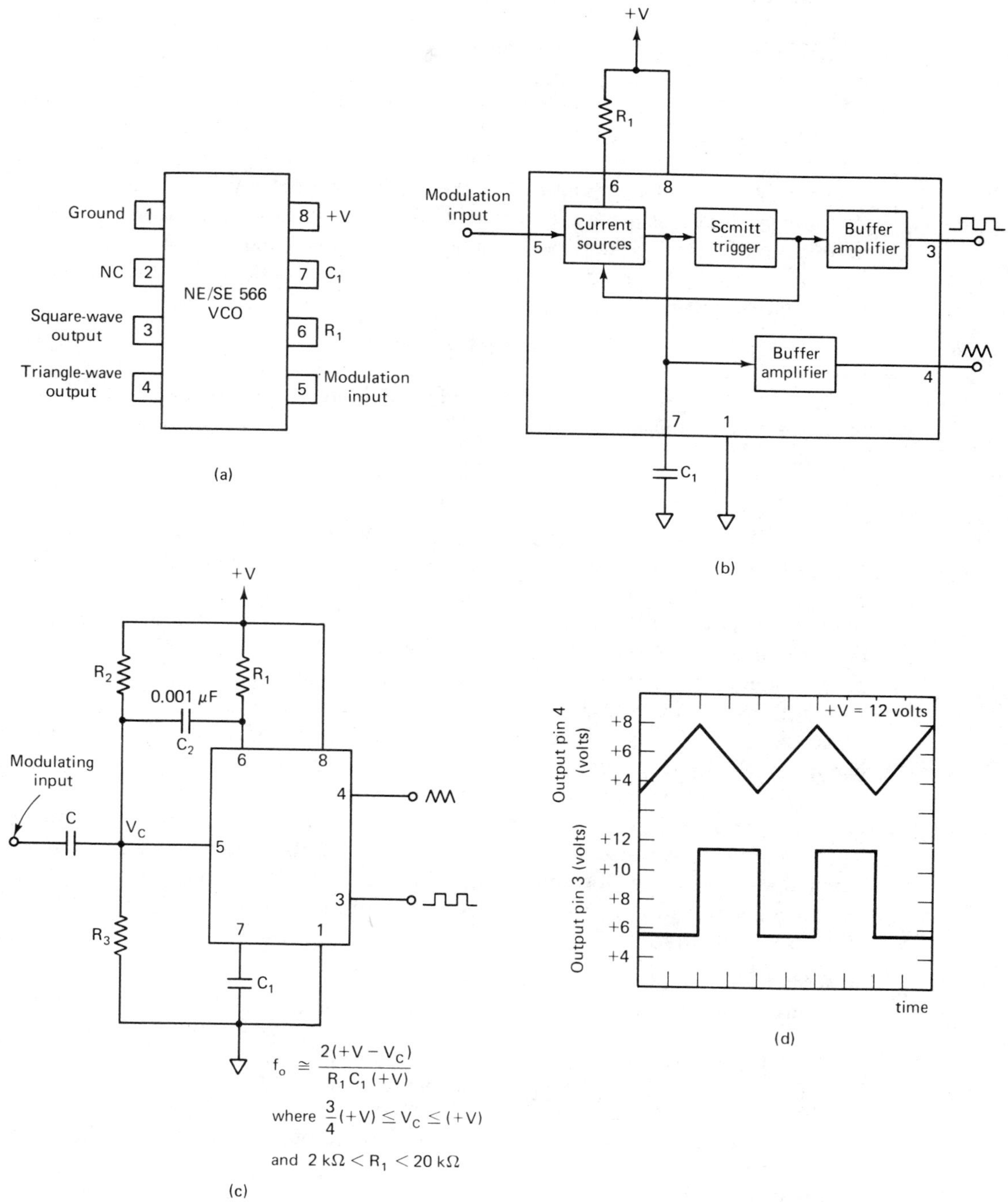

Figure 8-25 Voltage-controlled oscillator 566. (a) Pin configuration. (b) Block diagram. (c) Typical connection diagram. (d) Output waveforms. (Courtesy of Signetics Corporation.)

diagram of the 566. The frequency of oscillation is determined by an external resistor R_1 and capacitor C_1, and the voltage V_C applied to the control terminal 5 [see Figure 8.25(c)]. The triangular wave is generated by alternately charging the external capacitor C_1 by one current source and then linearly discharging it by another [see Figure 8-25(b)]. The charge–discharge levels are determined by Schmitt trigger action. The Schmitt trigger also provides the square-wave output. Both the output waveforms are buffered so that the output impedance of each is 50 Ω. The typical amplitude of the triangular wave is 2.4 V pp and that of the square wave 5.4 V pp.

Figure 8-25(c) is a typical connection diagram. In this arrangement, the R_1C_1 combination determines the free-running frequency, and the control voltage V_C at terminal 5 is set by the voltage divider formed with R_2 and R_3. The initial voltage V_C at terminal 5 must be in the range

$$\frac{3}{4}(+V) \leq V_C \leq +V \tag{8-31a}$$

where $+V$ is the total supply voltage. The modulating signal is ac coupled with the capacitor C and must be < 3 V pp. The frequency of the output waveforms is approximated by

$$f_o \simeq \frac{2(+V - V_C)}{R_1C_1(+V)} \tag{8-31b}$$

where R_1 should be in the range $2\text{ k}\Omega < R_1 < 20\text{ k}\Omega$. For a fixed V_C and constant C_1, the frequency f_o can be varied over a 10:1 frequency range by the choice of R_1 between 2 and 20 kΩ. Similarly, for a constant R_1C_1 product, the frequency f_o can be modulated over a 10:1 range by the control voltage V_C. In either case the maximum output frequency is 1 MHz. A small capacitor 0.001 μF should be connected between pins 5 and 6 to eliminate possible oscillations in the control current source.

If the VCO is to be used to drive standard logic circuitry, a dual supply of ±5 V is recommended so that the square-wave output has the proper dc levels for logic circuitry.

The VCO is commonly used in converting low-frequency signals such as electroencephalograms (EEG) or electrocardiograms (EKG) into an audio-frequency range. These audio signals can then be transmitted over telephone lines or two-way radio communication for diagnostic purposes or can be recorded on a magnetic tape for documentation or further reference. For more information on the VCO applications, refer to Section 10-4.

Example 8-17

In the circuit of Figure 8-25(c), $+V = 12$ V, $R_2 = 1.5\text{ k}\Omega$, $R_1 = R_3 = 10\text{ k}\Omega$, and $C_1 = 0.001\ \mu\text{F}$.

(a) Determine the nominal frequency of the output waveforms.

(b) Compute the modulation in the output frequencies if V_C is varied between 9.5 and 11.5 V.

(c) Draw the square-wave output waveform if the modulating input is a sine wave, as shown in Figure 8-26.

Solution

(a) Using the voltage-divider rules, the initial control voltage V_C at terminal 5 is

$$V_C = \frac{(10\text{ k}\Omega)(12)}{11.5\text{ k}\Omega} = 10.43\text{ V}$$

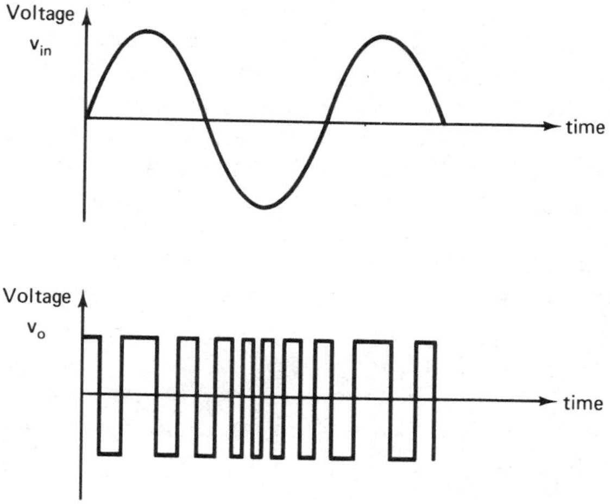

Figure 8-26 Input and Output waveforms for Example 8-17.

From Equation (8-31b), the approximate nominal frequency f_o is

$$f_o \cong \frac{(2)(12 - 10.43)}{(10^4)(10^{-9})(12)} = 26.17 \text{ kHz}$$

(b) The modulation in the output frequencies can be calculated from Equation (8-31b) by substituting for V_C, first 9.5 V and then 11.5 V, as follows:

$$f_o \cong \frac{(2)(12 - 9.5)}{(10^4)(10^{-9})(12)} = 41.67 \text{ kHz}$$

$$f_o \cong \frac{(2)(12 - 11.5)}{(10^4)(10^{-9})(12)} = 8.33 \text{ kHz}$$

Thus the change in the output frequencies is

$$41.67 \text{ kHz} - 8.33 \text{ kHz} = 33.34 \text{ kHz}$$

(c) During the positive half-cycle of the sine-wave input, the control voltage V_C will increase. Therefore, according to Equation (8-31b) the frequency of the output waveform will decrease and the time period will increase. Exactly the opposite action will take place during the negative half-cycle of the input, as shown in Figure 8-26.

SUMMARY

1. A filter is often a frequency selective circuit that passes a specified band of frequencies and attenuates signals of frequencies outside this band.
2. Filters may be classified in a number of ways: analog or digital, passive or active, audio or radio frequency.
3. The most commonly used filters are the following:
   (a) Low-pass filter
   (b) High-pass filter
   (c) Band-pass filter
   (d) Band-reject filter
   (e) All-pass filter

A low-pass filter has a constant gain from 0 Hz to a certain frequency called the cutoff frequency f_H, at which the gain is down by 3 dB. Above f_H, the gain decreases with an increase in frequency. On the other hand, a high-pass filter passes all the frequencies above a certain frequency called the low cutoff frequency f_L. The upper cutoff frequency of the high-pass filter, however, depends on the bandwidth of the op-amp. A band-pass filter has a passband between two cutoff frequencies f_H and f_L such that $f_H > f_L$. Any input frequency outside this passband is attenuated. The band-reject filter performs exactly opposite to the band-pass, in that it has a stopband between two cutoff frequencies f_H and f_L. Finally, the all-pass filter has input and output amplitudes equal at all frequencies; however, the phase shift between the two is a function of frequency.

4. The order of the filter indicates the rate at which the gain changes while the input frequency is approaching or exceeding the cutoff frequency of the filter. For example, in the first-order low-pass filter, for $f > f_H$, the gain decreases at the rate of 20 dB/decade, whereas in the first-order high-pass filter the gain increases at the rate of 20 dB/decade until $f = f_L$. Similarly, in the second-order filter the change in the rate of the gain is 40 dB/decade, in the third-order 60 dB/decade, and so on.
5. In order to convert a low-pass filter into a high-pass filter or vice versa, simply interchange frequency-determining components that is, resistors and capacitors.
6. Higher-order filters can be formed by simply using first- and second-order filters. For example, a third-order filter is constructed by cascading a first- and second-order filters; a fourth-order is formed by cascading two second-order filters; and so on.
7. There are two types of band-pass filters: wide band pass and narrow band pass. A wide band-pass filter can be formed simply by cascading high-pass and low-pass sections. On the other hand, the multiple-feedback filter is one of the narrow band-pass filters.
8. As with band-pass filters, band-reject filters can be either a wide band-reject or narrow band-reject. The wide band-reject filter is formed using a high-pass filter, a low-pass filter, and a summing amplifier. The outputs of low-pass and high-pass filters are added together through a summing amplifier. The narrow band-reject filter, often called the notch filter, uses the twin-T network. It is usually used for the rejection of a single frequency such as 60-Hz power line frequency hum.
9. The all-pass filter does just what its name suggests. It provides unity gain with predictable phase shifts for different input frequencies.
10. Basically, the function of an oscillator is to generate alternating current or voltage waveforms. Oscillators are classified according to the type of components used, the frequency of oscillation, and the type of waveform generated. There are two requirements for oscillation:

    (a) The magnitude of the loop gain A_vB must be greater than or equal to 1.

    (b) The total phase shift of the loop gain must be 0°.

    The Wien bridge, the phase shift, and the quadrature oscillators are the most commonly used sinusoidal oscillators. In all these oscillators the frequency of oscillation is a function of the RC time constant.
11. A square-wave output waveform is generated if the output of the op-amp is forced to swing repetitively between positive and negative saturation. One way to obtain the triangular wave is to integrate the square wave. Therefore, the triangular-wave generator can be formed by using a comparator and an integrator.
12. The sawtooth waveform has a rise time many times longer than the fall time, or vice versa. The triangular-wave generator using a comparator and an integrator can be converted to obtain a sawtooth waveform.

13. The oscillator whose output frequency depends on the amplitude of the input voltage is called a voltage-controlled oscillator (VCO) or voltage-to-frequency (V/F) converter. The VCO is used in phase-locked-loop circuits and for frequency modulation (FM), among other things.

QUESTIONS

8-1. Define a filter. How are the filters classified?

8-2. List the most commonly used filters.

8-3. What is a passband and a stopband for a filter?

8-4. What are the advantages of active filters over passive ones?

8-5. What is the Butterworth response?

8-6. What is an all-pass filter? Where and why is it needed?

8-7. Define an oscillator.

8-8. What are the two requirements for oscillation?

8-9. How are oscillators classified?

8-10. What is frequency stability? Explain its significance.

8-11. What is the difference between the sawtooth wave and the triangular wave?

8-12. In the sawtooth generator of Figure 8-24(a), how does the potentiometer R_4 affect the frequency and amplitude of the output waveform?

8-13. What is VCO? Give two applications that require VCO.

PROBLEMS

8-1. Design a first-order low-pass filter so that it has a cutoff frequency of 2 kHz and a passband gain of 1.

8-2. Convert the 2 kHz low-pass filter of Problem 8-1 to a cutoff frequency of 3 kHz.

8-3. Obtain the frequency response data similar to that in Table 8-1 for the filter of Problem 8-1. Construct the frequency response plot from this data.

8-4. Design a second-order low-pass filter at a cutoff frequency of 1.2 kHz.

8-5. Design a first-order high-pass filter at a cutoff frequency of 400 Hz and a passband gain of 1.

8-6. Obtain the frequency response data similar to that in Table 8-3 for the filter of Problem 8-5. Construct the frequency response plot from the data.

8-7. In the circuit of Figure 8-8, $C_2 = C_3 = 0.047\ \mu F$, $R_2 = R_3 = 3.3\ k\Omega$, $R_1 = 27\ k\Omega$, and $R_F = 15.8\ k\Omega$.

(a) Determine the low cutoff frequency f_L of the filter.

(b) Obtain the frequency response data and draw the frequency response plot for the filter.

8-8. Design a wide band-pass filter with $f_L = 400$ Hz, $f_H = 2$ kHz, and a passband gain $= 4$. Also draw an approximate frequency response plot for the filter.

8-9. (a) Design a narrow band-pass filter so that $f_C = 2$ kHz, $Q = 20$, and $A_F = 10$.
(b) Change the center frequency f_C to 1 kHz, keeping gain and bandwidth constant.

8-10. Design a wide band-reject filter using first-order high-pass and low-pass filters having $f_L = 2$ kHz and $f_H = 400$ Hz, respectively.

8-11. Design a 400-Hz active notch filter.

8-12. For the all-pass filter of Figure 8-16(a), determine the phase shift ϕ between the input and output at $f = 2$ kHz. To obtain a positive phase shift ϕ, what modifications are necessary in the circuit?

8-13. Design a phase shift oscillator so that $f_o = 1$ kHz.

8-14. Design a Wien bridge oscillator that will oscillate at 2 kHz.

8-15. In the circuit of Figure 8-20, $R_1 = R_2 = R_3 = 82\ \text{k}\Omega$ and $C_1 = C_2 = C_3 = 0.05\ \mu\text{F}$. Determine the frequency of oscillation.

8-16. In the square-wave generator of Figure 8-21(a), if $R_1 = 12\ \text{k}\Omega$, $R_2 = 13.92\ \text{k}\Omega$, $R = 100\ \text{k}\Omega$, and $C = 0.01\ \mu\text{F}$, what is the frequency of oscillation?

8-17. Design an integrator that can be used with the signal generator of Problem 8-16 so that the combination can be used as a triangular-wave generator.

8-18. In the triangular-wave generator of Figure 8-23(a), $R_2 = 1.2\ \text{k}\Omega$, $R_3 = 6.8\ \text{k}\Omega$, $R_1 = 120\ \text{k}\Omega$, and $C_1 = 0.01\ \mu\text{F}$. Determine (a) the peak-to-peak output amplitude of the triangular wave and (b) the frequency of the triangular wave.

8-19. For the VCO of Figure 8-25(c), determine the change in output frequency if V_C is varied between 9 V and 11 V. Assume that $+V = 12$ V, $R_2 = 15\ \text{k}\Omega$, $R_3 = 100\ \text{k}\Omega$, $R_1 = 6.8\ \text{k}\Omega$, and $C_1 = 75$ pF.

8-20. In the VCO of Figure 8-25(c), if $+V = 15$ V, $V_C = 13$ V, and $C_1 = 0.0068\ \mu\text{F}$, determine the approximate change in output frequency if R_1 is varied from 4 kΩ to 18 kΩ.

EXPERIMENT 8-1
FIRST-ORDER LOW-PASS AND HIGH-PASS FILTERS

Objectives

In this experiment you will verify the operation of first-order low-pass and high-pass filters. At the end of this experiment you should be able to:

1. Explain the difference between the low-pass and high-pass filters.
2. Construct a frequency response plot for the low-pass and high-pass filters from the experimental data.
3. Discuss the differences between the theoretical and experimental results.

Equipment

1. Oscilloscope
2. Audio-signal generator
3. ±15-V power supply

Materials

1. 741 or 351 op-amp
2. 3.3-kΩ resistor
3. Two 10-kΩ resistors
4. 0.05-μF capacitor
5. Semilog paper

Procedure

1. Connect the circuit as shown in Figure E8-1.1. Apply power to the circuit. Adjust the input voltage v_{in} to 1 V peak to peak and the frequency to 100 Hz.

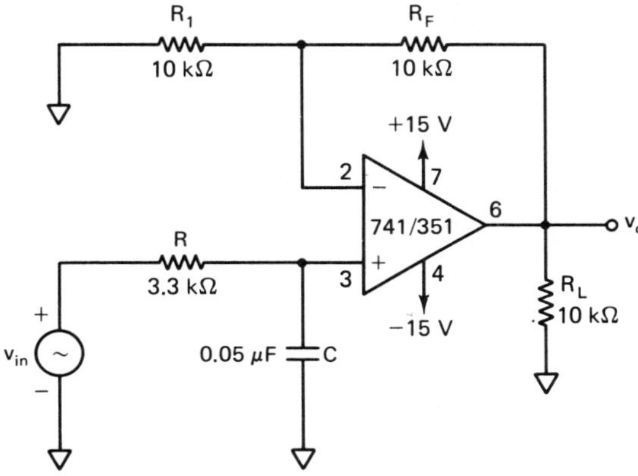

Figure E8-1.1 First-order low-pass filter. (Pin numbers refer to the 8-pin mini DIP.)

2. Using the oscilloscope, measure the output voltage and enter the measured value in Table E8-1.1.

TABLE E8-1.1 DATA FOR LOW-PASS FILTER

Frequency (Hz)	v_o pp (V)	v_o/v_{in}	20 log v_o/v_{in}
1. 100			
2. 300			
3. 700			
4. 1,000			
5. 2,000			
6. 5,000			
7. 8,000			
8. 10,000			

3. Vary the generators frequency and complete Table E8-1.1.
4. Calculate the cutoff frequency f_H and the passband gain using the following equations:

$$f_H = \frac{1}{2\pi RC}$$

$$\text{passband gain in (dB)} = 20 \log\left(1 + \frac{R_F}{R_1}\right)$$

5. Interchange the resistor R with capacitor C. Adjust the input voltage v_{in} to 1 V pp and frequency to 100 Hz. Measure output voltage using the scope, and enter the measured value in Table E8-1.2.
6. Vary the generator's frequency and complete Table E8-1.2.

TABLE E8-1.2 DATA FOR HIGH-PASS FILTER

Frequency (Hz)	v_o pp (V)	v_o/v_{in}	20 log v_o/v_{in}
1. 100			
2. 300			
3. 700			
4. 1,000			
5. 2,000			
6. 5,000			
7. 8,000			
8. 10,000			

7. Calculate the cutoff frequency f_L and the passband gain using the following equations:

$$f_L = \frac{1}{2\pi RC}$$

$$\text{passband gain (dB)} = 20 \log \left(1 + \frac{R_F}{R_1}\right)$$

8. Construct the frequency versus voltage gain (in dB) plot for the low-pass filter using the data of Table E8-1.1. Similarly, construct the frequency response plot for the high-pass filter using the data of Table E8-1.2. Remember that the frequency is assigned a logarithmic scale while amplitude is assigned a linear scale. From the graphs determine 3-dB frequencies f_H and f_L, respectively.

Questions

1. What is the difference between the low-pass and high-pass frequency response plots?
2. How do the theoretical cutoff frequencies of the low-pass and high-pass filters compare to their corresponding experimental cutoff frequencies? If they differ by more than 10%, suggest causes.

EXPERIMENT 8-2
WIEN BRIDGE OSCILLATOR

Objective

In this experiment you will construct the Wien bridge oscillator and verify its operation. At the end of this experiment you should be able to:

1. Measure the output voltage amplitude and frequency using an oscilloscope.
2. Determine the effect of different *RC* combinations on the frequency of oscillation.
3. Explain the use of zener diodes in limiting the output voltage amplitude.

Equipment

1. Oscilloscope
2. Digital multimeter
3. ±15 V power supply

Materials

1. 741 or 351 op-amp
2. 1N4735, 6.2-V zeners
3. Two 3.3-kΩ resistors
4. Two 10-kΩ resistors
5. 12-kΩ resistor
6. Two 0.01-μF capacitors
7. Two 0.05-μF capacitors

Procedure

1. Connect the circuit as shown in Figure E8-2.1. Make sure that the resistance R_F (50-kΩ potentiometer) is set to approximately 20 kΩ.

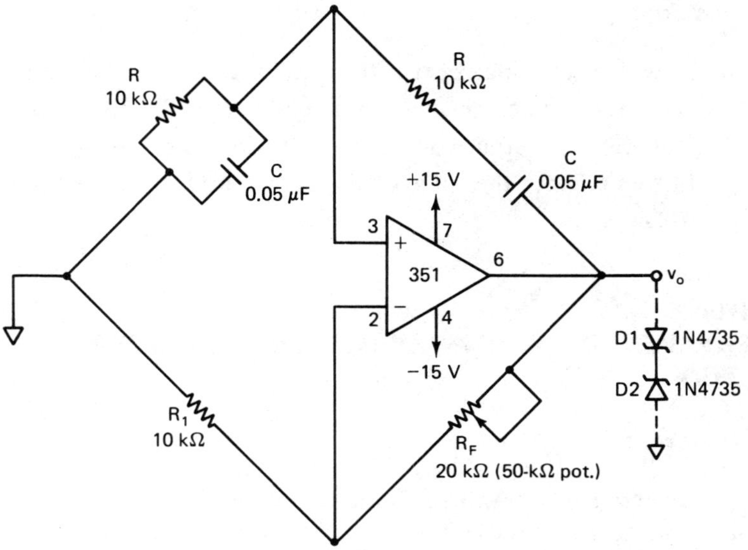

Figure E8-2.1 Wien bridge oscillator. (Pin numbers refer to the 8-pin mini DIP.)

2. Turn the dc power on and measure the amplitude and frequency of the output waveform using the oscilloscope. If the circuit fails to oscillate, gradually increase the value of R_F until the circuit starts oscillating. The output should be a sustained sine-wave. Enter the measured values in Table E8-2.1.
3. Connect the two 1N4735 zeners back to back between the output terminal of the op-amp and ground, and again measure the output amplitude. Enter the measured value in Table E8-2.1. The 1N4735 are 6.2-V zeners.

TABLE E8-2.1

			Output amplitude pp (V)		Frequency of the output waveform (Hz)	
	R (kΩ)	C (μF)	Without zeners	With zeners	Measured	Calculated $f_o = 0.159/RC$
1.	10	0.05				
2.	10	0.01				
3.	3.3	0.05				

4. Replace the 0.05-μF capacitors with the 0.01-μF capacitors. Repeat steps 2 and 3.
5. Replace the 10-kΩ resistors with the 3.3-kΩ resistors and the 0.01-μF capacitors with the 0.05-μF capacitors. Repeat steps 2 and 3.
6. Reduce the pot R_F slowly until the oscillations stop; then disconnect the pot and measure its value with the ohmmeter. The value of the pot R_F is = ______ Ω.

Questions

1. How does the value of the RC combination affect the frequency of oscillation?
2. What is the difference in output amplitude with and without the zeners?
3. How can the output voltage swing be limited to ± 5 V?
4. In step 6, when the value of R_F was reduced, the oscillations stopped. Explain why.

EXPERIMENT 8-3
SQUARE-WAVE, TRIANGULAR-WAVE, OR SAWTOOTH-WAVE GENERATOR

Objectives

In this experiment you will verify the operation of square-, triangular-, and sawtooth-wave generators. At the end of this experiment you should be able to:

1. Discuss the differences between the square, triangular, and sawtooth waveforms.

2. Explain the operation of the comparator.
3. Explain the operation of the integrator that converts a square wave into a triangular wave.
4. Discuss the effect of external components on the frequency of oscillation.

Equipment

1. Dual-trace oscilloscope
2. Digital multimeter
3. ±15 V power supply

Materials

1. 1458 or 353 op-amp
2. Two 4.7-kΩ resistors
3. Three 10-kΩ resistors
4. 50-kΩ potentiometer
5. 1 MΩ potentiometer
6. 0.05-μF capacitor
7. Linear graph paper

Procedure

1. Connect the circuit as shown in Figure E8-3.1. Make sure that the R_3 is set at 40 kΩ ($4R_2$).
2. Apply the dc power and monitor the output waveforms of A_1 and A_2 using the oscilloscope. Connect channel 1 and channel 2 of the scope to the output terminals of A_1 and A_2, respectively. Measure the amplitude and frequency of the output waveforms. Enter the measured values in Table E8-3.1.

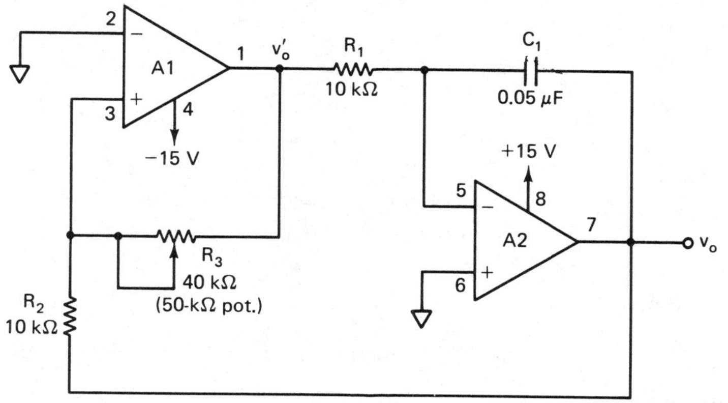

Figure E8-3.1 Square-, triangular or sawtooth-wave generator. (Pin numbers refer to the 8-pin mini DIP.)

TABLE E8-3.1

	R_2 (kΩ)	R_3 (kΩ)	Square-wave output, v_o'		Triangular-wave output, v_o	
			Amplitude	Frequency	Amplitude	Frequency
1.	10	40				
2.	4.7	18.8				

3. Draw the output waveforms using the same scale on the linear graph paper.
4. Switch the dc power off; then change the value of R_2 to 4.7 kΩ and reset the value of R_3 to 18.8 kΩ ($4R_2$).
5. Repeat steps 2 and 3.
6. Turn the dc power off. Disconnect the noninverting terminal of A_2 from ground. Now connect one terminal of the 1-MΩ potentiometer to the negative supply ($-V_{EE} = -15$ V), the other terminal to the positive supply, and the wiper to the noninverting input terminal of A_2.
7. Turn on the dc power. Connect voltmeter between the noninverting terminal of A_2 and ground. Also connect channels 1 and 2 of the scope to the output terminals of A_1 and A_2, respectively. Adjust the wiper in the 1-MΩ potentiometer so that the voltage at the noninverting terminal of A_2 is -5 V. Measure the amplitude and frequency of the output waveforms at this setting. Enter the measured values in Table E8-3.2. Complete Table E8-3.2.

TABLE E8-3.2

	Voltage at the noninverting terminal of A_2 (V)	Square-wave output, v_o'		Sawtooth-wave output, v_o	
		Amplitude	Frequency	Amplitude	Frequency
1.	-5				
2.	-10				
3.	$+5$				
4.	$+10$				

8. When the voltage at the noninverting terminal is -10 V, draw the output waveforms on linear graph paper. Use the same scale for both the waveforms.

Questions

1. Is the amplitude and frequency of the square wave a function of R_2 and R_3 values?
2. How do the R_2 and R_3 values affect the amplitude and frequency of the triangle wave?

3. Is there any difference in the amplitude and frequency of the square and triangular waveforms for a fixed value of R_2 and R_3? Explain your answer.
4. What is the difference between the square-wave outputs in steps 3 and 8? Explain why.
5. How does the voltage at the noninverting terminal of A_2 affect the sawtooth waveform?
6. What is the relationship between the rise times and fall times of the sawtooth waveform when the voltage at the noninverting terminal is -5 V and $+5$ V?

9

Comparators and Converters

9-1 INTRODUCTION

The amplifier, filter, and oscillator applications presented so far illustrate a fair sampling of typical op-amp uses. However, op-amps are used in many other circuits, where they are employed under specific names. Such functions include comparators, detectors, limiters, and digital interface devices, namely, converters. In this chapter we discuss comparators, limiters, detectors, and converters using a general-purpose op-amp. To obtain far better performance, we shall also look at integrated circuits designed specifically as comparators and converters.

A comparator, as its name implies, compares a signal voltage on one input of an op-amp with a known voltage called the *reference voltage* on the other input. In its simplest form, it is nothing more than an open-loop op-amp, with two analog inputs and a digital output; the output may be (+) or (−) saturation voltage, depending on which input is the larger. Comparators are used in circuits such as digital interfacing, Schmitt triggers, discriminators, voltage-level detectors, and oscillators.

9-2 BASIC COMPARATOR

Figure 9-1(a) shows an op-amp used as a comparator. A fixed reference voltage V_{ref} of 1 V is applied to the (−) input, and the other time-varying signal voltage v_{in} is applied to the (+) input. Because of this arrangement, the circuit is called the *noninverting comparator*. When v_{in} is less than V_{ref}, the output voltage v_o is at $-V_{sat}$

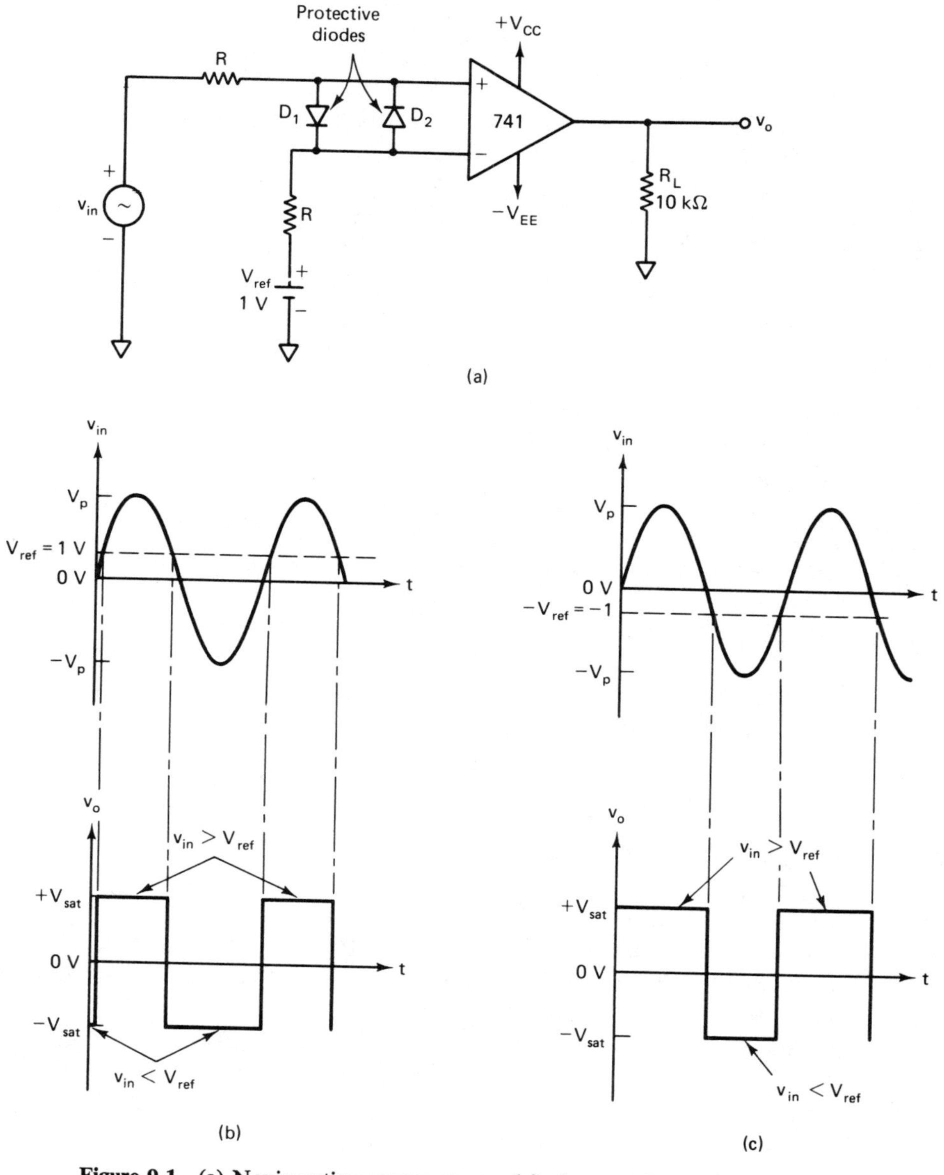

Figure 9-1 (a) Noninverting comparator and its input and output waveforms. (b) If V_{ref} is positive. (c) If V_{ref} is negative.

($\cong -V_{EE}$) because the voltage at the (−) input is higher than that at the (+) input. On the other hand, when v_{in} is greater than V_{ref}, the (+) input becomes positive with respect to the (−) input, and v_o goes to $+V_{sat}(\cong +V_{CC})$. Thus v_o changes from one saturation level to another whenever $v_{in} \cong V_{ref}$, as shown in Figure 9-1(b). In short, the comparator is a type of analog-to-digital converter. At any given time the v_o waveform shows whether v_{in} is greater or less than V_{ref}. The comparator is sometimes also called a *voltage-level detector* because for a desired value of V_{ref}, the voltage level of the input v_{in} can be detected.

In Figure 9-1(a) the diodes D1 and D2 protect the op-amp from damage due to excessive input voltage v_{in}. Because of these diodes, the difference input voltage v_{id} of op-amp is clamped to either 0.7 V or −0.7 V; hence the diodes are called *clamp diodes*. There are some op-amps with built-in input protection; in such op-amps the input diodes D1 and D2 are unnecessary. The resistance R in series with v_{in} is used to limit the current through D1 and D2. To reduce offset problems, a resistance $R_{OM} \cong R$ is connected between (−) input and V_{ref} [see Figure 9-1(a)].

If the reference voltage V_{ref} is negative with respect to ground, with sinusoidal signal applied to the (+) input, the output waveform will be as shown in Figure 9-1(c). When $v_{in} > V_{ref}$, v_o is at $+V_{sat}$; on the other hand, when $v_{in} < V_{ref}$, v_o is at $-V_{sat}$. Obviously, the amplitude of v_{in} must be large enough to pass through V_{ref} if the switching action is to take place.

Figure 9-2(a) shows an inverting comparator in which the reference voltage V_{ref} is applied to the (+) input and v_{in} is applied to the (−) input. In this circuit, V_{ref} is obtained by using a 10-kΩ potentiometer which forms a voltage divider with the dc supply voltages $+V_{CC}$ and $-V_{EE}$ and the wiper connected to the (+) input. As the wiper is moved toward $-V_{EE}$, V_{ref} becomes more negative, while if it is moved toward $+V_{CC}$, V_{ref} becomes more positive. Thus a V_{ref} of a desired amplitude and polarity can be obtained by simply adjusting the 10 kΩ potentiometer. With the sinusoidal input waveform, the output v_o has the waveform shown in Figure 9-2(b) or (c), depending on whether V_{ref} is positive or negative, respectively.

9-3 ZERO-CROSSING DETECTOR

An immediate application of the comparator is the *zero-crossing detector* or *sine wave-to-square wave converter*. The basic comparator of Figure 9-1(a) or Figure 9-2(a) can be used as the zero-crossing detector provided that V_{ref} is set to zero ($V_{ref} = 0$ V). Figure 9-3(a) shows the inverting comparator used as a zero-crossing detector. The output voltage v_o waveform in Figure 9-3(b) shows when and in what direction an input signal v_{in} crosses zero volts. That is, the output v_o is driven into negative saturation when the input signal v_{in} passes through zero in the positive direction. Conversely, when v_{in} passes through zero in the negative direction, the output v_o switches and saturates positively.

In some applications, the input v_{in} may be a slowly changing waveform, that is, a low-frequency signal. Therefore, it will take v_{in} more time to cross 0 V; therefore, v_o may not switch quickly from one saturation voltage to the other. On the other hand, because of the noise at the op-amp's input terminals, the output v_o may fluctuate between two saturation voltages $+V_{sat}$ and $-V_{sat}$, detecting zero reference crossings for noise voltages as well as v_{in}. Both of these problems can be cured with the use of regenerative or *positive feedback* that causes the output v_o to change faster and eliminate any false output transitions due to noise signals at the input.

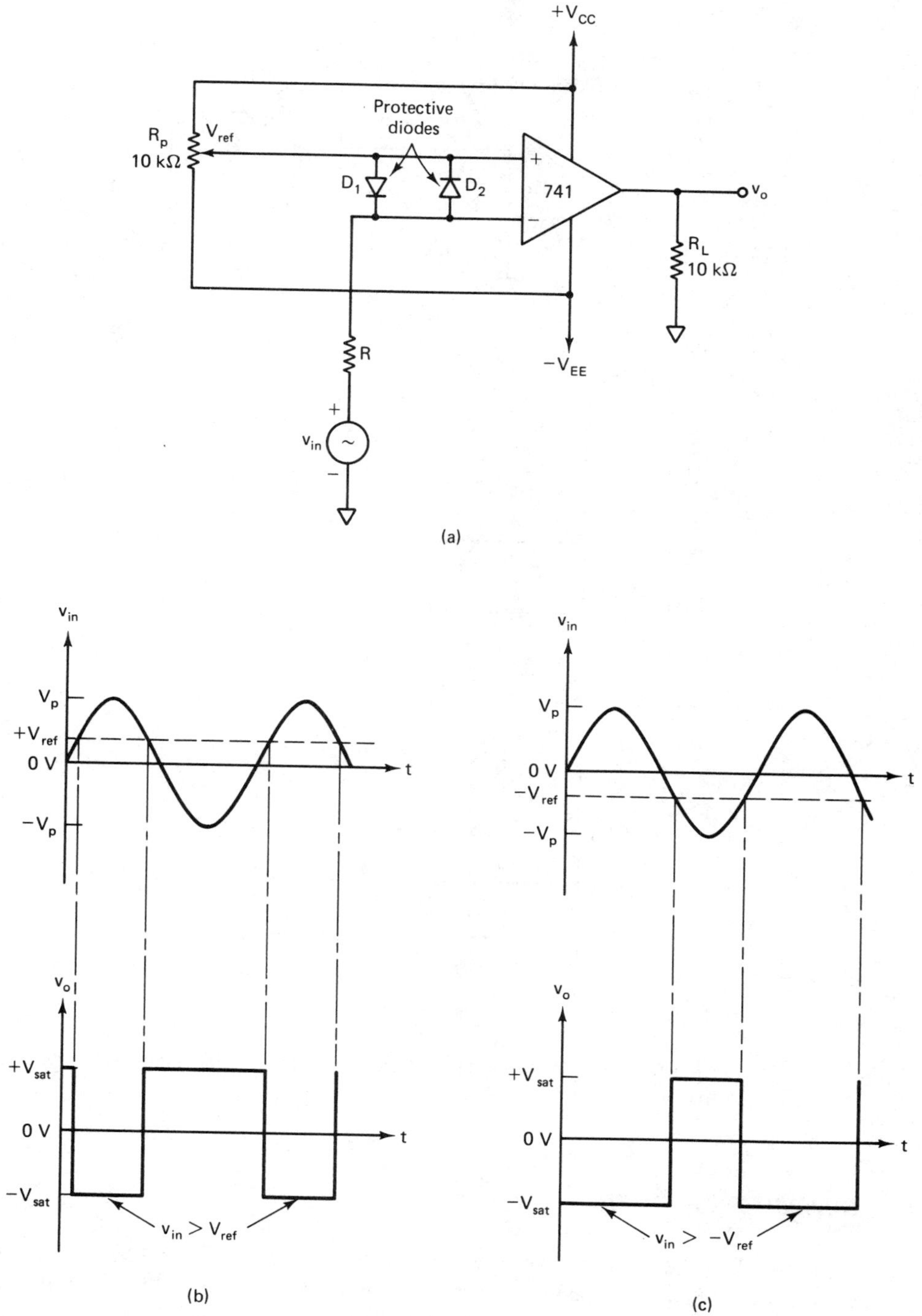

Figure 9-2 (a) Inverting comparator with input and output waveforms. (b) If V_{ref} is positive. (c) If V_{ref} is negative.

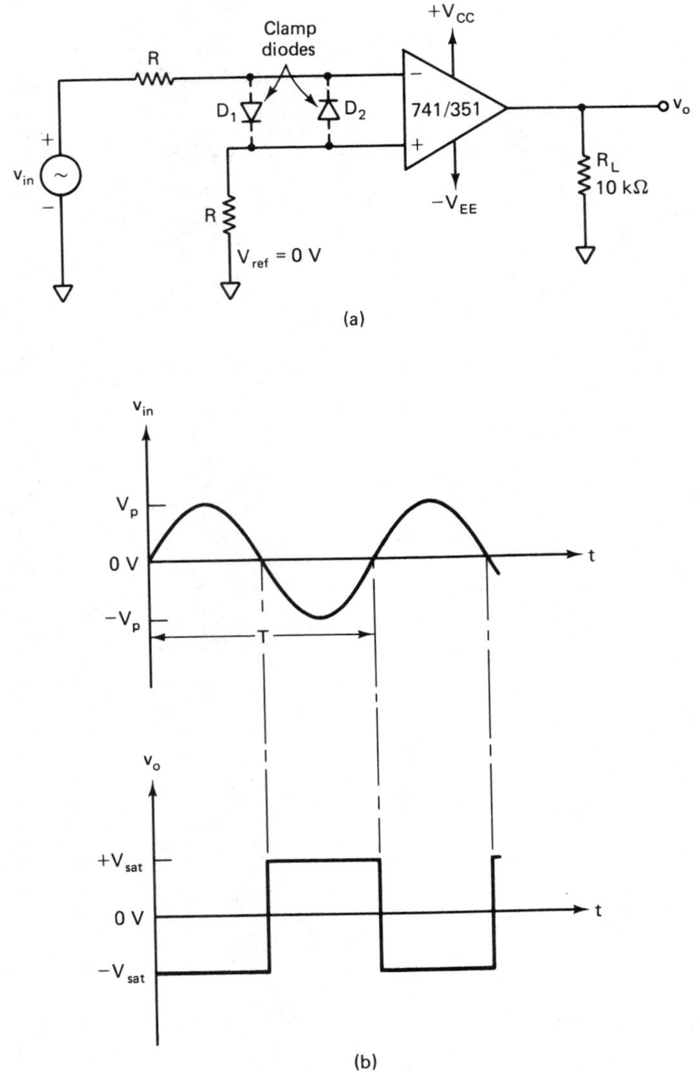

Figure 9-3 (a) Zero-crossing detector. (b) Its typical input and output waveforms.

9-4 SCHMITT TRIGGER

Figure 9-4(a) shows an inverting comparator with *positive feedback*. This circuit converts an irregular-shaped waveform to a square wave or pulse. The circuit is known as the *Schmitt trigger* or squaring circuit. The input voltage v_{in} triggers (changes the state of) the output v_o every time it exceeds certain voltage levels called upper threshold voltage V_{ut} and lower threshold voltage V_{lt}, as shown in Figure 9-4(b).

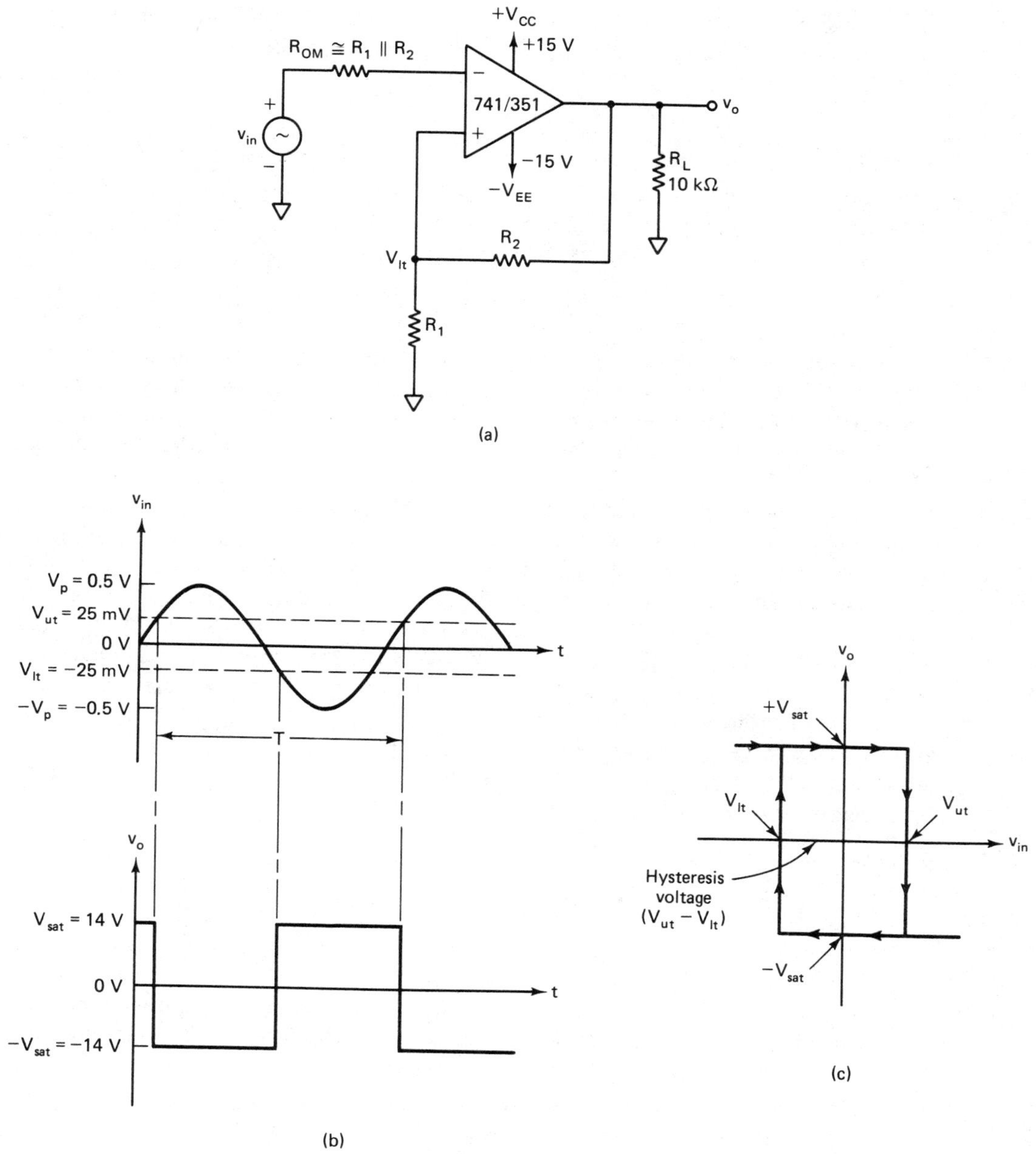

Figure 9-4 (a) Inverting comparator as Schmitt trigger. (b) Input and output waveforms of Schmitt trigger. (c) v_o versus v_{in} plot of the hysteresis voltage.

In Figure 9-4(a), these threshold voltages are obtained by using the voltage divider $R_1 - R_2$, where the voltage across R_1 is fed back to the (+) input. The voltage across R_1 is a variable reference threshold voltage which depends on the value and polarity of the output voltage v_o. When $v_0 = +V_{sat}$, the voltage across R_1 is called the *upper threshold voltage*, V_{ut}. The input voltage v_{in} must be slightly more positive than V_{ut} in order to cause the output v_o to switch from $+V_{sat}$ to $-V_{sat}$. As long as $v_{in} < V_{ut}$, v_o is at $+V_{sat}$. Using the voltage-divider rule,

$$V_{ut} = \frac{R_1}{R_1 + R_2}(+V_{sat}) \tag{9-1a}$$

On the other hand, when $v_o = -V_{sat}$, the voltage across R_1 is referred to as *lower threshold voltage*, V_{lt}. v_{in} must be slightly more negative than V_{lt} in order to cause v_o to switch from $-V_{sat}$ to $+V_{sat}$. In other words, for v_{in} values greater than V_{lt}, v_o is at $-V_{sat}$. V_{lt} is given by the following equation:

$$V_{lt} = \frac{R_1}{R_1 + R_2}(-V_{sat}) \tag{9-1b}$$

Thus if the threshold voltages V_{ut} and V_{lt} are made larger than the input noise voltages the positive feedback will eliminate the false output transitions. Besides that, the positive feedback, because of its regenerative action, will make v_o switch faster between $+V_{sat}$ and $-V_{sat}$. In Figure 9-4(a), resistance $R_{OM} \cong R_1 \| R_2$ is used to minimize the offset problems.

Figure 9-4(b) shows that the output of the Schmitt trigger is a square wave when the input is a sine wave. Recall that a slightly different version of the Schmitt trigger is used in the triangular-wave and sawtooth-wave generators of Figures 8-23 and 8-24, respectively. In these generators a noninverting comparator is used as a Schmitt trigger. When the input is a triangular wave, the output of the Schmitt trigger is a square wave, whereas if the input is a sawtooth wave, the output is a pulse waveform.

The comparator with positive feedback is said to exhibit *hysteresis*, a deadband condition. That is, when the input of the comparator exceeds V_{ut}, its output switches from $+V_{sat}$ to $-V_{sat}$ and reverts back to its original state, $+V_{sat}$, when the input goes below V_{lt} [see Figure 9-4(c)]. The hysteresis voltage is, of course, equal to the difference between V_{ut} and V_{lt}. Therefore,

$$\begin{aligned} V_{hy} &= V_{ut} - V_{lt} \\ &= \frac{R_1}{R_1 + R_2}[+V_{sat} - (-V_{sat})] \end{aligned} \tag{9-2}$$

Example 9-1

In the circuit of Figure 9-4(a), $R_1 = 100\ \Omega$, $R_2 = 56\ \text{k}\Omega$, $v_{in} = 1$ V pp sine wave, and the op-amp is type 741 with supply voltages $= \pm 15$ V. Determine the threshold voltages V_{ut} and V_{lt} and draw the output waveform.

Solution. For 741 the maximum output voltage swing is ± 14 V, that is, $+V_{sat} =$ 14 V, and $-V_{sat} = -14$ V. From Equations (9-1a) and (9-1b),

$$V_{ut} = \frac{100}{56{,}100}(14) = 25\ \text{mV}$$

$$V_{lt} = \frac{100}{56{,}100}(-14) = -25\ \text{mV}$$

The output v_o waveform is shown in Figure 9-4(b). From Equation (9-2), the hysteresis voltage $V_{hy} = 50$ mV.

9-5 COMPARATOR CHARACTERISTICS

The important characteristics of a comparator are these:

1. Speed of operation
2. Accuracy
3. Compatibility of output

The output of the comparator must switch rapidly between saturation levels and also respond instantly to any change of conditions at its inputs. This implies that the bandwidth of the op-amp comparator must be rather wide; in fact, the wider the bandwidth, the higher is the speed of operation. As discussed in Section 9-4, the speed of operation of the comparator is improved with positive feedback (hysteresis).

The accuracy of the comparator depends on its voltage gain, common-mode rejection ratio, input offsets, and thermal drifts. High voltage gain requires a smaller difference voltage (hysteresis voltage) to cause the comparator's output voltage to switch between saturation levels. On the other hand, a high CMRR helps to reject the common-mode input voltages, such as noise, at the input terminals. Finally, to minimize the offset problems, the input offset current and input offset voltage must be negligible; also, the *changes* in these offsets due to temperature variations should be very slight.

Since the comparator is a form of analog-to-digital converter, its output must swing between two logic levels suitable for a certain logic family such as transistor-transistor logic (TTL).

9-6 LIMITATIONS OF OP-AMPS AS COMPARATORS

A general-purpose op-amp such as the 741 can be used in relatively less critical comparator applications in which speed and accuracy are not major factors. As illustrated in Section 9-4, with positive feedback (hysteresis), the switching speed of the op-amp comparator can be improved and false transition due to noise can be eliminated. Besides that, an offset voltage compensating network and offset minimizing resistor can be used to minimize offset problems. However, the output voltage swing of an op-amp is relatively large because it is designed primarily as an amplifier. In other words, the output of an op-amp comparator is generally not compatible with a particular logic family such as the TTL, which requires input voltages of either approximately +5 V or 0 V. Therefore, to keep the output voltage swing within specific limits, op-amps are used with externally wired components such as zeners or diodes. The resulting circuits, in which the outputs are limited to predetermined values, are called *limiters*. The next section presents an analysis of typical limiters in order to suggest other such possibilities.

9-7 VOLTAGE LIMITERS

Several op-amp comparator circuits with output voltage limiting are shown in Figures 9-5 through 9-7. In the circuit of Figure 9-5(a), the zener diodes D_1 and D_2 are connected in the feedback path; this arrangement limits the positive and negative values of the output voltage v_o. When the input voltage v_{in} crosses 0 V and increases in the positive direction, the output voltage v_o increases in the negative direction until diode D_1 is forward biased and D_2 goes into avalanche conduction. Therefore, the maximum

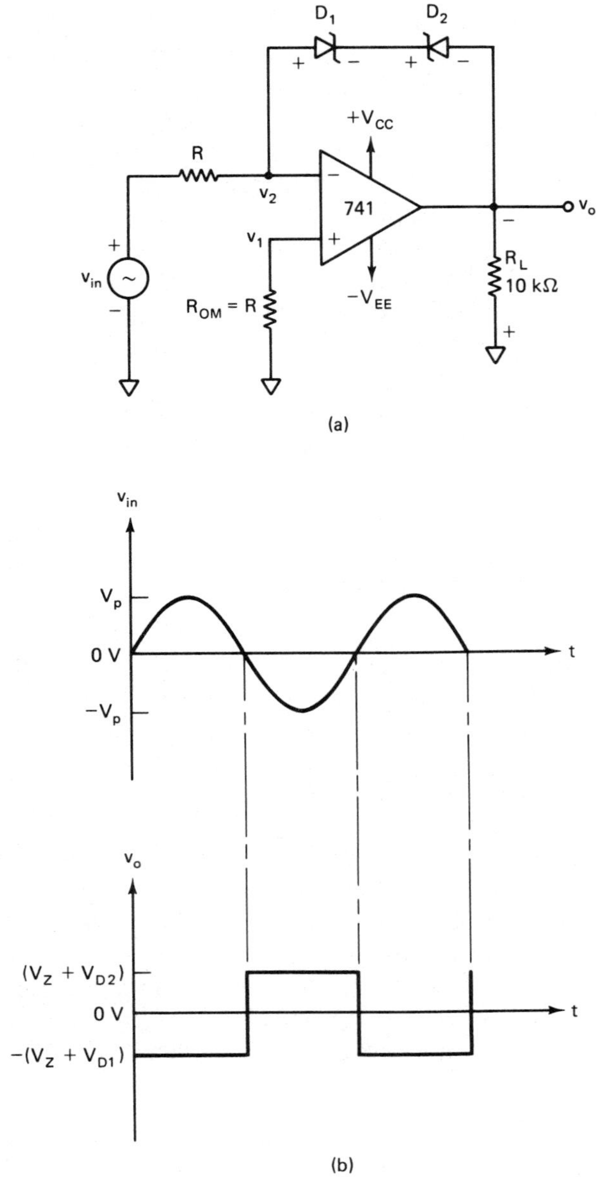

Figure 9-5 (a) Basic comparator with positive and negative output voltage limiting. (b) Its input and output waveforms.

negative value of v_o is equal to $(V_Z + V_{D1})$, where V_Z = zener voltage and V_{D1} = the voltage drop across the *forward*-biased zener D_1 (= 0.7 V typically). On the other hand, when v_{in} crosses 0 V and starts increasing in the negative direction, v_o starts increasing positively until diode D_2 is forward biased and D_1 goes into avalanche conduction. Thus the maximum positive v_o is equal to $V_Z + V_{D2}$, where V_Z is the zener voltage and V_{D2} the voltage drop across the forward-biased zener D_2 (= 0.7 V typically). Thus the output voltage swing is limited to $+(V_Z + 0.7)$ and $-(V_Z + 0.7.)$ [see Figure 9-5(b)].

Note that in the circuit of Figure 9-5(a), since the input terminals of the op-amp are at virtual ground ($v_1 = v_2 \cong 0$ V), the input voltage v_{in} appears across R, and v_o appears across D_1 and D_2. The resistance R_{OM} is used to minimize offset problems.

If there is a need to limit the swing of the output voltage v_o to a positive direction only, a combination of zener and rectifier diodes is used as shown in the circuit of Figure 9-6(a). When the input voltage v_{in} crosses 0 V and increases in the positive direction, the output voltage v_o is at $-V_{sat}$. This happens because diode D_2 is reverse biased, causing the op-amp to operate in the open-loop configuration. However, when v_{in} crosses 0 V and starts increasing in the negative direction, v_o starts increasing positively until D_2 is forward biased and D_1 goes into avalanche conduction. Therefore, the maximum positive $v_o = V_Z + V_{D2}$ [see Figure 9-6(b)].

In the circuit of Figure 9-6(a), if the position of D_1 and D_2 diodes is inter-

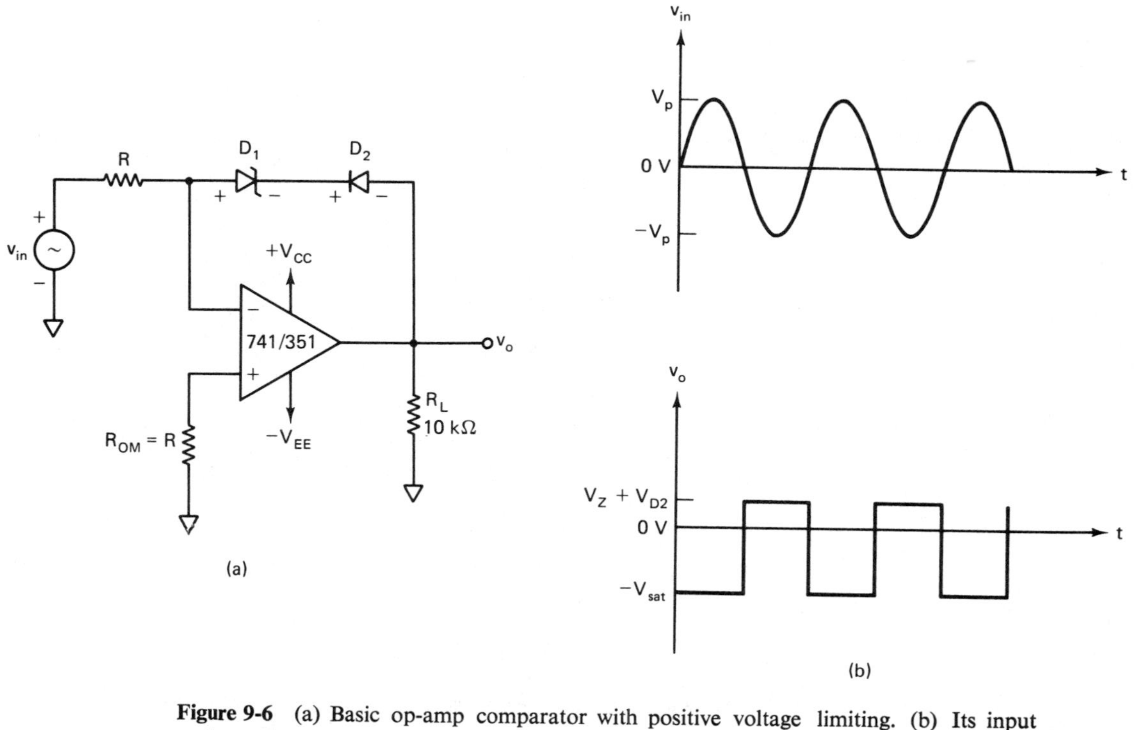

Figure 9-6 (a) Basic op-amp comparator with positive voltage limiting. (b) Its input and waveforms.

changed, the output voltage v_o will then be limited in a negative direction to $-(V_Z + V_{D2})$. However, the maximum positive $v_o = +V_{sat}$.

Finally, if only a single zener is used in the feedback path of an op-amp as shown in Figure 9-7(a), the output voltage v_o is limited to $+V_Z$ and $-V_D$, where V_Z in the zener voltage and V_D the voltage drop across the forward-biased zener $(= +0.7$ V typically) [see Figure 9-7(b)]. The exact opposite result can be obtained by reversing the direction of the zener diode. That is, v_o will be limited to $-V_Z$ and $+V_D$.

Voltage limiters are commonly used in communications devices such as TV and FM receivers.

Example 9-2

In the circuit of Figure 9-7(a), $v_{in} = 500$ mV peak 60-Hz sine wave, $R = 100\ \Omega$, 1N3826 zener with $V_Z = 5.1$ V, and the supply voltages $= \pm 15$ V. Determine the output voltage swing. Assume that the voltage drop across the forward-biased zener $=$ 0.7 V.

Solution. During the positive half cycle of the input waveform, the output voltage would be at $-V_D = -0.7$ V because the zener will be forward biased. However, during the negative half cycle of v_{in}, v_o would be at $+V_Z = +5.1$ V since the zener will be reverse biased into zener breakdown. Thus, the use of a zener diode in the feedback path limits v_o to $+5.1$ V and -0.7 V. This output voltage swing will, therefore, make the op-amp comparator of Figure 9-7(a) compatible with TTL.

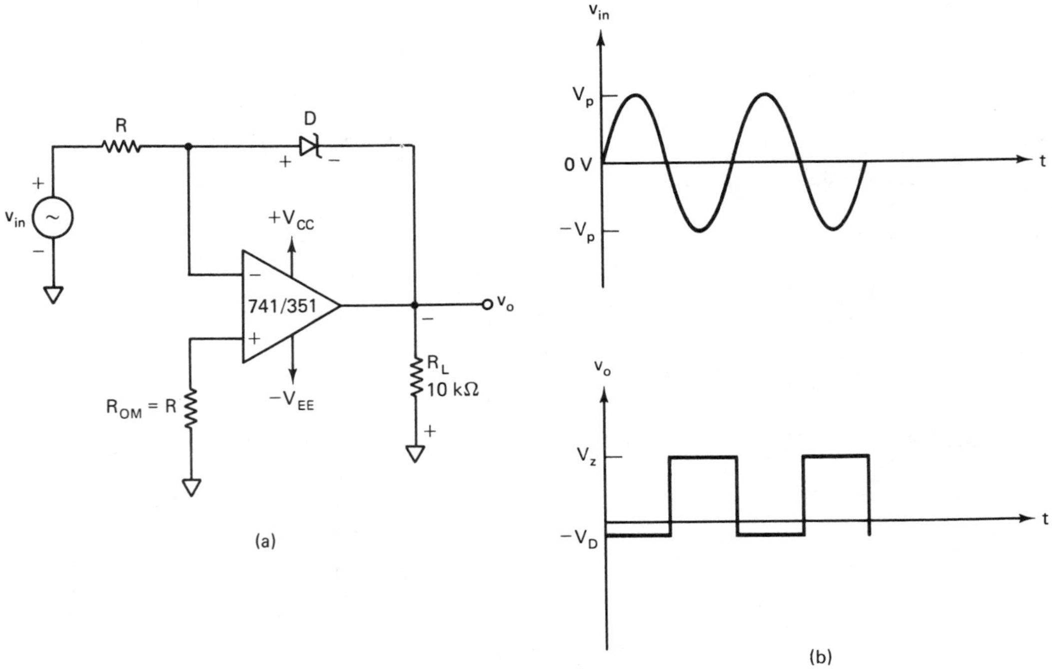

Figure 9-7 (a) Basic op-amp comparator with positive and negative limiting. (b) Its input and output waveforms.

9-8 HIGH-SPEED AND PRECISION-TYPE COMPARATORS

Although op-amps can be used as comparators, specially designed comparators outperform the op-amp types. The specially designed devices are optimaized for the most desirable parameters, such as speed and accuracy. Besides that, their output is compatible with RTL, DTL, TTL, and MOS logic. A variety of integrated comparators are available, including FET-input, low-power, low-offset voltage, high-speed, and precision voltage comparators. The information included on the comparator data sheets is identical to that on op-amp data sheets in most cases. A typical comparator data sheet includes a general description of the device, a schematic diagram, a connection diagram, absolute maximum ratings, electrical characteristics, typical performance curves, and applications. Electrical parameters such as response time, positive output level, negative output level, strobe current, strobe release time, and saturation voltage are typical of comparators. These parameters are defined as follows. The *response time* is the interval between the application of an input step function and the time when the output crosses the logic threshold voltage. The *positive output level* or *output high voltage* is the high output voltage level with a given load and the input drive equal to or greater than a specified value. The *negative output level* or *output low voltage* is the negative dc output voltage with the comparator saturated by a differential input equal to or greater than a specified voltage. The *strobe current* is the current out of the strobe terminal when it is at the zero logic level. The *strobe release time* is the time required for the output to rise to the logic threshold voltage after the strobe terminal has been driven from *zero* to the *one* logic level. The *saturation voltage* is the low-output voltage level with the input drive equal to or greater than a specified value. Furthermore, parameters such as strobe current and strobe release time are only listed for comparators with strobe capability.

The μA311 precision comparator is designed for low-level signal detection and high-level output drive capability. It can operate from $\pm$15-V op-amp supplies down to the single +5-V supply used for IC logic. In addition, its output can drive RTL, DTL, TTL, and MOS logic, as well as lamps or relays. Outputs can also be wire ORed. The input slew rate can be increased by increasing the input stage current. The μA311 has input offset voltage balancing and TTL strobe capability [see Figure 9-8(a)]. The strobe capability allows the comparator's output to either respond to input signals or be independent of input signals. The response time is typically 200 ns for a 100-mV input step with 5-mV overdrive.

The μAF311 is a FET input comparator that has input currents (I_{io} and I_B) more than 1000 times lower than for the μA311. Except for this difference, the μAF311 has the same characteristics and features as the μA311.

As shown in Figure 9-8(b), the LM1414 has two totally separate comparators with independent strobe capability. Compared to the 311's, it has lower response time (30 ns typical) at the expense of higher input currents.

The features of the ICL8001 include low input currents, low power consumption (30 mW), and 250 ns response time. An output stage enables the designer to control the output voltage swing. That is, the positive output level can be adjusted to suit a desired logic family by changing $V+$ [see Figure 9-8(c)].

The μA760 is a high-speed differential voltage comparator with *complementary*

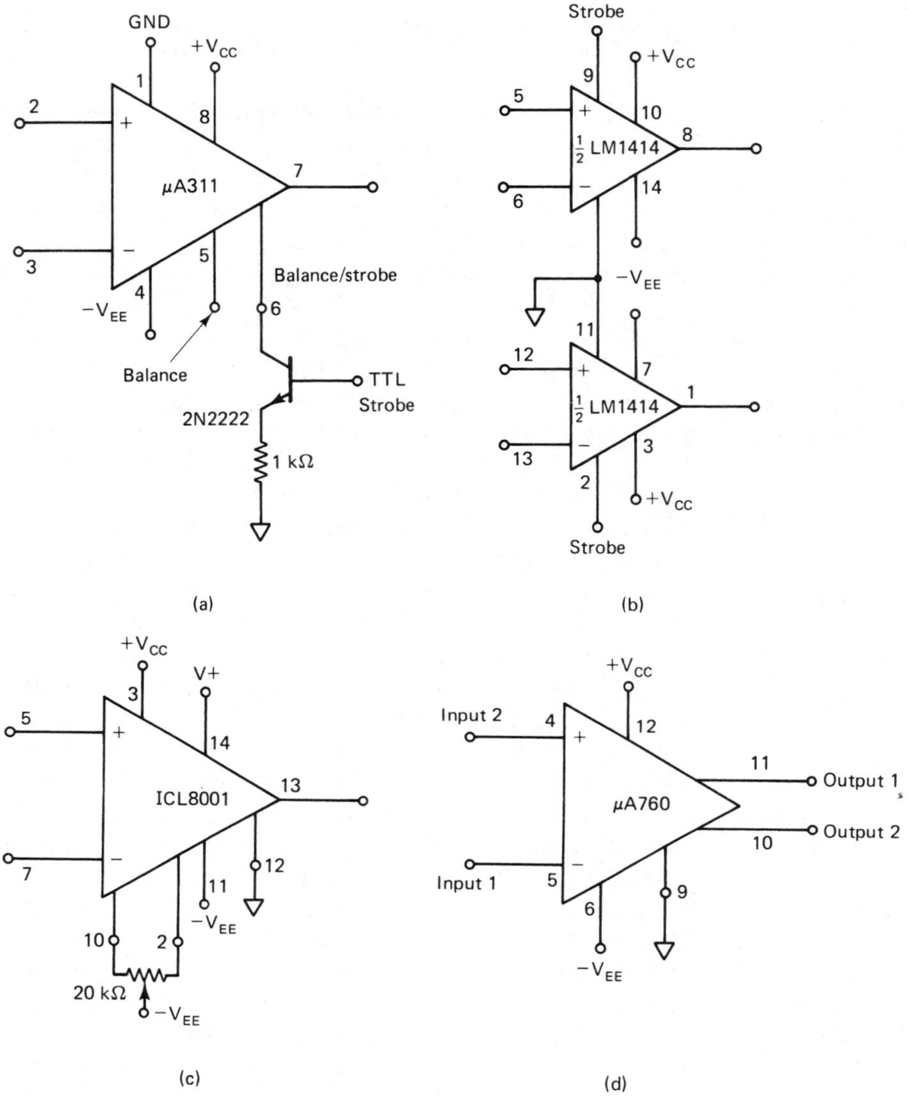

Figure 9-8 Connection diagrams of comparators. (a) μA311: when not used, strobe terminal (6) should either be open or connected to positive supply. (b) LM1414: pin 4 NC. (c) ICL8001: pin connections shown are for 14-pin DIP package. (d) μA760: pin connections shown are for 14-pin DIP package.

TTL outputs. It has a typical response time of 16 ns and operates from supplies of ±4.5 to ±6.5 V [see Figure 9-8(d)]. Its typical applications are high-speed analog-to-digital converters and zero-crossing detectors.

9-9 WINDOW DETECTOR

Sometimes there is a need to determine when an unknown input is between two precise reference thresholds V_{ut} and V_{lt}. This determination can be made by a circuit called the window detector. Figure 9-9(a) shows such a circuit using a dual comparator

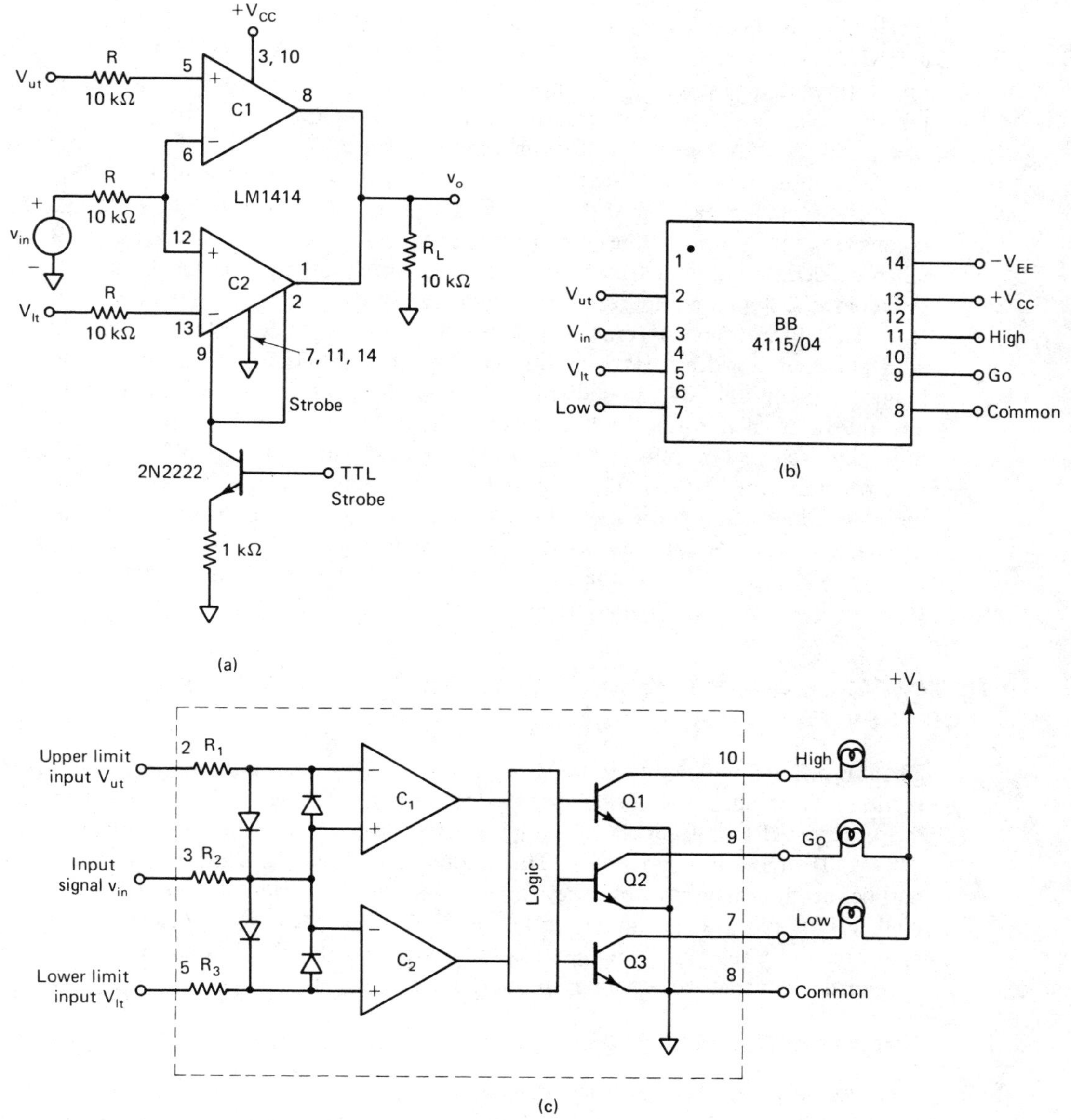

	$V_{ut} < v_{in}$	$V_{lt} < v_{in} < V_{lt}$	$v_{in} < V_{lt}$
High	On	Off	Off
Go	Off	On	Off
Low	Off	Off	On

(d.

Figure 9-9 (a) Window detector using LM1414. The BB 4415/04 window detector. (b) Pin diagram. (c) Block diagram. (d) Transfer characteristics. (Courtesy of Burr-Brown Research Corp.)

LM1414. An unknown voltage v_{in} is applied to the (−) input of comparator C_1 ($\frac{1}{2}$ LM1414) and to the (+) input of comparator C_2 (remaining $\frac{1}{2}$ LM1414). An upper threshold voltage V_{ut} is applied to the (+) input of C_1, whereas the lower threshold voltage V_{lt} is connected to the (−) input of C_2. In addition, the outputs of C_1 and C_2 are connected to form a single output v_o. When v_{in} is between V_{ut} and V_{lt}, the output v_o is *high*. However, when v_{in} goes above V_{ut} or drops below V_{lt}, the output v_o switches *low* or to 0 V. As a visual aid an appropriate lamp circuit may be connected at the output of the comparator. The window detector can also be "strobed" when needed. Window detectors are usually used in industrial alarms, level detectors and controls, digital computers, and production-line testing.

The Burr-Brown 4115/04 is a hybrid IC window detector that is available in a 14-pin double-width DIP. Figure 9-9(b) and (c) show the pin diagram and block diagram of the 4115/04. As shown in Figure 9-9(c), the inputs are diode protected and the value of the input voltages can be as high as ±15 V. The device has three mutually exclusive outputs, HIGH, LOW, and GO, and sinks up to 200 mA of current when an output is *on*. The required power supply voltages are ±15 V while the maximum load supply voltage $+V_L$ can be up to +30 V. For proper operation of the device, separate connections should be made from each power supply common ($+V_{CC}$, $-V_{EE}$, and $+V_L$) to the device common, that is, pin 8. The 4115/04 is designed to drive TTL and DTL logic devices as well as lamps and relays.

9-10 VOLTAGE-TO-FREQUENCY AND FREQUENCY-TO-VOLTAGE CONVERTERS

The SE/NE 566 voltage-to-frequency converter is discussed in Section 8-18. This section investigates the Teledyne 9400 series, which can be used as voltage-to-frequency (V/F) or frequency-to-voltage (F/V) converters. The series includes the 9400, 9401, and 9402 converters. These converters have the same internal circuitry and connection diagrams, but they differ slightly in electrical characteristics. A complete V/F or F/V system can be formed simply by using two external capacitors, three resistors, and a reference voltage. In addition, the 9400 series consists of CMOS and bipolar devices that can operate on dual or single supply voltages.

9-10.1 V/F Converter

The 9400 is designed for pulse- and square-wave outputs having a frequency range of 1 Hz to 100 kHz. Furthermore, the input can be either current or voltage, and the output can interface with most forms of logic. When it is used as a V/F converter, the equivalent circuit (inside the dashed box) and connection diagram of the 9400 is as shown in Figure 9-10. The equivalent circuit consists of an integrator, comparators, a delay network, a divide-by-2 network, and open-collector output transistors. The input current $I_{in} = V_{in}/R_{in}$ is converted to a charge by the integrating capacitor C_{int} and shows up as a linearly decreasing voltage V_A at the output of the op-amp integrator, as shown in Figure 9-10(b). In equation form,

$$V_A = -\left(\frac{I_{in}}{C_{int}}\right)t \tag{9-3}$$

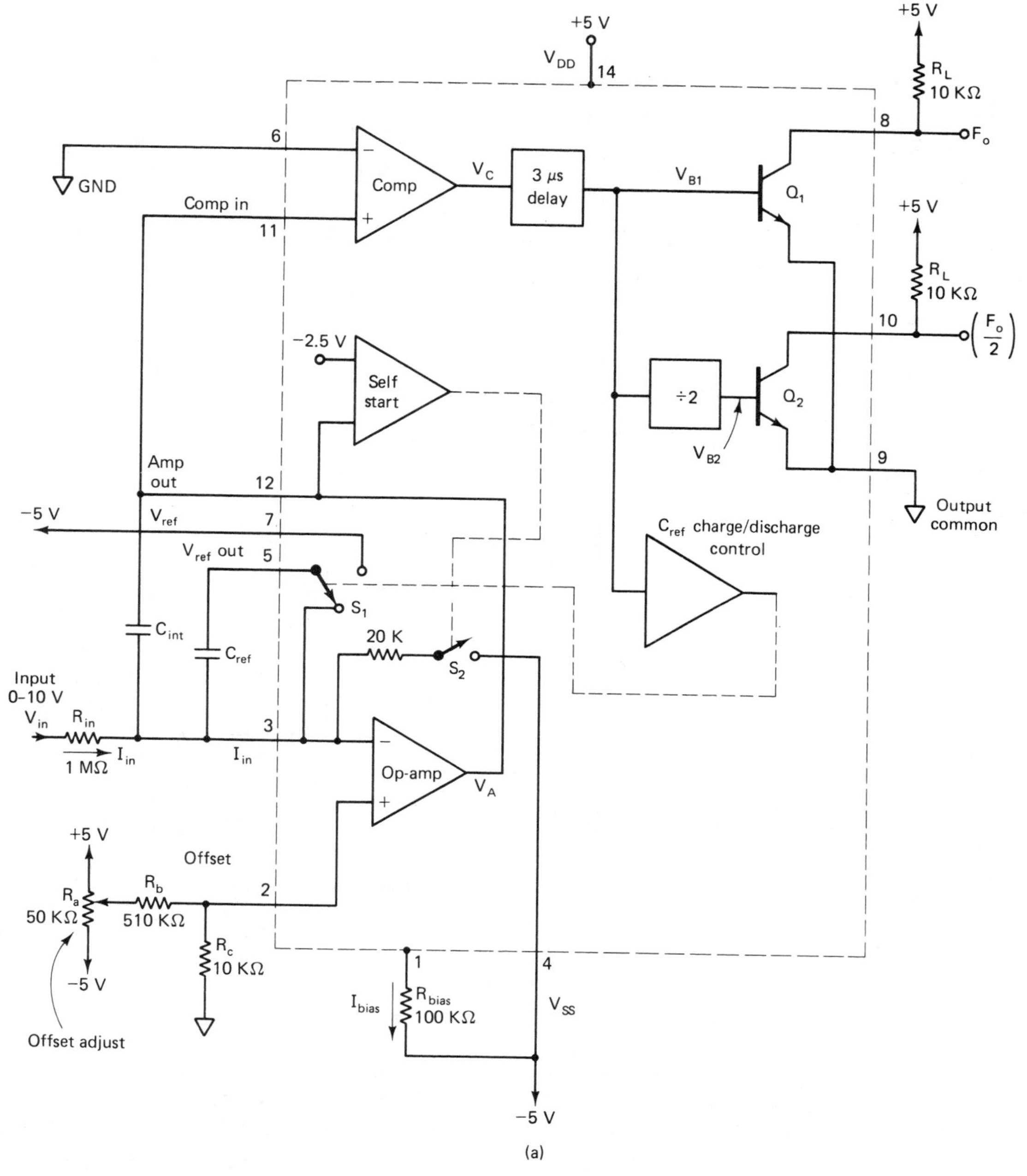

Figure 9-10 (a) 9400 V/F converter equivalent circuit and connection diagram. (b) Its waveforms. (Courtesy of Teledyne Semiconductor.)

where I_{in} = input current (amperes)
C_{int} = integrating capacitor (farads)
t = time (seconds)

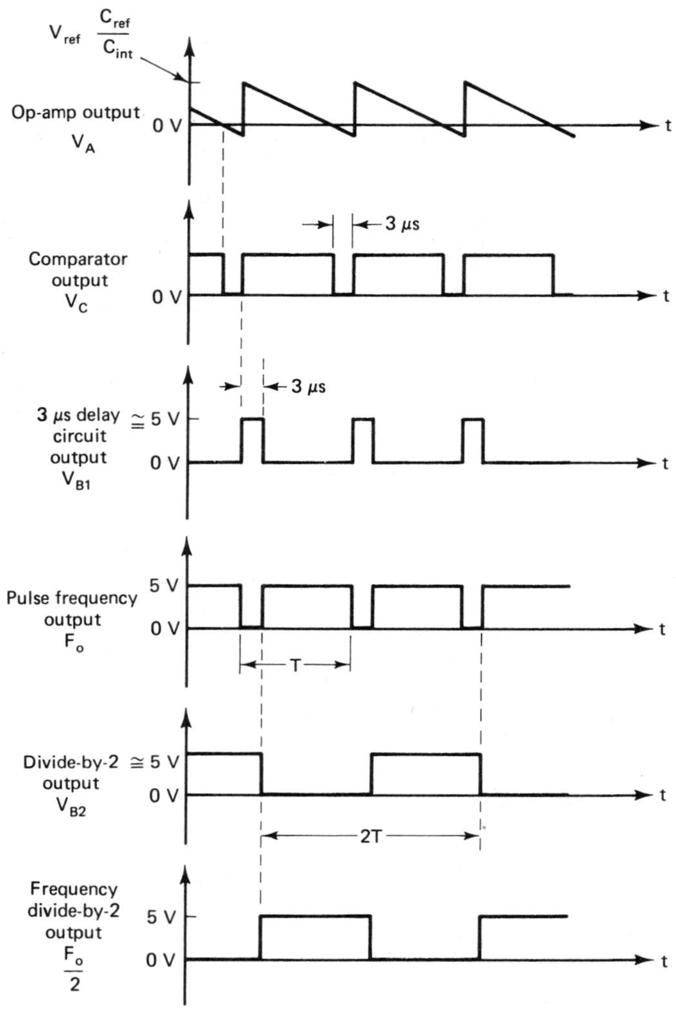

(b)

Figure 9-10 (cont.)

The output V_A of the integrator is sensed by the comparator. The output of the comparator is then applied via the 3 μs delay network to the output transistor Q_1, the divide-by-2 network, and the C_{ref} charge/discharge control circuit. The output of the divide-by-2 network drives the output transistor Q_2. When the output V_A of the op-amp is positive, the output V_C of the noninverting comparator is *high*, or $\cong +5$ V, and the output V_{B1} of the 3 μs delay network is *low* or 0 V. Since V_{B1} is *low*, the transistor Q_1 is *off*, and the output F_o is *high*, or $\cong +5$ V [see Figure 9-10(b)]. The divide-by-two network is a negative-edge-triggered flip-flop whose output voltage V_{B2} is a complement (inversion) of V_{B1}; hence the transistor Q_2 will be on, and output $F_o/2$ will be *low*. Finally, with V_{B1} *low*, the C_{ref} charge/discharge control circuit is disabled, and capacitor C_{ref} remains discharged (is shorted out). However, from Equation (9-3), the output V_A of the op-amp is linearly decreasing. Therefore, when V_A goes slightly below 0 V, the output V_C of the comparator switches "low" to 0 V. Three microseconds after V_C goes low, V_{B1} switches from low to high. This forces output F_o *low* and enables the C_{ref} charge/discharge control circuit which connects C_{ref} to the reference voltage V_{ref} of -5 V. At this instant the negative voltage at the ($-$) input causes the output of the op-amp to step up to a finite amount, as shown in Figure 9-10(b). The reference capacitor C_{ref} remains connected to V_{ref} for a time period long enough to virtually charge it to V_{ref}. The charging path for C_{ref} is through the output terminal of the op-amp, through C_{int}, through C_{ref}, and finally through the negative reference voltage. Now, since the output of the op-amp V_A is positive, the output of the comparator switches from low to high, and 3 μs later V_{B1} goes low. This disables the C_{ref} charge/discharge control circuit, and C_{ref} is shorted out, dissipating the stroed reference charge. When V_{B1} goes low, the output F_o goes high, the output of the divide-by-2 circuit goes low, and output $F_o/2$ therefore goes high [see Figure 9-10(b)]. The integrating capacitor C_{int} again begins converting the input current I_{in} to a charge, and in turn the output of the op-amp starts decreasing linearly [see Equation (9-3)]. When the output of the op-amp goes slightly below 0 V, the output of the comparator goes from high to low, and hence the cycle repeats. In short, the continued discharging of C_{int} by the input current I_{in} is balanced out by fixed charge from V_{ref}. As the input voltage V_{in} is increased, the rate of decrease of output voltage V_A also increases [see Equation (9-3)], causing output frequency F_o to also increase. Since each positive voltage increment in V_A is fixed, the increase in frequency with voltage is linear.

As shown in Figure 9-10(a), the 9400 contains a "self-start" circuit that assures proper V/F operation when power is first applied. If the op-amp output V_A is below the comparator threshold of 0 V and C_{ref} is already charged, then when the power is first applied, a positive voltage step will not occur. In this situation the output V_A of the op-amp continues to decrease until it crosses the -2.5 V threshold of the "self-start" comparator. When this happens, the self-start comparator connects through the 20-kΩ resistor a negative supply voltage V_{SS} to the ($-$) input of the op-amp, which in turn causes the output of the op-amp to quickly go positive. As soon as V_A goes positive, the self-start comparator is disabled, and the 9400 resumes its normal operating mode.

The 9400 is a 14-pin DIP with plastic and ceramic options. The pin diagram is shown in Figure 9-11. The functions of these pins are as follows:

Pin 11: Comparator input. This pin is the (+) input of the comparator and is connected to the output of the op-amp in the V/F mode. In the F/V mode, the input frequency is applied to the comparator input.

Pin 8: Pulse frequency output. At this pin a pulse waveform output is available whose frequency is proportional to the input voltage. Since this is an open-collector output, it requires a pull-up resistor and interfaces directly with all forms of logic.

Pin 10: Frequency divide-by-2 output. This pin provides a square-wave output which is one-half the frequency of the pulse output at pin 8. This output is an

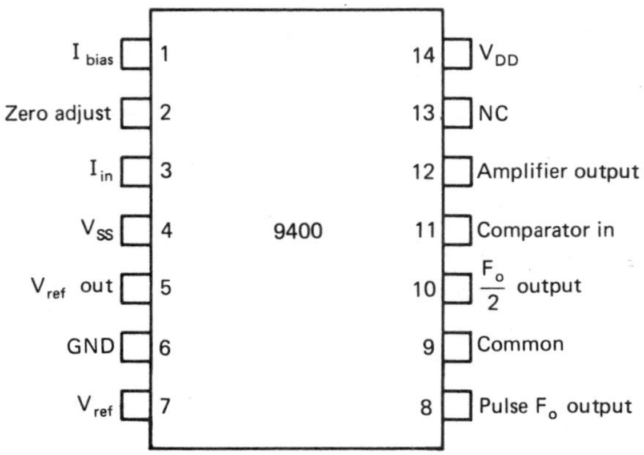

Figure 9-11 Pin diagram for the 9400.

open collector and hence requires a pull-up resistor. It is also compatible with all logic families.

Pin 9: Output common. The emitters of both the output transistors are connected to this pin. A unipolar or bipolar output voltage swing can be obtained by connecting this pin either to ground or to the negative supply V_{SS}, respectively. For high-performance applications, the output ground (pin 9) should be separated from the input ground (pin 6).

Pin 1: I_{bias}. A resistance R_{bias} is connected between pin 1 and pin 4. Specifications for the 9400 are based on $R_{\text{bias}} = 100\ \text{k}\Omega \pm 10\%$.

Pin 12: Amplifier out. This is the output of the op-amp which is a negative-going ramp signal in the V/F mode. In the F/V mode, a voltage proprotional to the frequency input is generated at this pin.

Pin 2: Zero adjust. The circuit that can be used to reduce the output initially to zero is connected to this pin. This low-frequency set point is determined by adjusting the voltage at this pin.

Pin 3: I_{in}. This pin is the (−) input of the op-amp, and the input current I_{in} is applied to it. I_{in} is 10 μA for nominal full scale with an over-range current up to 50 μA.

Pin 7: V_{ref}. A reference voltage from either a separate negative supply or the V_{SS} may be applied to this pin.

Pin 5: V_{ref} *OUT.* When C_{ref} is to be charged, the C_{ref} charge/discharge control circuit connects this pin to the reference voltage V_{ref} (pin 7).

Pin 13: NC. No connection.

Pin 6: GND. Input ground.

9-10.1(a) V/F design procedure. The output frequency F_o of the 9400 V/F converter is related to the analog input voltage V_{in} by the equation

$$F_o = \frac{V_{\text{in}}}{R_{\text{in}}} \frac{1}{(V_{\text{ref}})(C_{\text{ref}})} \tag{9-4}$$

The V/F converter shown in Figure 9-12 can be designed by using the following steps:

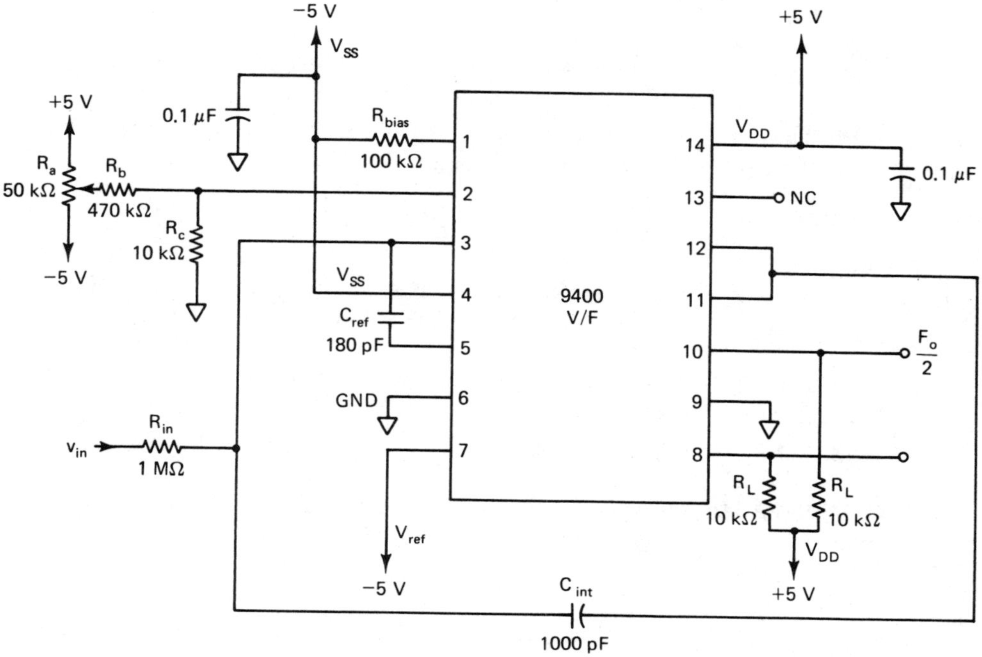

Figure 9-12 10 Hz to 10 kHz V/F converter. To increase $F_{o\,max}$ to 100 kHz, change C_{ref} to 27 pF and C_{int} to 75 pF. (Courtesy of Teledyne Semiconductor.)

1. Choose V_{DD} and V_{SS} such that

$$4\text{ V} \le V_{DD} \le 7.5\text{ V}$$
$$-7.5\text{ V} \le V_{SS} \le -4\text{ V}$$

   Generally, $V_{DD} = +5$ V and $V_{SS} = -5$ V. For high accuracy, 0.1-μF disk decoupling capacitors located near the pins 4 and 14 are recommended.

2. The reference voltage V_{ref} can be a separate precision source or can be the negative supply voltage V_{SS}. Usually, $V_{ref} = V_{SS} = -5$ V.
3. R_{in} is calculated from the equation

$$R_{in} = \frac{V_{in}\text{ full scale}}{I_{in}\text{ full scale}}$$

   For the 9400 V/F the specified values of V_{in} full scale = 10 V and I_{in} full scale = 10 μA.

   Hence $R_{in} = 1$ MΩ. Metal film resistors with ≤ 1% tolerance are recommended for high accuracy.

4. For the 9400 V/F, the specified value of $R_{bias} = 100$ kΩ.
5. The specified value of pull-up resistors R_L is 10 kΩ each (see Figure 9-12).
6. Although the exact value of C_{ref} is not critical, it should be <500 pF. This capacitor should be located as close as possible to pins 3 and 5. Glass film capacitors are recommended for high accuracy.

7. Choose the value of C_{int} such that

$$3C_{ref} \leq C_{int} \leq 10C_{ref}$$

However, for improved stability and linearity, C_{int} must be $\geq 4C_{ref}$. This capacitor should be located as close as possible to pins 3 and 12.

8. The exact values of offset adjust resistors R_a, R_b, and R_c are not critical; however, they are related by the relationship

$$R_c < R_a < R_b$$

(See Figure 9-12).

9-10.1(b) Adjustment/calibration procedure. For a 10-kHz full-scale output frequency F_o, the adjustment/calibration procedure is as follows:

1. Set V_{in} to 10 mV and adjust the R_a potentiometer to obtain a 10-Hz output frequency (see Figure 9-12).
2. Set V_{in} to 10 V and adjust R_{in}, V_{ref}, or C_{ref} to obtain a 10-kHz output frequency.

9-10.1(c) Single-supply operation. The 9400 V/F converter can also be operated using a single supply, as shown in Figure 9-13.

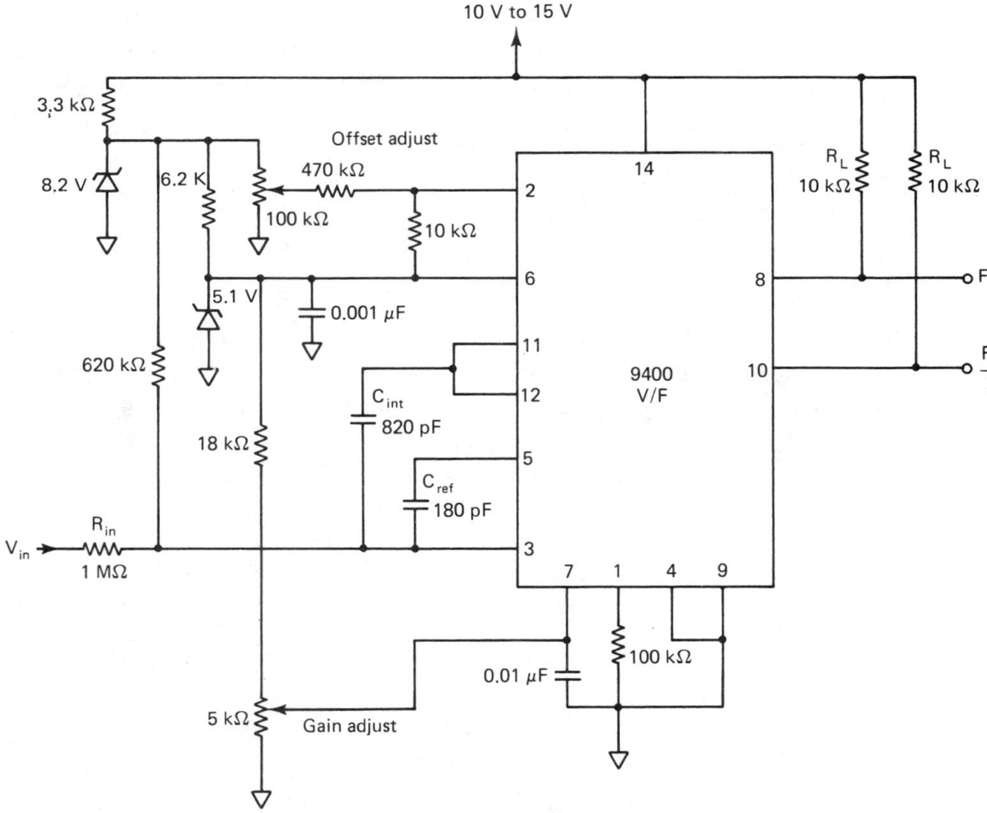

Figure 9-13 9400 V/F converter using single variable supply voltage with offset and gain adjust. (Courtesy of Teledyne Semiconductor.)

Example 9-3

The V/F converter of Figure 9-12 is initially adjusted for a 10-kHz full-scale output frequency. Determine the output frequencies F_o and $F_o/2$ if the input signal $V_{in} = 2$ V.

Solution. The output frequency F_o of the 9400 is directly proportional to the input voltage V_{in}. In addition, it has a guaranteed linearity of 0.05%. Since the circuit is initially adjusted for a 10-kHz full-scale output frequency (at $V_{in} = 10$ V), when the input $V_{in} = 2$ V, the output frequency F_o will be (2 kHz $\pm$ 0.05%) and $F_o/2$ will be 1 kHz $\pm$ 0.05%. The nature of these output waveforms will be as shown in Figure 9-10(b).

The V/F converter is used in instrumentation and control, digital, and communication systems. Typical applications that use V/F converters are in temperature sensing and control, transducer encoding, analog-to-digital (A/D) converters, microprocessor data acquisition, digital panel meters, phase-locked loops, and analog data transmission and recording. For specific application of the V/F converter, refer to Chapter 11.

9-10.2 F/V Converter

When used as an F/V converter, the 9400 generates an output voltage that is linearly proportional to the input frequency waveform. The features of the 9400 F/V converter include dc to 100 kHz operation, op-amp output, programmable scale factor, high input impendance (>10 MΩ), and, above all, its capability to accept any voltage wave shape. Figure 9-14(a) shows an equivalent circuit (inside the dashed box) and connection diagram of the 9400 when it is used as a F/V converter. The input frequency is applied to the (+) input of the comparator (pin 11). Since the comparator hysteresis voltage is ± 200 mV, the input signal amplitude must be greater than ± 200 mV in order to trip the comparator. If only a unipolar input signal is available, it can be converted into a bipolar waveform by using the offset circuit of Figure 9-15.

Each time the input signal crosses zero in the negative direction, the output of the comparator goes low. Three microseconds later the C_{ref} *charge/discharge control circuit* is enabled, which instantaneously connects the reference capacitor C_{ref} to the reference voltage V_{ref} [see Figure 9-14(a)]. This action charges C_{int} each time with a precise amount of voltage until the voltage across it can no longer increase. The charging path is through the output terminal of the op-amp, through C_{int}, through C_{ref}, and finally through V_{ref}. On the other hand, each time the input waveform crosses zero in the positive direction, the output of the comparator switches high. This disables the C_{ref} *charge/dischagre control circuit*, and C_{ref} is shorted out. However, the voltage across C_{int} is retained because the only discharge path for C_{int} is through R_{int}, which is very large (1 MΩ). The voltage across C_{int} is the output voltage V_o.

The amount of ripple on V_o is inversely proportional to C_{int} and the frequency of the input F_{in}. This means that for low frequencies C_{int} can be increased in the range 1 to 100 μF to reduce the ripple. To eliminate the ripple on V_o, an extra op-amp that is operating in the common-mode configuration shown in Figure 9-16 can be connected to the output of the F/V converter of Figure 9-14(a). Because of the common-

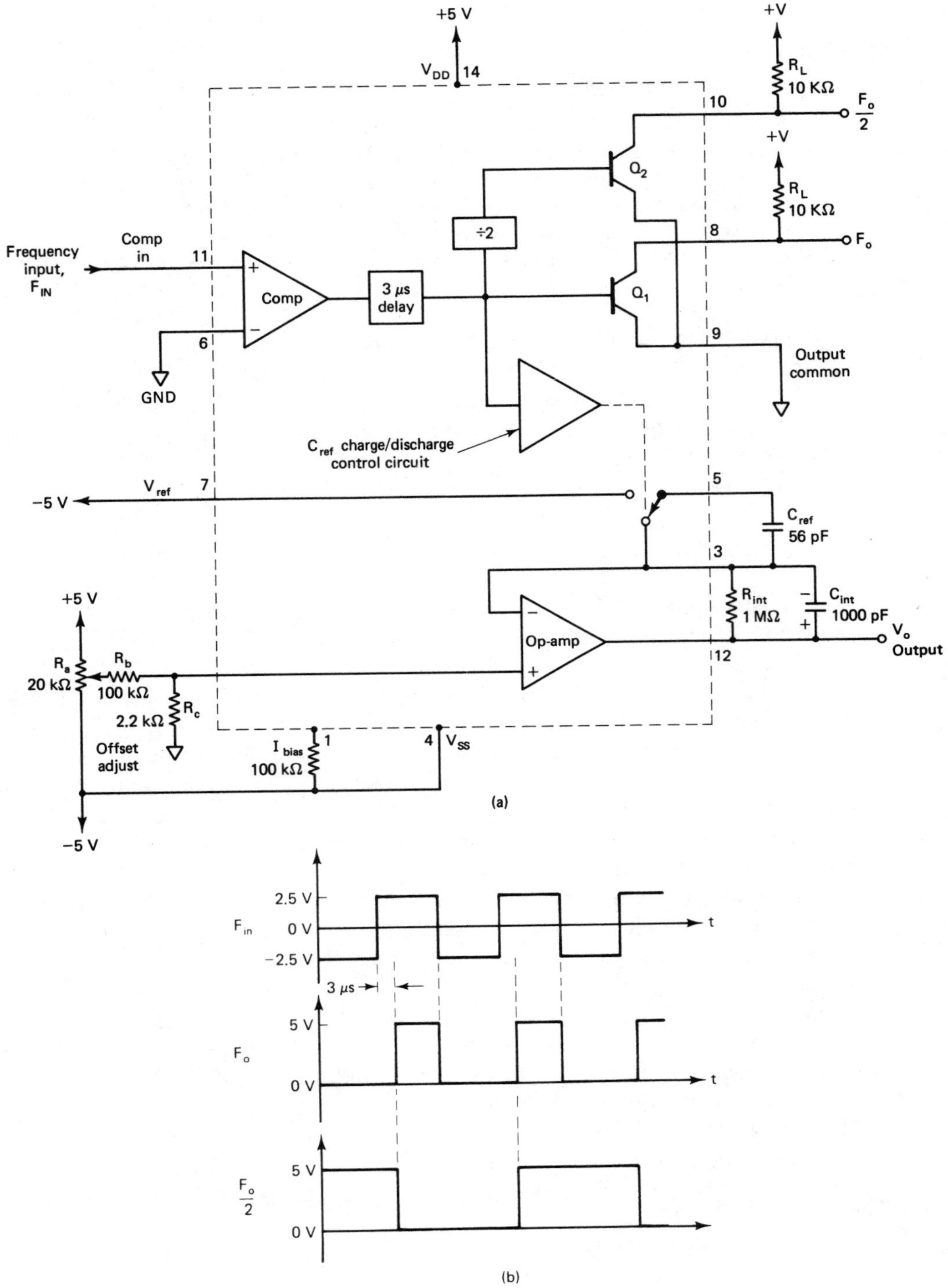

Figure 9-14 (a) Dc to 10 kHz F/V converter. $F_o/2$ and F_o are optional if buffer is needed. When not used, pins 8, 9, and 10 may be connected to ground. (b) F/V digital outputs. (Courtesy of Teledyne Semiconductor.)

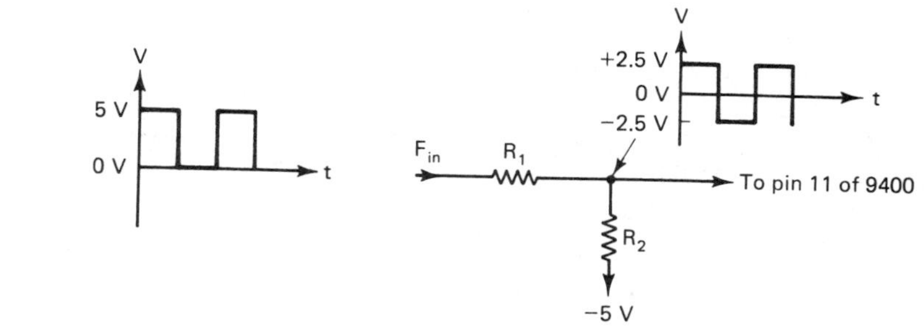

Figure 9-15 Offset circuit to convert unipolar waveform into bipolar waveform.

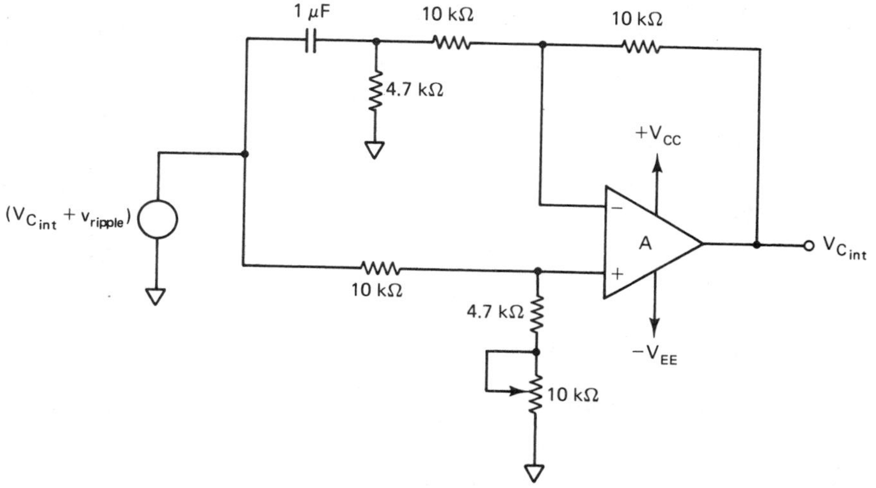

Figure 9-16 F/V ripple eliminator.

mode configuration, the ac ripple is canceled at the output of an op-amp provided that both (+) and (−) inputs have the same gain. The 10 kΩ potentiometer is used to make the gain of (+) and (−) inputs equal. Besides that, the circuit in Figure 9-16 has a *dc* gain of unity, so that output voltage is equal to input.

In the F/V converter of Figure 9-14(a), outputs F_o and $F_o/2$ are optional because these outputs are useful in only some applications. The output F_o follows the input frequency waveform, with a 3-μS delay on the rising edge [see Figure 9-14(b)]. $F_o/2$ is a square wave with a frequency one half of F_o. If F_o and $F_o/2$ outputs are not used, pins 8, 9, and 10 may be connected to ground.

Although the F/V converter accepts any input wave shape, the circuit will work only if the positive pulse width of the input waveform is at least 5 μs and the negative pulse width is ≥0.5 μs. When $F_{\text{in max}}$ is less than 1 kHz, the duty cycle (= pulse width/time period × 100) should be greater than 20% to ensure that C_{ref} is fully charged and discharged.

9-10.2(a) F/V design procedure. The output voltage V_o of the F/V converter is related to the input frequency F_{in} by the equation

$$V_o = (V_{\text{ref}} C_{\text{ref}} R_{\text{int}}) F_{\text{in}} \tag{9-5}$$

The response time to a change in F_{in} is equal to $R_{int}C_{int}$. The F/V converter can be designed by following these steps:

1. Choose V_{DD} and V_{SS} such that

$$4\text{ V} \leq V_{DD} \leq 7.5\text{ V}$$
$$-7.5\text{ V} \leq V_{SS} \leq -4\text{ V}$$

   (Usually, $V_{DD} = +5$ V and $V_{SS} = -5$ V.)
2. Choose $V_{ref} = V_{SS} = -5$ V.
3. Choose $R_{int} = 1$ MΩ for $F_{in\ max}$ of 10 kHz. However, for $F_{in\ max}$ of 100 kHz, R_{int} should be decreased to 100 kΩ.
4. $R_{bias} = 100$ kΩ is the specified value for the 9400 F/V converter.
5. Choose pull-up resistors $R_L = 10$ kΩ.
6. Choose $C_{ref} \cong 56$ pF for a $F_{in\ max}$ of 10 kHz. However, C_{ref} should be increased for a lower $F_{in\ max}$.
7. Choose $C_{int} = 1{,}000$ pF for $F_{in\ max}$ of 10 kHz. However, for a lower $F_{in\ max}$, C_{int} can be increased in the range 1 to 100 μF.
8. Choose offset voltage compensating network resistors R_a, R_b, and R_c such that

$$R_c < R_a < R_b$$

   [See Figure 9-14(a)].

9-10.2(b) Adjustment/calibration procedure. For a 10-kHz maximum input frequency $F_{in\ max}$, the adjustment/calibration procedure is as follows:

1. When no input signal is applied ($F_{in} = 0$), adjust the offset null circuit to obtain a 0 V dc output voltage.
2. Set $F_{in} = 10$ kHz and adjust C_{ref} so that V_o is approximately 2.5 to 3 V.

9-10.2(c) Single-supply operation. The 9400 F/V converter can also be operated using a single supply, as shown in Figure 9-17. In this figure, V_{ref} is defined as the voltage difference between pins 2 and 7. The comparator's (−) input is referenced to 6.2 V; therefore, the input signal must satisfy the requirement

$$4\text{ V} < \text{input voltage} < V_{DD}$$

The output voltage in turn is also referenced to 6.2 V. Additionally, if the input signal is ac-coupled, a 100-kΩ to 10-MΩ resistor must be connected between (+) and (−) inputs (pins 11 and 6) of the comparator.

The 9400 F/V converter can be used in applications such as frequency meters/tachometers, speedometers, rpm (revolutions per minute) indicators, FM demodulation frequency multipliers/dividers, and motor control.

Example 9-4

The F/V converter of Figure 9-14(a) is initially adjusted for $V_o = 2.8$ V at $F_{in\ max}$ of 10 kHz. Determine the output voltage V_o if $F_{in} = 1$ kHz.

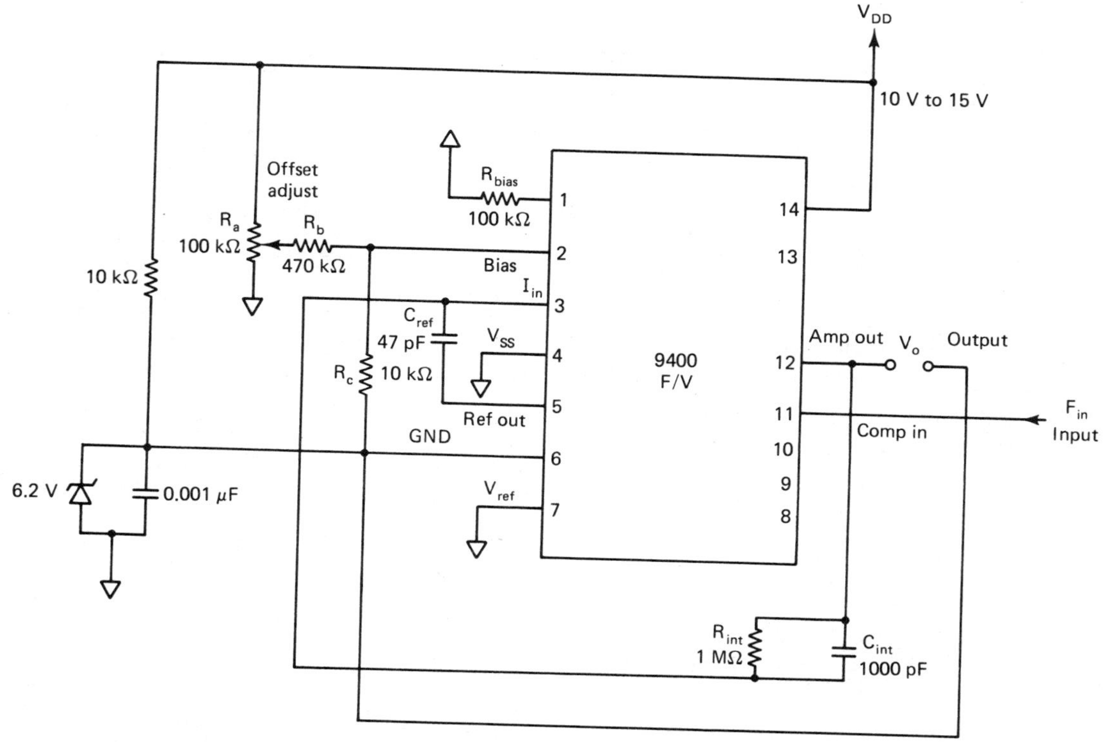

Figure 9-17 9400 F/V converter using single supply. (Courtesy of Teledyne Semiconductor.)

Solution. The 9400 F/V converter generates an output voltage that is linearly proportional to the input frequency F_{in}. Therefore, at $F_{in} = 1$ kHz, $V_o = 2.8$ V/10 $= 0.28$ V.

9-11 ANALOG-TO-DIGITAL AND DIGITAL-TO-ANALOG CONVERTERS

Digital systems are used in ever more applications because of their increasingly efficient, reliable, and economical operation. With the development of the microprocessor, data processing has become an integral part of various systems. Data processing involves transfer of data to and from the microcomputer via an input/output devices. Since digital systems such as microcomputers use a *binary system* of "ones" and "zeros," the data to be put in to the microcomputer have to be converted from analog form to digital form. The circuit that performs this conversion is called an *analog-to-digital* (*A*/*D*) *converter*. On the other hand, a *digital-to-analog* (*D*/*A*) *converter* is used when a binary output from a digital system must be converted to some equivalent analog voltage or current. The binary output (a sequence of 1's 0's and) from a digital system is difficult to interpret; however, a D/A converter makes the interpretation easier. The function of a D/A converter is exactly opposite to that of an A/D converter. This section presents both types of converters. The

D/A converter is presented first because (1) it is simpler than the A/D converter and (2) it can also be used to form the A/D converter.

9-11.1 D/A Converters

A D/A converter in its simplest form uses an op-amp and either *binary-weighted resistors* or R and $2R$ resistors (see Figures 9-18 and 9-19).

9-11.1(a) D/A converter with binary-weighted resistors. Figure 9-18 shows a D/A converter using an op-amp and binary-weighted resistors. Although in this figure the op-amp is connected in the inverting mode, it can also be connected in the noninverting mode. Since the number of binary inputs is four, the converter is called a 4-*bit* (*b*inary dig*it*) converter. Because there are 16 (2^4) combinations of binary inputs for b0 through b3, an analog output should have 16 possible corresponding values. In Figure 9-18, four switches (b0 to b3) are used to simulate the binary inputs; in practice, a 4-bit binary counter such as the 7493 may be used instead. When switch b0 is closed (connected to +5 V), the voltage across R is 5 V because $V_2 = V_1 = 0$ V. Therefore, the current through R is 5 V/10 kΩ = 0.5 mA. However, the input bias current I_B is negligible; hence the current through feedback resistor R_F is also 0.5 mA. which, in turn produces an output voltage of −(1 kΩ)(0.5 mA) = −0.5 V. Note that the op-amp is working as a current-to-voltage converter. Now suppose that switch b1 is closed and b0 is opened. This action connects $R/2$ to the positive supply of +5 V, causing twice as much current (1 mA) to flow through R_F, which in turn doubles the output voltage. Thus the output voltage V_o is −1 V when switch b1 is closed. Similarly, if both switches b0 and b1 are closed, the current through R_F will be 1.5 mA which will be converted to an output voltage of −[(1 kΩ(1.5 mA)] = −1.5 V.

Thus, depending on whether switches b0 to b3 are open or closed, the binary-weighted currents will be set up in input resistors. The sum of these currents is equal to the current through R_F, which in turn is converted to a proportional output voltage. When all the switches are closed, obviously the output will be maximum. The output voltage equation is given by

$$V_o = -R_F\left(\frac{b0}{R} + \frac{b1}{R/2} + \frac{b2}{R/4} + \frac{b3}{R/8}\right) \tag{9-6}$$

where each of the inputs b3, b2, b1, and b0 may either be high (+5 V) or low (0 V).

Figure 9-18(b) shows analog outputs versus possible combinations of inputs. The output is a negative-going staircase waveform with 15 steps of −0.5 V each. In practice, however, the steps may not all be the same size because of the variations in *logic high* voltage levels. Notice that the size of the steps depends on the value of R_F. Therefore, a desired step can be obtained by selecting a proper value of R_F, provided that the maximum output voltage does not exceed the saturation levels of an op-amp. For accurate operation of the D/A converter, precision metal film resistors are recommended.

The problem with the D/A converter of Figure 9-18 is that it requires binary-weighted resistors, which may not be readily available, especially if the number of

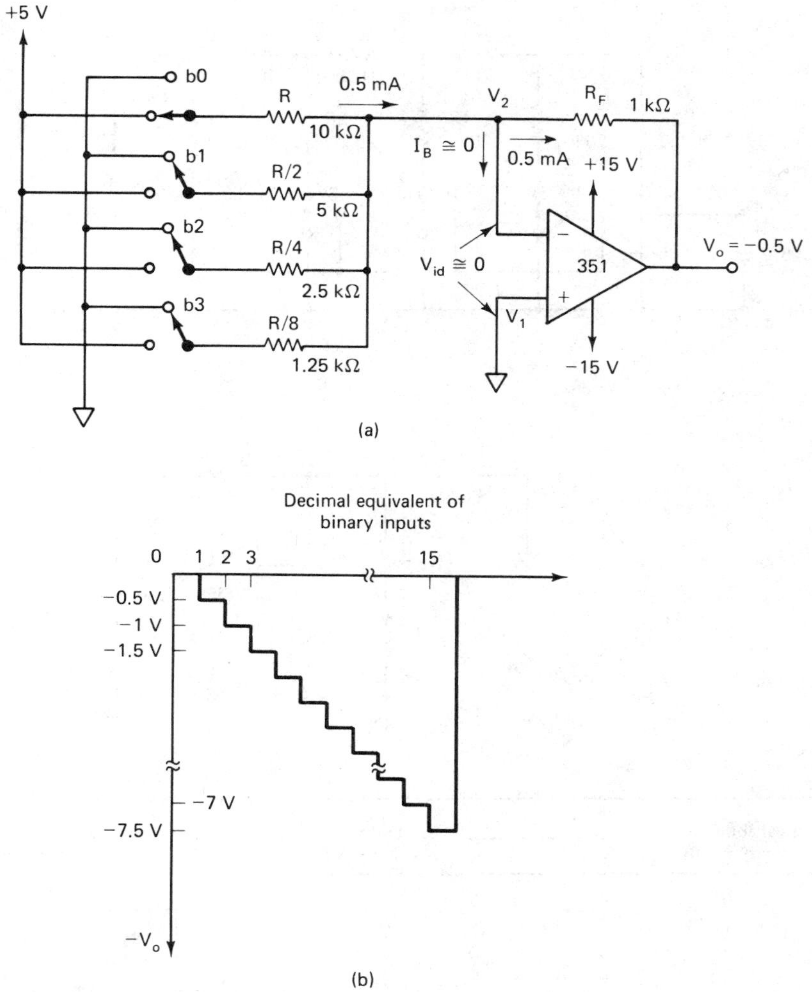

Figure 9-18 (a) D/A converter with binary-weighted resistors. (b) Graph of output versus inputs.

inputs is more than four. An attractive alternative is to use R and $2R$ resistors for the D/A converter since it requires only two sets of precision resistance values.

9-11.1(b) D/A converter with *R* and 2*R* resistors. Figure 9-19(a) shows a D/A converter with R and $2R$ resistors. As before, the binary inputs are simulated by switches b0 through b3, and the output is proportional to the binary inputs. Binary inputs can either be in the *high* (+5 V) or *low* (0 V) state. Assume that the most significant bit (MSB) switch b3 is connected to +5 V and other switches are connected to ground as in Figure 9-19(a). Thévenizing the circuit to the left of switch b3, Thévenin's equivalent resistance R_{TH} is

$$R_{TH} = [\{[(2R \| 2R + R) \| 2R] + R\} \| 2R] + R$$
$$= 2R = 20 \text{ k}\Omega \qquad (9\text{-}7a)$$

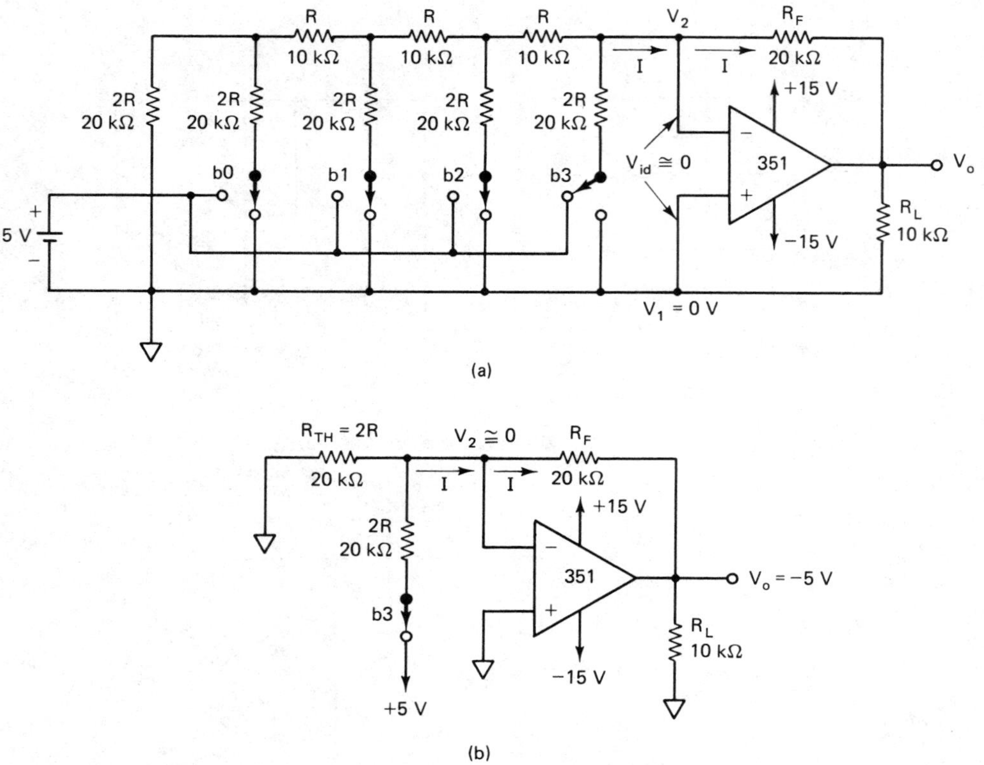

Decimal equivalent of binary inputs	Input (V) b3	b2	b1	b0	Output voltage (V)
0	0	0	0	0	0
1	0	0	0	5	−0.625
2	0	0	5	0	−1.25
3	0	0	5	5	−1.875
4	0	5	0	0	−2.50
5	0	5	0	5	−3.125
6	0	5	5	0	−3.750
7	0	5	5	5	−4.375
8	5	0	0	0	−5.0
9	5	0	0	5	−5.625
10	5	0	5	0	−6.25
11	5	0	5	5	−6.875
12	5	5	0	0	−7.50
13	5	5	0	5	−8.125
14	5	5	5	0	−8.875
15	5	5	5	5	−9.375

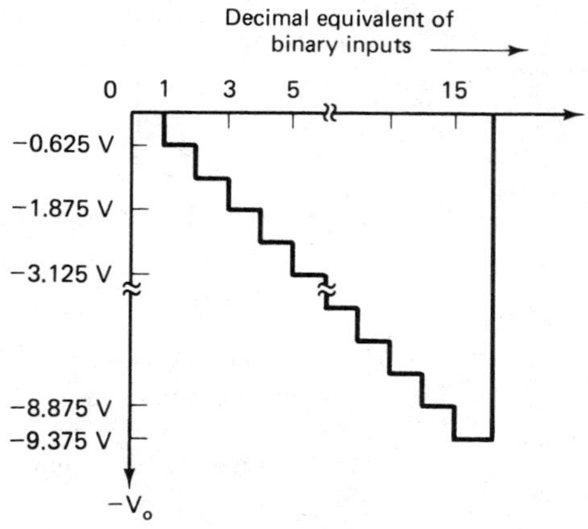

(c)

Figure 9-19 (a) D/A converter with *R*/2*R* resistors. (b) Equivalent circuit when b3 is *high* and b0, b1, and b2 are *low*. (c) Output versus inputs.

The resultant circuit is shown in Figure 9-19(b). In this figure the (−) input is at virtual ground ($V_2 \cong 0$ V); therefore, the current through R_{TH} (= 2R) is zero. However, the current through 2R connected to +5 V is 5 V/20 kΩ = 0.25 mA. The same current flows through R_F and in turn produces the output voltage

$$V_o = -(20\ \text{k}\Omega)(0.25\ \text{mA}) = -5\ \text{V} \tag{9-7b}$$

Using the same analysis, the output voltage corresponding to all possible combinations of binary inputs can be calculated as shown in Figure 9-19(c). The maximum or full-scale output of −9.375 V is obtained when all the inputs are high. The output voltage equation can be written as

$$V_o = -R_F\left(\frac{b3}{2R} + \frac{b2}{4R} + \frac{b1}{8R} + \frac{b0}{16R}\right) \tag{9-8}$$

where each of the inputs b3, b2, b1, and b0 may be either high (+5 V) or low (0 V).

The great advantage of the D/A converter of Figure 9-19(a) is that is requires only two sets of precision resistance values; nevertheless, it requires more resistors and is also more difficult to analyze than the binary-weighted resistor type. As the number of binary inputs is increased beyond four, both D/A converter circuits get complex and their accuracy degenerates. Therefore, in critical applications an integrated circuit specially designed as a D/A converter should be used.

9-11.1(c) Monolithic/hybrid D/A converters. At present, 8-, 10-, 12-, 14-, and 16-bit D/A converters are available with either a current output, a voltage output, or both current and voltage outputs. The MC1408L is a common 8-bit monolithic D/A converter with a current output which can be converted to a voltage type by using an I-to-V converter op-amp [see Figure 9-20(a)]. On the other hand, the SE/NE 5018 is a typical 8-bit D/A converter with voltage output. Figure 9-20(b) shows the SE/NE 5018 configured for unipolar output (0 to 10 V). For 12 bits of resolution as well as current and voltage outputs, hybrid D/A converters such as the DATEL DAC-HZ series are an excellent choice. To make an intelligent selection from a variety of D/A converters to suit a given application, it is necessary to know the key specifications of D/A converters, such as resolution, nonlinearity, gain error, and settling time.

Resolution. Resulution is determined by the number of input bits of the D/A converter. For example, an 8-bit converter has 2^8 (= 256) possible output levels; therefore, its resolution is 1 part in 256. In short, resolution is a value of the LSB.

Nonlinearity or linearity error. This is the difference between the actual output of the DAC and its ideal straight-line output. The error is normally expressed as a percentage of the full-scale range.

Gain error and offset error. Gain error is usually caused by the deviations in the feedback resistor on the I-to-V converter. On the other hand, offset error implies that the output of the DAC is not zero when the binary inputs are all zero. This error stems from the input offsets (V and I) of the op-amp as well as the DAC.

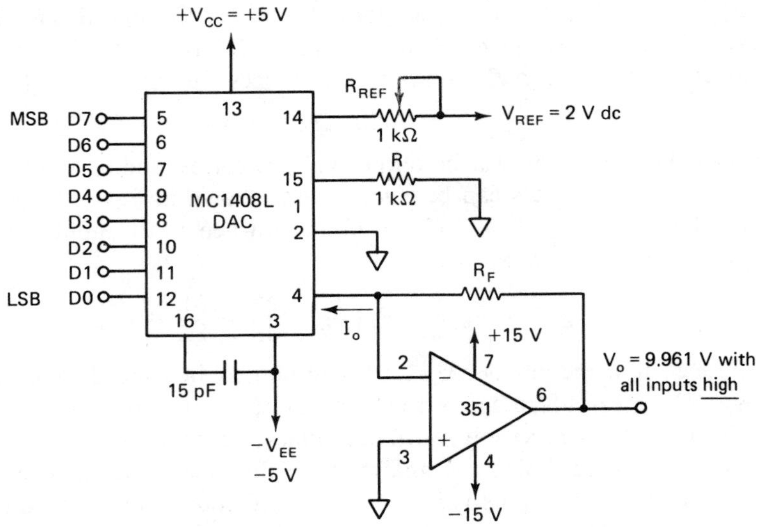

(a)

$$V_o = \frac{V_{ref}}{R_{ref}} \quad (R_F) \quad \left\{ \frac{D7}{2} + \frac{D6}{4} + \frac{D5}{8} + \frac{D4}{16} + \frac{D3}{6} + \frac{D2}{64} + \frac{D1}{128} + \frac{D0}{256} \right\}$$

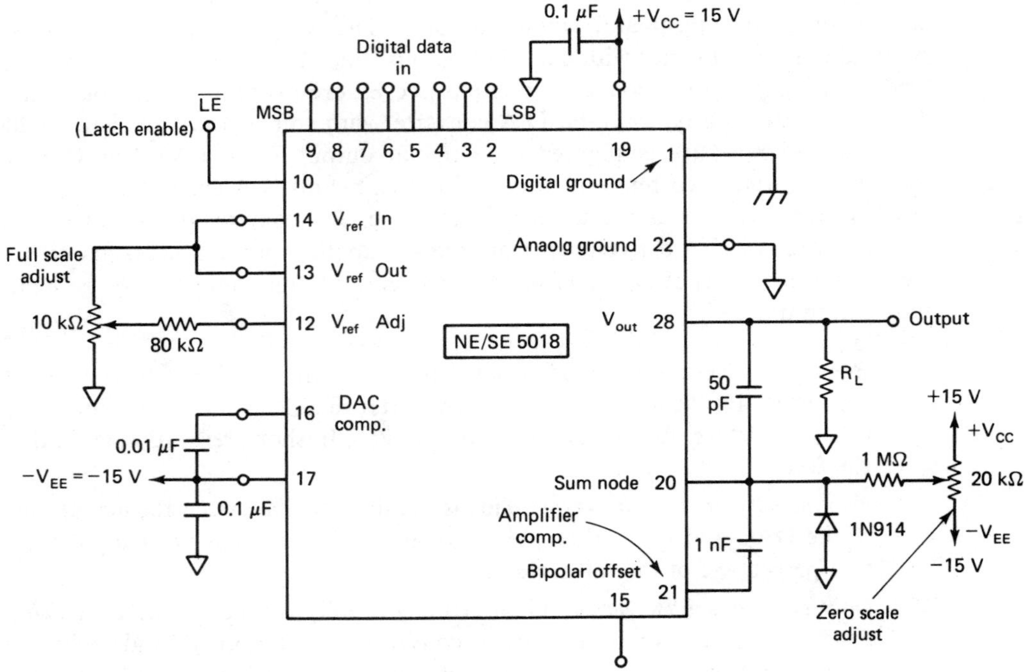

Figure 9-20 (a) MC1408 D/A converter with current output. (Courtesy of Motorola Semiconductor.) (b) NE/SE 5018 D/A converter with voltage output. (Courtesy of Signetics Corporation.)

Settling time. This is the time required for the output of the DAC to settle to within $\pm\frac{1}{2}$ LSB of the final value for a given digital input, that is, zero to full scale.

Typical applications for D/A converters include microcomputer interfacing, CRT graphics generation, programmable power supplies, digitally controlled gain circuits, digital filters, and others.

9-11.2 A/D Converters

A/D converters convert an analog voltage to the digital output that best represents the input. As in the case of D/A converters, analog converters are also specified as 8, 10, 12, or 16 bit. There are many types of A/D converters: single-ramp integrating, dual-ramp integrating, single counter, tracking, and successive approximation. This section discusses only the successive-approximation type, which uses a comparator, a successive-approximation register, output latches, and a D/A converter.

9-11.2(a) Successive-approximation A/D converter. Figure 9-21 shows a successive-approximation type of A/D converter. The heart of the circuit is an 8-bit successive-approximation register (SAR), whose output is applied to an 8-bit D/A converter. The analog output (V_a) of the D/A converter is then compared

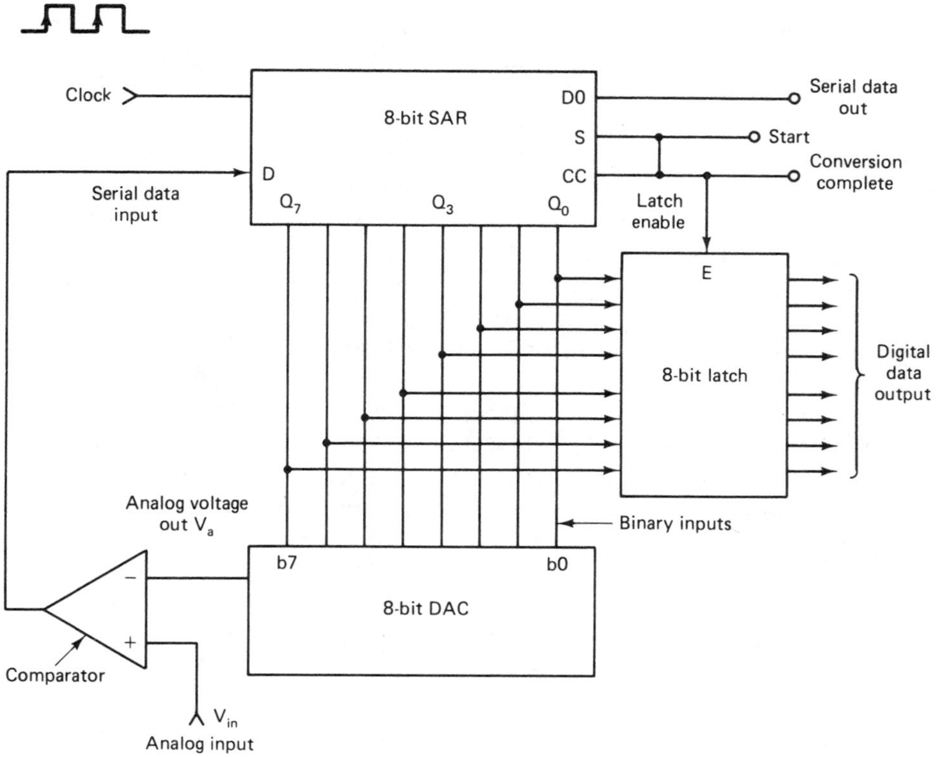

Figure 9-21 Successive-approximation-type A/D converter.

to an analog input signal V_{in} by the comparator. The output of the comparator is a serial data input to the SAR. The SAR then adjusts its digital output (8 bits) until it is equivalent to analog input V_{in}. The 8-bit latch at the end of conversion holds onto the resultant digital data output. The circuit works as follows. At the start of a conversion cycle, the SAR is reset by holding the start (S) signal high. On the first clock pulse LOW-to-HIGH transition, the most significant output bit Q_7 of the SAR is set. The D/A converter then generates an analog equivalent to the Q_7 bit, which is compared with the analog input V_{in}. If the comparator output is *low*, the D/A output $> V_{in}$ and the SAR will clear its MSB Q_7. On the other hand, if the comparator output is *high*, the D/A output $< V_{in}$ and the SAR will keep the MSB Q_7 set. In any case, on the next clock pulse LOW-to-HIGH transition, the SAR will set the next MSB Q_6. Depending on the output of the comparator, the SAR will then either keep or reset the bit Q_6. This process is continued until the SAR tries all the bits. As soon as the LSB Q_0 is tried, the SAR forces the conversion complete (CC) signal HIGH to indicate that the parallel output lines contain valid data. The CC signal in turn enables the latch, and digital data appear at the output of the latch. Digital data are also available serially as the SAR determines each bit. To cycle the converter continuously, the CC signal may be connected to the start conversion input (see Figure 9-21). The advantage of the successive approximation A/D converter is its high speed and excellent resolution. For example, the 8-bit successive approximation A/D converter of Figure 9-21 requires only eight clock pulses.

9-11.2(b) Monolithic/hybrid A/D converters. There are many monolithic A/D converters, such as the integrating A/D, integrating A/D with three-state outputs, and the tracking A/D with latched outputs. In addition, the outputs of A/D are coded in straight binary, binary-coded decimal (BCD), complementary binary (1's or 2's), or sign-magnitude binary. Figure 9-22 shows the Teledyne 8703, an 8-bit monolithic CMOS A/D converter with three-state output. The converter is microprocessor-compatible, exhibits high stability over a full temperature range, contains all required active elements, and has latched parallel binary outputs with strobed or free-running conversion. It has an infinite input range since any positive voltage can be applied via a scaling register R_{in}.

As in the case of monolithics, there are many hybrid A/D converters, such as the successive-approximation A/D with input buffer amplifier, low-power CMOS A/D, the fast A/D with sample-and-hold, and the ultrafast A/D with input buffer amplifier. Datel-Intersil's hybrid ADC-815MC is a very high speed 8-bit successive-approximation A/D converter. It is capable of 8-bit resolution in only 600 ns. It has a six analog-input voltage range with parallel or serial outputs and requires no calibration. Datel's ADC-MC8B is an 8-bit monolithic multifunction A/D–D/A converter which operates on a single +5-V supply. It is a complete D/A converter which can be configured as an A/D converter by using an external comparator and a quad two-input Schmitt trigger NAND gate.

As with D/A converters, specifications such as resolution or nonlinearity are also used for A/D converters. Another important parameter for A/D is *conversion time*, the time required to convert an analog input into valid digital outputs.

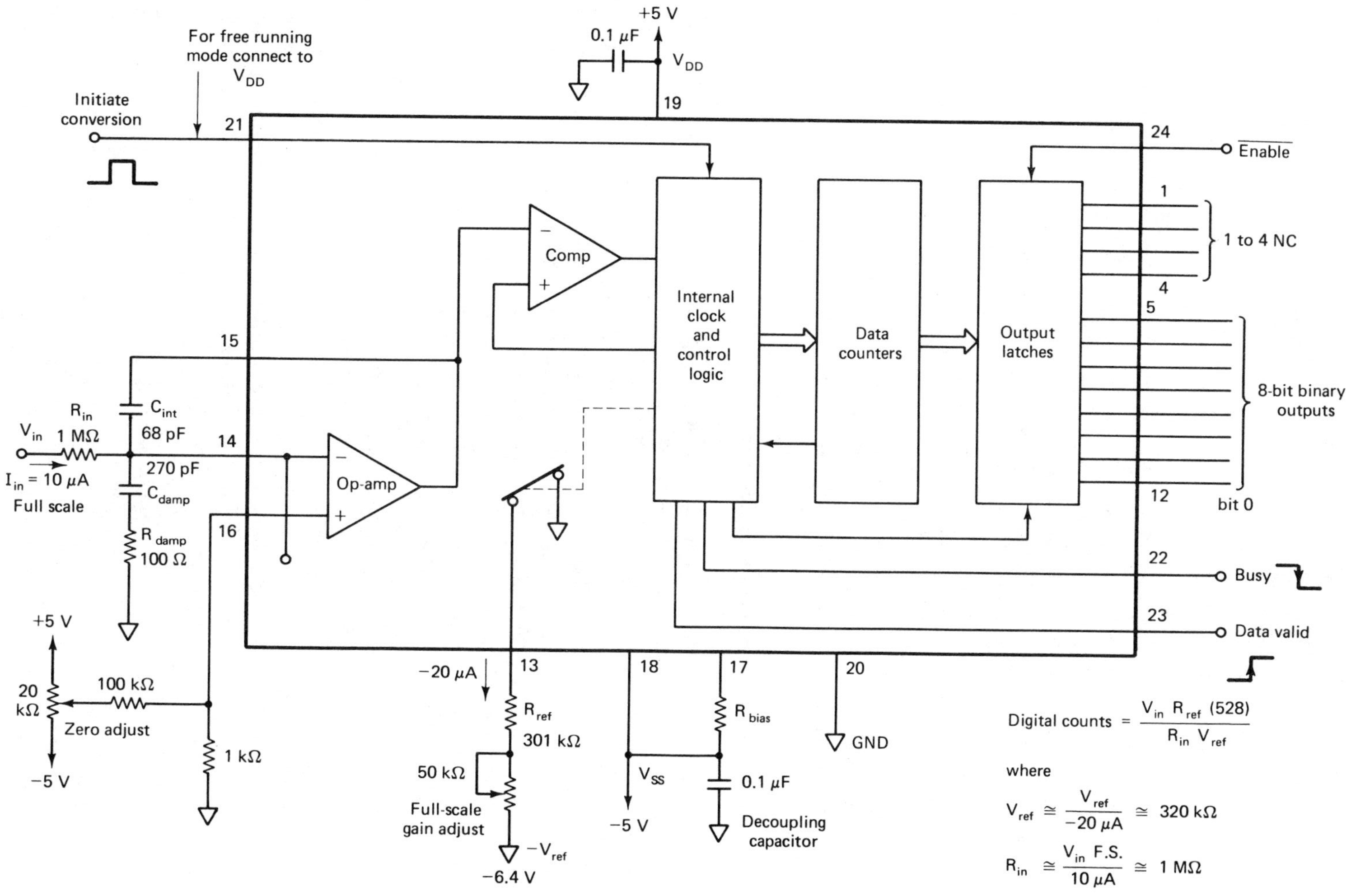

$$\text{Digital counts} = \frac{V_{in}\ R_{ref}\ (528)}{R_{in}\ V_{ref}}$$

where

$$V_{ref} \cong \frac{V_{ref}}{-20\ \mu A} \cong 320\ k\Omega$$

$$R_{in} \cong \frac{V_{in}\ F.S.}{10\ \mu A} \cong 1\ M\Omega$$

Figure 9-22 Teledyne 8703 8-bit monolithic CMOS A/D converter: a single 24-pin DIP package. (Courtesy of Teledyne Semiconductor.)

Typical applications of A/D converters include microprocessor interfacing, data printing and recording, digital voltmeters, and control of LED or LCD displays.

9-12 CLIPPERS AND CLAMPERS

Waveshaping circuits are commonly used in digital computers and communications such as TV and FM receivers. Waveshaping techniques include limiting, clipping, and clamping. Voltage limiting was discussed in Section 9-7. The present section presents typical circuits that use clipping and clamping techniques. In op-amp clipper circuits a rectifier diode may be used to clip off a certain portion of the input signal to obtain a desired output waveshape. An op-amp half-wave rectifier and an absolute-value output circuit use diodes to remove or modify certain portions of the input waveform. In the applications described above the diode works as an *ideal diode* (switch) because when *on*, the voltage drop across the diode is divided by the open-loop gain of the op-amp. On the other hand, when reverse biased (off), the diode is an open circuit. In an op-amp clamper circuit, however, a predetermined dc level is deliberately inserted in the output voltage. For this reason, the clamper is sometimes called a *dc inserter*.

9-12.1 Positive and Negative Clippers

A *positive clipper*, a circuit that removes positive parts of the input signal, can be formed by using an op-amp with a rectifier diode as shown in Figure 9-23(a). In this circuit the op-amp is basically used as a voltage follower with a diode in the feedback path. The clipping level is determined by the reference voltage V_{ref}, which should be less than the input voltage range of the op-amp. Additionally, since V_{ref} is derived from the positive supply voltage ($+V_{CC}$), the dc supply voltages must be well regulated. As shown in Figure 9-23(b) the output voltage has portions of the positive half cycles above V_{ref} clipped off.

The circuit works as follows. During the positive half cycle of the input, the diode D_1 conducts only until $v_{in} = V_{ref}$. This happens because when $v_{in} < V_{ref}$, the voltage (V_{ref}) at the (−) input is higher than that at the (+) input; hence the output voltage v'_o of the op-amp becomes sufficiently negative to drive D_1 into conduction. When D_1 conducts, it closes the feedback loop and the op-amp operates as a voltage follower; that is, output v_o follows input v_{in} until $v_{in} = V_{ref}$. However, when v_{in} is slightly higher than V_{ref}, the output v'_o of the op-amp becomes sufficiently positive to drive D_1 into cutoff. This opens the feedback loop and the op-amp operates open-loop; therefore, it further drives its output v'_o toward positive saturation ($\cong +V_{CC}$). With D_1 reverse biased, the output voltage $v_o = V_{ref}$. Thus when $v_{in} > V_{ref}$, $v'_o \cong +V_{CC}$ and $v_o = V_{ref}$ [see Figure 9-23(b)].

When v_{in} drops below V_{ref}, the output of the op-amp v'_o again becomes sufficiently negative to drive D_1 into conduction. This closes the feedback loop; hence the output follows the input. Thus diode D_1 is *on* for $v_{in} < V_{ref}$ and *off* for $v_{in} > V_{ref}$. The output follows the input only when the diode is *on*. The op-amp alternates between open-loop and closed-loop operations as the diode D_1 is turned *off* and *on*, respec-

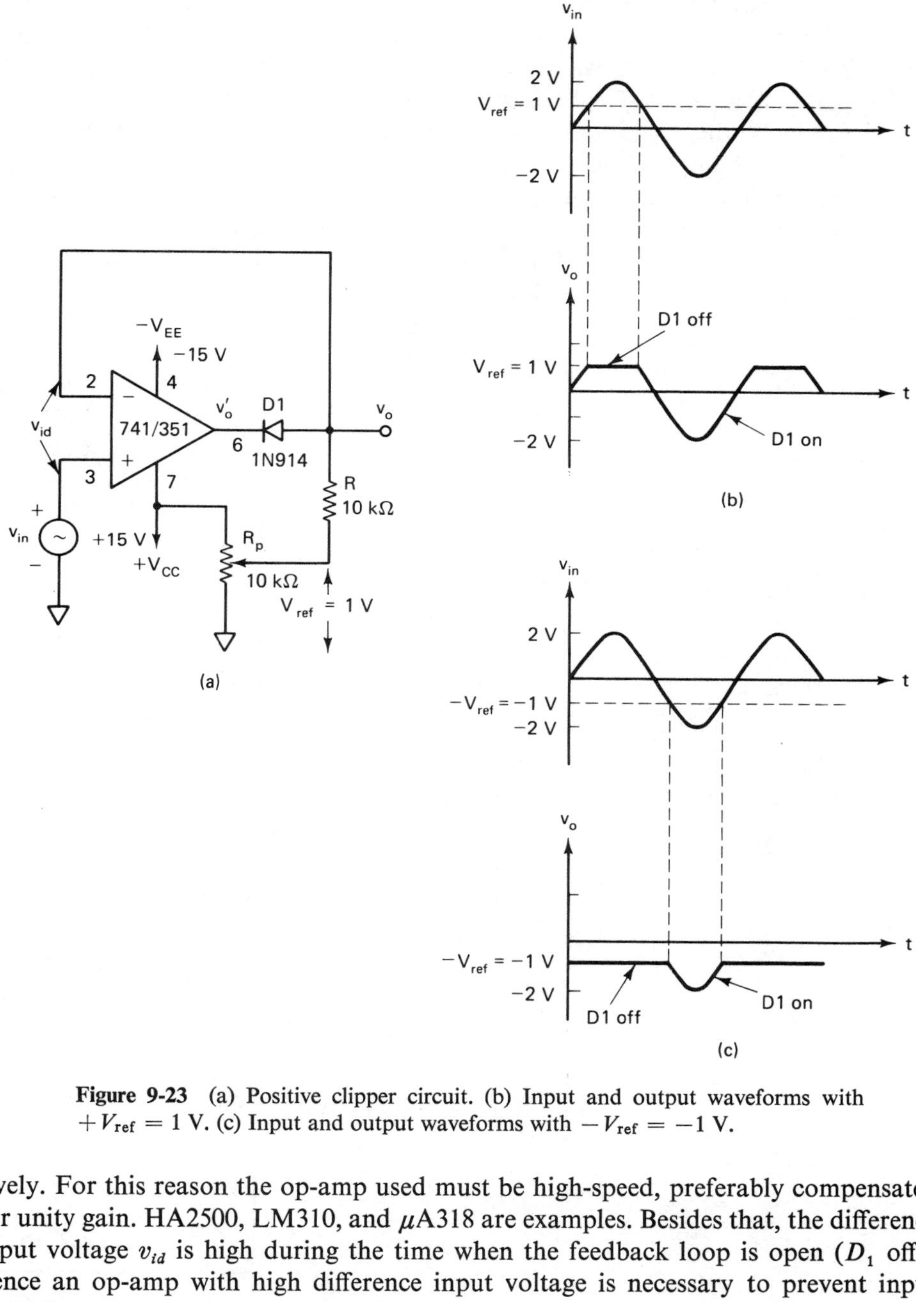

Figure 9-23 (a) Positive clipper circuit. (b) Input and output waveforms with $+V_{ref} = 1$ V. (c) Input and output waveforms with $-V_{ref} = -1$ V.

tively. For this reason the op-amp used must be high-speed, preferably compensated for unity gain. HA2500, LM310, and μA318 are examples. Besides that, the difference input voltage v_{id} is high during the time when the feedback loop is open (D_1 off); hence an op-amp with high difference input voltage is necessary to prevent input breakdown.

In Figure 9-23(a), if pot R_P is connected to the negative supply $-V_{EE}$ instead of $+V_{CC}$, the reference voltage V_{ref} will be negative. This will cause the entire output waveform above $-V_{ref}$ to be clipped off, as shown in Figure 9-23(c). The output follows the input only when $v_{in} < -V_{ref}$.

The positive clipper of Figure 9-23(a) is converted into a negative clipper by simply reversing diode D_1 and changing the polarity of reference voltage V_{ref}. The resultant circuit is shown in Figure 9-24(a). The negative clipper clips off the negative parts of the input signal below the reference voltage [see Figure 9-24(b)]. Diode D_1 conducts when $v_{in} > -V_{ref}$, and therefore during this period output v_o follows the input v_{in}. However, the negative portion of the output voltage below $-V_{ref}$ is clipped off because D_1 is off for $v_{in} < -V_{ref}$. In the circuit of Figure 9-24(a), if $-V_{ref}$ is changed to $+V_{ref}$ by connecting the potentiometer R_p to the $+V_{CC}$, the output voltage below $+V_{ref}$ will be clipped off, as shown in Figure 9-24(c). The diode D_1 must be *on* for $v_{in} > +V_{ref}$ and *off* for $v_{in} < +V_{ref}$.

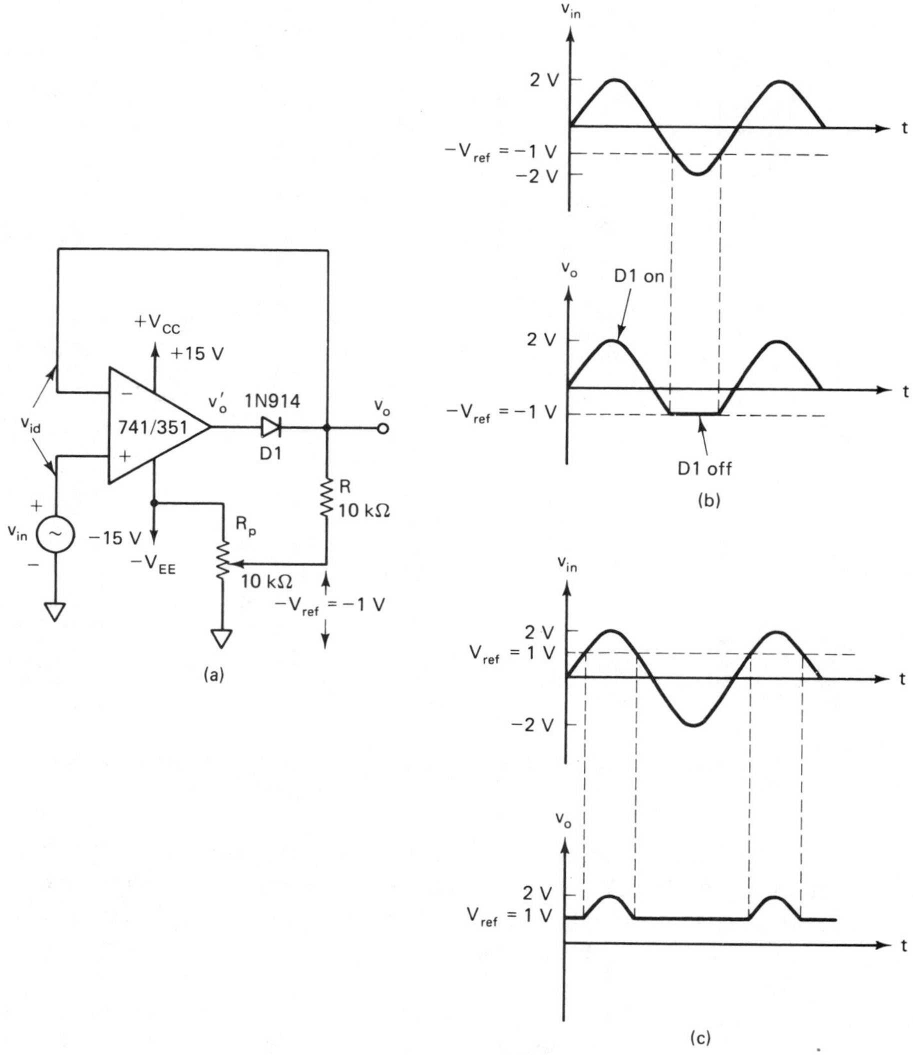

Figure 9-24 (a) Negative clipper circuit. (b) Input and output waveforms with $-V_{ref} = -1$ V. (c) Input and output waveforms with $+V_{ref} = +1$ V.

9-12.2 Small-Signal Half-Wave Rectifiers

The circuit of Figure 9-24(a) can be used as a positive small-signal half-wave rectifier provided that $-V_{ref} = 0$ V. Shown in Figure 9-25(a), the resultant circuit can rectify signals with peak values down to a few millivolts, unlike conventional diodes. This is

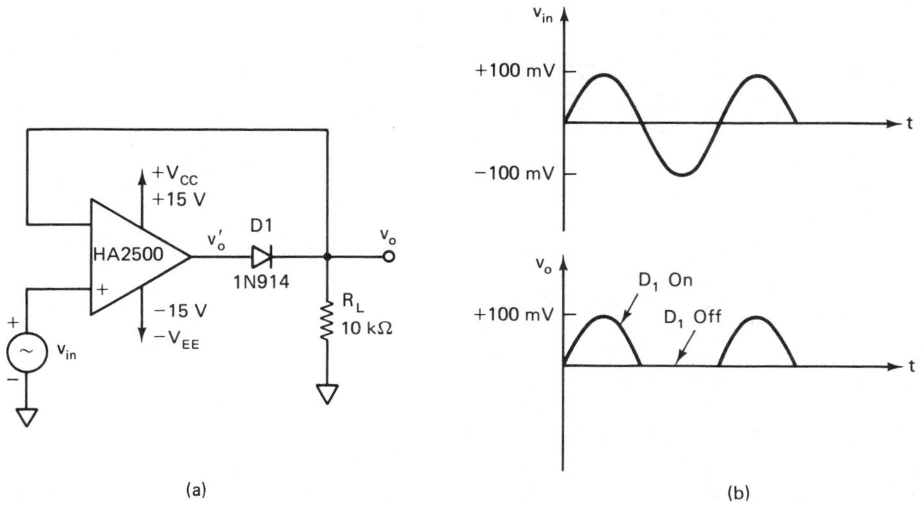

Figure 9-25 (a) Positive small-signal half-wave rectifier circuit. (b) Its input and output waveforms.

possible because the high open-loop gain of the op-amp automatically adjusts the voltage drive to the diode D_1 so that the rectified output peak is the same as the input [see Figure 9-25(b)]. In fact, the diode acts as an *ideal diode* (switch), since the voltage drop across the *on* diode is divided by the open-loop gain of the op-amp. As v_{in} starts increasing in the positive direction, the v_o' also starts increasing positively until diode D_1 is forward biased. When D_1 is forward biased, it closes a feedback loop and the op-amp works as a voltage follower. Therefore, the output voltage v_o follows the input voltgae v_{in} during the positive half-cycle, as shown in Figure 9-25(b). However, when v_{in} starts increasing in the negative direction, v_o' also increases negatively until it is equal to the negative saturation voltage ($\cong -V_{EE}$). This reverse biases diode D_1 and opens the feedback loop. Therefore, during the negative half-cycle of the input signal, v_o is 0 V.

The op-amp in the circuit of Figure 9-25(a) must be a high-speed op-amp since it alternates between open-loop and closed-loop operations. μA318, HA2500, and LM310 are typical examples of high-speed op-amps.

Figure 9-26(a) shows a negative small-signal half-wave rectifier. This circuit, in fact, can be obtained from the circuit of Figure 9-23(a) by setting $V_{ref} = 0$ V. During the positive alternation of v_{in}, D_1 is reverse biased; therefore, $v_o = 0$ V. On the other hand, during the negative alternation, D_1 is forward biased; hence v_o follows v_{in}.

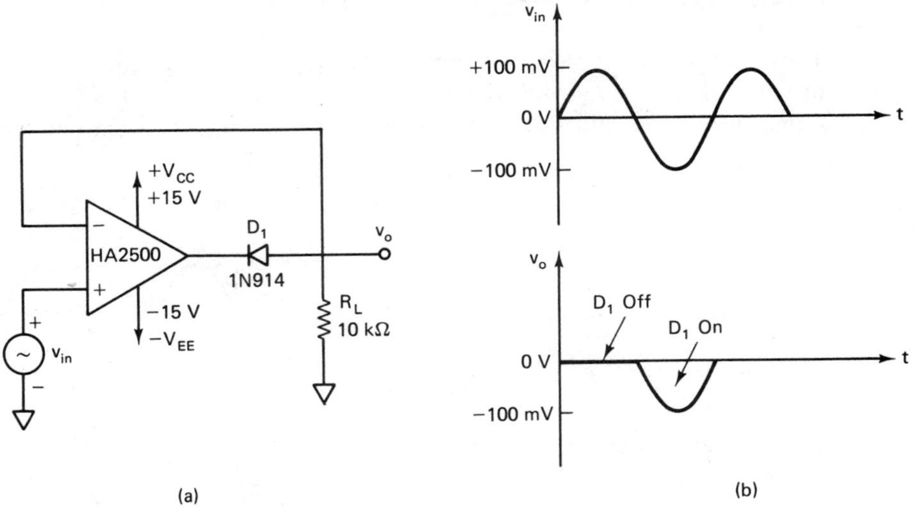

Figure 9-26 (a) Negative small-signal half-wave rectifier circuit. (b) Its input and output waveforms.

Yet another negative half-wave rectifier is shown in Figure 9-27(a). In this circuit two diodes are used so that the output v'_o of the op-amp does not saturate. This minimizes the response time and increases the operating frequency range of the op-amp. However, notice that the op-amp is used in the inverting configuration, and the output is measured at the anode of diode D_1 with respect to ground. Besides that, the output resistance is nonuniform since it depends on the *state* of diode D_1. In other words, the output impedance is low when D_1 is on and high ($\cong R_F$) when D_1 is off. This problem, however, can be cured by connecting a voltage follower

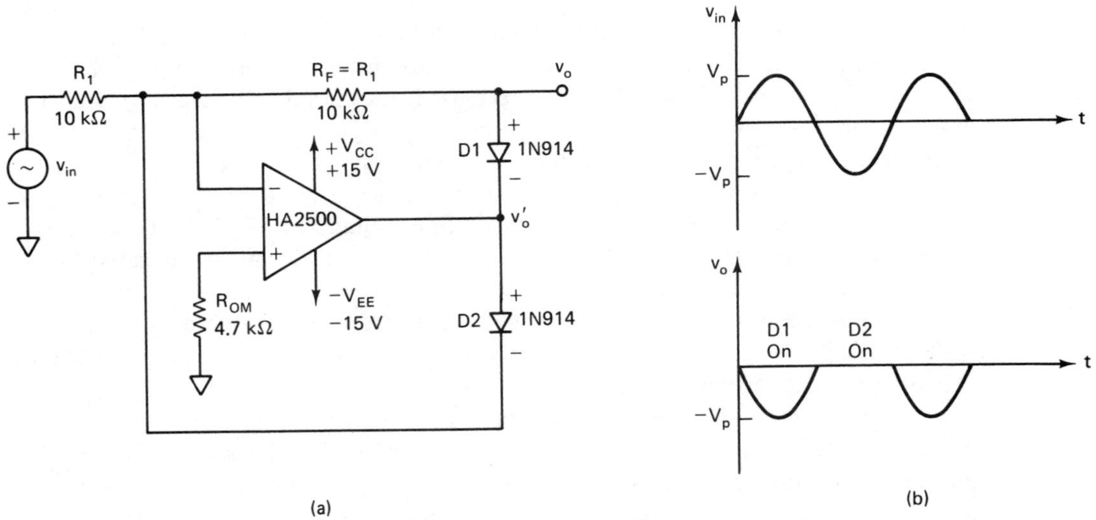

Figure 9-27 (a) Negative half-wave rectifier. (b) Its input and output waveforms.

stage at the output. During the positive half-cycle of v_{in}, output v_o' is negative, which forward biases diode D_1 and closes the feedback loop through R_F. Since $R_1 = R_F$, $v_o = v_{in}$. However, on the negative alternation of v_{in}, output v_o' is positive; hence diode D_2 is forward biased. In fact, it is this diode which prevents the op-amp from going into positive saturation. Since diode D_1 is *off*, output $v_o = 0$ V. To obtain positive half-wave rectified outputs, diodes D_1 and D_2 must be reversed.

Example 9-5

In the circuit of Figure 9-25(a), $v_{in} = 200$ mV peak-to-peak sine wave at 100 Hz. Briefly describe the operation of the circuit and draw the output waveform.

Solution. During the positive half-cycle of the input, diode D_1 conducts; therefore, output follows the input; that is, $v_o = 100$ mV peak. During the negative half-cycle of v_{in}, diode D_1 is reverse biased; hence $v_o = 0$ V. The output waveform is shown in Figure 9-25(b).

9-12.3 Positive and Negative Clampers

In clamper circuits a predetermined dc level is added to the output voltage. In other words, the output is clamped to a desired dc level. If the clamped dc level is positive, the clamper is a *positive clamper*. On the other hand, if the clamped dc level is negative, the clamper is called a *negative clamper*. Other, equivalent terms used for clamper are *dc inserter* or *restorer*.

Recall that op-amps designed to operate from a single power supply, such as the 124 series, employ a dc insertion technique. Inverting and noninverting amplifiers that use this technique are repeated here for convenience [see Figure 9-28(a) and (b)]. The value of the capacitors in these circuits depends on different input rates and pulse widths. Notice that in both amplifier circuits the dc level added to the output voltage is approximately equal to $V_{CC}/2$. This fixed positive dc level is needed to obtain a maximum undistorted symmetrical sine wave. A clamper with a variable positive dc level is shown in Figure 9-29(a). However, this circuit functions differently from that of Figure 9-28(a) in that the input waveform peak is clamped at V_{ref} [see Figure 9-29(b)]. For this reason the circuit is called the *peak clamper*.

The output voltage of the clamper is a net result of ac and dc input voltages applied to (−) and (+) input terminals, respectively. Therefore, in order to understand the circuit operation, each input must be considered separately. First consider the input voltage V_{ref} at the (+) input. Since this voltage is positive, v_o' is also positive, which forward biases diode D_1. This closes the feedback loop and the op-amp operates as a voltage follower. This is possible because C_1 is an open circuit for dc voltage. Therefore, $v_o = +V_{ref}$. As far as voltage v_{in} at the (−) input is concerned, during its negative half-cycle, diode D_1 conducts, charging C_t to the negative peak value of V_P. However, during the positive half-cycle of v_{in}, diode D_1 is reverse biased; hence, the voltage V_P across the capacitor acquired during the negative half-cycle is retained. Since this voltage V_P is in series with the positive peak voltage V_P, the output voltage $v_o = 2V_P$. Thus the net output is V_{ref} plus $2V_P$, so that the negative peak of $2V_P$ is at V_{ref} [see Figure 9-29(b)]. For precision clamping,

$$C_t R_d \ll \frac{T}{2}$$

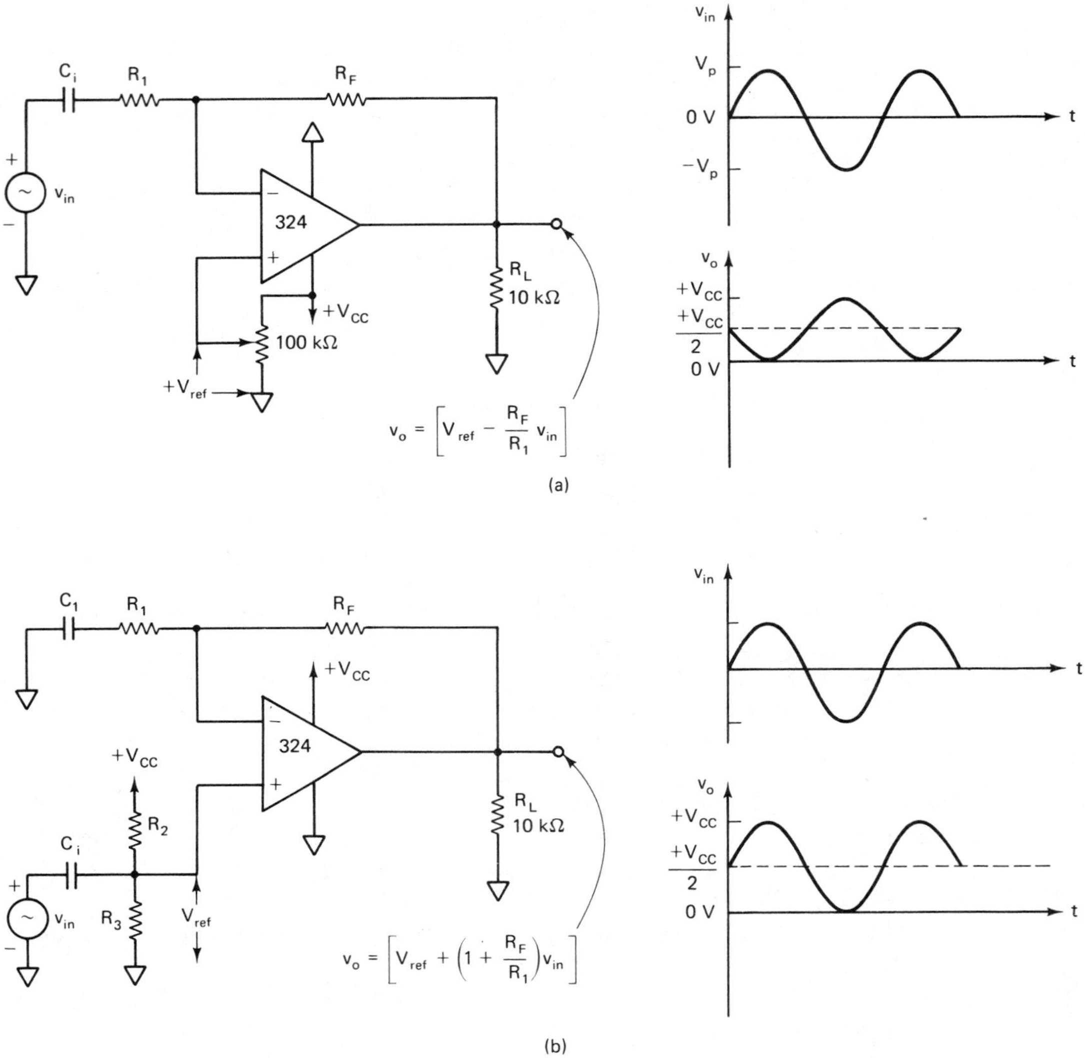

Figure 9-28 (a) Inverting amplifier with dc level. (b) Noninverting amplifier with dc level.

where R_d = resistance of diode D_1 when it is forward biased
= 100 Ω typically
T = time period of the input waveform

Resistor R is used to protect the op-amp against excessive discharge currents from capacitor C_i, especially when the dc supply voltages are switched off. A positive-peak clamping is accomplished by reversing D_1 and using negative reference voltage ($-V_{ref}$) [see waveform in Figure 9-29(c)].

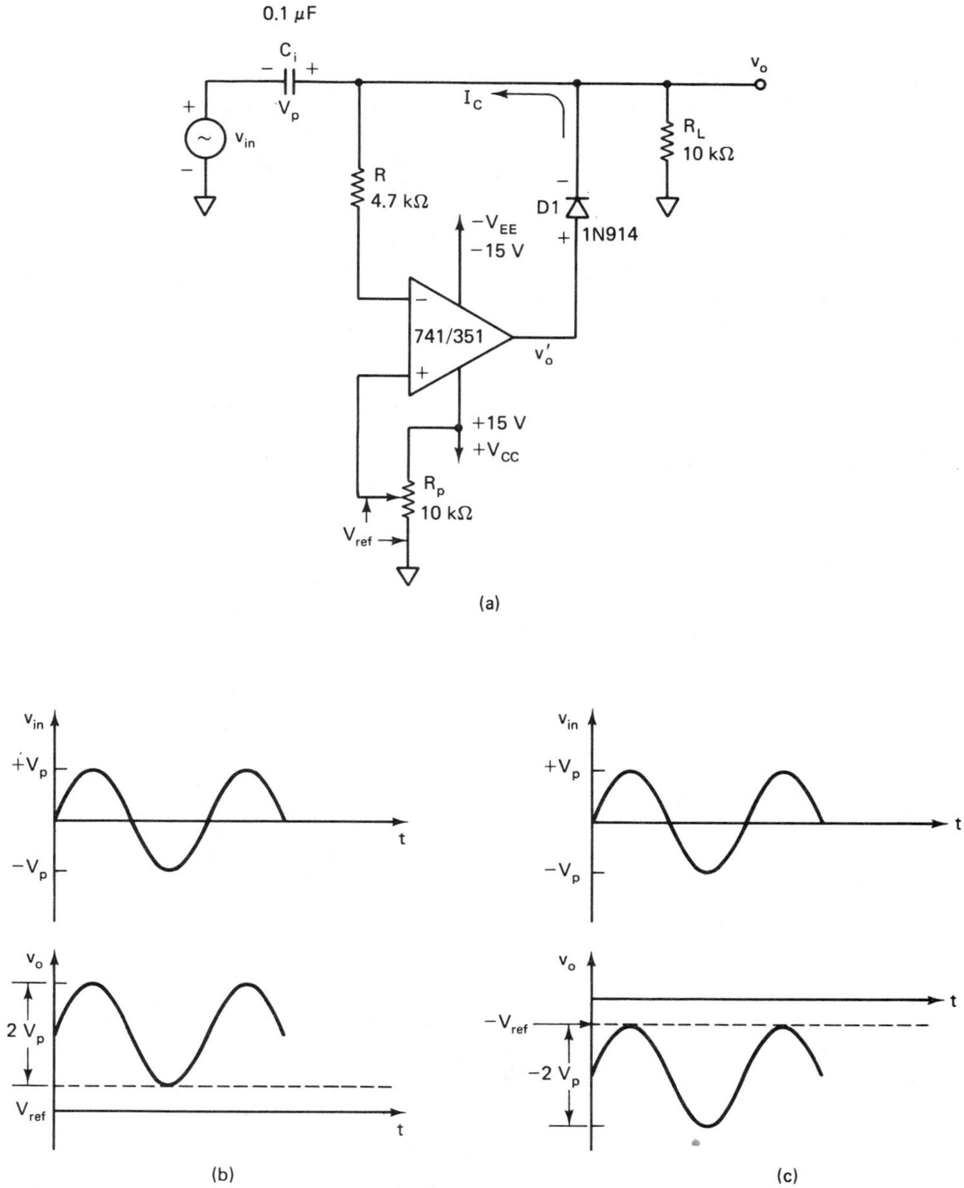

Figure 9-29 (a) Peak clamper circuit. (b) Input and output waveforms with $+V_{ref}$. (c) Input and output waveforms with $-V_{ref}$.

9-13 ABSOLUTE-VALUE OUTPUT CIRCUIT

Another important circuit used in waveshaping is the *absolute-value output circuit*. This circuit produces an output signal which swings positively only, regardless of the polarity of the input signal. Figure 9-30 shows an absolute-value output circuit with

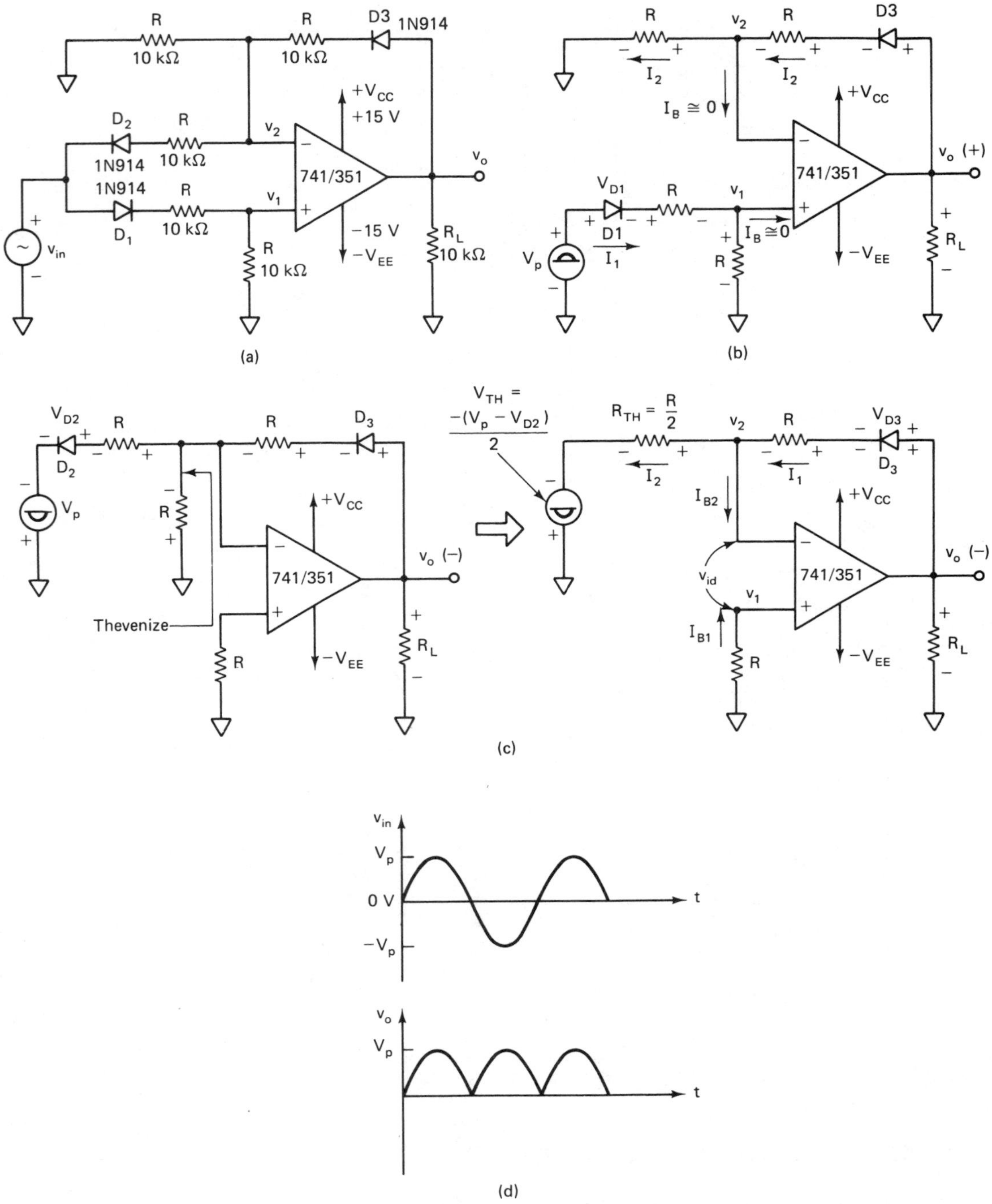

Figure 9-30 (a) Absolute-value output circuit, equivalent circuit during (b) positive half-cycle and (c) negative half-cycle. (d) Its input and output waveforms.

its input and output waveforms. Because of the nature of its output waveform, the circuit may be used as a full-wave rectifier if the input is a sine wave. During the positive half-cycle of v_{in}, diode D_1 is forward biased and D_2 is reverse biased; therefore, the equivalent circuit is as shown in Figure 9-30(b). In this circuit the voltage at the (+) input is

$$v_1 = \frac{V_P - V_{D1}}{2} \tag{9-9a}$$

where V_{D1}, the voltage drop across $D_1 = 0.7$ V. Similarly, voltage v_2 at the (−) input is

$$v_2 = \frac{v_o(+) - V_{D3}}{2} \tag{9-9b}$$

where $v_o(+)$ = output voltage during the positive half-cycle
V_{D3} = voltage drop across $D_3 = 0.7$ V
Since $v_{id} \cong 0$ V, $v_1 = v_2$. Therefore, from Equations (9-9a) and (9-9b),

$$\frac{V_P - V_{D1}}{2} = \frac{v_o(+) - V_{D3}}{2}$$

or

$$v_o(+) = V_P \tag{9-10}$$

On the other hand, during the negative half-cycle of v_{in}, diode D_1 is reverse biased and D_2 is forward biased; hence the equivalent circuit is as shown in Figure 9-30(c). This circuit can further be simplified by Thévenizing to the left of (−) input. Thévenin's voltage and resistance are

$$V_{\text{TH}} = -\left(\frac{V_P - V_{D2}}{2}\right) \tag{9-11a}$$

$$R_{\text{TH}} \cong \frac{R}{2} \tag{9-11b}$$

[See Figure 9-30(c)]. Now output voltage $v_o(-)$ can be obtained by writing Kirchhoff's current equation at node v_2:

$$I_1 = I_2 + I_{B2}$$

$$I_1 \cong I_2 \qquad \text{since } I_{B2} \cong 0 \text{ A}$$

$$\frac{(v_o(-) - V_{D3}) - v_2}{R} = \frac{v_2 - (V_{\text{TH}})}{R/2} \tag{9-11c}$$

However, $v_{id} = 0$ V and $v_1 = 0$. Therefore, $v_2 = 0$ and

$$\frac{v_o(-) - V_{D3}}{R} = \frac{-V_{\text{TH}}}{R/2}$$

Substituting for V_{TH} from Equation (9-11a), we get

$$\begin{aligned} v_o(-) - V_{D3} &= V_P - V_{D2} \\ v_o(-) &= V_P \end{aligned} \tag{9-12}$$

Thus, regardless of the polarity of the input signal, output is always positive going, hence the name *absolute-value output circuit*. Note that the gain of the circuit is 1;

therefore, the positive peak amplitudes are the same as input peak amplitudes. Besides that, diode D_3 compensates for the voltage drop across D_1 or D_2.

9-14 PEAK DETECTOR

Square, triangular, sawtooth, and pulse waves are typical examples of nonsinusoidal waveforms. A conventional *ac voltmeter* cannot be used to measure these nonsinusoidal waveforms because it is designed to measure the rms value of the pure sine wave. One possible solution to this problem is to measure the peak values of the nonsinusoidal waveforms.

Figure 9-31 shows a *peak detector* that measures the positive peak values of the square wave input. During the positive half cycle of v_{in}, the output of the op-amp drives D_1 on, charging capacitor C to the positive peak value V_P of the input voltage v_{in}. Thus when D_1 is forward biased, the op-amp operates as a voltage follower. On the other hand, during the negative half-cycle of v_{in}, diode D_1 is reverse biased, and voltage across C is retained. The only discharge path for C is through R_L since the input bias current I_B is negligible. For proper operation of the circuit, charging time constant (CR_d) and discharging time constant (CR_L) must satisfy the following conditions:

$$CR_d \leq \frac{T}{10} \tag{9-13a}$$

where R_d = resistance of the forward-biased diode, 100 Ω, typically
T = time period of the input waveform

and

$$CR_L \geq 10T \tag{9-13b}$$

where R_L is the load resistor.

If R_L is very small so that Equation (9-13b) cannot be satisfied, use a buffer (voltage follower) between capacitor C and load resistor R_L. Although a 741-type op-amp is used in the circuit of Figure 9-31, a high-speed, precision-type op-amp

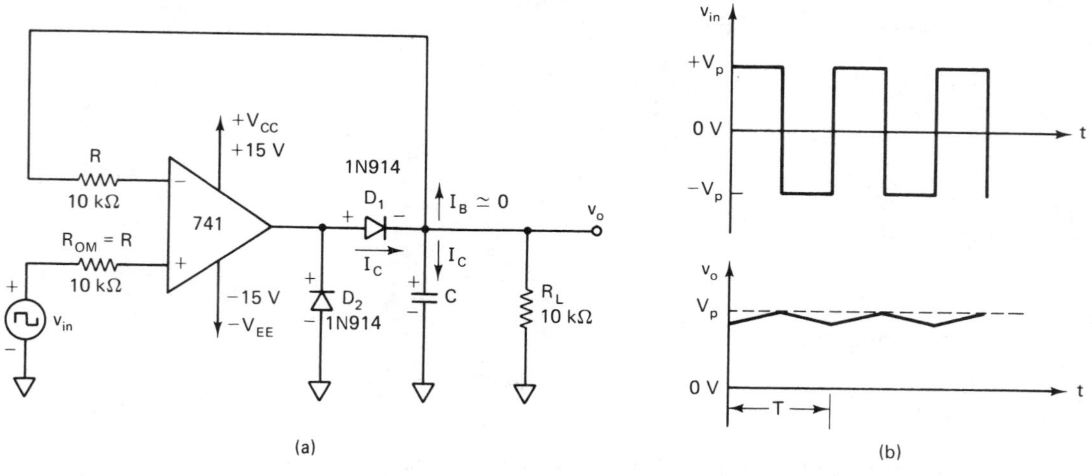

Figure 9-31 (a) Peak detector circuit. (b) Its input and output waveforms.

such as the μA771 or μA714 may be desirable in critical applications. The resistor R is used to protect the op-amp against excessive discharge currents, especially when the power supply is switched off. The resistor $R_{OM} = R$ minimizes the offset problems caused by input currents. In addition, diode D_2 conducts during the negative half-cycle of v_{in} and hence prevents the op-amp from going into negative saturation. This in turn helps to reduce the recovery time of the op-amp. Negative peaks of input signal v_{in} can be detected simply by reversing diodes D_1 and D_2.

9-15 SAMPLE-AND-HOLD CIRCUIT

The sample-and-hold circuit, as its name implies, samples an input signal and holds on to its last sampled value until the input is sampled again. Figure 9-32 shows a sample-and-hold circuit using an op-amp with an E-MOSFET. In this circuit the E-MOSFET works as a switch that is controlled by the sample-and-hold control voltage V_S, and the capacitor C serves as a storage element. The circuit operates as follows. The analog signal v_{in} to be sampled is applied to the drain and sample-and-hold control voltage V_S is applied to the gate of the E-MOSFET. During the positive portion of V_S, the E-MOSFET conducts and acts as a closed switch. This allows input voltage to charge capacitor C. In other words, input voltage appears across C and in turn at the output, as shown in Figure 9-32(b). On the other hand, when V_S is zero, the E-MOSFET is *off* (nonconductive) and acts as an open switch.

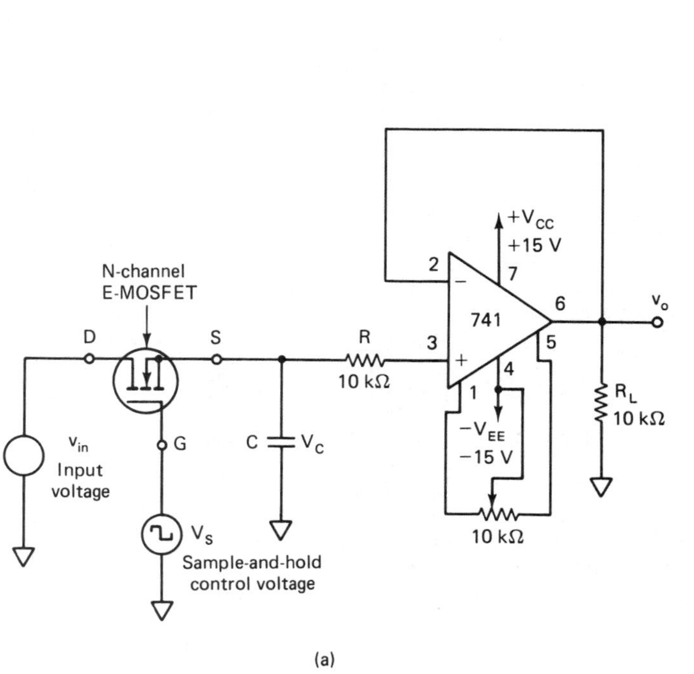

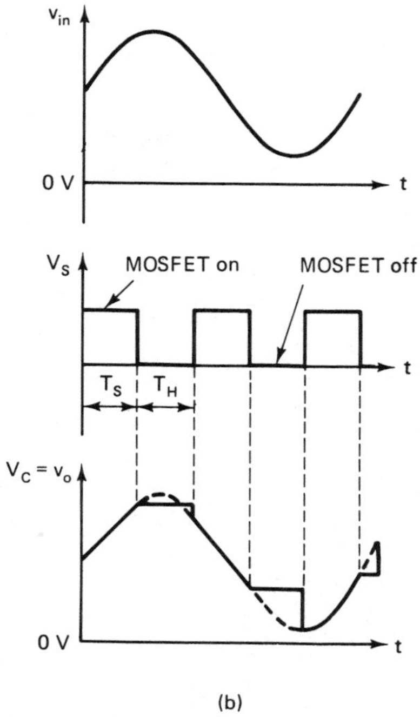

Figure 9-32 (a) Sample-and-hold circuit. (b) Its input and output waveforms.

The only discharge path for C is, therefore, through the op-amp. However, the input resistance of the op-amp voltage follower is also very high; hence the voltage across C is retained. The time periods T_S of the sample-and-hold control voltage V_S during which the voltage across the capacitor is equal to the input voltage are called *sample periods*. The time periods T_H of V_S during which the voltage across the capacitor is constant are called *hold periods* [see Figure 9-32(b)]. The output of the op-amp is usually processed/observed during hold periods. To obtain close approximation of the input waveform, the frequency of the sample-and-hold control voltage must be significantly higher than that of the input. In critical applications a precision and/or high-speed op-amp is helpful. If possible, choose a low-leakage capacitor such as Teflon or polyethylene.

A significant reduction in size and an improved performance can be achieved by using a specially designed sample-and-hold IC such as the LF398. The functional circuit of the LF398 shown in Figure 9-33 requires only an external storage capacitor.

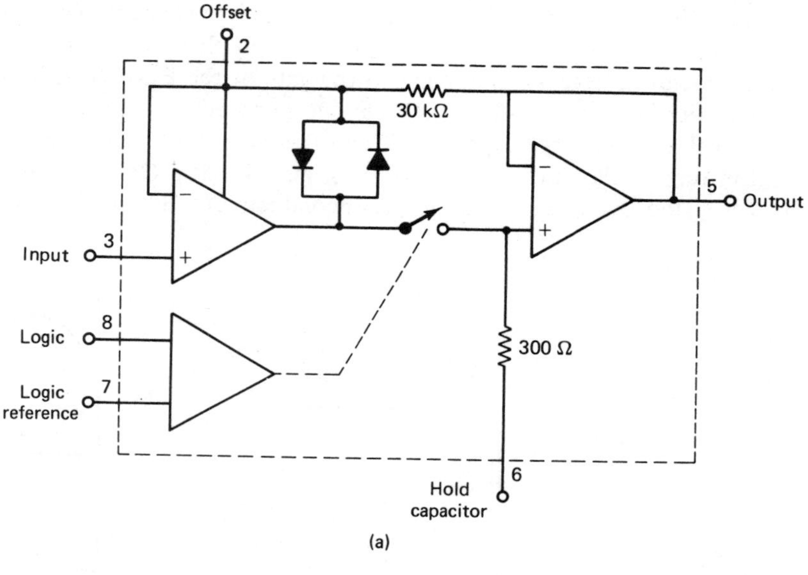

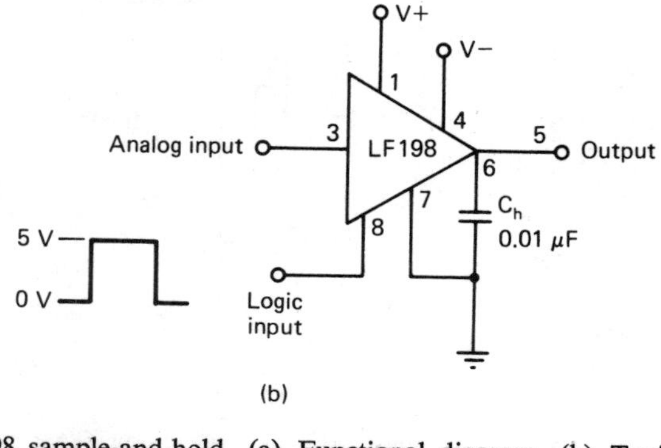

Figure 9-33 LF398 sample-and-hold. (a) Functional diagram. (b) Typical connection diagram. (Courtesy of National Semiconductor.)

The sample-and-hold circuit is commonly used in digital interfacing and communications such as analog-to-digital and pulse modulation systems.

SUMMARY

1. Together with such functions as amplifier, filter, and oscillator, the op-amp can also be used as a comparator, limiter, detector, clipper, clamper, and converter.
2. In its simplest form, a comparator is nothing more than an open-loop op-amp with two analog inputs and a digital output, depending on which input is the larger. It is used in digital interfacing, Schmitt triggers, analog-to-digital converters, oscillators, and others.
3. An immediate application of the comparator is the zero-crossing detector, in which the reference voltage $V_{ref} = 0$ V.
4. The Schmitt trigger is a comparator with positive feedback that converts an irregular waveform to a square or pulse waveform. In the Schmitt trigger the input voltage triggers the output every time it exceeds certain voltage levels called upper threshold V_{ut} and lower threshold V_{lt}.
5. Switching speeds, accuracy, and compatibility of outputs are the limiting factors to the use of op-amp comparators in critical applications. The output of an op-amp can be limited to a predetermined value by using external components such as zeners and diodes. Such circuits are called voltage limiters.
6. The window detector uses two comparators and two threshold levels to determine when an unknown input is between these levels. When the input is between two predetermined limits, the output is high. However, when the input goes above or below the set limits, the output is low.
7. The Teledyne 9400 can be used as a V/F or F/V converter simply by using two external capacitors, three resistors, and a reference voltage. The V/F converter is used in such applications as temperature sensing and control, analog-to-digital converters, digital panel meters, and phase-locked loops. The F/V converter is used for applications such as frequency meters, speedometers, frequency multipliers/dividers, and motor controls.
8. The analog-to-digital (A/D) converter changes an analog input into a digital output while a digital-to-analog (D/A) converter changes a digital input into an analog output. The D/A converter can be formed by using an op-amp and either binary-weighted resistors or an R and $2R$ ladder network. Integrated D/A converters are available as 8, 10, 12, 14, and 16 bits with voltage or current outputs. Microcomputer interfacing, CRT graphics generation, programmable power supplies, and digital filters are typical applications of D/A converters. On the other hand, the A/D converter can be a single-ramp or double-ramp integrating type, single-counter type, tracking type, or successive approximation type. As with D/A converters, there are many monolithic/hybrid A/D converters. Typical applications of A/D converters include microprocessor interfacing, digital voltmeters, and LED/LCD displays.
9. A clipper is a circuit that removes certain parts of the input waveform. The clipper can be formed by using an op-amp with a rectifier diode. On the other hand, in an op-amp clamper a predetermined dc level is deliberately inserted in the output voltage. A half-wave rectifier is a form of clipper. The outputs of the amplifiers that operate on a single power supply are clamped to a desired value to avoid output distortion. Yet another waveshaping circuit is the absolute value output circuit, in which the output swings positively only. Such a circuit is formed by using an op-amp with rectifier diodes.

10. The peak detector is generally used to detect the positive peak values of nonsinusoidal waveforms such as triangular, sawtooth, or pulse.
11. The sample-and-hold circuit samples an input signal and holds on to its last sampled value until the input is sampled again. It can be constructed by simply using an op-amp with a E-MOSFET as a switch. However, a monolithic sample-and-hold such as the LF398 requires only an external capacitor. The sample-and-hold circuit is used in digital interfacing and communications.

QUESTIONS

9-1. What is a comparator?

9-2. What is the difference between a basic comparator and the Schmitt trigger?

9-3. List the important characteristics of the comparator.

9-4. What is voltage limiting, and why is it needed?

9-5. What is a window detector?

9-6. Name and then briefly describe one application of V/F and F/V converters.

9-7. Briefly describe the operation of the F/V ripple eliminator shown in Figure 9-16.

9-8. For the 9400 F/V converter, what is the limitation on input pulse widths?

9-9. What is the difference between A/D and D/A converters? Give one application of each.

9-10. What is the major difference between 1408 and 5018 D/A converters?

9-11. Define the following terms: resolution, settling time, conversion time.

9-12. What is the difference between clippers and clampers? Give one application of each.

9-13. What is an absolute-value output circuit? How can it be used as a full-wave rectifier?

9-14. What is the name of the circuit that is used to detect the peak value of the nonsinusoidal input waveforms? Briefly explain its operation.

9-15. What is a sample-and-hold circuit? Why is it needed?

PROBLEMS

9-1. For the basic comparator of Figure 9-1(a), $v_{in} = 2$ V pp sine wave at 1 kHz, $V_{ref} = 500$ mV, $R = 100\ \Omega$ and supply voltages $= \pm 15$ V. Draw the output waveform.

9-2. In the circuit of Figure 9-4(a), $R_1 = 150\ \Omega$, $R_2 = 68\ \text{k}\Omega$, $v_{in} = 500$ mV pp sine wave, and the saturation voltages $= \pm 14$ V.

(a) Determine the threshold voltages V_{ut} and V_{lt}.

(b) What is the value of hysteresis voltage V_{hy}?

9-3. In the circuit of Figure 9-5, $v_{in} = 100$ mV peak sine wave at 100 Hz, $R = 1\ \text{k}\Omega$, and D_1 and D_2 are 6.2-V zeners. The op-amp is a 741 with supply voltages $= \pm 12$ V. Draw the output voltage waveform.

9-4. Repeat Probem 9-3 for the circuit shown in Figure 9-6(a). Assume that the voltage drop across the forward-biased diode is 0.7 V.

9-5. In the 9400 V/F converter shown in Figure 9-10(a), $V_{in} = 5$ V, $R_{in} = 1\ \text{M}\Omega$, $V_{ref} = -5$ V, and $C_{ref} = 180$ pF.

(a) Calculate the value of output frequency F_o.
(b) Draw the F_o and $F_o/2$ waveforms.
(c) In order to obtain bipolar outputs, what modification is required in the output circuitry?

9-6. Calculate the output voltage V_o for the 9400 F/V converter of Figure 9-14(a) if F_{in} = 2.5 kHz. What modifications are required in the external components if F_{in} is changed to 100 kHz?

9-7. Referring to the D/A converter with binary weighted resistors shown in Figure 9-18(a), determine the size of each step if $R_F = 1.2\ k\Omega$. What is the output voltage when inputs b0 through b3 are at 5 V?

9-8. For the D/A converter using an R–$2R$ ladder network shown in Figure 9-19(a):
(a) Determine the size of each step if $R_F = 27\ k\Omega$.
(b) Calculate the output voltage when the inputs b0, b1, b2, and b3 are at 5 V.
(c) What is the advantage of this type of D/A converter over the one with binary-weighted resistors?

9-9. For the small-signal half-wave rectifier of Figure 9-25(a), draw the output waveform if v_{in} is a 300-mV peak sine wave at 1 kHz.

9-10. Repeat Problem 9-9 with D_1 reversed.

9-11. For the peak clamper shown in Figure 9-29(a), draw the output voltage waveform if v_{in} = 500 mV pp sine wave at 100 Hz and V_{ref} = 50 mV.

9-12. In the peak detector of Figure 9-31, $C = 0.01\ \mu F$, $R_L = 1\ M\Omega$, and v_{in} = 2 V pp square wave at 1 kHz. Draw the approximate output voltage waveform. (Assume that the resistance of the forward-biased diode $R_d = 100\ \Omega$.)

EXPERIMENT 9-1
COMPARATOR AND SCHMITT TRIGGER

Objectives

In this experiment you will see how an op-amp can be used as a comparator and a Schmitt trigger. At the end of this experiment you should be able to:

1. Verify that an open-loop op-amp is a basic comparator.
2. Verify the operation of a zero-crossing detector.
3. Draw the output waveform of a comparator for a given value of reference voltage.
4. Explain the difference between the comparator and the Schmitt trigger.
5. Select the appropriate components for desired values of thresholds.

Equipment

1. Dual-trace oscilloscope
2. Audio-signal generaltor
3. Digital voltmeter
4. ±15-V power supply

Materials

1. 741 or 351 op-amp
2. Two 1N914 diodes
3. Two 1N3826 (5.1-V) zeners
4. Two 1-kΩ resistors
5. 10-kΩ resistor
6. 100-kΩ resistor
7. 180-kΩ resistor
8. 10-kΩ potentiometer

Procedure

1. Connect the circuit of Figure E9-1.1(a).

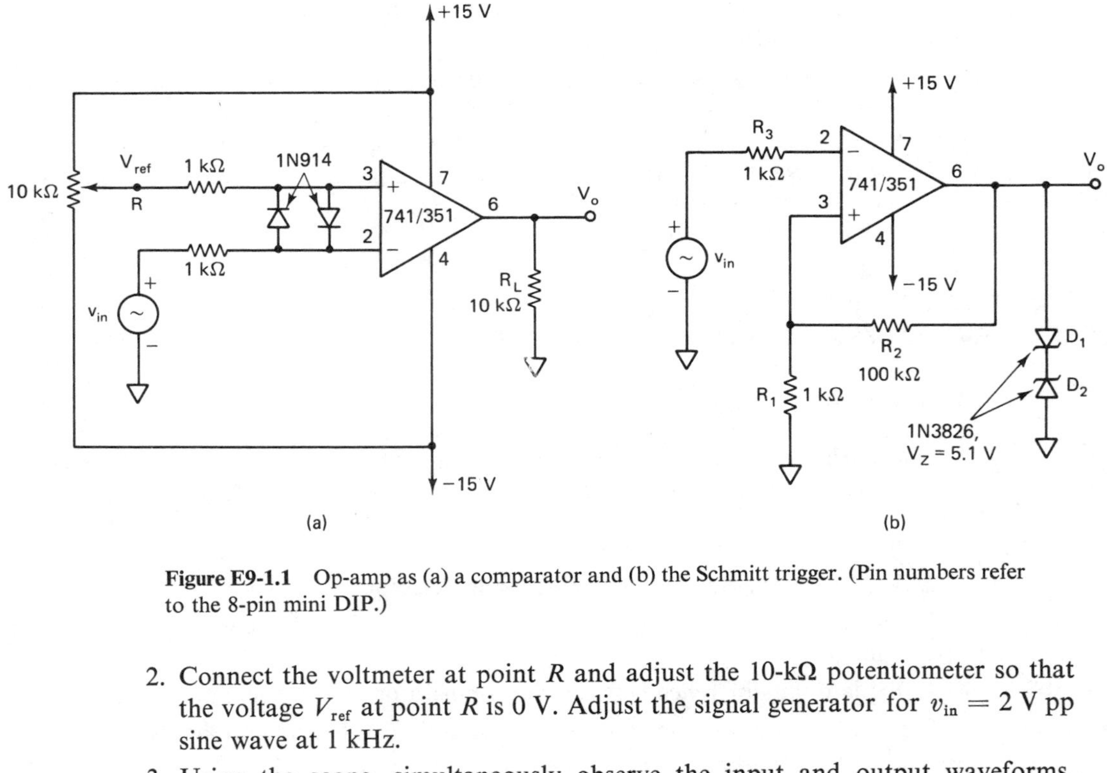

Figure E9-1.1 Op-amp as (a) a comparator and (b) the Schmitt trigger. (Pin numbers refer to the 8-pin mini DIP.)

2. Connect the voltmeter at point *R* and adjust the 10-kΩ potentiometer so that the voltage V_{ref} at point *R* is 0 V. Adjust the signal generator for $v_{in} = 2$ V pp sine wave at 1 kHz.
3. Using the scope, simultaneously observe the input and output waveforms. Draw the output waveform in Figure E9-1.2(a).
4. Reconnect the voltmeter to point *R* and adjust the 10-kΩ potentiometer so that voltage $V_{ref} = +0.5$ V. Repeat step 3 [Figure E9-1.2(b)].
5. Connect the voltmeter to point *R* again and adjust the 10-kΩ potentiometer so that $V_{ref} = -0.5$ V. Repeat step 3 [Figure E9-1.2(c)].

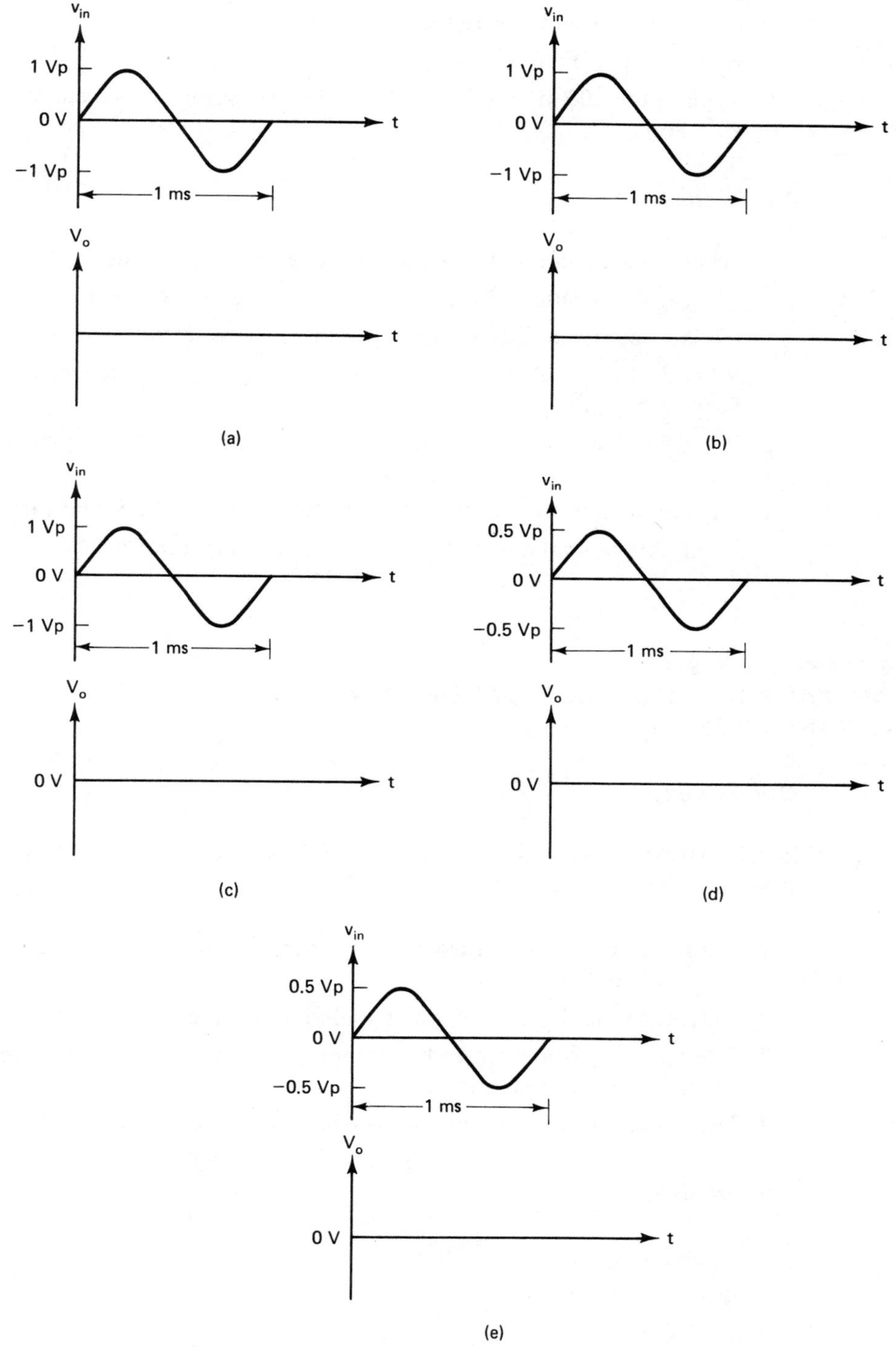

Figure E9-1.2 Waveforms for (a) step 3, (b) step 4, (c) step 5, (d) step 7, (e) step 8.

6. Connect the circuit of Figure E9-1.1(b). Adjust the signal generator so that $v_{in} = 1$ V pp sine wave at 1 kHz.
7. Repeat step 3 [Figure E9-1.2(d)].
8. Replace the 100-kΩ resistor with a 180-kΩ resistor and repeat step 3 [Figure E9-1.2(e)].

Questions

1. What is the function of the rectifier diodes in Figure E9-1.1(a)?
2. What is the function of the zener diodes in Figure E9-1.1(b)?
3. Explain the significance of the V_{ref} value in comparator operation.
4. What is the significance of the R_1 and R_2 values in the operation of the Schmitt trigger?
5. What parameter is most important in the operation of an op-amp as a comparator?
6. Explain the differences between the comparator and the Schmitt trigger.
7. What are the limitations of an op-amp as a comparator?

EXPERIMENT 9-2 DIGITAL-TO-ANALOG CONVERTER USING THE R–2R LADDER NETWORK

Objectives

In this experiment you will learn how the D/A converter works. At the end of this experiment you should be able to:

1. Apply the concepts learned in this experiment to construct a D/A converter with more digital inputs.
2. Determine the size of each step (resolution) for a given D/A converter.
3. Select a D/A converter with an appropriate number of digital inputs for a desired step size (resolution).
4. Discuss the advantages and disadvantages of an op-amp as a D/A converter.

Equipment

1. Digital multimeter
2. +5-V power supply
3. ±15-V power supply

Materials

1. 741 or 351 op-amp
2. 100-Ω resister
3. Four 10-kΩ resistors (±1% or better tolerance)
4. Six 20-kΩ resistors (±1% or better tolerance)
5. 100-kΩ resistor
6. 20-kΩ potentiometer
7. Miniature four-slide or rocker switch assembly
8. Linear graph paper

Procedure

1. Connect the circuit as shown in Figure E9-2.1. (The op-amp pin numbers indicated in this figure refer to the 8-pin mini DIP).

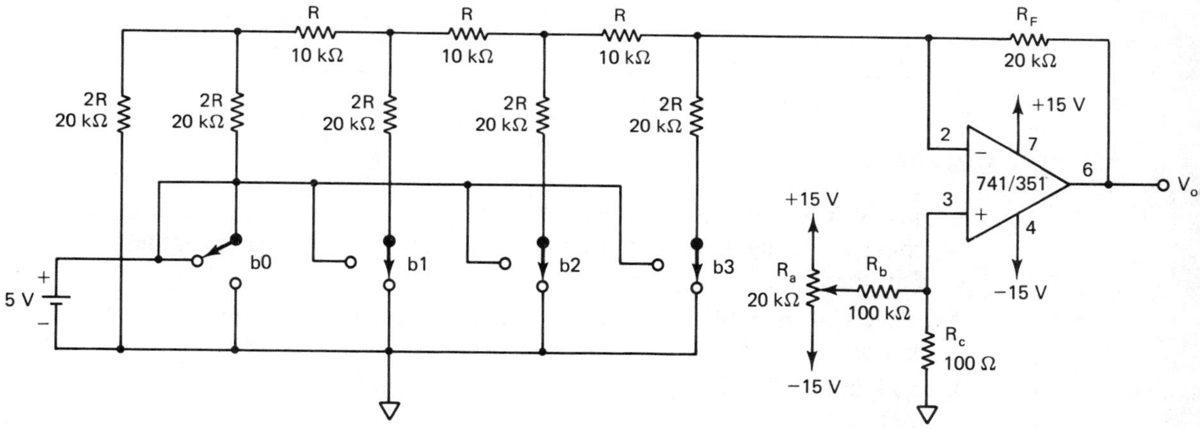

Figure E9-2.1 Op-amp used as a 4-bit D/A converter.

2. Connect the voltmeter to the output terminal. With all switches (inputs b0 to b3) shorted to ground, adjust the 20-kΩ potentiometer until the output is reduced to 0 V. If you have a problem in reducing output to 0 V, you may try a different value for the 100-Ω resistor.
3. Connect the switch corresponding to input b0 to +5 V (logic 1) and measure the output voltage. Enter the measured value in Table E9-2.1.
4. Connect the switch corresponding to input b1 to +5 V, and short all other remaining switches to ground. Measure the output voltage and enter the result in Table E9-2.1.
5. Complete Table E9-2.1 for the indicated input conditions.
6. Using the data in Table E9-2.1, plot the graph of decimal equivalents of binary inputs versus output voltage.

TABLE E9-2.1

Decimal equivalents of binary inputs	Inputs[a]				Output voltage, V_o
	b3	b2	b1	b0	
0	0 V	0 V	0 V	0 V	0 V
1	0 V	0 V	0 V	5 V	
2	0 V	0 V	5 V	0 V	
3	0 V	0 V	5 V	5 V	
4	0 V	5 V	0 V	0 V	
5	0 V	5 V	0 V	5 V	
6	0 V	5 V	5 V	0 V	
7	0 V	5 V	5 V	5 V	
8	5 V	0 V	0 V	0 V	
9	5 V	0 V	0 V	5 V	
10	5 V	0 V	5 V	0 V	
11	5 V	0 V	5 V	5 V	
12	5 V	5 V	0 V	0 V	
13	5 V	5 V	0 V	5 V	
14	5 V	5 V	5 V	0 V	
15	5 V	5 V	5 V	5 V	

[a]0 V = logic 0 and 5 V = logic 1.

Questions

1. In the D/A converter of Figure E9-2.1, if the number of inputs is increased from four to eight, what is the size of each step and what is the output voltage if all the inputs are logic 1 (+5 V)?
2. In the converter of Figure E9-2.1, what will be the size of each step if R_F is changed to 10-kΩ?
3. In Figure E9-2.1, if $R_F = 10$ kΩ, what will the output voltage be when all inputs are logic 1?
4. Determine the number of inputs required for a D/A converter if the size of each step is equal to 9.766 mV. Assume that $R = 10$ kΩ, $2R = 20$ kΩ, and $R_F = 20$ kΩ.
5. Is the graph of decimal equivalents of binary inputs versus output voltage linear? If not, give the reasons.
6. What are the advantages and disadvantages of an op-amp D/A converter?

10

Specialized IC Applications

10-1 INTRODUCTION

Integrated circuits with improved capabilities are appearing in ever-increasing numbers. Innovative design methods and fabrication procedures have not only helped produce a large variety of new integrated circuits but have also improved the old ones. Often the use of specialized ICs produces a simpler and more accurate circuit, such as Datel's universal filter FLT-U2, Signetics' phase-locked loop (PLL) SE/NE 565 and timer SE/NE 555, and others. This chapter presents a sampling of specialized integrated circuits and their applications to such devices as universal filters, timers, phase-locked loops, power amplifiers, voltage regulators, and voltage references.

10-2 UNIVERSAL ACTIVE FILTER

Various filter networks using general-purpose op-amps are discussed in Chapter 8. However, in critical applications specially designed filter ICs are preferred. Besides being more accurate, the specially designed IC filters are simpler, easier to use, and more flexible. Datel's FLT-U2 is a typical example of such specialized IC filters. Datel's FLT-U2 is a universal filter which uses the state-variable active filter principle to implement second-order low-pass, high-pass, and band-pass output functions. These output functions are simultaneously available at the output of three committed op-amps (pins 3, 13, and 5) as shown in the FLT-U2 block diagram of Figure 10-1(a). A fourth uncommitted op-amp can be used as a gain stage, buffer amplifier, or to raise the order of the low-pass, high-pass, or band-pass functions. The uncommitted

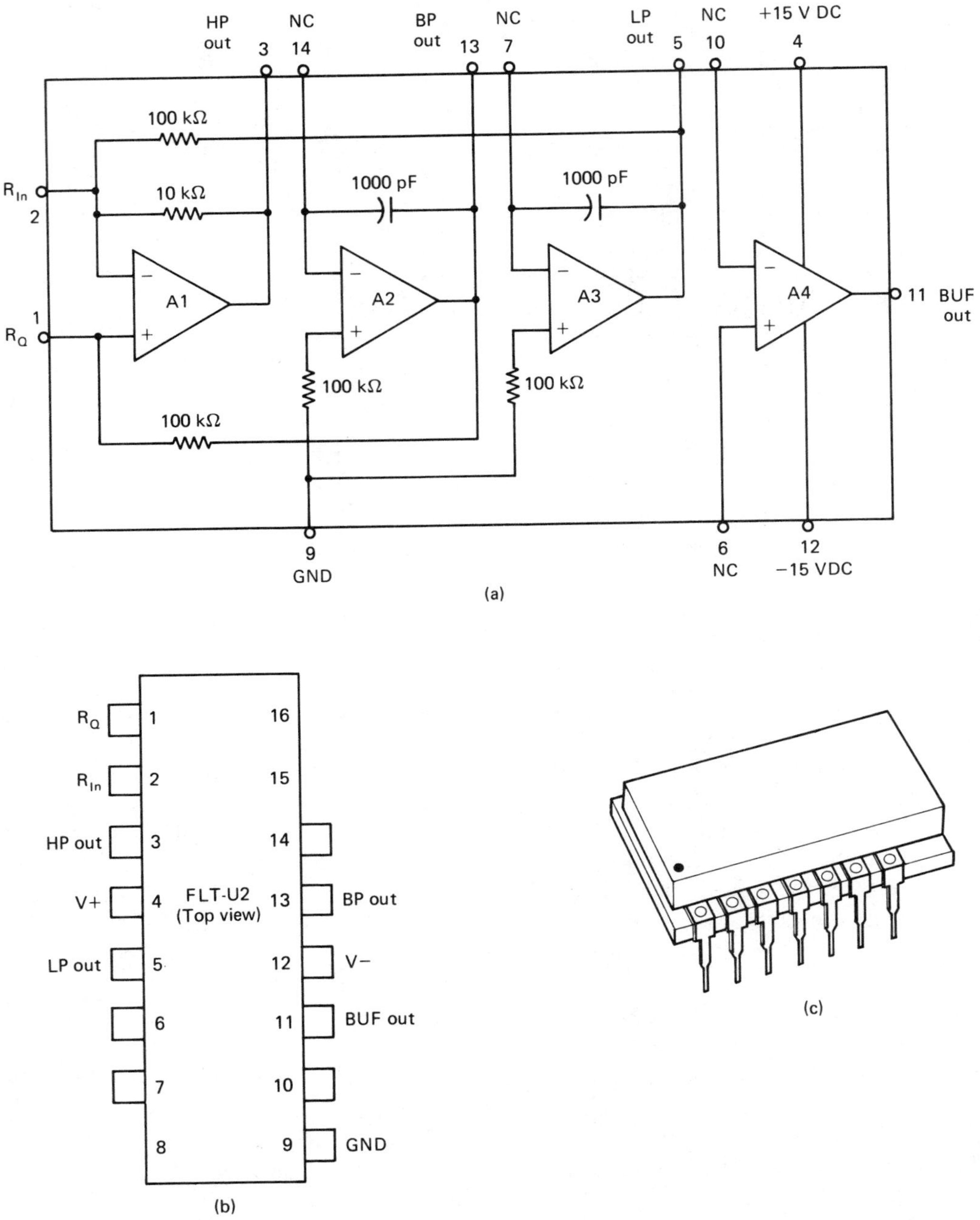

Figure 10-1 (a) FLT-U2 universal filter block diagram. (b) Connection diagram. (c) Ceramic 16-pin DIP. (Courtesy of Datel-Intersil.)

op-amp can also be used to realize the *notch* and *all-pass* functions. The important characteristics of the FLT-U2 are the following:

Frequency range	0.001 Hz to 200 kHz
Figure of merit (Q) range	0.1 to 1000
Frequency accuracy	$\pm 5\%$
Voltage gain	0.1 to 1000
Input impedance	5 MΩ
Operating voltage range	± 5 to ± 18 V
Unity gain bandwidth	3 MHz
Slew rate	1 V/μs

Frequency tuning is accomplished by using two external resistors and Q tuning by using a third resistor. In addition, by using the FLT-U2 any of the filter types, such as Butterworth, Chebyshev, or Bessel, may be designed by proper selection of external components.

10-2.1 Design Procedure

In the design procedure that follows, it is assumed that the gain of the desired filter function is unity. For simplicity, Q is assumed to be larger than 1. Figure 10-2 shows the relative gains of the three simultaneous filter outputs assuming that the band-pass gain is set to unity (0 dB). From this figure it is clear that the low-pass gain is always 10 dB higher than band-pass gain, and high-pass gain is always 10 dB lower than band-pass gain.

In following the design steps listed below, refer to Figure 10-3.

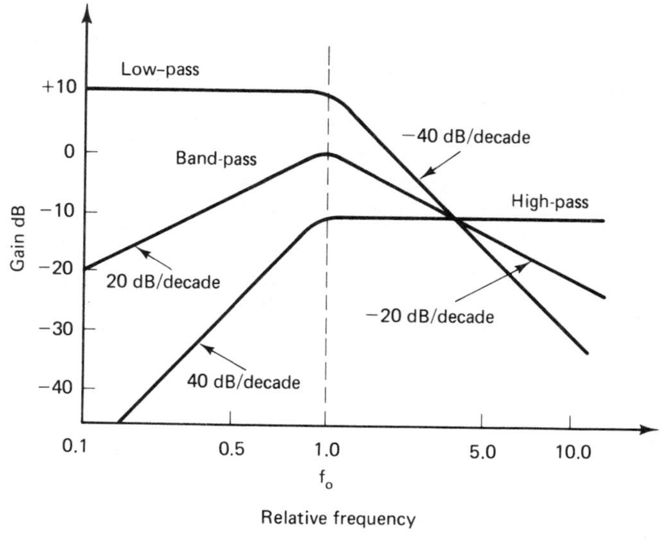

Figure 10-2 Relative gains of simultaneous filter outputs ($Q = 1$).

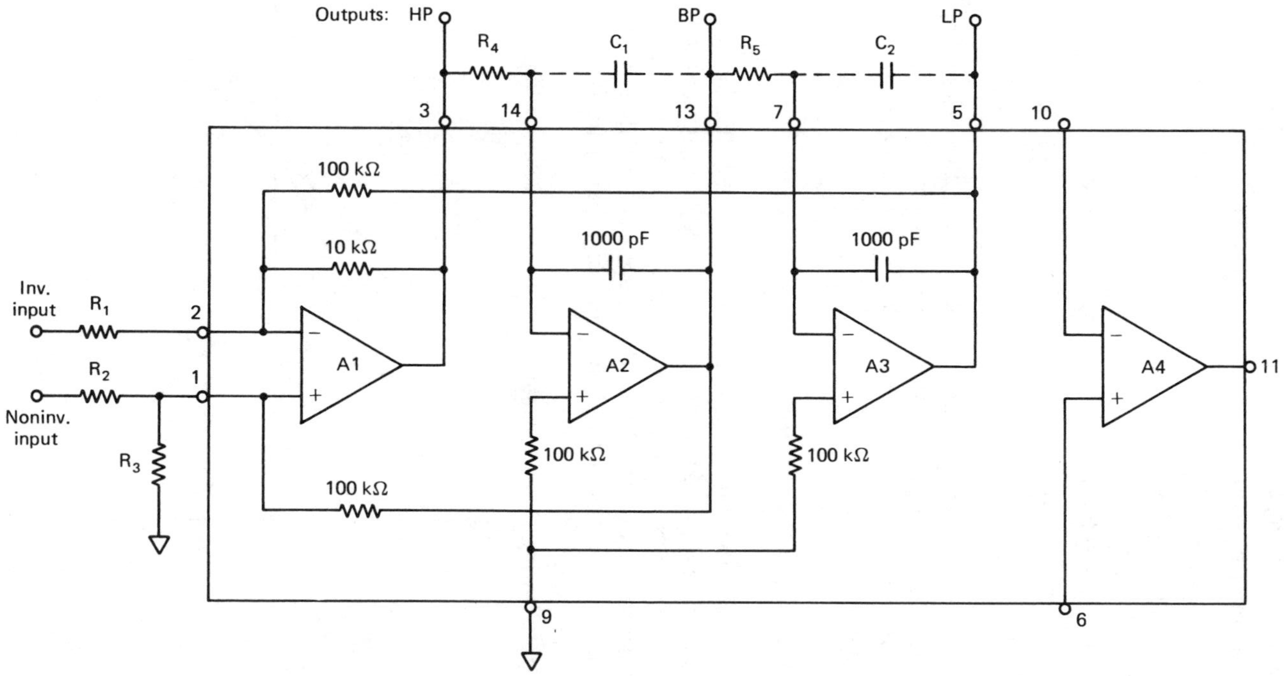

Figure 10-3 FLT-U2 as second-order low-pass, high-pass, and band-pass filters. (Courtesy of Datel-Intersil.)

1. For a desired function—low-pass, high-pass, or band-pass—use Table 10-1 to select an appropriate filter configuration for inverted or noninverted output.

TABLE 10-1 FILTER CONFIGURATIONS

Configuration	LP	HP	BP
Inverting input Noninverting input	Inverted Noninverted	Inverted Noninverted	Noninverted Inverted

Note that the low-pass, high-pass outputs are in phase.

For an inverting configuration, follow this step; otherwise, go to step 3. Using the desired value of Q, calculate R_1 and R_3 from Table 10-2. (R_2 is open.)

TABLE 10-2 INVERTING CONFIGURATIONS

	R_1	R_3
Low-pass	100 kΩ	$\frac{100\text{ k}\Omega}{(3.80)(Q)-1}$
High-pass	10 kΩ	$\frac{100\text{ k}\Omega}{(6.64)(Q)-1}$
Band-pass	$(Q)(31.6\text{ k}\Omega)$	$\frac{100\text{ k}\Omega}{(3.48)(Q)}$

3. Follow this step if a noninverting configuration is to be used; otherwise, go to step 4. Calculate R_2 and R_3 values from Table 10-3, using the desired value of Q. (R_1 is open.)

TABLE 10-3 NONINVERTING CONFIGURATIONS

	R_2	R_3
Low-pass	$\frac{316\text{ k}\Omega}{Q}$	$\frac{100\text{ k}\Omega}{(3.16)(Q)-1}$
High-pass	$\frac{31.6\text{ k}\Omega}{Q}$	$\frac{100\text{ k}\Omega}{(0.316)(Q)-1}$
Band-pass	100 kΩ	$\frac{100\text{ k}\Omega}{(3.48)(Q)-1}$

4. Using the desired value of resonant frequency f_1, which is the center frequency for band-pass and the cutoff frequency for low-pass or high-pass, calculate R_4 and R_5 values from the equation

$$R_4 = R_5 = \frac{(5.03)(10^7)}{f_1} \quad \text{ohms} \tag{10-1}$$

where f_1 is either the center frequency or cutoff frequency expressed in hertz.

For a band-pass filter the center frequency can be varied without affecting bandwidth by varying R_5, with R_4 fixed.

5. If $f_1 < 50$ Hz, the internal 1000-pF capacitors should be shunted with equal value external capacitors across pins 5 and 7 and 13 and 14 (see Figure 10-3). R_4 and R_5 values are then computed from the equation

$$R_4 = R_5 = \frac{(5.03)(10^{10})}{f_1 C} \quad \text{ohms} \tag{10-2}$$

where C is the total capacitance in pF, that is, the sum of the external capacitance across pins 5 and 7 or across pins 13 and 14 in pF, and the internal 1000-pF capacitor. The value f_1 is in hertz.

6. The procedure described above is based on unity gain output for the desired function. However, if additional gain is required, the fourth uncommitted op-amp should be used as an inverting or noninverting gain stage following the selected output (see Figure 10-4). Also, as shown in Figure 10-5, the order of the filter function can be raised by adding a capacitor to the gain stage.

The resonant frequency tuning is done by varying R_4 or R_5 resistors while the figure of merit Q tuning is done by adjusting input resistors R_1, R_2, or R_3, depending on the filter configuration desired (see Figure 10-3). When tuning the filter and checking

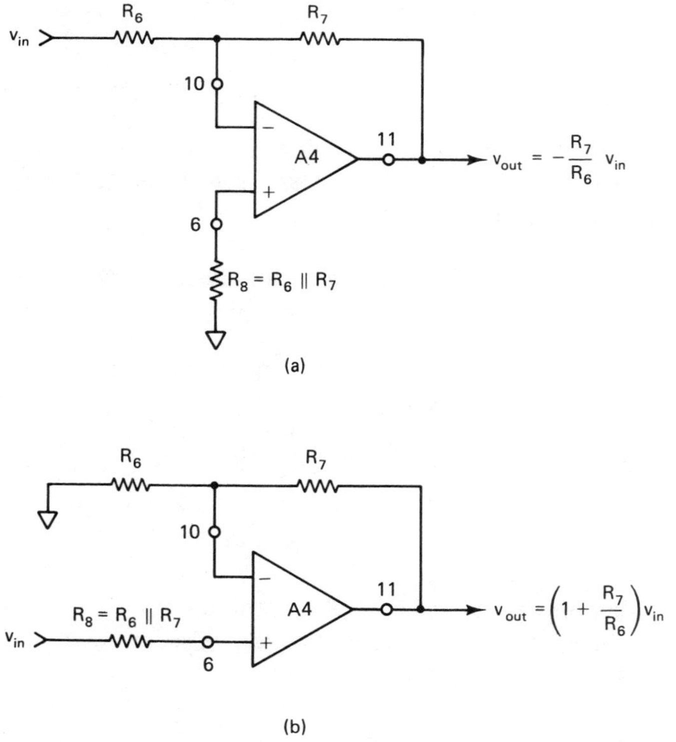

Figure 10-4 Uncommitted op-amp gain configurations. (a) Inverting. (b) Non-inverting.

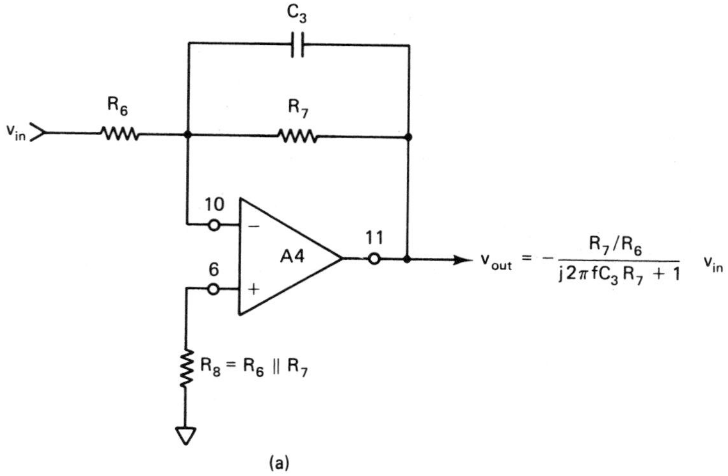

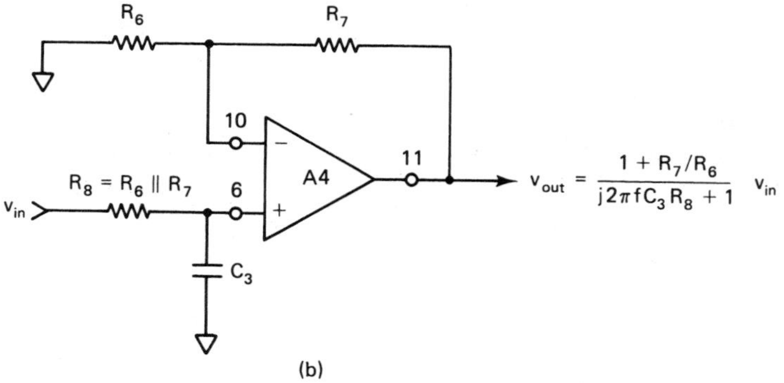

Figure 10-5 Using the uncommitted op-amp to raise the order of low-pass function. (a) Inverting configuration. (b) Noninverting configuration.

it over its frequency range, care should be exercised so that clipping does not occur in the filter. If clipping occurs, the input signal level should be reduced. For best performance, all external resistors should be 1% metal film, and capacitors should be a stable type such as tantalum and Mylar. Higher-order filters can be made by using cascaded FLT-U2 stages. To construct a notch filter, the simplest method is to use the FLT-U2 as an inverting band-pass filter and then to sum the output of the band-pass filter with the input signal by means of the uncommitted op-amp.

The FLT-U2 universal filter may be used in audio tone signaling, data acquisition, and feedback control systems.

Example 10-1

The FLT-U2 is to be used as a second-order inverting Butterworth low-pass filter with a *dc* gain of 5, cutoff frequency of 2 kHz, and $Q = 10$. Determine the values of the external components.

Solution. The values of external components can be determined by following these design steps:

1. According to Table 10-1, the inverting configuration would normally be used to give an inverting low-pass output. However, to obtain a gain of 5, an inverting uncommitted op-amp has to be used; hence the noninverting filter configuration must be used.
2. From Table 10-3, using $Q = 10$,

$$R_2 = \frac{316\ \text{k}\Omega}{10} = 31.6\ \text{k}\Omega$$

$$R_3 = \frac{100\ \text{k}\Omega}{(3.16)(10) - 1} = 3.27\ \text{k}\Omega \quad (\text{use } 3.3\ \text{k}\Omega)$$

$$R_1 = \text{open}$$

3. Substituting $f_1 = 2$ kHz in Equation (10-1), we get

$$R_4 = R_5 = \frac{(5.03)(10^7)}{(2)(10^3)} = 25.15\ \text{k}\Omega$$

(Use $R_4 = R_5 = 24.9\ \text{k}\Omega$.)

4. The final step is to use the uncommitted op-amp as an inverting amplifier with a gain of 5 [see Figure 10-4(a)]. Let $R_6 = 1.8\ \text{k}\Omega$; then $R_7 = (5)(1.8\ \text{k}\Omega) = 9.0\ \text{k}\Omega$ (use a 10-kΩ potentiometer). $R_8 = R_6 \| R_7 = 1.5\ \text{k}\Omega$. The complete circuit is shown in Figure 10-6.

Example 10-2

Using the FLT-U2, design a second-order inverting Butterworth band-pass filter with center frequency $f_1 = 5$ kHz and $Q = 10$.

Solution. The values of external components needed to use the FLT-U2 as a band-pass filter can be determined by following these design steps:

1. According to Table 10-1, the noninverting configuration is chosen to obtain an inverted band-pass output.
2. Using $Q = 10$, the values of R_2 and R_3 from Table 10-3 are

$$R_2 = 100\ \text{k}\Omega$$

$$R_3 = \frac{100\ \text{k}\Omega}{(3.48)(10) - 1} = 2.96\ \text{k}\Omega$$

(use $R_3 = 2.7\ \text{k}\Omega$). However,

$$R_1 = \text{open}$$

3. Since $f_1 = 5$ kHz, from Equation 10-1,

$$R_4 = R_5 = \frac{(5.03)(10^7)}{(5)(10^3)} \cong 10\ \text{k}\Omega$$

For accurate circuit operation, R_4 may be a fixed 10-kΩ resistor while R_5 may be a 10-kΩ potentiometer, so that R_5 can be adjusted for the exact center frequency of 5 kHz.

The complete circuit is shown in Figure 10-7.

Example 10-3

Using the FLT-U2, design a notch filter with 5-kHz notch-out frequency and $Q = 10$.

Solution. As mentioned earlier, the FLT-U2 can be used as a notch filter by summing the inverted output of the band-pass filter designed in Example 10-2 with the input

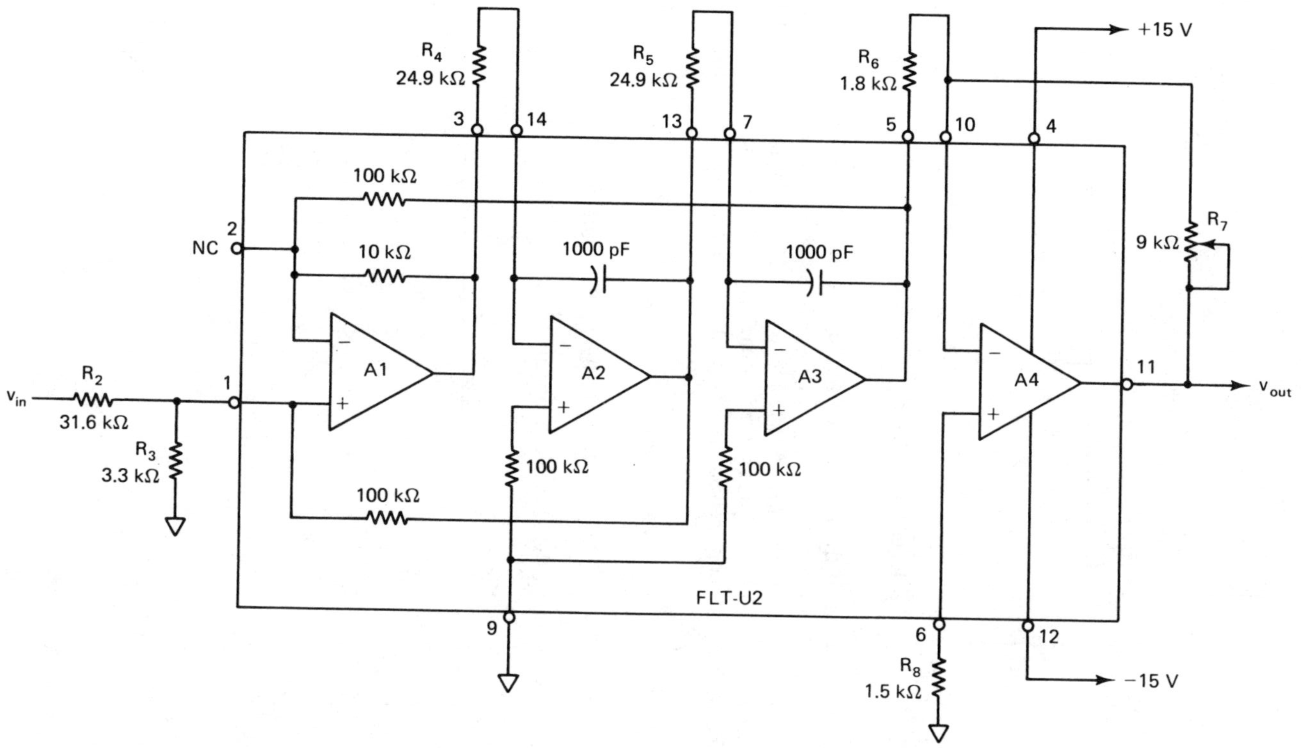

Figure 10-6 Second-order inverting Butterworth low-pass filter of Example 11-1.

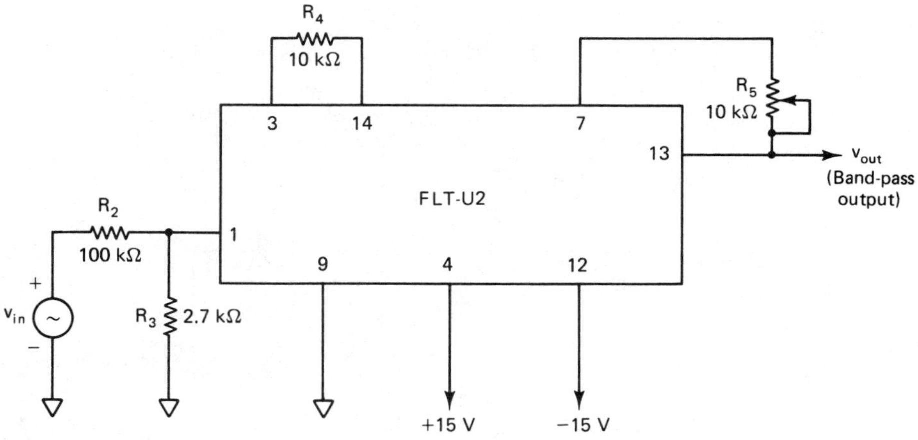

Figure 10-7 Band-pass filter of Example 10-2.

signal by means of the uncommitted op-amp (see Figure 10-8). The use of the band-pass filter of Example 10-2 is possible here because its 5-kHz center frequency and Q value are equal to the 5-kHz notch-out frequency and the Q value of the notch filter. Since no extra gain is required, the feedback resistance and the input resistances of the uncommitted op-amp used as the summing amplifier must be equal in value. Therefore, let

$$R_6 = R_7 = R_8 = 10 \text{ k}\Omega$$

Then

$$R_9 \cong R6 \,||\, R7 \,||\, R8 = 3.3 \text{ k}\Omega$$

Figure 10-8 shows the complete diagram of the notch filter.

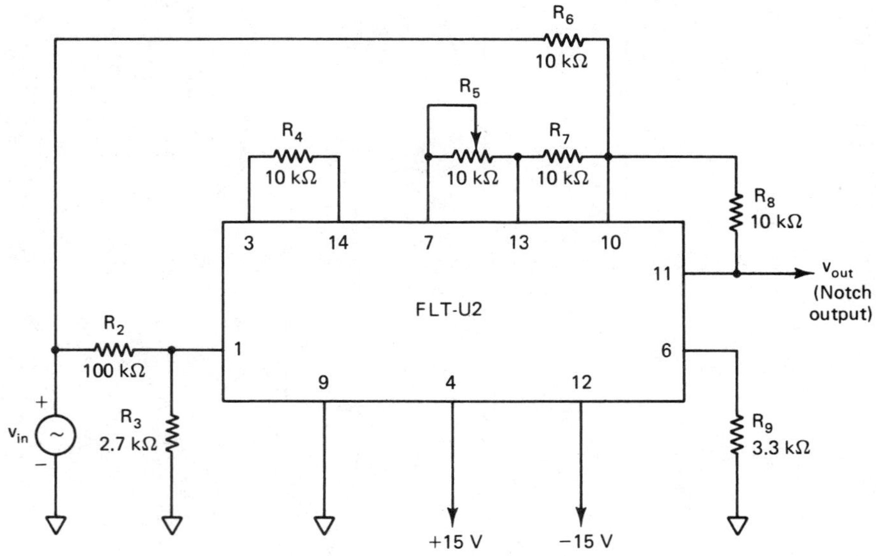

Figure 10-8 Notch filter of Example 10-3.

10-3 THE 555 TIMER

One of the most versatile linear integrated circuits is the 555 timer. Signetics Corporation first introduced this device as the SE/NE 555 in early 1970. Since its debut, the device has been used in a number of novel and useful applications. A sample of these applications includes monostable and astable multivibrators, dc–dc converters, digital logic probes, waveform generators, analog frequency meters and tachometers, temperature measurement and control, infrared transmitters, burglar and toxic gas alarms, voltage regulators, electric eyes, and many others. The 555 is a monolithic timing circuit that can produce accurate and highly stable time delays or oscillation. In other words, the timer basically operates in one of the two modes: either as a monostable (one-shot) multivibrator or as an astable (free-running) multivibrator. The device is available as an 8-pin metal can, as an 8-pin mini DIP, or a 14-pin DIP. Figure 10-9 shows the connection diagram and the block diagram of the SE/NE 555 timer. The SE555 is designed for the operating temperature range −55 to +125°C, while the NE555 operates over a temperature range of 0 to +70°C. The important features of the 555 timer are these: it operates on +5 to +18 V supply voltage in both free-running (astable) and one-shot (monostable) modes; it has an adjustable duty cycle; timing is from microseconds through hours; it has a high current output; it can source or sink 200 mA; the output can drive TTL and has a temperature stability of 50 parts per million (ppm) per degree Celsius change in temperature, or equivalently 0.005%/°C. Like general-purpose op-amps, the 555 timer is reliable, easy to use, and low cost.

The next several sections explain the operation of the 555 timer as a monostable and astable multivibrator; a few simple applications using these two modes are then presented.

10-3.1 The 555 as a Monostable Multivibrator

A monostable multivibrator, often called a *one-shot* multivibrator, is a pulse-generating circuit in which the duration of the pulse is determined by the *RC* network connected externally to the 555 timer. In a stable or standby state the output of the circuit is approximately zero or at logic-low level. When an external trigger pulse is applied, the output is forced to go *high* ($\cong V_{CC}$). The time the output remains high is determined by the external *RC* network connected to the timer. At the end of the timing interval, the output automatically reverts back to its logic-low stable state. The output stays low until the trigger pulse is again applied. Then the cycle repeats. The monostable circuit has only one stable state (output low), hence the name *monostable*. Normally, the output of the monostable multivibrator is low.

Figure 10-10(a) shows the 555 configured for monostable operation. To better explain the circuit's operation, the internal block diagram is included in Figure 10-10(b).

Before proceeding with the operation of the 555 timer as a monostable multivibrator, it is important to examine its pin functions. The pin numbers used in the following discussion refer to the 8-pin mini DIP and 8-pin metal can packages [see Figure 10-9(a)].

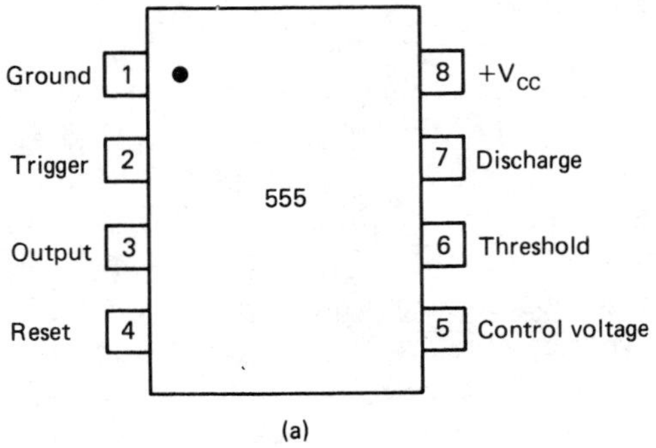

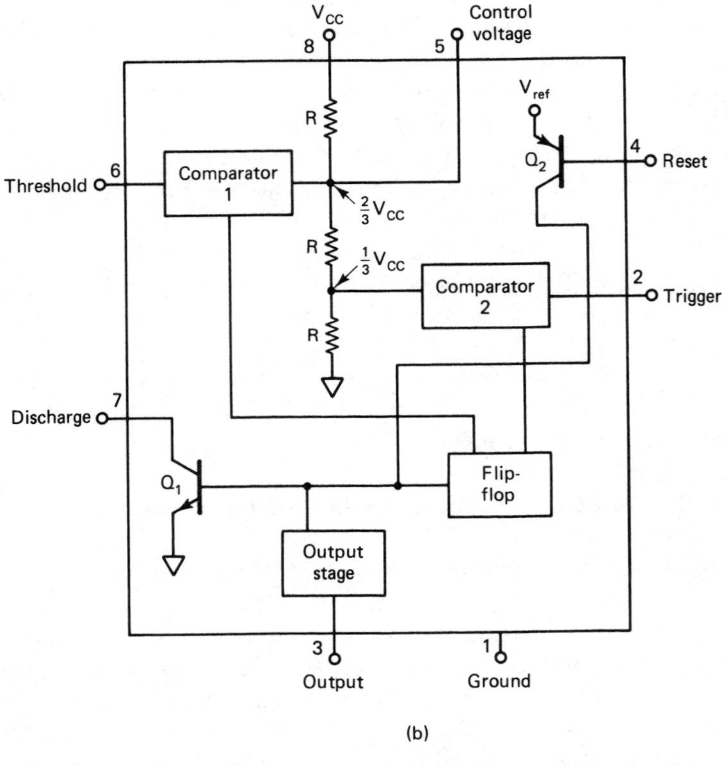

Figure 10-9 (a) 555 timer connection diagram. (b) Block diagram. (Courtesy of Signetics Corp.)

Pin 1: Ground. All voltages are measured with respect to this terminal.

Pin 2: Trigger. The output of the timer depends on the amplitude of the external trigger pulse applied to this pin. The output is low if the voltage at this pin is greater than $\frac{2}{3}V_{CC}$. However, when a negative-going pulse of amplitude larger than $\frac{1}{3}V_{CC}$ is applied to this pin, the comparator 2 output goes low, which in turn switches the output of the timer high [see Figure 10-9(b)]. The output remains high as long as the trigger terminal is held at a low voltage.

Pin 3: Output. There are two ways a load can be connected to the output

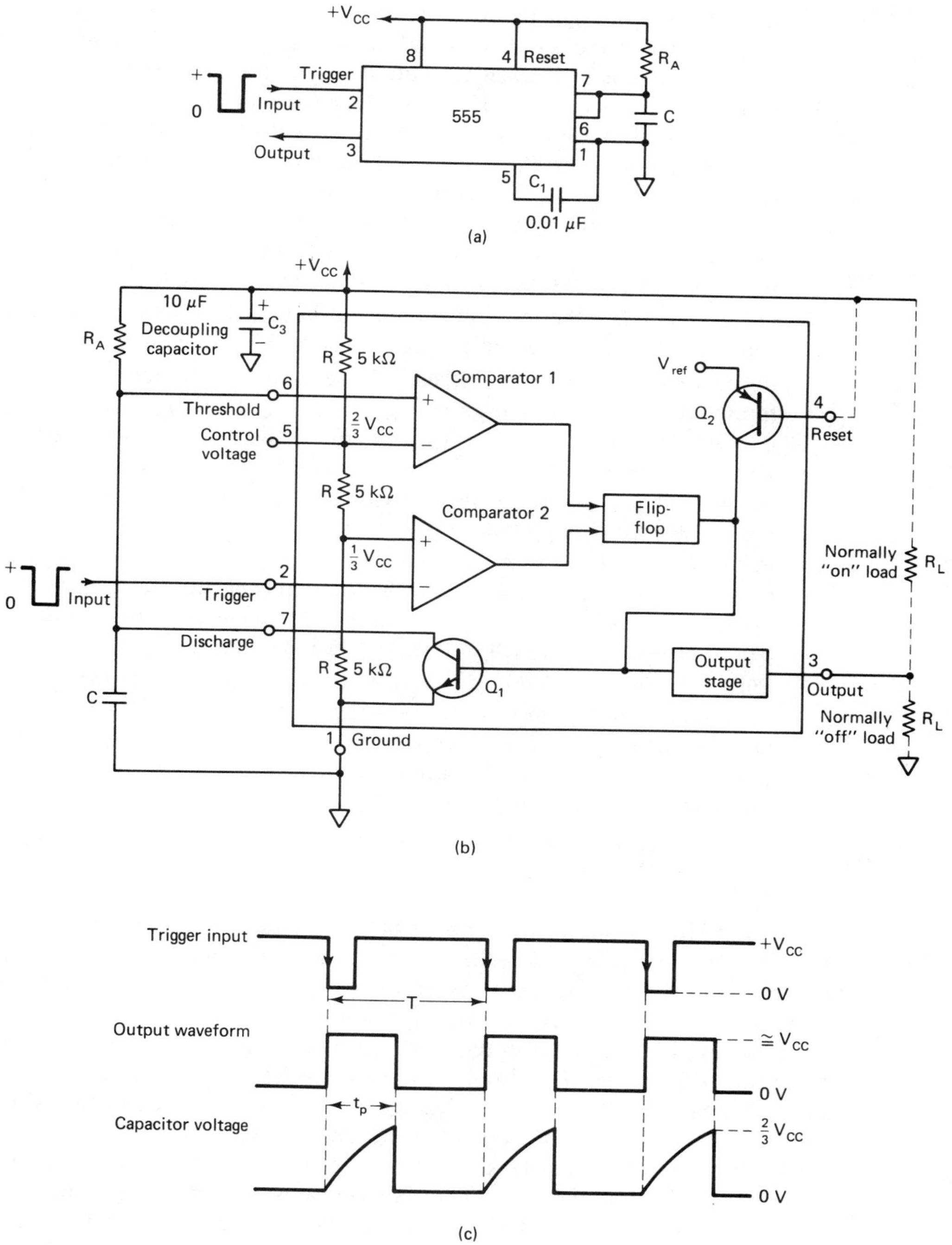

Figure 10-10 (a, b) 555 connected as a monostable multivibrator. (c) Input and output waveforms.

terminal; either between pin 3 and ground (pin 1) or between pin 3 and supply voltage $+V_{CC}$ (pin 8) [see Figure 10-10(b)]. When the output is low, the load current flows through the load connected between pin 3 and $+V_{CC}$ into the output terminal and is called the *sink* current. However, the current through the grounded load is zero when the output is low. For this reason, the load connected between pin 3 and $+V_{CC}$ is called the *normally-on load* and that connected between pin 3 and ground is called the *normally-off load.* On the other hand, when the output is high, the current through the load connected between pin 3 and $+V_{CC}$ (normally-on load) is zero. However, the output terminal supplies current to the normally-off load. This current is called the *source* current. The maximum value of sink and source current is 200 mA.

Pin 4: Reset. The 555 timer can be reset (disabled) by applying a negative pulse to this pin. When the reset function is not in use, the reset terminal should be connected to $+V_{CC}$ to avoid any possibility of false triggering.

Pin 5: Control voltage. An external voltage applied to this terminal changes the threshold as well as the trigger voltage [see Figure 10-10(b)]. In other words, by imposing a voltage on this pin or by connecting a pot between this pin and ground, the pulse width of the output waveform can be varied. When not used, the control pin should be bypassed to ground with a 0.01-μF capacitor to prevent any noise problems.

Pin 6: Threshold. This is the noninverting input terminal of comparator 1, which monitors the voltage across the external capacitor [see Figure 10-10(b)]. When the voltage at this pin is $\geq$ threshold voltage $\frac{2}{3}V_{CC}$, the output of comparator 1 goes high, which in turn switches the output of the timer low.

Pin 7: Discharge. This pin is connected internally to the collector of transistor Q_1, as shown in Figure 10-10(b). When the output is high, Q_1 is off and acts as an open circuit to the external capacitor connected across it. On the other hand, when the output is low, Q_1 is saturated and acts as a short circuit, shorting out the external capacitor to ground.

Pin 8: $+V_{CC}$. The supply voltage of +5 V to +18 is applied to this pin with respect to ground (pin 1).

10-3.1(a) Monostable operation. According to Figure 10-10(b), initially when the output is low, that is, the circuit is in a stable state, transistor Q_1 is on and capacitor C is shorted out to ground. However, upon application of a negative trigger pulse to pin 2, transistor Q_1 is turned off, which releases the short circuit across the external capacitor C and drives the output high. The capacitor C now starts charging up toward V_{CC} through R_A. However, when the voltage across the capacitor equals $\frac{2}{3}V_{CC}$, comparator 1's output switches from low to high, which in turn drives the output to its low state via the output of the flip-flop. At the same time, the output of the flip-flop turns transistor Q_1 on, and hence capacitor C rapidly discharges through the transistor. The output of the monostable remains low until a trigger pulse is again applied. Then the cycle repeats. Figure 10-10(c) shows the input trigger, output voltage, and capacitor voltage waveforms. As shown here, the pulse width of the input trigger must be smaller than the expected pulse width of the output waveform.

Also, the trigger pulse must be a negative-going input signal with an amplitude larger than $\frac{1}{3}V_{CC}$.

The time during which the output remains high is given by

$$t_p = 1.1R_AC \qquad \text{seconds} \tag{10-3}$$

where R_A is in ohms and C is in farads. Figure 10-11 shows a graph of the various combinations of R_A and C necessary to produce desired time delays. Note that this graph can only be used as a guideline and gives only the approximate value of R_A and C for a given time delay.

Once triggered, the circuit's output will remain in the high state until the set time t_p elapses. The output will not change its state even if an input trigger is applied again during this time interval t_p. However, the circuit can be reset during the timing cycle by applying a negative pulse to the reset terminal. The output will then remain in the low state until a trigger is again applied.

Often in practice a decoupling capacitor (10 μF) is used between $+V_{CC}$ (pin 8) and ground (pin 1) to eliminate unwanted voltage spikes in the output waveform. Sometimes, to prevent any possibility of mistriggering the monostable multivibrator on positive pulse edges, a waveshaping circuit consisting of R, C_2, and diode D is connected between the trigger input pin 2 and V_{CC} pin 8, as shown in Figure 10-12. The

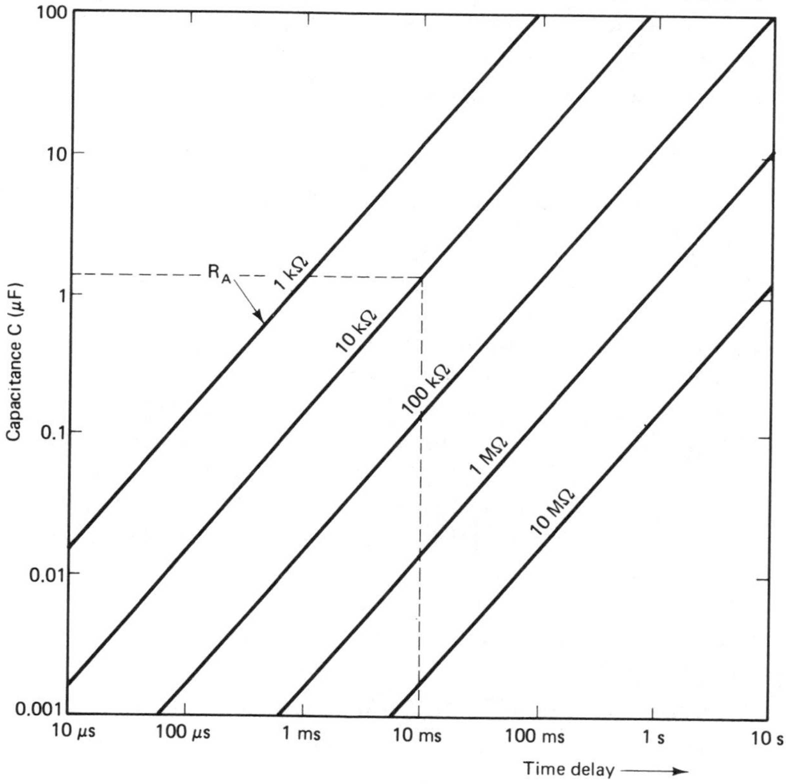

Figure 10-11 Determining R_A and C values for various time delays.

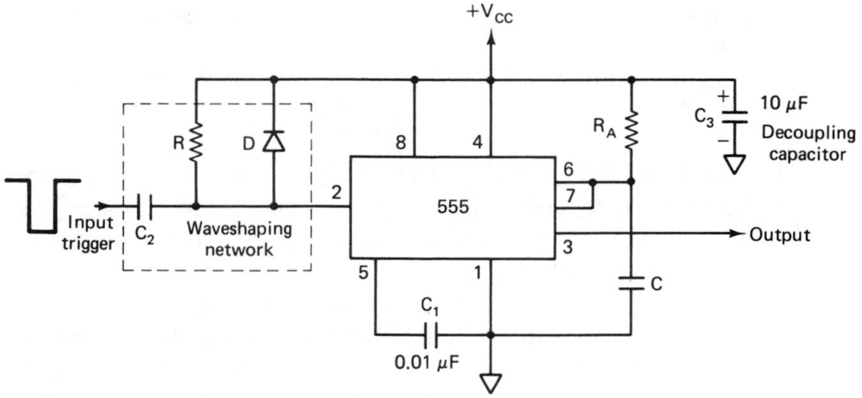

Figure 10-12 Monostable multivibrator with waveshaping network to prevent positive pulse edge triggering.

values of R and C_2 should be selected so that the time constant RC_2 is smaller than the output pulse width t_p.

Example 10-4

In the circuit of Figure 10-10(a), $R_A = 10\ \text{k}\Omega$, and the output pulse width $t_p = 10$ ms. Determine the value of C.

Solution. Rearranging Equation (10-3), we get

$$C = \frac{(10)(10^{-3})}{(1.1)(10^4)} = 0.909\ \mu\text{F} \cong 1\ \mu\text{F}$$

(Notice that approximately the same value can be obtained for C from the time delay graph of Figure 10-11.)

10-3.2 Monostable Multivibrator Applications

10-3.2(a) Frequency divider. The monostable multivibrator of Figure 10-10(a) can be used as a frequency divider by adjusting the length of the timing cycle t_p with respect to the time period T of the input trigger signal applied to pin 2. In order to use the monostable multivibrator as a divide-by-2 circuit, the timing interval t_p must be slightly larger than the time period T of the input trigger signal, as shown in Figure 10-13. By the same concept, in order to use the monostable multivibrator as a divide-by-3 circuit, t_p must be slightly larger than twice the period of the input trigger signal, and so on.

The frequency-divider application is possible because the monostable multivibrator cannot be retriggered during the timing cycle.

Example 10-5

The circuit of Figure 10-10(a) is to be used as a divide-by-2 network. The frequency of the input trigger signal is 2 kHz. If the value of $C = 0.01\ \mu$F, what should be the value of R_A?

Solution. For a divide-by-2 network, t_p should be slightly larger than T. Let

$$t_p = 1.2T$$

Therefore,

$$t_p = (1.2)\frac{1}{2\text{ kHz}} = 0.6\text{ ms}$$

Rearranging Equation (10-3) yields

$$R_A = \frac{(0.6)(10^{-3})}{(1.1)(10^{-8})} = 54.5\text{ k}\Omega$$

A combination of a 10-kΩ fixed resistor and a 50-kΩ potentiometer may be used for R_A. The input and output waveforms are shown in Figure 10-13.

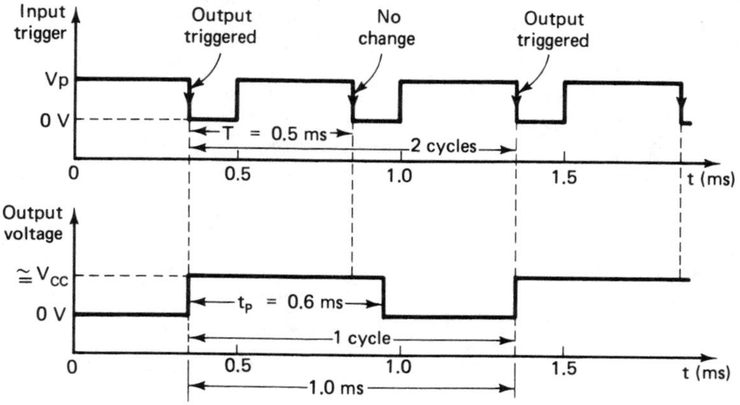

Figure 10-13 Input and output waveforms of the monostable multivibrator as a divide-by-2 network.

10-3.2(b) Pulse stretcher. This application makes use of the fact that the output pulse width (timing interval) of the monostable multivibrator is of longer duration than the negative pulse width of the input trigger. As such, the output pulse width of the monostable multivibrator can be viewed as a stretched version of the narrow input pulse, hence the name *pulse stretcher*. Often, narrow-pulse-width signals are not suitable for driving an LED display, mainly because of their very narrow pulse widths. In other words, the LED may be flashing but is not visible to the eye because its *on* time is infinitesimally small compared to its *off* time. The 555 pulse stretcher can be used to remedy this problem.

Figure 10-14 shows a basic monostable used as a pulse stretcher with an LED indicator at the output. The LED will be on during the timing interval $t_p = 1.1R_AC$, which can be varied by changing the value of R_A and/or C.

10-3.3 The 555 as an Astable Multivibrator

An astable multivibrator, often called a *free-running* multivibrator, is a rectangular-wave generating circuit. Unlike the monostable multivibrator, this circuit does not require an external trigger to change the state of the output, hence the name *free-running*. However, the time during which the output is either high or low is determined by the two resistors and a capacitor, which are externally connected to the 555 timer.

10-3.3(a) Astable operation. Figure 10-15(a) shows the 555 timer connected as an astable multivibrator. Initially, when the output is high, capacitor

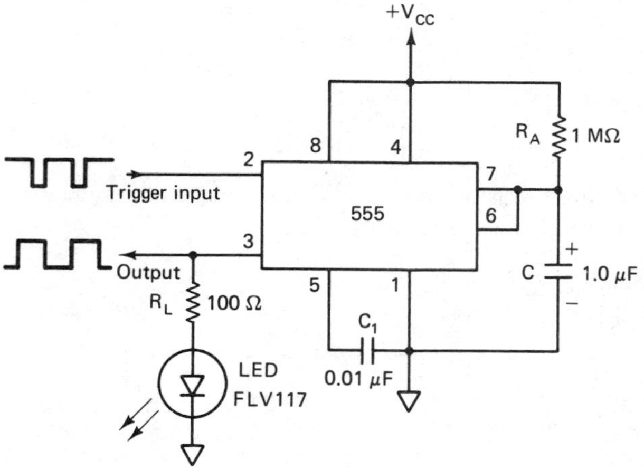

Figure 10-14 Monostable multivibrator as a pulse stretcher.

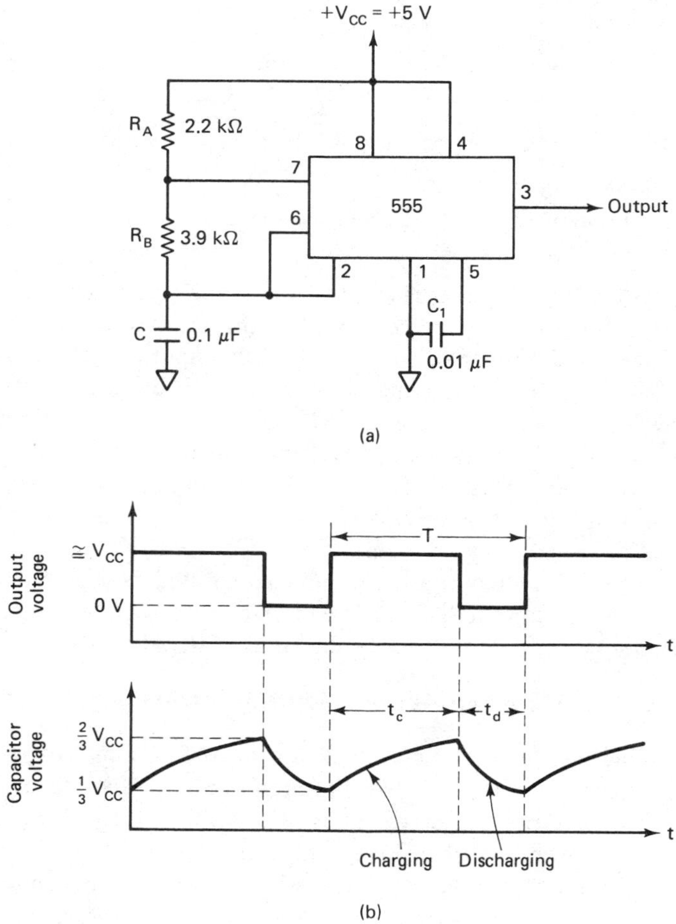

Figure 10-15 555 as an astable multivibrator. (a) Circuit. (b) Voltage across capacitor and output voltage waveforms.

C starts charging toward V_{CC} through R_A and R_B. However, as soon as voltage across the capacitor equals $\frac{2}{3}V_{CC}$, comparator 1 triggers the flip-flop, and the output switches low [see Figure 10-15(b)]. Now capacitor C starts discharging through R_B and transistor Q_1. When the voltage across C equals $\frac{1}{3}V_{CC}$, comparator 2's output triggers the flip-flop, and the output goes high. Then the cycle repeats. The output voltage and capacitor voltage waveforms are shown in Figure 10-15(b).

As shown in this figure, the capacitor is periodically charged and discharged between $\frac{2}{3}V_{CC}$ and $\frac{1}{3}V_{CC}$, respectively. The time during which the capacitor charges from $\frac{1}{3}V_{CC}$ to $\frac{2}{3}V_{CC}$ is equal to the time the output is high and is given by

$$t_c = 0.69(R_A + R_B)C \tag{10-4a}$$

where R_A and R_B are in ohms and C is in farads. Similarly, the time during which the capacitor discharges from $\frac{2}{3}V_{CC}$ to $\frac{1}{3}V_{CC}$ is equal to the time the output is low and is given by

$$t_d = 0.69(R_B)C \tag{10-4b}$$

where R_B is in ohms and C is in farads. Thus the total period of the output waveform is

$$T = t_c + t_d = 0.69(R_A + 2R_B)C \tag{10-5}$$

This, in turn, gives the frequency of oscillation as

$$f_o = \frac{1}{T} = \frac{1.45}{(R_A + 2R_B)C} \tag{10-6}$$

Equation (10-6) indicates that the frequency f_o is independent of the supply voltage V_{CC}. The frequency of oscillation (free-running frequency) f_o can also be found by Figure 10-16.

Often the term *duty cycle* is used in conjunction with the astable multivibrator. The duty cycle is the ratio of the time t_d during which the output is low to the total time period T. It is generally expressed as a percentage. In equation form,

$$\begin{aligned} \%\ \text{duty cycle} &= \frac{t_d}{T} \times 100 \\ &= \frac{R_B}{R_A + 2R_B}(100) \end{aligned} \tag{10-7}$$

The duty cycle of the square wave is 50%. This means that, according to Equation (10-7), the astable multivibrator shown in Figure 10-15(a) will not produce square-wave output unless the resistance $R_A = 0\ \Omega$. However, there is a danger in shorting resistance R_A to zero. With $R_A = 0\ \Omega$, terminal 7 is connected directly to $+V_{CC}$. When the capacitor discharges through R_B and Q_1 (pin 7), an extra current is supplied to Q_1 by V_{CC} through a short between terminal 7 and $+V_{CC}$, which may damage Q_1 and hence the timer. Fortunately, an alternative is available which is explained in the section on astable multivibrator applications.

Example 10-6

In the astable multivibrator of Figure 10-15(a), $R_A = 2.2\ \text{k}\Omega$, $R_B = 3.9\ \text{k}\Omega$, and $C = 0.1\ \mu\text{F}$. Determine the positive pulse width t_c, negative pulse width t_d, and free-running frequency f_o.

Solution. From Equations (10-4a) and (10-4b),

$$t_c = (0.69)(2.2\text{ k}\Omega + 3.9\text{ k}\Omega)(0.1)(10^{-6})$$
$$= 0.421\text{ ms}$$
$$t_d = (0.69)(3.9\text{ k}\Omega)(0.1)(10^{-6})$$
$$= 0.269\text{ ms}$$

Therefore,

$$f_o = \frac{1}{(0.421 + 0.269)(10^{-3})} = 1.45\text{ kHz}$$

Using $C = 0.1\ \mu\text{F}$ and $R_A + 2R_B = 10\text{ k}\Omega$, the free-running frequency can also be found from the graph in Figure 10-16, as indicated by point A. Note that the graph uses the log-log scale; that is, vertical and horizontal axes are marked in logarithmic scales.

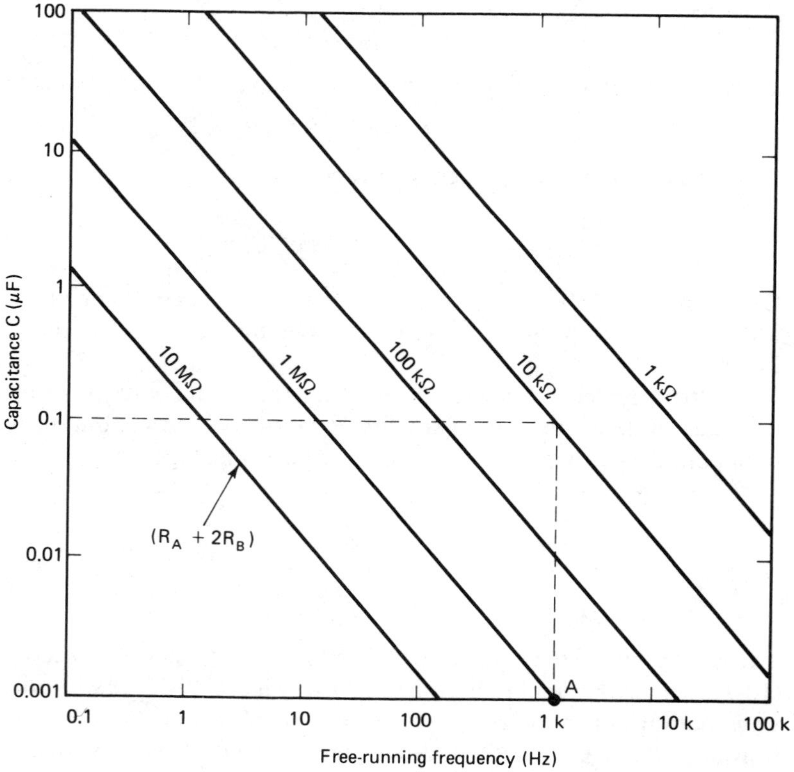

Figure 10-16 Free-running frequency versus R_A, R_B, and C.

10-3.4 Astable Multivibrator Applications

10-3.4(a) Square-wave oscillator. Without reducing $R_A = 0\ \Omega$, the astable multivibrator can be used to produce a square-wave output simply by connecting diode D across resistor R_B, as shown in Figure 10-17. The capacitor C charges through R_A and diode D to approximately $\frac{2}{3}V_{CC}$ and discharges through R_B and

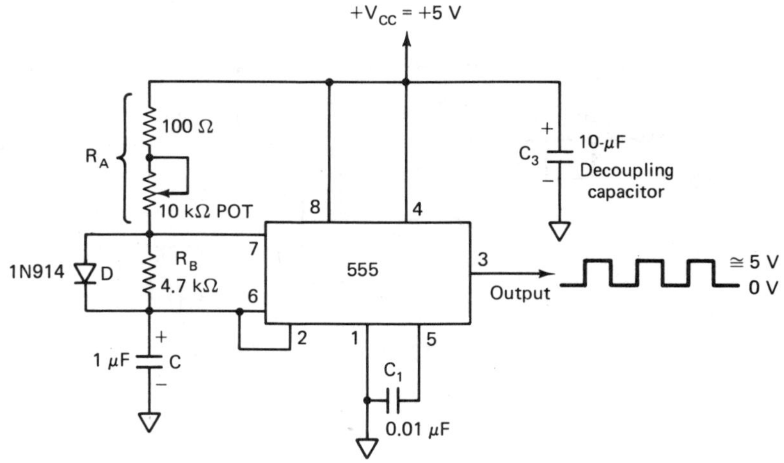

Figure 10-17 Astable multivibrator as a square-wave oscillator.

terminal 7 (transistor Q_1) until the capacitor voltage equals approximately $\frac{1}{3}V_{CC}$; then the cycle repeats. To obtain a square-wave output (50% duty cycle), R_A must be a combination of a fixed resistor and potentiometer, so that the potentiometer can be adjusted for the exact square wave.

10-3.4(b) Free-running ramp generator. The astable multivibrator can be used as a free-running ramp generator when resistors R_A and R_B are replaced by a current mirror. Figure 10-18(a) shows an astable multivibrator configured to perform this function. The current mirror starts charging capacitor C toward V_{CC} at a constant rate. When voltage across C equals $\frac{2}{3}V_{CC}$, comparator 1 turns transistor Q_1 on and C rapidly discharges through transistor Q_1. Refer to Figure 10-9(b). However, when the discharge voltage across C is approximately equal to $\frac{1}{3}V_{CC}$, comparator 2 switches transistor Q_1 off, and then capacitor C starts charging up again. Thus the charge–discharge cycle keeps repeating. The discharging time of the capacitor is relatively negligible compared to its charging time; hence, for all practical purposes, the time period of the ramp waveform is equal to the charging time and is approximately given by

$$T = \frac{(V_{CC})(C)}{3I_C} \tag{10-8a}$$

where $I_C = (V_{CC} - V_{BE})/R$ = constant current in amperes and C is in farads. Therefore, the free-running frequency of the ramp generator is

$$f_o = \frac{3I_C}{(V_{CC})(C)} \tag{10-8b}$$

Figure 10-18(b) shows the generator's output waveform. For the analysis and design of the current mirror, the reader may refer to Section 1-11.

Example 10-7

Referring to the circuit of Figure 10-18(a), determine the frequency of the free-running ramp generator if R is set at 10 kΩ. Assume that the $V_{BE} = V_{D1} = 0.7$ V.

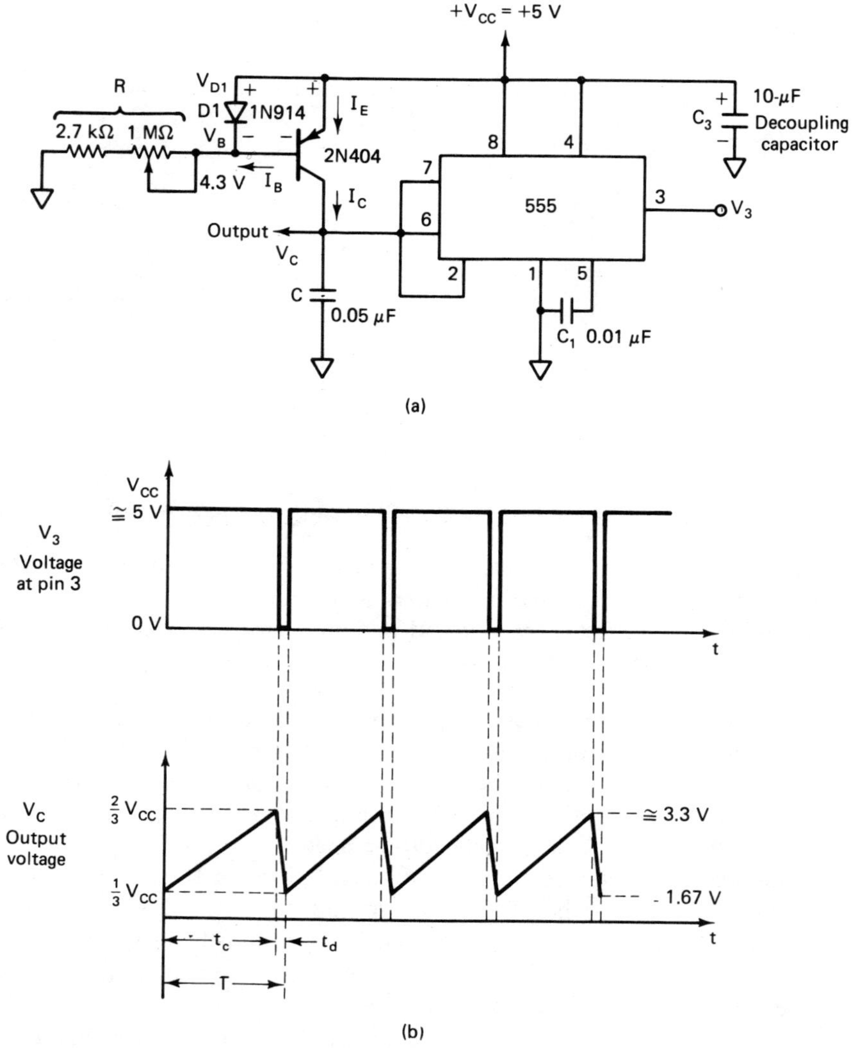

Figure 10-18 (a) Free-running ramp generator circuit. (b) Output waveform.

Solution: By Equation (10-8b),

$$f_o = \frac{(3)\left[\dfrac{5 - 0.7}{10\text{ k}}\right]}{(5)(5)(10^{-8})} = 5.16 \text{ kHz}$$

10-4 PHASE-LOCKED LOOPS

Although the evolution of the phase-locked loop began in the early 1930s, its cost outweighed its advantages at first. With the rapid development of integrated-circuit technology, however, the phase-locked loop has emerged as one of the fundamental building blocks in electronics technology. The phase-locked-loop principle has been used in applications such as FM (frequency modulation) stereo decoders, motor speed controls, tracking filters, frequency synthesized transmitters and receivers, FM demodulators, frequency shift keying (FSK) decoders, and a generation of local oscillator frequencies in TV and in FM tuners. Today the phase-locked loop is even available as a single package, typical examples of which include the Signetics' SE/NE 560 series (the 560, 561, 562, 564, 565, and 567). However, for more economical operation, discrete ICs can be used to construct a phase-locked loop.

10-4.1 Operating Principles

Figure 10-19 shows the phase-locked loop in its basic form. As illustrated in this figure, the phase-locked loop consists of:

1. A phase detector
2. A low-pass filter
3. A voltage-controlled oscillator

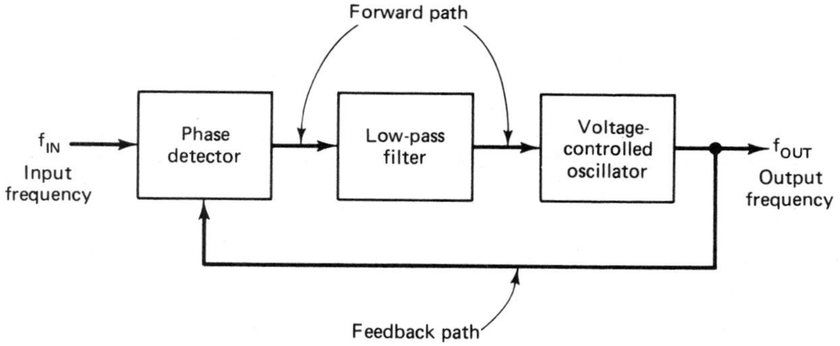

Figure 10-19 Block diagram of a phase-locked loop.

The phase detector, or comparator, compares the input frequency f_{IN} with the feedback frequency f_{OUT}. The output of the phase detector is proportional to the phase difference between f_{IN} and f_{OUT}. The output voltage of a phase detector is a dc voltage and therefore is often referred to as the error voltage. The output of the phase detector is then applied to the low-pass filter, which removes the high-frequency noise and produces a dc level. This dc level, in turn, is the input to the voltage-controlled oscillator (VCO). The filter also helps in establishing the dynamic characteristics of the PLL circuit. The output frequency of the VCO is directly proportional to the input dc level. The VCO frequency is compared with the input frequencies and adjusted until it is equal to the input frequencies. In short, the phase-locked loop goes through three states: *free-running*, *capture*, and *phase lock*.

Before the input is applied, the phase-locked loop is in the free-running state. Once the input frequency is applied, the VCO frequency starts to change and the phase-locked loop is said to be in the capture mode. The VCO frequency continues to change until it equals the input frequency, and the phase-locked loop is then in the phase-locked state. When phase locked, the loop tracks any change in the input frequency through its repetitive action.

Before studying the specialized phase-locked-loop IC, we shall consider the discrete phase-locked loop, which may be assembled by combining a phase detector, a low-pass filter, and a voltage-controlled oscillator (see Figure 10-19).

10-4.1(a) Phase detector. The phase detector compares the input frequency and the VCO frequency and generates a *dc* voltage that is proportional to the phase difference between the two frequencies. Depending on the analog or digital type of phase detector used, the PLL is either called an analog or digital type, respectively. Even though most of the monolithic PLL integrated circuits use analog phase detectors, the majority of discrete phase detectors in use are of the digital type mainly because of its simplicity. For this reason, we shall consider only digital-type phase detectors here.

A double-balanced mixer is a classic example of an analog phase detector. On the other hand, examples of digital phase detectors are these:

1. An exclusive—OR phase detector
2. An edge-triggered phase detector
3. A monolithic phase detector (such as type 4044)

Figure 10-20(a) shows the exclusive-OR phase detector that uses an exclusive-OR gate usch as CMOS type 4070. The output of the exclusive-OR gate is *high* only when f_{IN} *or* f_{OUT} is high, as shown in Figure 10-20(b).

In this figure, f_{IN} is leading f_{OUT} by ϕ (phi) degrees; that is, the phase difference between f_{IN} and f_{OUT} is ϕ degrees. The *dc* output voltage of the exclusive-OR phase detector is a function of the phase difference between its two inputs. Figure 10-20(c) shows dc output voltage as a function of the phase difference between f_{IN} and f_{OUT}. This graph indicates that the maximum dc output voltage occurs when the phase difference is π radians or 180°. The slope of the curve between 0 and π radians is the conversion gain k_p of the phase detector. For example, if the exclusive-OR gate uses a supply voltage $V_{CC} = 5$ V, the conversion gain k_p is

$$k_p = \frac{5\text{ V}}{\pi} = 1.59\text{ V/rad}$$

The exclusive-OR type of phase detector is generally used if the f_{IN} and f_{OUT} are square waves. The edge-triggered phase detector, on the other hand, is preferred if the f_{IN} and f_{OUT} are pulse waveforms with less than 50% duty cycles.

Figure 10-21(a) shows the edge-triggered type of phase detector using an R-S (reset–set) flip-flop. The R-S flip-flop, in turn, is formed from a pair of cross-coupled NOR gates, such as the CD4001. The R-S flip-flop is triggered; that is, the output of the detector changes its logic state on the positive (leading) edge of the inputs f_{IN} and f_{OUT} [see Figure 10-21(b)]. The graph of dc output voltage versus phase difference between f_{IN} and f_{OUT} is shown in Figure 10-21(c).

The advantages of the edge-triggered type phase detector over the exclusive-OR type of detector are: (1) the dc output voltage is linear over 2π radians or 360°, as opposed to π radians or 180° in the case of the exclusive-OR detector; and (2) the

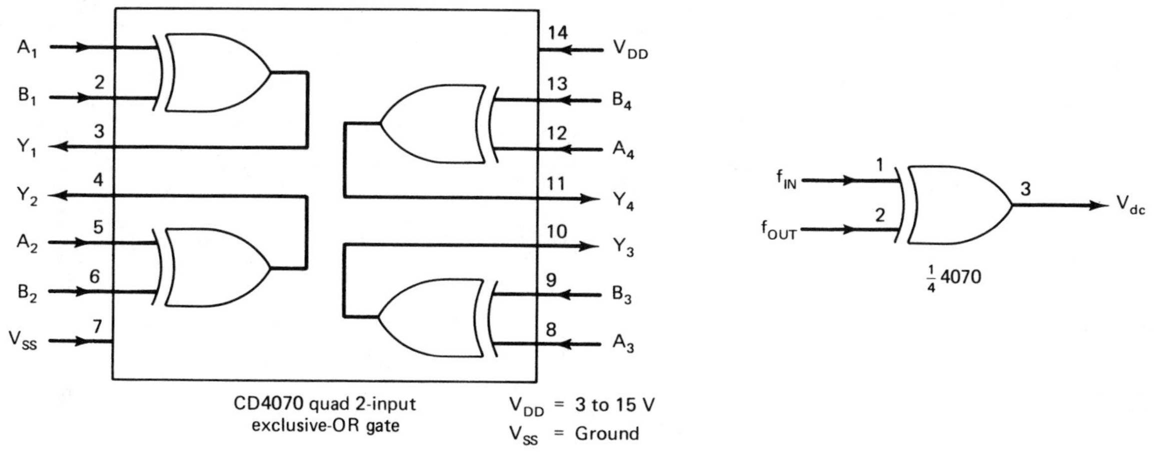

Truth table

Inputs		Outputs
A	B	Y
L	L	L
L	H	H
H	L	H
H	H	L

L = Low
H = High

(a)

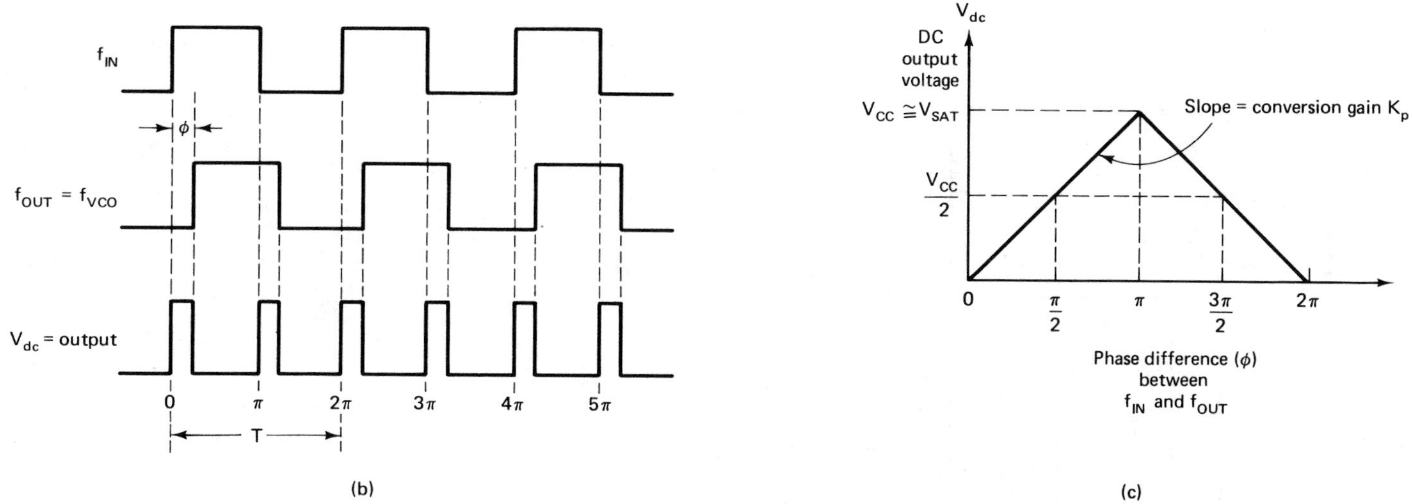

Figure 10-20 (a) Exclusive-OR phase detector connection and logic diagram (Courtesy of National Semiconductor) (b) Input and output waveforms. (c) Average output voltage versus phase difference between f_{IN} and f_{OUT} curve.

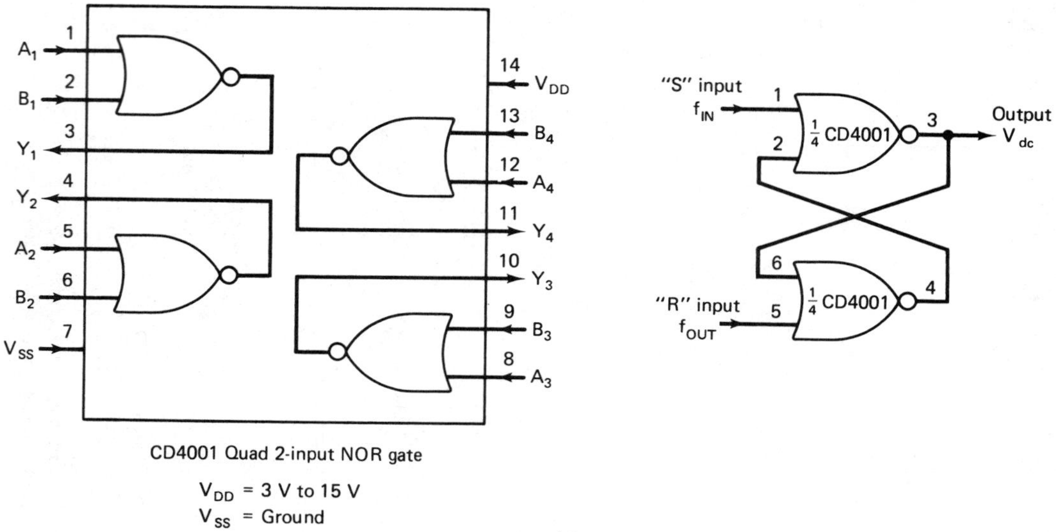

(a)

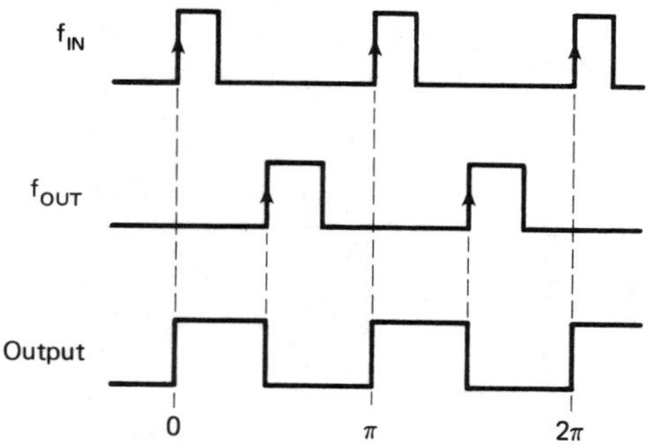

(b)

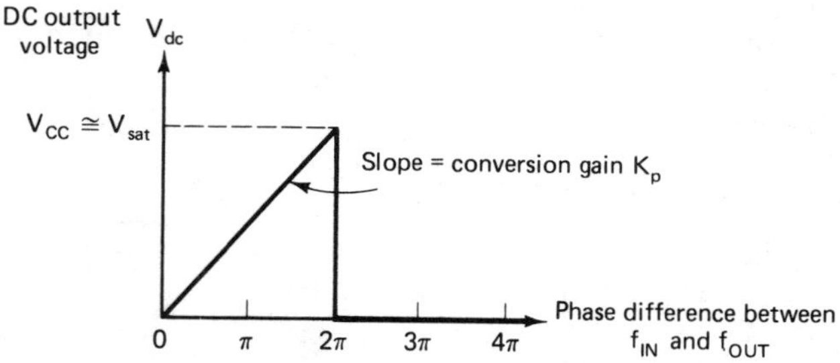

(c)

Figure 10-21 Edge-triggered type phase detector. (a) NOR gate R-S flip-flop connection diagram. (b) Input and output waveforms. (c) Dc output voltage versus phase difference between f_{IN} and f_{OUT}.

edge-triggered detector also exhibits better capture, tracking, and locking characteristics than the exclusive-OR detector. However, both types of detectors are sensitive to hormonics of the input signal and changes in the duty cycles of f_{IN} and f_{OUT}.

In a monolithic phase detector IC such as CMOS type 4044, the harmonic sensitivity and duty cycle problems are absent, since the circuit responds only to transitions in the input signals. In other words, the phase error and hence the output error voltage of the monolithic phase detector is independent of the variations in the amplitude and duty cycle of the input waveforms. Therefore, in critical applications the monolithic phase detector is preferred over both the exclusive-OR and edge-triggered types of phase detectors.

Figure 10-22(a) shows the block diagram of the MC4344/4044 phase detector. As shown in this figure, the MC4344/4044 consists of two digital phase detectors, a charge pump, and an amplifier. Phase detector 1 is used in applications that require zero frequency and phase difference at lock. On the other hand, if quadrature lock is desired, phase detector 2 can be used. When detector 1 is used in the main loop, detector 2 can also be used to indicate whether the main loop is in lock or out of lock.

The input/output transfer characteristic curve of phase detector 1 is shown in Figure 10-22(b). As shown here, the curve is linear over 4π radians or 720° and has a conversion gain k_p of $1.5\ V/4\pi = 0.12\ V/rad$.

10-4.1(b) Low-Pass filter. The function of the low-pass filter in the phase-locked loop is to remove the high-frequency components in the output of the phase detector and to remove high-frequency noise. More important, the low-pass filter controls the dynamic characteristics of the phase-locked loop. These characteristics include capture and lock ranges, bandwidth, and transient response. The lock range is defined as the range of frequencies over which the PLL system follows the changes in the input frequency f_{IN}. An equivalent term for lock range is tracking range. On the other hand, the capture range is the frequency range in which the PLL acquires phase lock. Obviously, the capture range is always smaller than the lock range.

As the filter bandwidth is reduced, its response time increases. However, reduced bandwidth reduces the capture range of the PLL. Nevertheless, reduced bandwidth helps to keep the loop in lock through momentary losses of signal and also minimizes noise.

The loop filter used in the PLL may be one of the three types shown in Figure 10-23. With the passive filters of Figure 10-23(a) and (b), an amplifier is generally used for gain. On the other hand, the active filter of Figure 10-23(c) includes the gain.

10-4.1(c) Voltage-controlled oscillator. A third section of PLL is the voltage-controlled oscillator. The VCO generates an output frequency which is directly proportional to its input voltage. The block diagram of the VCO is shown in Figure 10-24.

While the SE/NE 566 VCO is discussed in Section 8-18, a typical connection diagram is repeated in Figure 10-25 for convenience. The maximum output frequency of the NE/SE 566 is 500 kHz. For higher output frequency, therefore, integrated circuits such as Motorola's MC4324/4024 and MC1648 may be used.

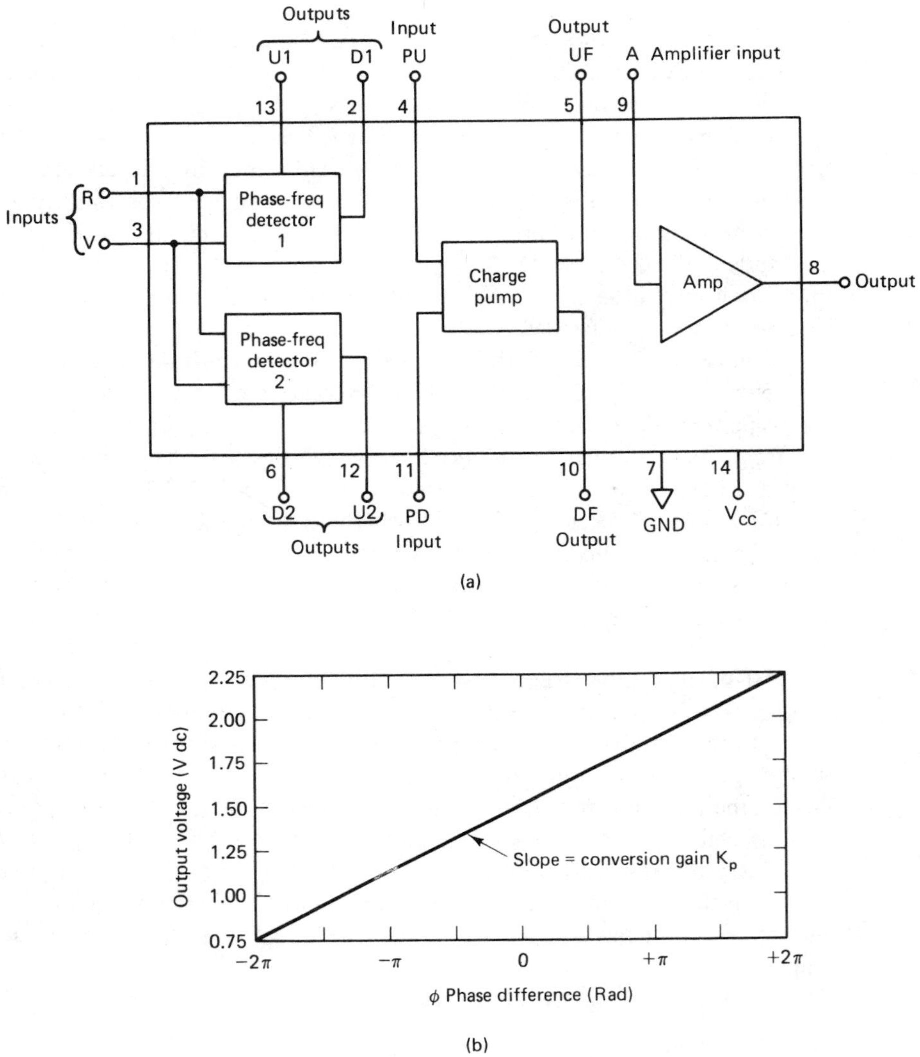

Figure 10-22 (a) MC4344/4044 phase detector block diagram. R, reference input; V, variable or feedback input; PU, pump-up signal; PD, pump-down signal; UF, up-frequency output signal; DF, down-frequency output signal. (b) Its input/output transfer characteristic curve. (Courtesy of Motorola Semiconductor.)

10-4.2 Monolithic Phase-Locked Loops

Signetics' SE/NE 560 series are monolithic phase-locked loops. The SE/NE 560, 561, 562, 564, 565, and 567 differ mainly in operating frequency range, power supply requirements, and frequency and bandwidth adjustment ranges. Only the SE/NE 565 phase-locked loop is discussed here because it is one of the most commonly used devices of the 560 series.

Figure 10-26 shows the block diagram and connection diagram of the 565 PLL.

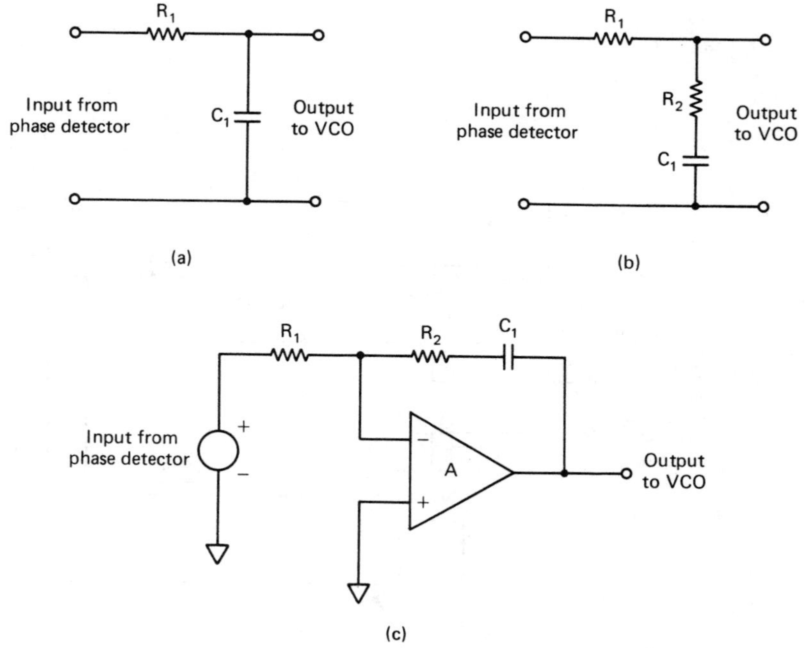

Figure 10-23 Low-pass filters. (a, b) Passive filters. (c) Active filter.

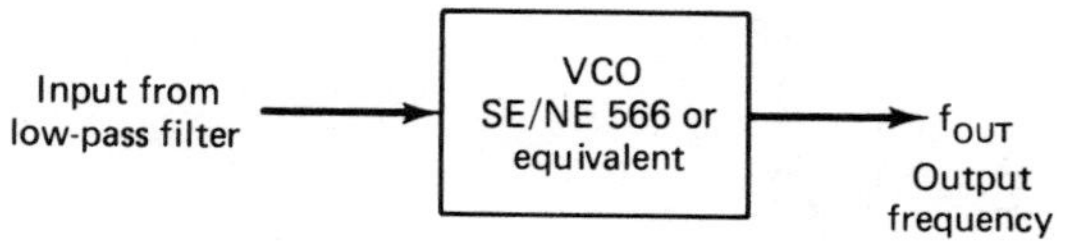

Figure 10-24 VCO block diagram.

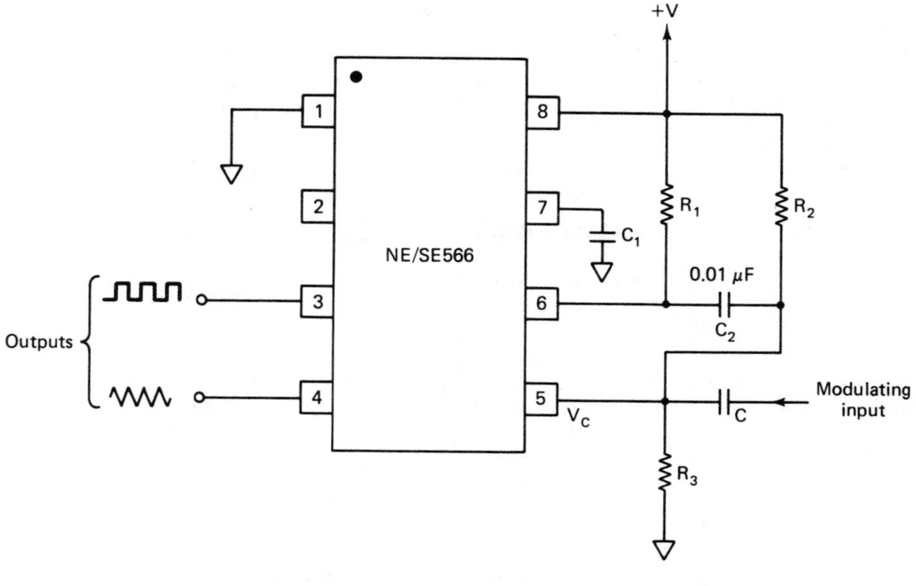

Figure 10-25 SE566 VCO connection diagram.

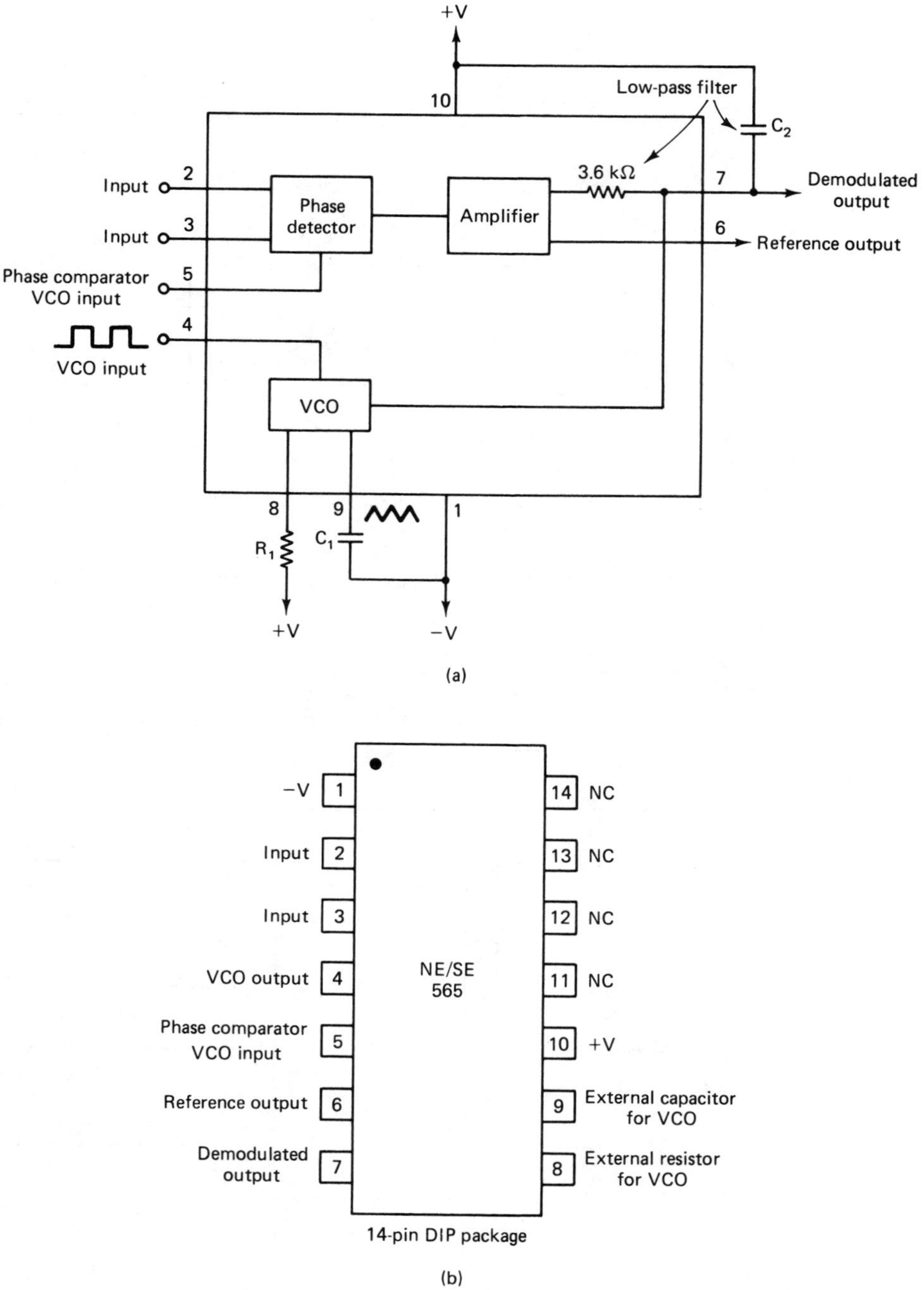

Figure 10-26 (a) NE/SE 565 PLL block diagram. (b) Connections diagram. (Courtesy of Signetics Corp.)

The device is available as a 14-pin DIP package and as 10-pin metal can package. The important electrical characteristics of the 565 PLL are

Operating frequency range: 0.001 Hz to 500 kHz
Operating voltage range: ± 6 to ± 12 V
Input level required for tracking: 10 mV rms minimum to 3 V peak-to-peak maximum
Input impedance: 10 kΩ typically
Output sink current: 1 mA, typically
Output source current: 10 mA, typically
Drift in VCO center frequency (f_{OUT}) with temperature: 300 ppm/°C, typically
Drift in VCO center frequency with supply voltage: 1.5%/V maximum
Triangle-wave amplitude: typically 2.4 V pp at ± 6 V (see block diagram, pin 9)
Square-wave amplitude: typically 5.4 V pp at ± 6 V (see block diagram, pin 4)
Bandwidth adjustment range: $\langle \pm 1$ to$\rangle$ $\pm 60\%$

The center frequency of the PLL is determined by the free-running frequency of the VCO, which is given by the equation

$$f_{OUT} \cong \frac{1.2}{4R_1C_1} \quad \text{Hz} \tag{10-9}$$

where R_1 and C_1 are an external resistor and a capacitor connected to pins 8 and 9, respectively [see block diagram of Figure 10-26(a).] The VCO free-running frequency f_{OUT} is adjusted externally with R_1 and C_1 to be at the center of the input frequency range. Although C_1 can be any value, R_1 must have a value between 2 and 20 kΩ. A capacitor C_2 connected between pin 7 and the positive supply (pin 10) forms a first-order low-pass filter with an internal resistance of 3.6 kΩ. The filter capacitor C_2 should be large enough to eliminate variations in the demodulated output voltage at pin 7 in order to stabilize the VCO frequency.

The 565 PLL can lock to and track an input signal over typically $\pm 60\%$ bandwidth with respect to f_{OUT} as the center frequency. The lock-range f_L and capture range f_C of the PLL is given by the following equations:

$$f_L = \pm \frac{8f_{OUT}}{V} \quad \text{Hz} \tag{10-10}$$

where f_{OUT} = free-running frequency of VCO (hertz)
$V = (+V) - (-V)$ (volts)

and

$$f_C = \pm \left[\frac{f_L}{(2\pi)(3.6)(10^3)(C_2)} \right]^{1/2} \tag{10-11}$$

where C_2 is in farads.

The lock range usually increases with an increase in input voltage but decreases with an increase in supply voltages. Pins 2 and 3 are the input terminals of the 565 PLL, and an input signal can be direct-coupled, provided that there is no dc voltage difference between the pins and dc resistances seen from pins 2 and 3 are equal. A short between pins 4 and 5 connects the VCO output (f_{OUT}) to the phase comparator and enables the comparator to compare f_{OUT} with the input signal f_{IN}.

In frequency multiplication applications a digital frequency divider is inserted between pins 4 and 5, as will be shown below. A dc reference voltage at pin 6 is approximately equal to the dc potential of the demodulated output at pin 7. In applications such as frequency shift keying (FSK), the dc reference voltage at pin 6 is used as an input to the comparator. [Refer to Section 10-4.3(b) for an example.] The lock range of PLL can be decreased with little change in the free-running frequency of the VCO by connecting a resistance between pins 6 and 7.

Example 10-8

Referring to the circuit of Figure 10-27(a), determine the free-running frequency f_{OUT}, the lock range f_L, and the capture range f_C.

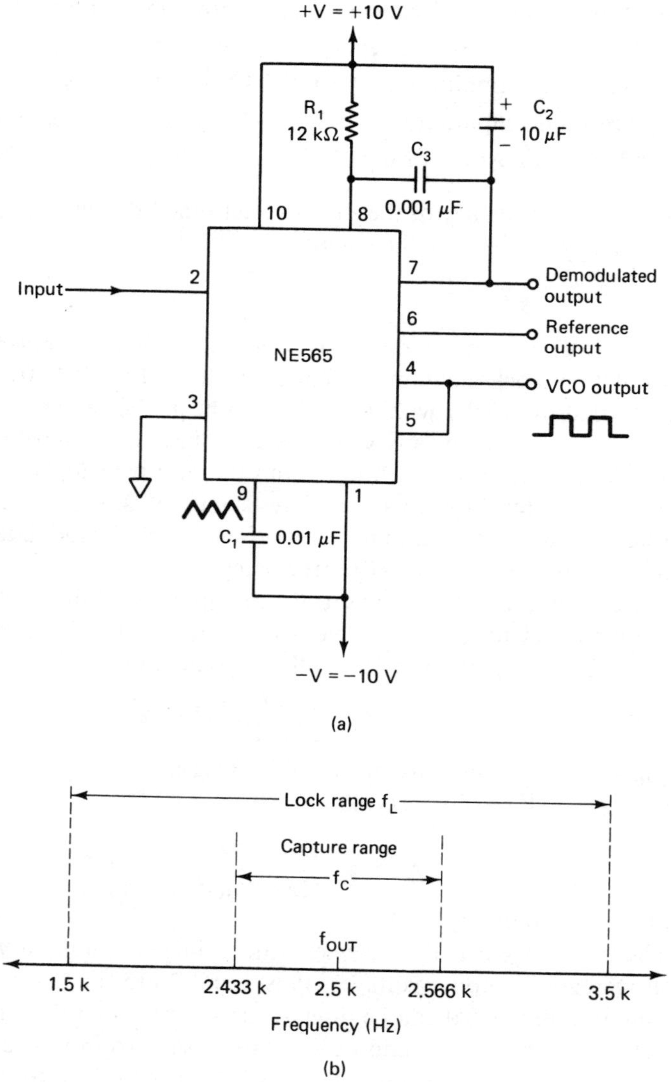

Figure 10-27 (a) Circuit of Example 10-8. (b) Relationship between f_{OUT}, f_L, and f_C.

Solution. From Equations (10-9) through (10-11),

$$f_{OUT} = \frac{1.2}{(4)(12)(10^3)(10^{-8})} = 2.5 \text{ kHz}$$

$$f_L = \pm\frac{8(2.5)(10^3)}{(20)} = \pm 1 \text{ kHz}$$

$$f_C = \pm\left[\frac{(10^3)}{(2\pi)(3.6)(10^3)(10)(10^{-6})}\right]^{1/2} = \pm 66.49 \text{ Hz}$$

The relationship between f_{OUT}, f_L, and f_C is depicted in Figure 10-27(b).

10-4.3 565 PLL Applications

10-4.3(a) Frequency multiplier. Figure 10-28(a) is a block diagram of a frequency multiplier using the 565 PLL. As shown in this diagram, the frequency divider is inserted between the VCO and the phase comparator. Since the output of the divider is locked to the input frequency f_{IN}, the VCO is actually running at a multiple of the input frequency. The desired amount of multiplication can be obtained by selecting a proper divide-by-N network, where N is an integer. For example, in order to obtain the output frequency $f_{OUT} = 5f_{IN}$, a divide-by-$N = 5$ network is needed. Figure 10-28(b) shows this function performed by a 7490 (4-bit binary counter) configured as a divide-by-5 circuit. In this figure, transistor Q_1 is used as a driver stage to increase the driving capability of the NE565.

To verify the operation of the circuit, one must determine the input frequency range and then adjust the free-running frequency f_{OUT} of the VCO by means of R_1 and C_1 so that the output frequency of the 7490 divider is midway within the predetermined input frequency range. The output of the VCO now should be $5f_{IN}$. In Figure 10-28(b), the output frequency f_{OUT} can be adjusted from 1.5 kHz to 15 kHz by varying potentiometer R_1 [see Equation (10-9)]. This means that the input frequency f_{IN} range has to be within 300 Hz to 3 kHz. In addition, the input waveform can either be sine or square and may be applied to input pins 2 or 3. Figure 10-28(c) shows the input-output waveforms for $f_{OUT} = 5f_{IN}$.

Even though supply voltages of ± 10 V are used in Figure 10-28(b), the NE565 can be operated on ± 5-V supply voltages instead. A small capacitor, typically 1000 pF, is connected between pins 7 and 8 to eliminate possible oscillations. Also, capacitor C_2 should be large enough to stabilize the VCO frequency.

10-4.3(b) Frequency shift keying (FSK) demodulator. In computer peripheral and radio (wireless) communications, the binary data or code is transmitted by means of a carrier frequency which is shifted between two preset frequencies. Since a carrier frequency is shifted between two preset frequencies, the data transmission is said to use a frequency shift keying (FSK) technique.

A very useful application of the 565 PLL is as a FSK demodulator. In the 565 PLL the frequency shift is usually accomplished by driving a VCO with the binary data signal so that the two resulting frequencies correspond to the logic 0 and logic 1 states of the binary data signal. The frequencies corresponding to logic 1 and logic 0 states are commonly called the *mark* and *space* frequencies. Several standards are used to set the mark and space frequencies. For example, when transmitting teletype-

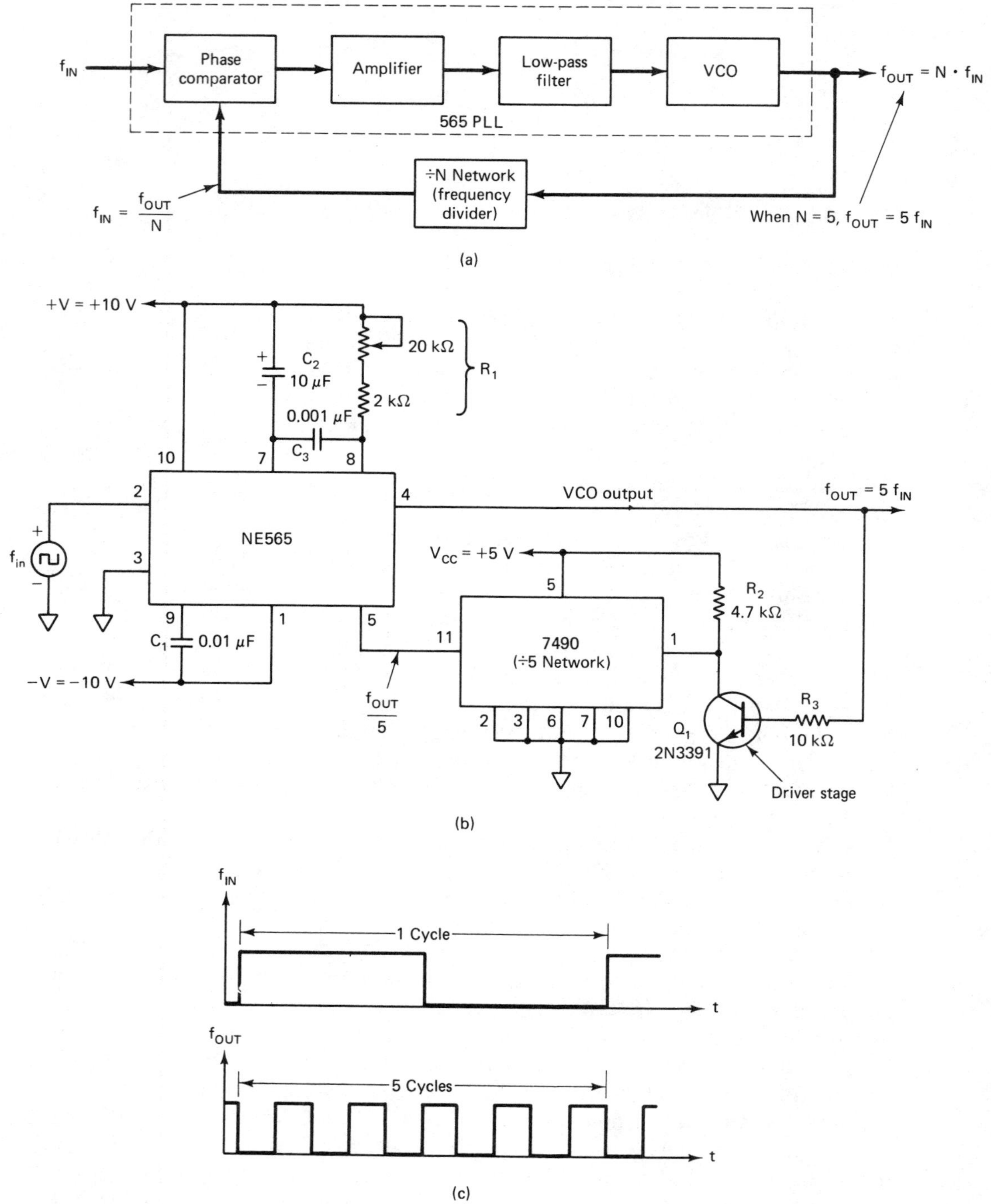

Figure 10-28 Frequency multiplier using the 565. (a) Block diagram. (b) Connection diagram for multiply-by-5. (c) Input-output waveforms.

writer information using a modulator–demodulator system (modem for short), a 1070 Hz–1270 Hz (mark-space) pair represents the originate signal, while a 2025 Hz–2225 Hz (mark-space) pair represents the answer signal.

Figure 10-29 shows a typical block diagram of the AM (amplitude modulation) and FM (frequency modulation) transmitters and receivers. A simplified diagram using only an FSK generator and FSK demodulator is shown in Figure 10-30. The FSK generator is formed by using a 555 as an astable multivibrator whose frequency is controlled by the state of transistor Q_1. In other words, the output frequency of the FSK generator depends on the logic state of the digital data input. 150 Hz is one of the standard frequencies at which the data are commonly transmitted. When the input is logic 1, transistor Q_1 is off. Under these conditions, the 555 works in its normal mode as an astable multivibrator; that is, capacitor C charges through R_A and R_B to $\frac{2}{3}V_{CC}$ and discharges through R_B to $\frac{1}{3}V_{CC}$. Thus capacitor C charges and discharges alternately between $\frac{2}{3}V_{CC}$ and $\frac{1}{3}V_{CC}$ as long as the input is at a logic 1 state. The frequency of the output waveform is given by Equation (10-6):

$$f_o = \frac{1.45}{(R_A + 2R_B)C} = 1070 \text{ Hz}$$

The values of R_A, R_B, and C are selected so that f_o represents a mark frequency (1070 Hz). On the other hand, when the input is logic 0, Q_1 is on (saturated), which in turn connects the resistance R_C across R_A. This action reduces the charging time of the capacitor and increases the output frequency which is given by Equation (10-6):

$$f_o = \frac{1.45}{(R_A \| R_C + 2R_B)C} = 1270 \text{ Hz}$$

By proper selection of resistance R_C, this frequency is adjusted to equal the space frequency of 1270 Hz. The difference between FSK signals of 1070 Hz and 1270 Hz is 200 Hz; this difference is called the *frequency shift*.

As shown in Figure 10-30, the output of the 555 FSK generator is then applied to the 565 FSK demodulator. Capacitive coupling is used at the input to remove a dc

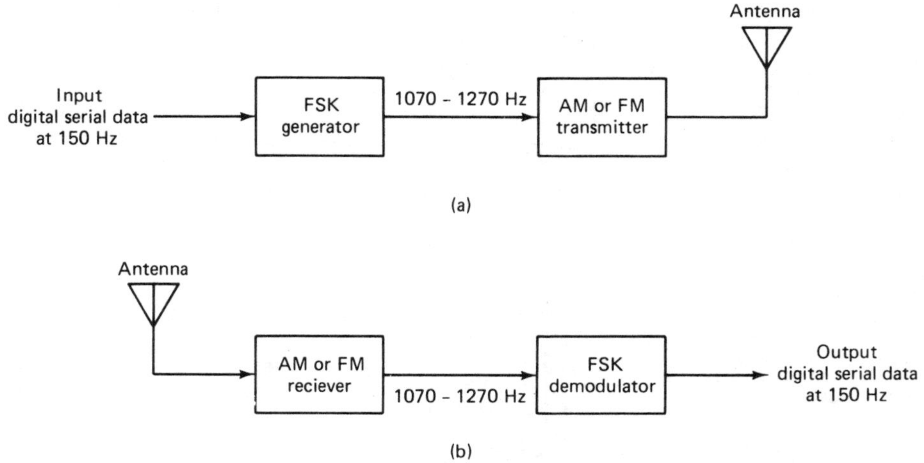

Figure 10-29 (a) Block diagram of AM/FM transmitter. (b) Receiver.

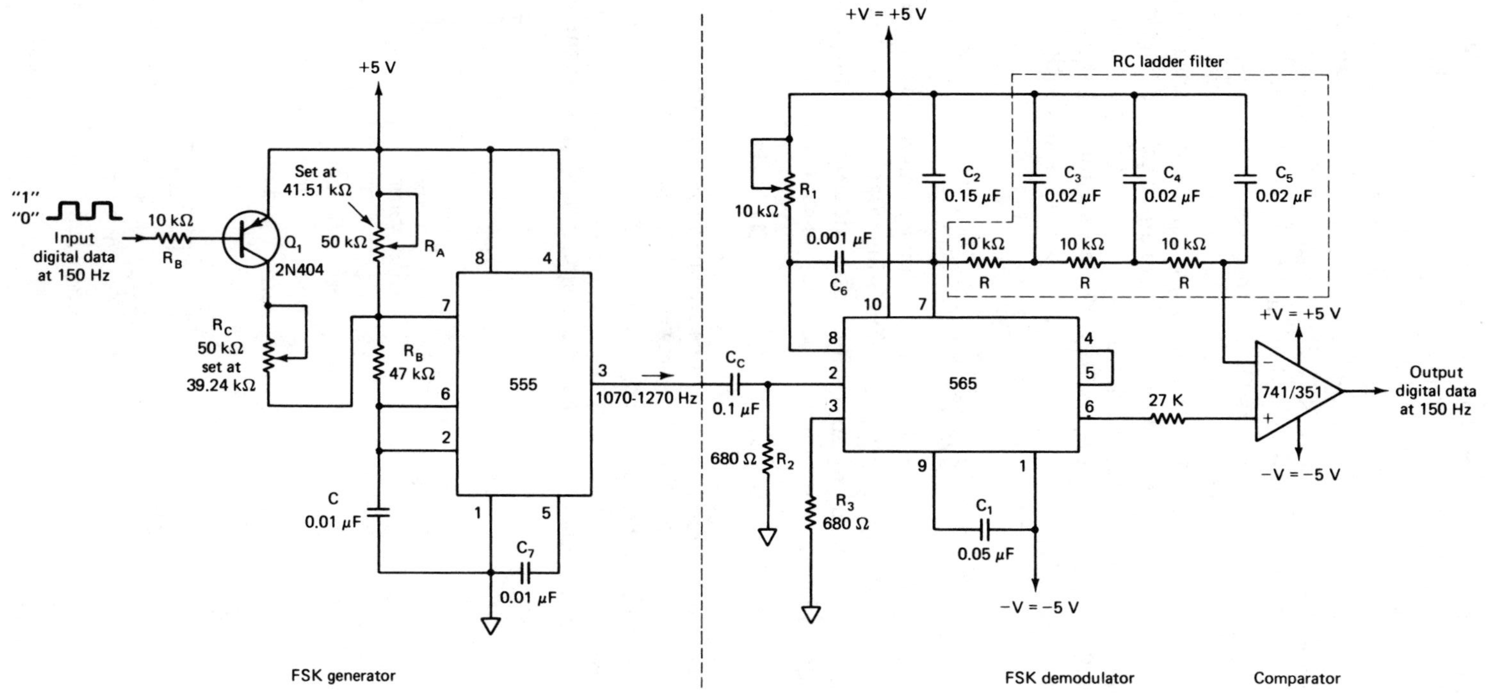

Figure 10-30 565 as an FSK demodulator.

level. As the signal appears at the input of the 565, the loop locks to the input frequency and tracks it between the two frequencies with a corresponding dc shift at the output. Resistor R_1 and capacitor C_1 determine the free-running frequency of the VCO, while C_2 is a loop filter capacitor that establishes the dynamic characteristics of the demodulator. Here C_2 must be chosen smaller than usual to eliminate overshoot on the output pulse. A three-stage *RC* ladder (low-pass) filter is used to remove the carrier component from the output. The high cutoff frequency ($f_H = 1/2\pi RC$) of the ladder filter is chosen to be approximately halfway between the maximum keying rate of 150 Hz and twice the input frequency, that is, approximately 2200 Hz. The output signal of 150 Hz can be made logic compatible by connecting a voltage comparator between the output of the ladder filter and pin 6 of the PLL. The VCO frequency is adjusted with R_1 so that at $f_{IN} = 1070$ Hz a slightly positive voltage is obtained at the output.

10-5 POWER AMPLIFIERS

Small-signal amplifiers are essentially voltage amplifiers that supply their loads with larger amplified signal voltages. On the other hand, large signal or power amplifiers supply a large-signal current to current-operated loads such as speakers and motors.

In previous discussion the general-purpose op-amp has been emphasized as a voltage amplifier because it is essentially a small-signal amplifier with a limited output current capability. For instance, the maximum output current supplied by the 741-type op-amp is equal to its *short-circuit* ($R_L = 0\ \Omega$) current, namely 25 mA ideally. Obviously, the current supplied by the 741 will decrease with an increase in load. In audio applications, however, the amplifier is called upon to deliver much higher current than that supplied by general-purpose op-amps. This means that loads such as speakers and motors requiring substantial currents cannot be driven directly by the output of general-purpose op-amps. However, there are two possible solutions available. One method is to use discrete or monolithic power transistors called *power boosters* at the output of the op-amp; a second method is to use specialized ICs designed as power amplifiers.

10-5.1 Power Amplifiers Using Power Boosters

A simple method of increasing the output current of a general-purpose op-amp is to connect a *power booster* in series with the op-amp. Obviously, the output characteristics of this arrangement, such as output current and output impendance, are then controlled by the choice of the power booster and the configuration in which it is used. Usually, a discrete power booster is an emitter-follower stage formed by using a power transistor, as in Figure 10-31(a). (Transistors with power dissipation larger than $\frac{1}{2}$ W are called power transistors.) Because of its unity gain, the emitter follower helps to retain the voltage-gain characteristics of the general-purpose op-amp. In addition, the emitter follower offers a relatively high input impedance; therefore, it also works as an isolation stage (buffer) between the op-amp and the load. The input impedance and output current of the emitter follower may be further increased

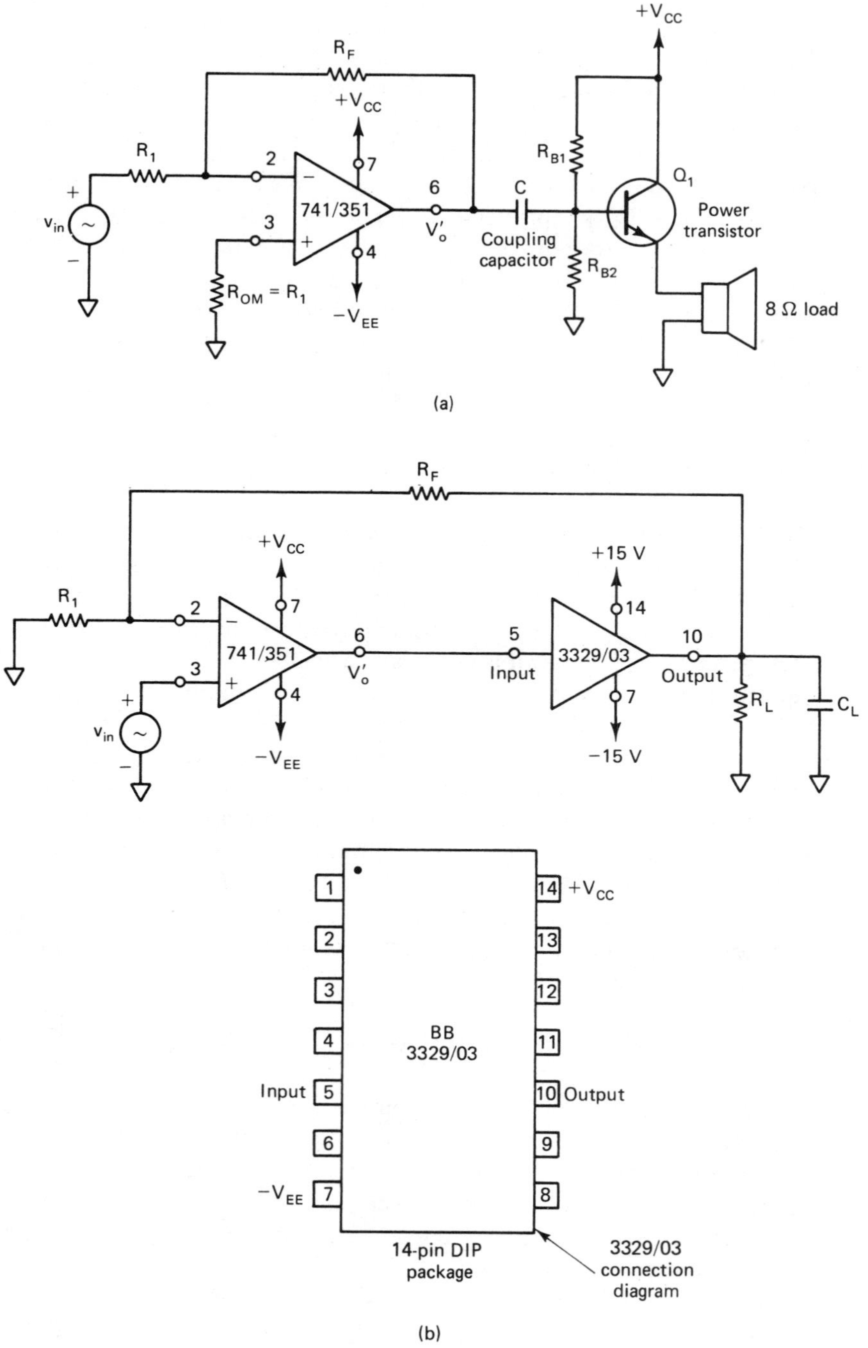

Figure 10-31 Power amplifiers using general-purpose op-amp input stage and (a) discrete power booster. (b) 3329/03 IC power booster. (Courtesy of Burr-Brown Research Corporation.)

by the use of Darlington power transistors. For increased efficiency, the power booster is often configured as a push-pull Class B amplifier.

Figure 10-31(b) shows a typical connection diagram for the power amplifier using a general-purpose op-amp and the Burr-Brown 3329/03 monolithic power booster. The 3329/03, a unity-gain power booster amplifier which does not require any external components, is designed to be used inside the feedback loop of an op-amp. It operates over a power supply range of ± 12 to ± 18 V. The device has a Class B output stage which provides output current of ± 100 mA at 20 V peak-to-peak output voltage swing when operated on supplies of ± 15 V. In addition, the device will operate over the temperature range -40 to $+85$°C without a heat sink. Above all, when the 3329/03 power booster is used with an op-amp, it does not degrade the frequency response and output characteristics of the op-amp, mainly because of the following electrical characteristics:

Full power frequency = 1 MHz
Small-signal bandwidth = 5 MHz
Input impedance = 10 kΩ
Open-loop output impedance = 10 Ω

Another power booster with increased output current, slew rate, and bandwidth is the Burr-Brown 3553. It has an output current of ± 200 mA with a 300-MHz bandwidth and a slew rate of 2000 V/μs. The 3553 is therefore ideally suited for line driving applications (50-Ω load) in which fast pulses or wideband signals are involved. For applications such as stepper motor drivers, servomotor drivers, power supplies, and power DACs requiring output voltage up to ± 30 V and full output current ± 2 A, the Intersil ICL8063 (16-pin DIP) monolithic power transistor driver and amplifier is most suitable. The ICL8063 will operate from the outputs of most of op-amps and devices such as timers and comparators. When used in conjunction with general-purpose op-amps, external complementary power transistors, and 8 to 10 passive components (resistors and capacitors), the ICL8063 can deliver more than 50 W to external loads. It has built-in safe area protection, short-circuit proof protection, and built-in ± 13-V regulators.

10-5.2 Monolithic Power Amplifiers

A variety of monolithic as well as hybrid power amplifiers are commercially available. Although power amplifiers differ from general-purpose op-amps in delivering various amounts of power, they are nearly as compact. Because of its wide acceptance, National Semiconductor's LM380 audio power amplifier is introduced here and followed by a brief discussion of hybrid power amplifiers with output power up to 100 W.

10-5.2(a) LM380 power audio amplifier. The National Semiconductor's LM380 is a power audio amplifier designed to deliver a minimum of 2.5 W (rms) to an 8-Ω load and hence is ideal for consumer applications. Because of the following features, the LM380 requires a minimum number of external components:

An internally fixed gain of 50 (34 dB)

Output automatically self-centering to one half of the supply voltage

Output short-circuit proof with internal thermal limiting

A unique input stage allowing inputs to be ground referenced or *ac* coupled

In addition, the LM380 has a wide supply voltage range (5 to 22 V), high peak current capability (1.3 A maximum), high impedance (150 kΩ), low (0.2%) total harmonic distortion (THD), and is available in a standard DIP package. Also, the LM380 has a bandwidth of 100 kHz typically at $P_{OUT} = 2$ W and $R_L = 8\ \Omega$.

Figure 10-32 shows the connection diagram, block diagram, and schematic diagram of the LM380. A copper lead frame used with the center three pins (3, 4, 5, 10, 11, and 12) on either side of the DIP package comprises a heat sink [see Figure 10-32(a)]. Therefore, there is no need to use a separate heat sink for the audio amplifier.

LM380 Circuit Description. The schematic diagram of the LM380 shown in Figure 10-32(c) is composed of four stages: PNP emitter follower, differential amplifier, common emitter, and quasi-complementary emitter follower. The input stage is an emitter follower composed of PNP transistors Q_1 and Q_2, which drives the PNP Q_3–Q_4 differential pair. The choice of PNP input transistors Q_1 and Q_2 allows the input to be referenced to ground; that is, input can be direct coupled to either the inverting (pin 6) or the noninverting (pin 2) terminals of the amplifier.

The current in the PNP differential pair Q_3–Q_4 is established by Q_7, R_3, and $+V$. The current mirror formed by transistors Q_7, Q_8, and associated resistors then establishes the collector current of Q_9. Transistors Q_5 and Q_6 constitute collector loads for the PNP differential pair. The output of the differential amplifier is taken at the junction of Q_4 and Q_6 transistors and is applied as an input to the common-emitter voltage gain stage.

The common-emitter amplifier stage is formed by transistor Q_9 with D_1, D_2, and Q_8 as a current source load. The capacitor C between the base and collector of Q_9 provides internal compensation and helps to establish the upper cutoff frequency of 100 kHz at 2 W for 8-Ω loads. Since Q_7 and Q_8 form a current mirror, the current through D_1 and D_2 is approximately the same as the current through R_3. In addition, D_1 and D_2 are temperature-compensating diodes for transistors Q_{10} and Q_{11} in that D_1 and D_2 have the same characteristics as the base–emitter junctions of Q_{10} and Q_{11}. Therefore, the current through Q_{10} and (Q_{11}–Q_{12}) is approximately equal to the current through diodes D_1 and D_2.

The output stage is a quasi (false)-complementary pair emitter follower formed by NPN transistors Q_{10} and Q_{12}. In fact, the combination of PNP transistor Q_{11} and NPN transistor Q_{12} has the power capability of an NPN transistor but the characteristics of a PNP transistor.

Because of the arrangement of the output stage, the quiescent output voltage is half the supply voltage ($+V$). Furthermore, the negative *dc* feedback applied through R_5 balances the differential amplifier so that the dc output voltage is stabilized at $+V/2$. To decouple the input stage from the supply voltage $+V$, a bypass capacitor on the order of microfarads should be connected between the bypass terminal (pin 1) and ground (pin 7). The overall internal voltage gain of the amplifier is fixed at 50.

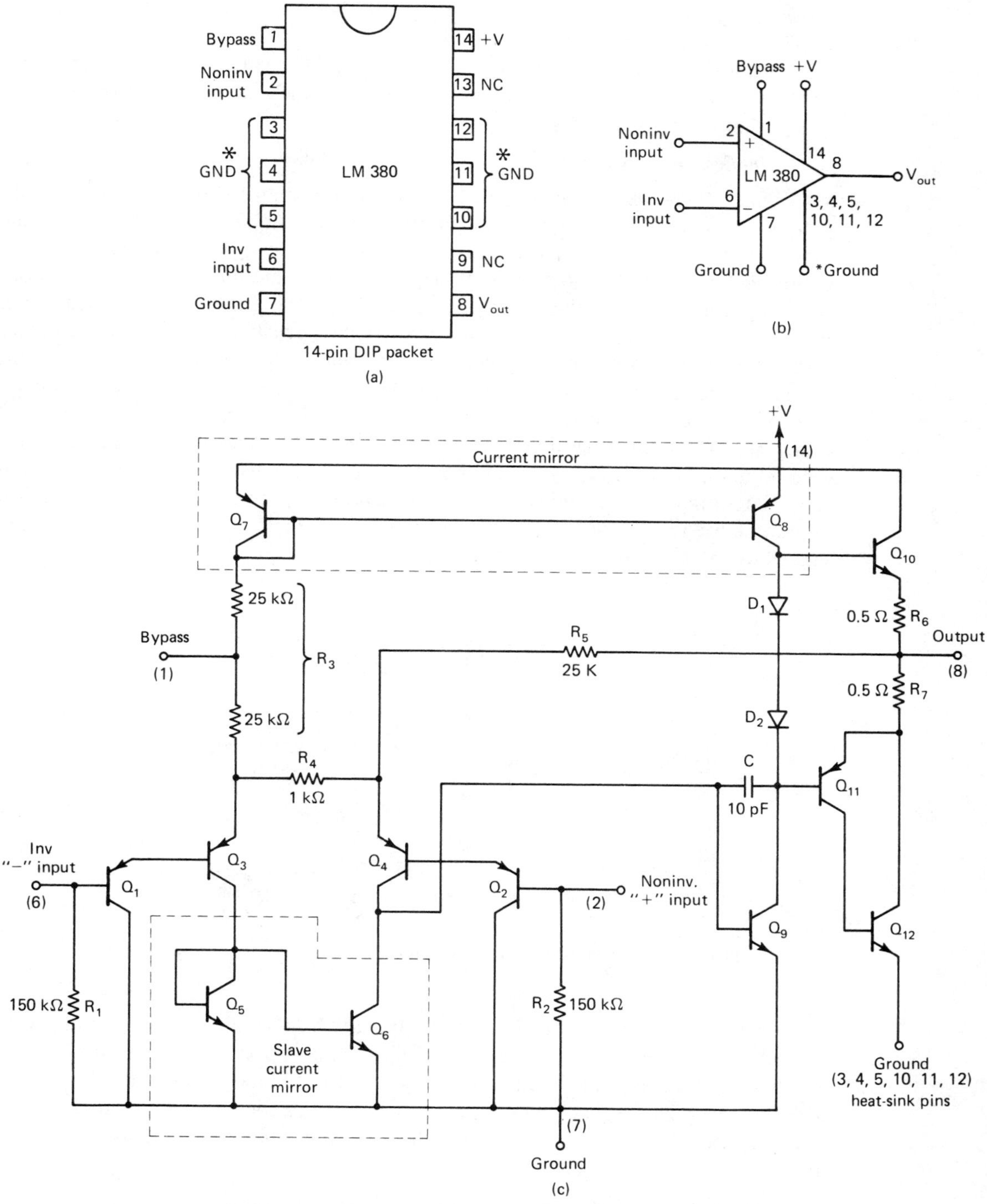

Figure 11-32 (a) LM380 power audio amplifier connection diagram. An asterisk indicates ground-heat sink pins. (b) Block diagram. (c) Schematic diagram. (Numbers in parentheses are pin numbers.) (Courtesy of National Semiconductor.)

However, gain can be increased by using positive feedback, as illustrated in the applications discussed in the next section.

Another commonly used audio amplifier is the LM384. The schematic diagram of the LM384 is the same as that of the LM380, which is shown in Figure 10-32, except that the former is designed to deliver 5 W.

Applications

Audio Power Amplifier. Figure 10-33 shows the simplest and most basic application of the LM380—as an audio power amplifier. As shown in this figure, the amplifier requires very few external components because of the internal biasing, compensation, and the fixed gain. When the power amplifier is used in the non-inverting configuration, the inverting terminal may be either shorted to ground, connected to ground through a resistor or capacitor, or left open as shown in Figure 10-33. Similarly, when the power amplifier is used in the inverting mode, the non-inverting terminal may be either shorted to ground or returned to ground through a resistor or capacitor. Usually, a capacitor is connected between the inverting terminal and ground if the input has a high internal impedance. Nevertheless, in either configuration the supply voltage $+V$ should be decoupled by connecting a capacitor between the $+V$ terminal (pin 14) and ground. As a precautionary measure, an *RC* combination should be used at the output terminal (pin 8) to eliminate 5- to 10-MHz oscillation, especially in an RF-sensitive environment.

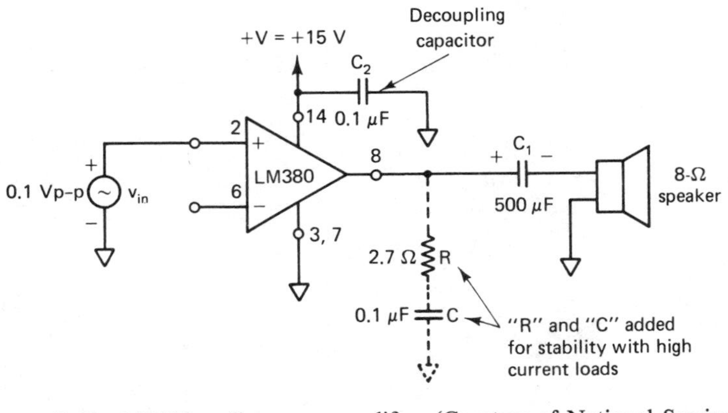

Figure 10-33 LM380 audio power amplifier. (Courtesy of National Semiconductor.)

Although the gain of the LM380 is internally fixed at 50, it can be changed with the use of external components. Specifically, gains up to 300 are possible when positive feedback is used. Figure 10-34(a) shows the LM380 configured for a gain of 200 using positive feedback. On the other hand, variable gains up to 50 are obtained with the use of a potentiometer across the two input terminals, as shown in Figure 10-34(b).

Bridge Power Audio Amplifier. For applications requiring more power than is provided by the single LM380 amplifier, two LM380s can be used in the bridge configuration shown in Figure 10-35. In this arrangement the maximum output voltage

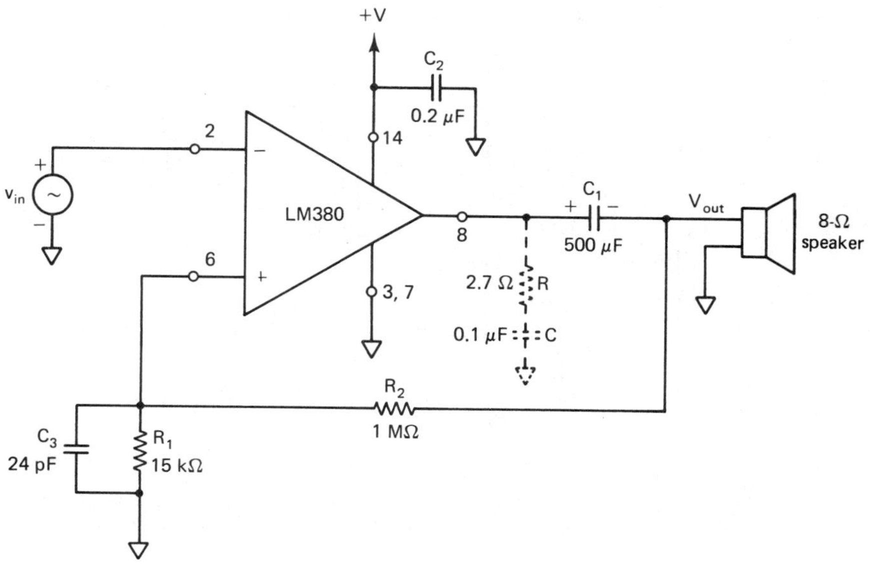

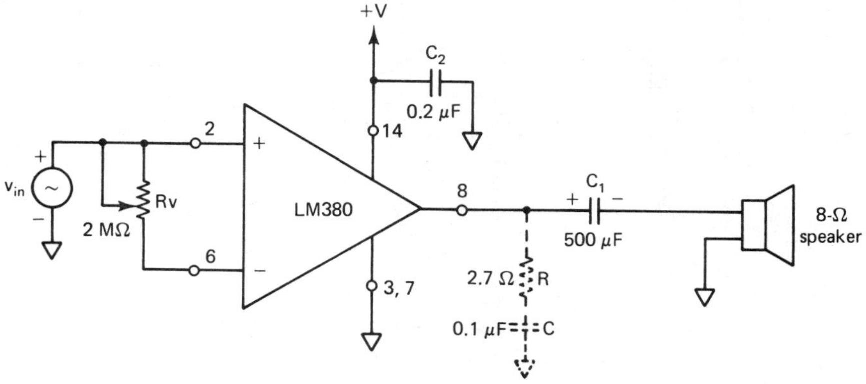

Figure 10-34 (a) LM380 configured for a gain of 200 using positive feedback. (b) LM380 with a variable gain (volume control) up to 50 using a potentiometer across the two input terminals. (Courtesy of National Semiconductor.)

swing will be twice that of a single LM380 amplifier; therefore, the power delivered to the load will be four times as much. For improved performances, potentiometer R_4 should be used to balance the output offset voltages of the LM380s.

Intercom System. A simple and inexpensive intercom system can be formed by using the LM380 as shown in Figure 10-36. The speakers used in this figure are permanent-magnet types and hence act as microphones as well. The talk and listen modes are defined with reference to a master station. When the switch is in the talk

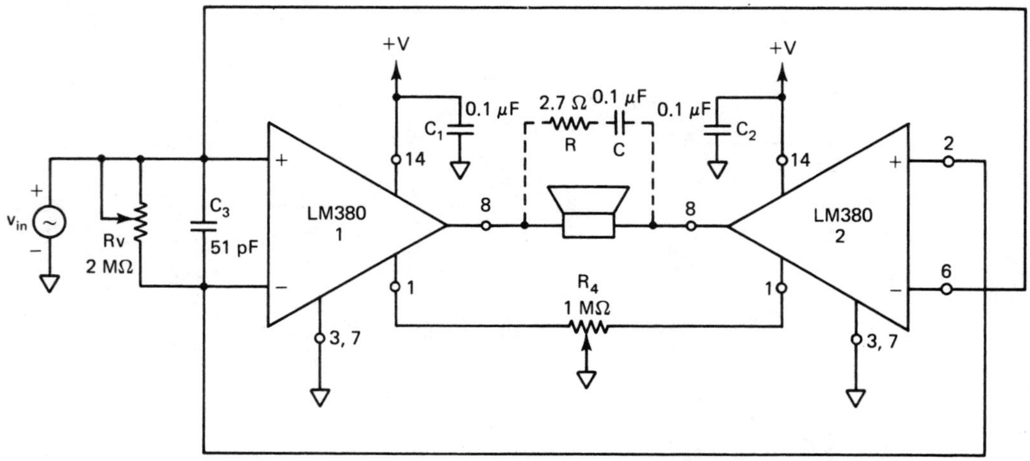

Figure 10-35 LM380s used in bridge configuration to provide more power. (*R* and *C* for stability with high current loads.) (Courtesy of National Semiconductor.)

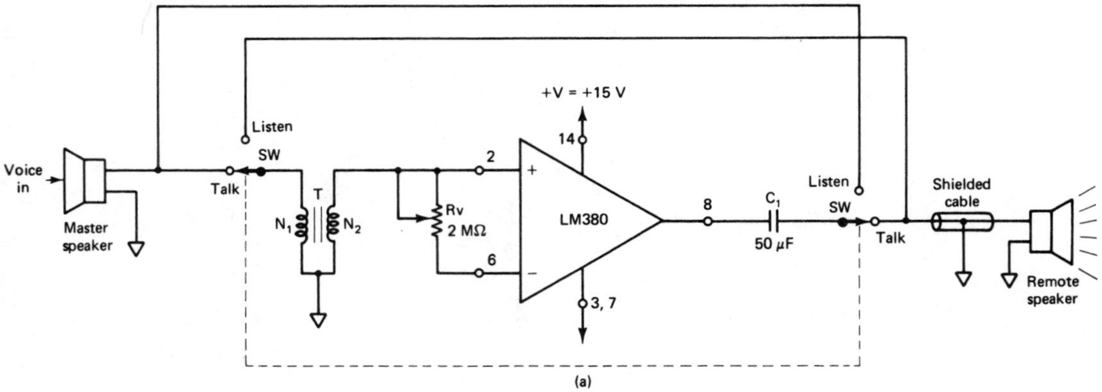

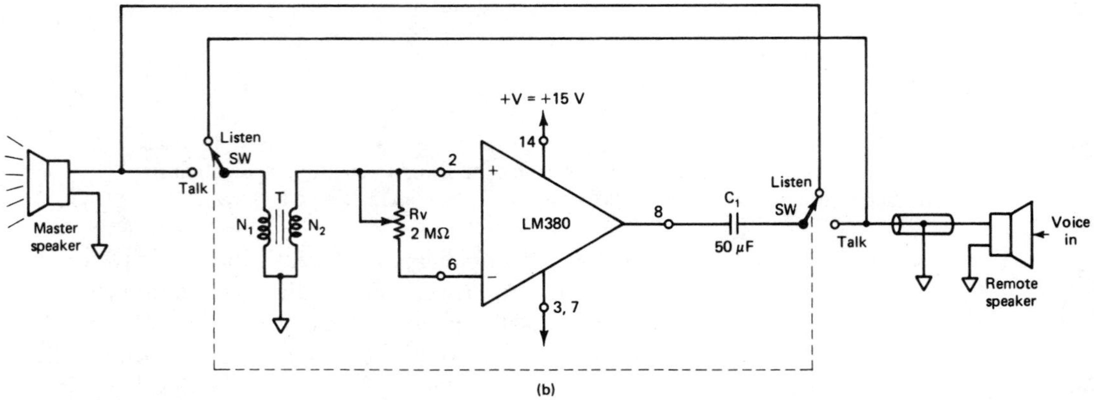

Figure 10-36 Intercom system using the LM380. (a) Talk mode. (b) Listen mode. (Courtesy of National Semiconductor.)

position, the master speaker acts as the microphone [see Figure 10-36(a)]. On the other hand, when the switch is in the listen mode, the remote speaker acts as the microphone [see Figure 10-36(b)]. In either position the overall gain of the circuit is the same and depends on the turns ratio of the transformer T as well as the internal gain of the LM380. For example, if the turns ratio $N1/N2 = 20$, the overall gain of the circuit will be $50 \times 20 = 1000$. However, the internal gain of the LM380 can be controlled with the use of potentiometer R_V.

A sample of monolithic power audio amplifiers with their output power ratings is listed in Table 10-4. In these power ICs, the heat-sink requirements are met without the use of separate heatsinks. Because of their compact size, power audio amplifiers are more convenient than any of the discrete forms of audio amplifiers. Duel power amplifiers such as the LM377, which delivers 2 W/channel, are used for stereo phonographs, tape players and recorders, and AM–FM stereo receivers.

TABLE 10-4 TYPICAL MONOLITHIC POWER AUDIO AMPLIFIERS

Output power (W)	IC type
1	MC1454
2	LM377 (dual)
4	LM378 (dual)
5	μA706
6	LM379 (dual)
8	TDA2002

For applications requiring more output power than that supplied by the monolithic power ICs, hybrid power amplifiers are commonly used. For example, the Intersil's ICH8510/8520/8530 is a family of hybrid power amplifiers designed to deliver 1, 2, and 2.8 A, respectively, at 24-V output levels. All amplifiers are protected against inductive kickback with internal power limiting, and against shorts to ground. Each device has a dc gain of 10^5 (100 dB), internal frequency compensation, and an electrically isolated package style that allows easy heat sinking. These devices are used to drive electronic values, push-pull solenoids, and dc and ac motors.

Another classic example of a hybrid power amplifier is the Burr-Brown's 3573, which is designed to deliver 100 W peak or 40 W continuous output power. When operated from $\pm$28-V power supplies, the device delivers $\pm$5 A peak minimum at $\pm$20 V (40 V peak to peak) to the load. The 3573 is internally frequency compensated with good input and distortion characteristics and is low in cost. It can be used to drive dc and ac motors, electronic valves, and push-pull solenoids.

10-6 VOLTAGE REGULATORS

A *voltage regulator* is a circuit that supplies a constant voltage regardless of changes in load currents. Although voltage regulators can be designed using op-amps, it is quicker and easier to use IC voltage regulators. Furthermore, IC voltage regulators are versatile, relatively inexpensive, and are available with features such as a programmable output, current/voltage boosting, internal short-circuit current limiting, thermal shutdown, and floating operation for high-voltage applications. IC voltage regulators are of the following types:

Fixed output voltage regulators: positive and/or negative output voltage

Variable output voltage regulators: positive or negative output voltage

Switching regulators

Special regulators

Except for the switching regulators, all other types of regulators are called *linear regulators*. The impedance of a linear regulator's active element may be continuously varied to supply a desired current to the load. On the other hand, in the *switching regulator* a switch is turned on and off at a rate such that the regulator delivers the desired average current in periodic pulses to the load. Because the switching element dissipates negligible power in either the *on* or *off* state, the switching regulator is more efficient than the linear regulator. Nevertheless, in switching regulators the power dissipation is substantial during the switching intervals (on to off or off to on). In addition, most loads cannot accept the average current in periodic pulses. Therefore, most practical regulators are of the linear type.

Voltage regulators are commonly used for on-card regulation and laboratory-type power supplies. Voltage regulators, especially the switching type, are used as control circuits in pulse width modulation (PWD), push-pull bridges, and series-type switchmode supplies. Almost all power supplies use some type of voltage regulator IC because voltage regulators are simple to use, reliable, low in cost, and above all, available in a variety of voltage/current ratings. A vast number of voltage regulators is available; data sheets and application notes provided by the manufacturers contain information on the design/use of these devices. The next section presents a selection of voltage regulators to illustrate design and construction features of these devices.

10-6.1 Positive Voltage Regulator Series with Seven Voltage Options

The 7800 series consists of three-terminal positive voltage regulators with seven voltage options [see Figure 10-37(a)]. These ICs are designed as fixed-voltage regulators and with adequate heat sinking can deliver output currents in excess of 1 A. Although these devices do not require external components, such components can be used to obtain adjustable voltages and currents. These ICs also have internal thermal overload protection and internal short-circuit current limiting. As shown in Figure 10-37(c), proper operation requires a common ground between input and output voltages. In addition, the difference between input and output voltages ($V_{in} - V_o$), called *dropout voltage*, must be typically 2.0 V even during the low point on the input ripple voltage. Further, the capacitor C_i is required if the regulator is located an appreciable distance from a power supply filter. Even though C_o is not needed, it may be used to improve the transient response of the regulator.

Typical performance parameters for voltage regulators are line regulation, load

Device type	Output voltage (V)	Maximum input voltage (V)
7805	5.0	
7806	6.0	
7808	8.0	
7812	12.0	35
7815	15.0	
7818	18.0	
7824	24.0	40

(a)

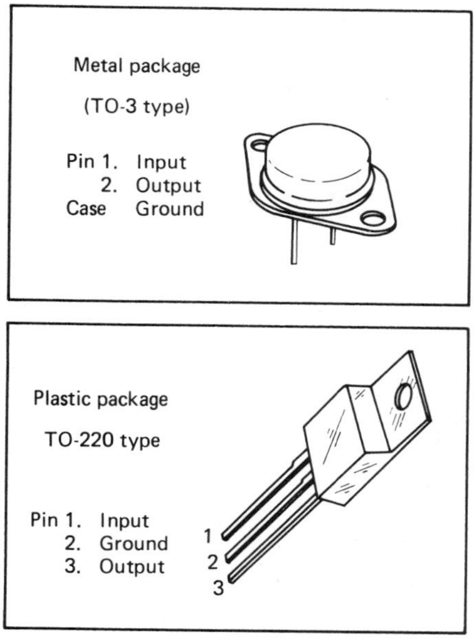

(b)

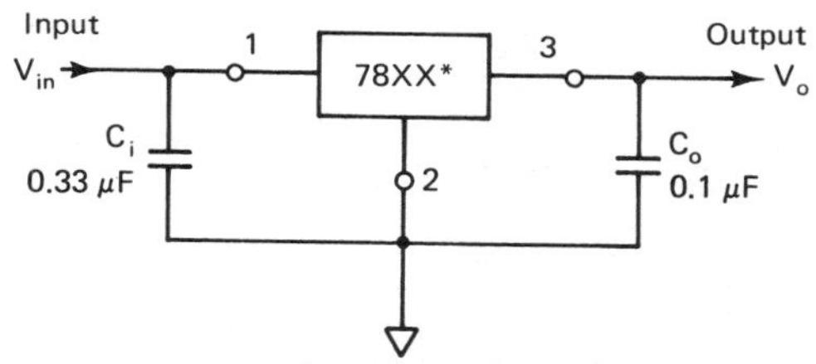

*XX: These two numbers of the type number indicate output voltage

(c)

Figure 10-37 7800 series regulators. (a) Voltage options. (b) Package types. (c) Standard application. (Courtesy of Motorola Semiconductor.)

regulation, temperature stability, and ripple rejection. *Line or input regulation* is defined as the change in output voltage for a change in the input voltage and is usually expressed in millivolts or as a percentage of output voltage V_o. *Load regulation* is the change in output voltage for a change in load current and is also expressed in millivolts or as a percentage of V_o. *Temperature stability* or *average temperature coefficient of output voltage* (TCV_0) is the change in output voltage per unit change in temperature and is expressed in either millivolts/°C or parts per million (ppm)/°C. *Ripple rejection* is the measure of a regulator's ability to reject ripple voltages. It is usually expressed in dB. The smaller the values of line regulation, load regulation, and temperature stability, the better the regulator.

The 7800 regulators can also be used as current sources. Figure 10-38 shows a typical connection diagram of the 7805C as a 0.5-A current source. The current supplied to the load is given by the equation

$$I_L = \frac{V_R}{R} + I_Q \tag{10-12a}$$

where I_Q = quiscent current (amperes)
= 4.3 mA typically for the 7805C

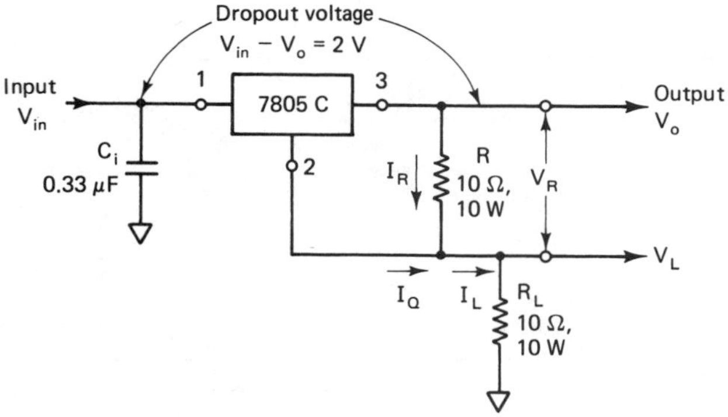

Figure 10-38 7805C as a 0.5-A current source.

Referring to Figure 10-38, $V_R = V_{23} = 5$ V and $R = 10\ \Omega$; therefore,

$$I_L \cong 0.5\text{ A}$$

The output voltage V_o with respect to ground is

$$V_o = V_R + V_L \tag{10-12b}$$

where $V_L = I_L R_L$. The load resistance $R_L = 10\ \Omega$; hence $V_L = 5$ V. Therefore, $V_o =$ 10 V. Since the dropout voltage for the 7805C is 2 V, the minimum input voltage required is given by the equation

$$\begin{aligned} V_{in} &= V_o + (\text{dropout voltage}) \\ &= 12\text{ V} \end{aligned} \tag{10-12c}$$

In short, a current source circuit using a voltage regulator can be designed for a desired value of load current (I_L) simply by selecting an appropriate value for R. Note, however, that V_{in} depends on the size of R_L and also the dropout voltage of the regulator [see Equation (10-12c)].

Example 10-9

Using the 7805C voltage regulator, design a current source that will deliver 0.25 A current to the 48-Ω 10-W load.

Solution. Determine the value of R and find the minimum value of V_{in} needed by using Equations (10-12a) through (10-12c):

$$R \cong \frac{5\text{ V}}{0.25\text{ A}} = 20\ \Omega$$

$$V_o = 5.0 + (48)(0.25) = 17\text{ V}$$

$$V_{in} = 17 + 2 = 19\text{ V}$$

The 7900 series of fixed output negative voltage regulators are complements to the 7800 series devices. These negative regulators are available in the same seven-voltage options as the 7800 devices. In addition, two extra voltage options, −2 V, and −5.2 V, are also available in the negative 7900 series, as shown in Figure 10-39(a). Figure 10-39(b) shows the package types in which the 7900 series voltage regulators are available.

Device type	Output voltage (V)	Maximum input voltage (V)
7902	−2.0	
7905	−5.0	
7905.2	−5.2	
7906	−6.0	−35
7908	−8.0	
7912	−12.0	
7915	−15.0	
7918	−18.0	
7924	−24.0	−40

(a)

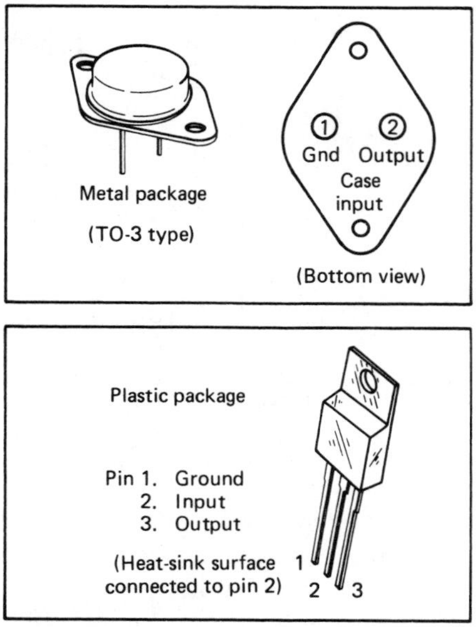

(b)

Figure 10-39 7900 series voltage regulators. (a) Voltage options. (b) Package types. (Courtesy of Motorola Semiconductor.)

10-6.2 Switching Regulators

An example of a general-purpose regulator is Motorola's MC1723. It can be used in many different ways, for example, as a fixed positive or negative output voltage regulator, variable output voltage regulator, or switching regulator. Because of its flexibility, it has become a standard type in the electronics industry. Although it is designed to deliver load current to 150 mA, the current capability can be increased to several amperes through the use of one or more external pass transistors. Figure 10-40 shows the connections for switching-regulator action of the 723 type regulator for positive output. The regulator requires an external transistor and a 1-mH choke. To minimize its power dissipation during switching, the external transistor used must be a switching power transistor. The 1-mH choke smooths out the current pulses delivered to the load while capacitor C holds output voltage at a constant dc level.

10-6.3 Voltage References

Voltage references are a special type of voltage regulators that are used for reference purposes as reference voltages. Typical applications of voltage references include D/A and A/D converters that do not have internal references, amplifier biasing, low-

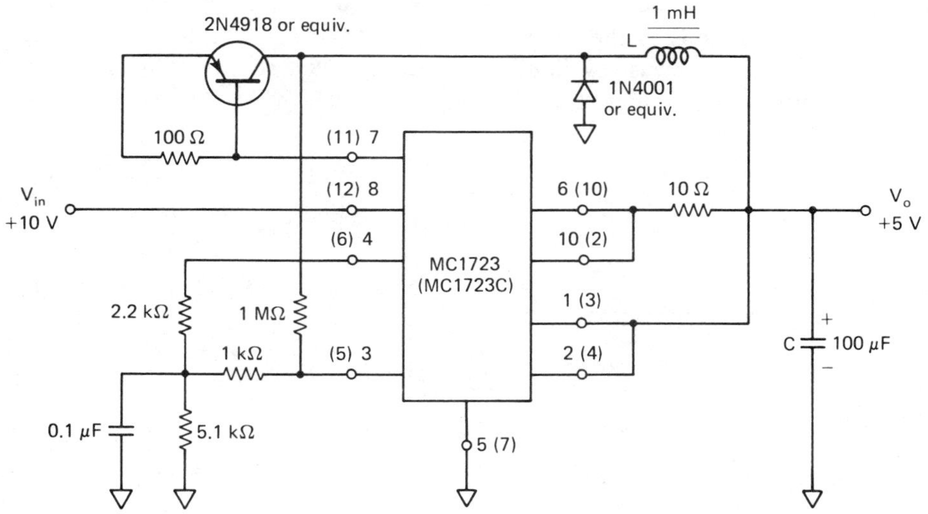

Figure 10-40 Motorola MC1723 as a 5-V 1-A switching regulator. (Pin numbers adjacent to terminals are for the metal packages and those in parentheses are for the DIP packages.) (Courtesy of Motorola Semiconductor.)

temperature-coefficient zener replacements, high-stability current references, comparator circuits, and voltmeter system references. Typical examples of voltage references include the following: The ICL8069 is a 1.2-V temperature-compensated voltage reference; the 9495 is a Teledyne's 5-V reference; and the MC1403 is a Motorola's 2.5-V reference.

10-6.3(a) MC1403 used as a voltage reference for DAC MC1408. Figure 10-41 shows a D/A converter using the MC1403 voltage reference. A stable current reference of nominally 2.0 mA required for the MC1408 is obtained from the MC1403 with the addition of series resistors R_1 and R_2. Besides that, resistor R_3 improves temperature performance and capacitor C_o decouples any noise present on the reference line. A single MC1403 can provide the required current input for up to five of the D/A converters. For complete connection diagram of MC1408, refer to Figure 9-20(a).

10-7 VOLTAGE INVERTER

Datel's VI-7660 is a monolithic CMOS voltage inverter that provides −1.5 to −10 V from +1.5 to +10 V supplies with the addition of only two noncritical external capacitors. The block diagram of the VI-7660 is shown in Figure 10-42(a), which contains a dc voltage regulator, *RC* oscillator, voltage-level translator, four output-power MOS switches, and a logic network. The logic network senses the most negative voltage in the device and ensures that the output N-channel switches are not forward biased. This assures latch-up-free operation. When unloaded, the oscillator oscillates at a nominal frequency of 10 kHz for an input supply voltage of 5 V. Because of noise or other considerations, it may be desirable to increase the oscillator

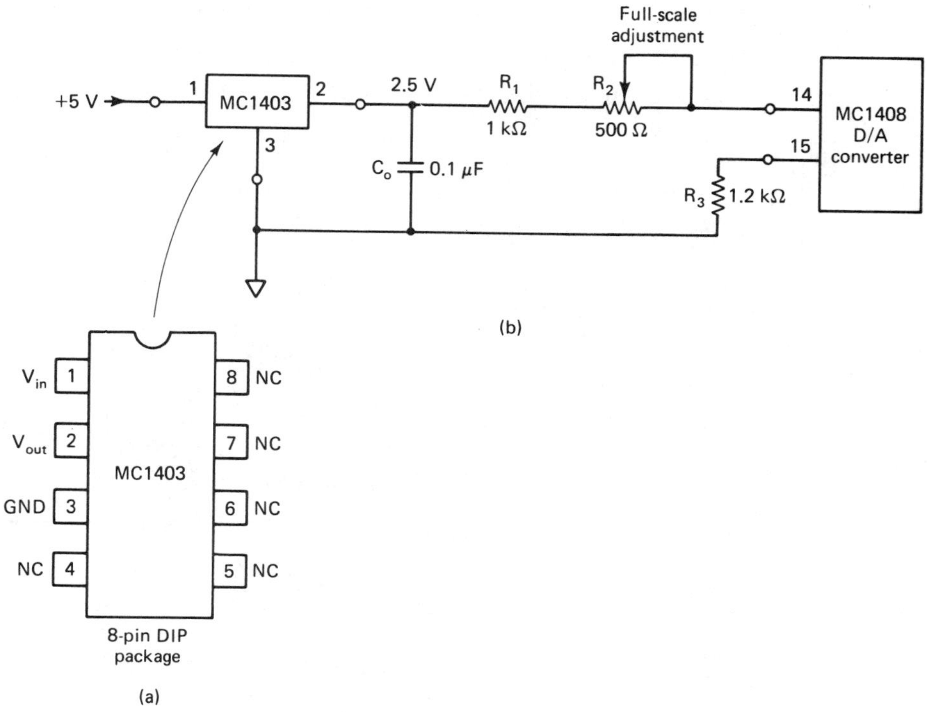

Figure 10-41 (a) MC1403 voltage reference pin diagram. (b) MC1403 providing the reference current for the MC1408. (Courtesy of Motorola Semiconductor.)

frequency in some applications. This may be done by overdriving the oscillator from an external clock. On the other hand, to maximize the conversion efficiency it may be necessary to lower the oscillator frequency. This is achieved by connecting an additional capacitor (typically 100 pF) between pins 7 and 8 [see Figure 10-42(b)].

Typical applications for the VI-7660 include data acquisition and microprocessor-based systems in which a positive supply is available and an additional negative supply is required. The VI-7660 is also ideally suited as on-board negative supply for up to 64 dynamic RAMs (random-access memory ICs).

10-7.1 VI-7660 Applications

10-7.1(a) Simple negative converter. Figure 10-43 shows typical connections when the VI-7660 is used as a simple negative converter. To improve the low-voltage (LV) operation, that is, when supply voltage $(+V) < 3.5$ V, the LV pin may be grounded. However, for $(+V) \geq 3.5$ V, the LV pin is left open to prevent device latch-up. Also, an additional diode D_1 must be included for proper operation at higher supply voltages $(+V > 6.5$ V) and/or elevated temperatures. When capacitor C_{osc} is used to maximize the conversion efficiency of the device, the oscillator frequency decreases and hence the reactances of C_1 and C_2 will increase (see Figure 10-43). Therefore, to overcome the increase in their reactances, the values of C_1 and C_2 must be increased by the same factor that the frequency has been reduced. For example,

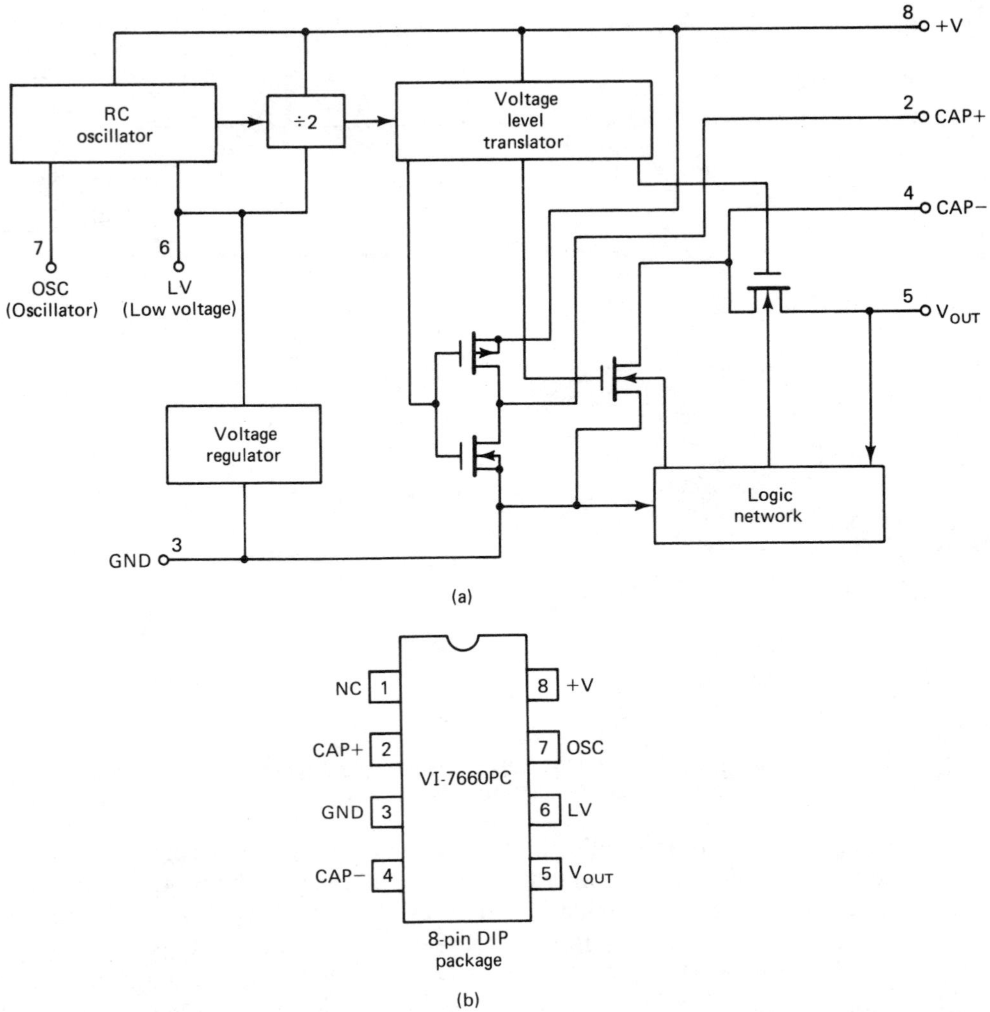

Figure 10-42 VI-7660 voltage inverter. (a) Block diagram. (b) Connection diagram. (Courtesy of Datel-Intersil Inc.)

the addition of $C_{osc} = 100$ pF between pins 7 and 8 will lower the oscillator frequency from 10 kHz to 1 kHz and will thereby call for an increase in the values of C_1 and C_2 from 10 μF to 100 μF. The output voltage equation is

$$V_{OUT} = -(+V) \qquad \text{for } 1.5 \leq +V \leq 6.5\ V \qquad (10\text{-}13a)$$

$$V_{OUT} = -(+V - V_{D1}) \qquad \text{for } 6.5 \leq +V \leq 10\ V \qquad (10\text{-}13b)$$

where V_{D1} = forward voltage drop of diode D_1
= 0.7 V typically

10-7.1(b) Negative-voltage multiplier. To produce larger negative multiplication of the initial supply voltage, the VI-7660's may be cascaded as shown in Figure 10-44. The practical limit to the number of devices that can be cascaded is

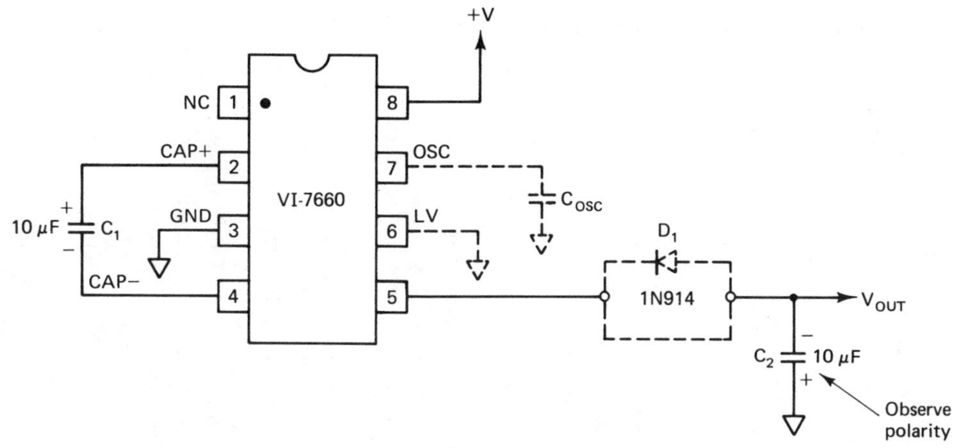

Figure 10-43 VI-7660 as a simple negative converter. (Courtesy of Datel-Intersil Inc.)

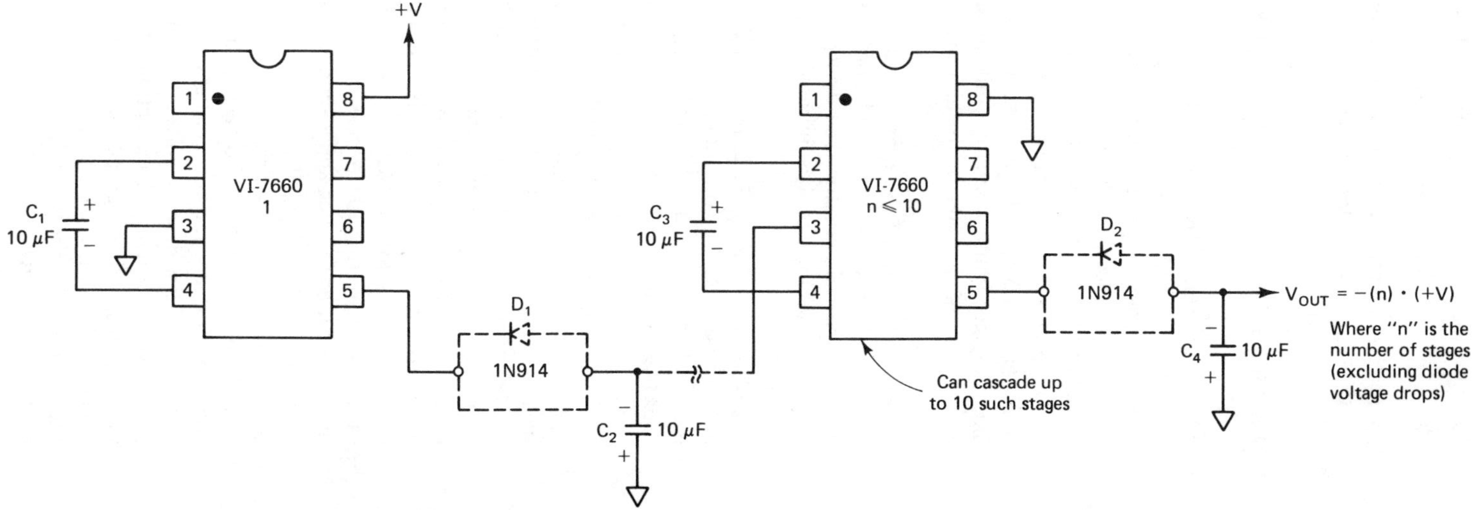

Figure 10-44 Cascading VI-7660s for increased negative output voltage. (Courtesy of Datel-Intersil Inc.)

10 for light loads. The output voltage is given by the equation

$$V_{OUT} = -(n)(+V) \tag{10-14}$$

where n = number of devices cascaded and must be ≤ 10
$+V$ = input supply voltage

SUMMARY

1. Although a general-purpose op-amp may work satisfactorily in certain applications, the use of specialized ICs results in simpler, easier, more economical, as well as more accurate circuits. In some other applications general-purpose op-amps will not work at all. Under such circumstances, the designer is forced to use specialized ICs. The best source of such specialized ICs are the manufacturers, data books.
2. Datel's FLT-U2 is a universal filter which provides second-order low-pass, high-pass, and band-pass output functions simultaneously. It can also be used as a notch or all-pass filter. An uncommitted op-amp available in the FLT-U2 can be used to either increase the overall gain of the filter or to raise the order of the low-pass or high-pass filter to the third order.
3. Signetics' NE/SE 555 is a monolithic timing circuit that can produce accurate and highly stable time delays or oscillation. Thus the 555 may be used as either a monostable or an astable multivibrator. This device can be used in such applications as waveform generators, digital logic probes, infrared transmitters, burglar alarms, toxic gas alarms, and electric eyes.
4. The phase-locked loop (PLL) basically consists of a phase detector, an amplifier/low-pass filter, and a VCO. The output f_{OUT} of the VCO is fed back to the phase detector, where it is compared with the input frequency f_{IN}. When $f_{IN} = f_{OUT}$, the loop is said to be locked. Signetics' SE/NE 560 series—560, 561, 562, 564, 565, and 567—are monolithic PLLs. The PLLs are used in such applications as FM stereo decoders, motor speed controls, FSK decoders, and FM demodulators.
5. Power amplifiers are intended to supply a large-signal current to current-operated loads such as speakers and motors. The output current capability of general-purpose op-amps can be increased by using a power transistor called a power booster in series with it. In some applications, however, monolithic/hybrid power amplifiers are preferred, mainly because of their compact size and simpler and more reliable operation. National Semiconductor's LM380 is a monolithic power audio amplifier designed to deliver a minimum of 2.5 W (rms) to an 8-Ω load, while Burr-Brown's 3573 hybrid power amplifier is capable of delivering 100 W peak or 40 W continuous output power. Power audio amplifiers can be used for such applications as simple phonograph amplifiers, alarms, TV sound systems, and intercoms.
6. A voltage regulator is a circuit that supplies a constant voltage regardless of changes in load currents. Monolithic voltage regulators are available in a variety of different output voltage ratings and are also quicker and easier to use. The 7800 is a fixed-positive-voltage regulator series with seven voltage options, while the 7900 is a fixed-negative-voltage regulator series with nine voltage options. Voltage references are special regulators designed for reference voltage purposes. A voltage inverter such as Datel's VI-7660 converts positive voltages from +1.5 to +10 V to −1.5 to −10 V. Voltage regulators are mainly used in power supplies, while voltage references are used with comparators, amplifiers, and D/A converters.

QUESTIONS

10-1. What is the function of the FLT-U2? List important features of this device.

10-2. What maximum-order low-pass (or high-pass) filter can be realized using just one FLT-U2?

10-3. What is the function of the uncommitted op-amp of the FLT-U2?

10-4. Can a notch filter be realized using FLT-U2? Explain how.

10-5. To form a fourth-order low-pass, high-pass, or band-pass filter, how many FLT-U2s are needed? Explain the connection procedures required to form a fourth-order low-pass filter.

10-6. List one application in which you will use the FLT-U2.

10-7. List important features of the 555 timer.

10-8. What are the two basic modes in which the 555 timer operates?

10-9. Briefly explain the differences between the two operating modes of the 555 timer.

10-10. What must the relationship be between the pulse width t_p and the period T of the input trigger signal if the 555 is to be used as a divide-by-4 network?

10-11. List one application each in which the 555 can be used as a monostable and astable multivibrator.

10-12. What is a phase-locked loop?

10-13. List the basic building blocks of the discrete PLL.

10-14. What is the major difference between digital and analog PLLs?

10-15. What is the major advantage of a monolithic phase detector such as the MC4344 over an exclusive-OR and an edge-triggered phase detector?

10-16. Briefly explain the roles of a low-pass filter and VCO in PLLs.

10-17. What are the advantages and disadvantages of monolithic PLLs over discrete PLLs?

10-18. List one application of the PLL and then briefly describe the role of the PLL in that application.

10-19. What is the major difference between small-signal and power amplifiers?

10-20. What is a power booster? Why is it needed?

10-21. List the differences between the monolithic power amplifiers and the power amplifiers using power boosters.

10-22. List important features of the LM380 power audio amplifier.

10-23. What is the major difference between monolithic and hybrid power amplifiers?

10-24. What is a voltage regulator? List four different types of voltage regulators.

10-25. How many voltage options are available with each of the 7800 and 7900 voltage regulators?

10-26. What is a voltage reference? Why is it needed?

10-27. List two important features of VI-7660 voltage inverter.

PROBLEMS

10-1. Using the FLT-U2, design a second-order noninverting Butterworth low-pass filter with 10 kHz cutoff frequency, a gain of 10, $Q = 5$, and $V_S = \pm 15$ V.

10-2. Repeat Problem 10-1 for a high-pass filter.

10-3. Design a band-pass filter with 2 kHz center frequency, $Q = 20$, and noninverted output. Use FLT-U2 with supply voltages $= \pm 15$ V.

10-4. Repeat Problem 10-3 for inverted output.

10-5. Using the FLT-U2, design a notch filter with 2 kHz notch-out frequency and $Q = 20$.

10-6. Using the 555 timer, design a monostable multivibrator having an output pulse width of 100 ms. Verify the designed values of R_A and C with the graph of Figure 10-11.

10-7. In the circuit of Figure 10-10(a), $C = 0.1\ \mu$F and the output pulse width is 1 ms. Determine the value of R_A.

10-8. The monostable multivibrator of Figure 10-10(a) is to be used as a divide-by-3 network. The frequency of the input trigger is 12 kHz. If the value of $C = 0.05\ \mu$F, what should be the value of R_A?

10-9. For the astable multivibrator of Figure 10-15(a), $R_A = 4.7$ kΩ, $R_B = 1$ kΩ, and $C = 1\ \mu$F. Determine the positive pulse width, the negative pulse width, and the free-running frequency. What is the duty cycle of the output waveform?

10-10. Design an astable multivibrator having an output frequency of 10 kHz with a duty cycle of 25%.

10-11. In the ramp generator of Figure 10-18(a), R is set at 100 kΩ, keeping all the other values the same. What is the approximate frequency of the output ramp?

10-12. Design a ramp generator having an output frequency of approximately 5 kHz.

10-13. Referring to Figure 10-27(a), determine f_{OUT}, f_L, and f_C if $C_1 = 0.001\ \mu$F.

10-14. Repeat Problem 10-13 with $C_1 = 470$ pF.

10-15. Referring to the 565 PLL frequency multiplier of Figure 10-28(b), determine the range for the output frequency f_{OUT} if R_1 is varied from 2 to 20 kΩ.

10-16. In the circuit of Figure 10-28(b), if $f_{IN} = 500$ Hz, what is the value of R_1 needed?

10-17. In the intercom system of Figure 10-36, an overall gain of 200 is needed. What must be the turns ratio N_1/N_2 of the transformer if potentiometer R_V is open?

10-18. Using the 7805C voltage regulator, design a current source that will deliver 150 mA current to the 8-Ω, 10-W load.

10-19. The VI-7660 is used as a simple negative converter, as shown in Figure 10-43. Determine the output voltage V_{OUT} if (a) $+V = 5$ V and (b) $+V = 9$ V.

10-20. The VI-7660 voltage inverters are to be used to obtain an output voltage of -15 V. How many VI-7660s are needed if the supply voltage $+V = 5$ V? Draw the complete connection diagram.

EXPERIMENT 10-1
THE 555 TIMER AS A MONOSTABLE AND ASTABLE MULTIVIBRATOR

Objectives

In this experiment you will learn the operation of a 555 timer as a monostable and astable multivibrator. At the end of this experiment you should be able to:

1. Select appropriate timing components for a desired output pulse width when the 555 is used as a monostable multivibrator.

2. Verify that for proper operation of the monostable multivibrator, the width of the negative-going trigger pulse must be smaller than the desired output pulse width.
3. Understand the operation of reset and control pins when the 555 is used as a monostable multivibrator.
4. Select a proper combination of external components for a desired output frequency using the 555 as an astable multivibrator.
5. Determine the duty cycle of a given waveform.

Equipment

1. Dual-trace oscilloscope
2. Audio square-wave generato
3. +5 V power supply

Materials

1. NE555
2. 1N914
3. 1-kΩ resistor
4. 2.2-kΩ resistor
5. Two 3.3-kΩ resistors
6. 4.7-kΩ resistor
7. 6.8-kΩ resistor
8. 10-kΩ resistor
9. 12-kΩ resistor
10. 20-kΩ resistor
11. 100-kΩ resistor
12. 10-kΩ potentiometer
13. 2.5-MΩ potentiometer
14. Two 0.01-μF capacitors
15. 1.0-μF capacitor
16. 10.0-μF capacitor
17. SPDT switch

All resistors should be with ±1% tolerance and all capacitors should be Mylar.

Procedure

1. Connect the monostable multivibrator as shown in Figure E10-1.1.
2. Set the square-wave generator for 5 V pp output at 50 Hz. Connect channels 1 and 2 of the scope to the input and output terminals, respectively, of the monostable multivibrator. Make sure that switch SW_1 is in position *A*. (Pin 4 connected to +5 V.)

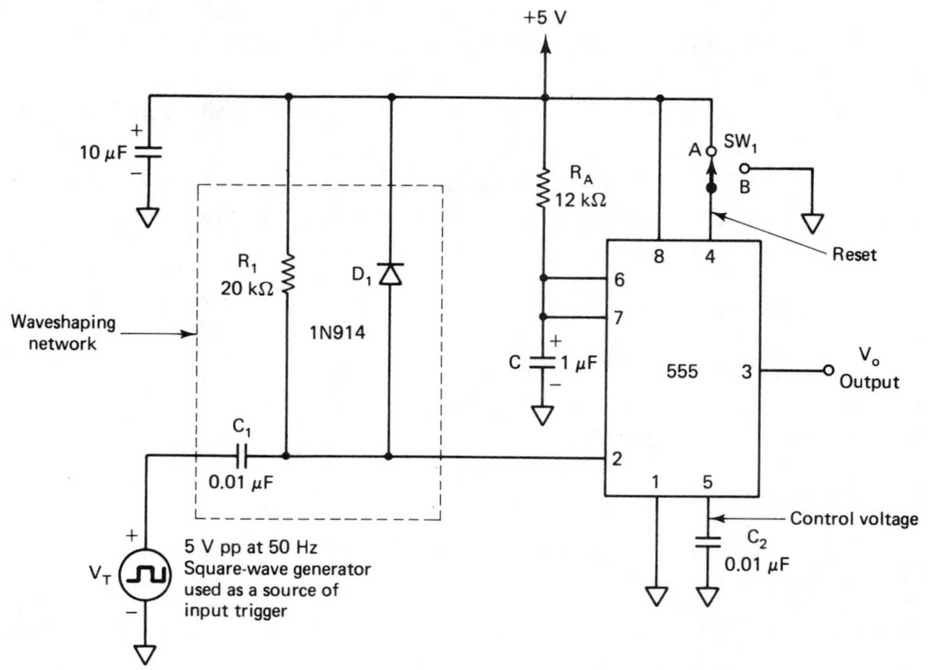

Figure E10-1.1 555 used as a monostable multivibrator.

3. Measure the pulse width of the output waveform and enter the measured value in Table E10-1.1(a).

TABLE E10-1.1
(a)

C (μF)	R_A (kΩ)	Output pulse width (s): Measured	Output pulse width (s): Calculated ($t_p = 1.1R_AC$)
1	12		
1	6.8		
1	4.7		
1	2.2		

4. Momentarily switch off the generator and the power supply, and replace the 12-kΩ resistor by the 6.8-kΩ resistor.
5. Turn the generator and the power supply back on and repeat step 3.
6. Complete Table E10-1.1(a) for the remaining values of R_A.
7. With $R_A = 2.2$ kΩ, $C = 1$ μF, and the dual-trace scope connected to the input and output terminals of the multivibrator, place switch SW_1 in position *B*.

Observe the output waveform. Enter the amplitude and frequency of the output waveform in Table E10-1.1(b).

TABLE E10-1.1
(b)

Output waveform	
Amplitude (V)	Frequency (Hz)

8. Return switch SW_1 to its original position, A. Connect the series combination of the 10-kΩ resistor and the 2.5-MΩ potentiometer between the control voltage pin 5 and ground (across C_2). Vary the 2.5 MΩ potentiometer and simultaneously observe the output waveform. In Table E10-1.1(c), enter the amplitude and pulse-width of the output waveform for the minimum and maximum values of the potentiometer. Remove the 10-kΩ resistor and 2.5-MΩ potentiometer connected across C_2.

TABLE E10-1.1
(c)

2.5-MΩ potentiometer setting	Output waveform	
	Amplitude (V)	Pulse-width (ms)
Minimum (0 Ω)		
Maximum (2.5 MΩ)		

9. Remove the waveshaping components C_1, R_1, and D_1 from the circuit of Figure E10-1.1 and apply the 50-Hz square wave directly to the trigger pin 2. Using the scope, simultaneously observe the input and output waveforms. Enter the amplitude and pulse-width of the output waveform in Table E10-1.1(d).

TABLE E10-1.1
(d)

Input waveform		Output waveform	
Amplitude (V)	Time period (ms)	Amplitude (V)	Pulse-width (ms)

10. Connect the astable multivibrator of Figure E10-1.2.
11. Connect the scope to the output terminal and measure the frequency and duty cycle of the output waveform. Enter the measured values in Table E10-1.2(a).

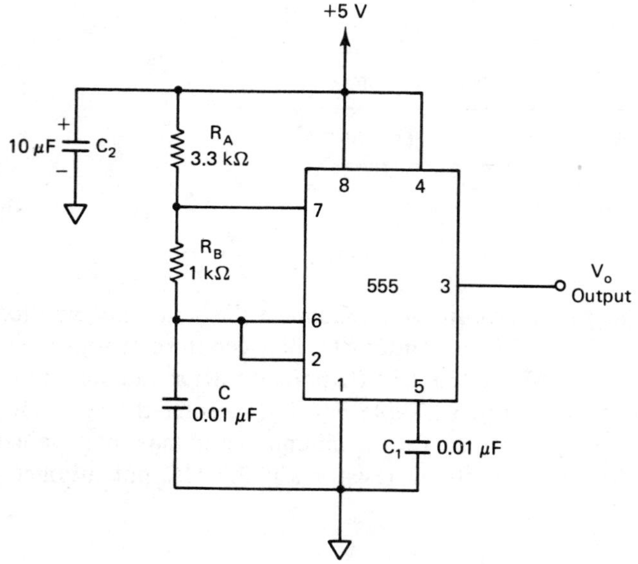

Figure E10.1.2 555 as an astable multivibrator.

TABLE E10-1.2
(a)

C (μF)	R_A (kΩ)	R_B (kΩ)	% Duty cycle: Measured	% Duty cycle: Calculated $\frac{100R_B}{R_A + 2R_B}$	Output frequency: Measured	Output frequency: Calculated $f = \frac{1.45}{(R_A + 2R_B)C}$
0.01	3.3	1				
0.01	3.3	10				
0.01	3.3	100				
0.01	3.3	3.3				

12. Momentarily switch off the power supply and replace R_B of 1 kΩ by 10 kΩ. Turn the power supply back on and repeat step 11.
13. Complete the table for the remaining values of R_B.
14. Switch off the power supply again, and replace the 3.3-kΩ value of R_A by a series combination of a 1-kΩ resistor and a 10-kΩ potentiometer. Also, connect a 1N914 diode across R_B such that the anode of the diode is connected to the junction of R_B and R_A and the cathode to the junction of R_B and C. Connect the scope to the output terminal of the astable multivibrator and turn on the power

supply. Adjust the 10-kΩ potentiometer until the output is a square wave. Measure the frequency and the duty cycle of the output waveform. Enter the total value of R_A in Table E10-1.2(b).

(b)

C (μF)	R_B (kΩ)	R_A (kΩ)	% Duty cycle $= \frac{100t_p}{T}$	Output frequency
0.01	3.3			

Questions

1. Are the measured pulse widths within ±10% of calculated pulse widths? If not, give the reasons.
2. Explain the result of step 7.
3. Explain the result of step 8.
4. What is the function of the waveshaping circuit in Figure E10-1.1?
5. Why did the monostable multivibrator not work in step 9?
6. For the astable multivibrator, is the percent duty cycle the same at different output frequencies? Explain your answer.
7. In order to produce a square-wave output (50% duty cycle), a diode is used across R_B in the astable multivibrator of Figure E10-1.2. Explain briefly how the circuit works.
8. For a square-wave output, what is the relationship between R_A and R_B? Explain your answer.

EXPERIMENT 10-2
NE565 PHASE-LOCKED LOOP AS A FREQUENCY MULTIPLIER

Objectives

In this experiment you will learn the operation of the NE565 phase-locked loop. You will also use the NE565 as a multiplier. At the end of this experiment you should be able to:

1. Understand the relationship between the free-running frequency, lock range, and capture range of the PLL.
2. Verify that the capture range depends on the value of the low-pass filter capacitor.
3. Select an appropriate resistor–capacitor combination for a desired VCO free-running frequency.
4. Use a PLL as a frequency multiplier for a desired output frequency.

Equipment

1. Dual-trace oscilloscope
2. Audio-square-wave generator
3. ±6-V power supply

Materials

1. NE565 PLL
2. 74LS90 decade counter
3. 2N2222 switching transistor
4. 2-kΩ resistor
5. 4.7-kΩ resistor
6. 6.8-kΩ resistor
7. 10-kΩ resistor
8. 12-kΩ resistor
9. 20-kΩ potentiometer
10. Two 0.001-μF capacitors
11. 5-μF capacitor
12. 10-μF capacitor

All resistors are with 1% tolerance and all capacitors are Mylar.

Procedure

1. Connect the 565 PLL as shown in Figure E10-2.1. Do not apply the input voltage V_{in} yet. The numbers indicated in the figure represent the pins.
2. Connect one channel of the scope to the VCO output pin 4, and measure the free-running frequency of the VCO. Enter the measured value in Table E10-2.1.
3. Turn on the audio generator and set it for 1 V pp square wave at 1 kHz. Connect another channel of the scope to the input terminal pin 2 of the PLL. Make sure that the scope is properly triggered.
4. Observe the scope display and gradually increase the input frequency until the loop is phase-locked onto the input frequency. This frequency is the lower end of the capture range. Enter it in Table E10-2.1. As the input frequency is increased further, the loop tracks the input until f_2, the upper end of the lock range. For input frequencies greater than f_2, the loop is unlocked. Enter the frequency f_2 at which the loop unlocks in Table E10-2.1.
5. Gradually decrease the frequency until the loop is again phase-locked onto the input frequency. This is the frequency f_3, the upper end of the capture range. Enter the value of f_3 in Table E10-2.1. Keep on decreasing the input frequency until the loop is unlocked. The frequency f_4 at which this happens is the lower end of the lock range. Enter the value of f_4 in Table E10-2.1.
6. Momentarily switch the generator and the power supply off and replace 6.8 kΩ

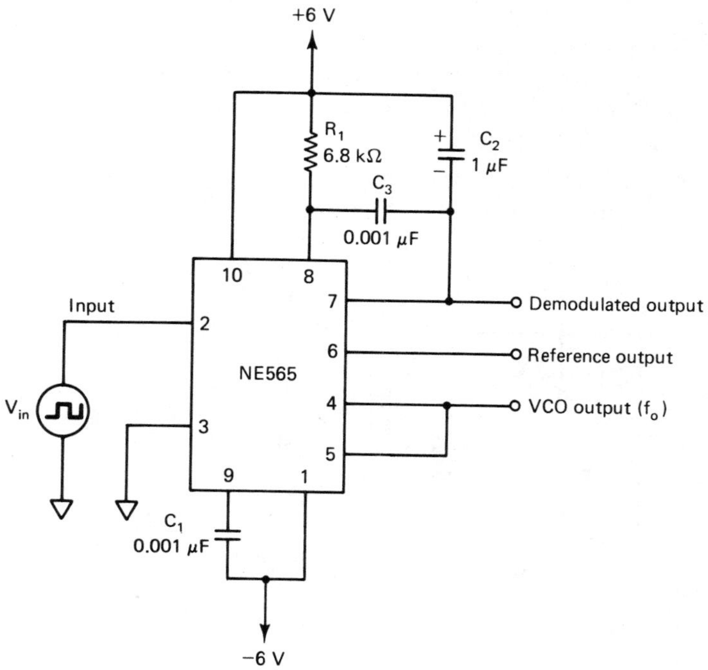

Figure E10-2.1 NE565 PLL.

TABLE E10-2.1

	R_1 (kΩ)	C_1 (μF)	C_2 (μF)	VCO free-running frequency, f_{OUT}: Measured	VCO free-running frequency, f_{OUT}: Calculated $f_{OUT} = \frac{0.3}{R_1 C_1}$	Lower end of the capture range, f_1 (Hz)	Upper end of the lock range, f_2 (Hz)	Upper end of the capture range, f_3 (Hz)	Lower end of the lock range, f_4 (Hz)
1.	6.8	0.001	1						
2.	12	0.001	1						
3.	12	0.001	5						

	Lock range f_L: Measured $(f_2 - f_4)$	Lock range f_L: Calculated $\pm\frac{8 f_{OUT}}{12}$	Capture range f_C: Measured $(f_3 - f_1)$	Capture range f_C: Calculated $\pm\left[\frac{f_L}{(2\pi)(3.6)C_2}\right]^{1/2}$
1.				
2.				
3.				

(R_1) by 12 kΩ. Switch the generator and the power supply back on and repeat steps 4 and 5.

7. Repeat step 6, but this time change the value of C_2 from 1 μF to 5 μF instead of changing the value of R_1.
8. Complete Table E10-2.1.
9. Connect the 565 PLL as shown in Figure E10-2.2. Connect channels 1 and 2 of the scope to the input and output terminals.

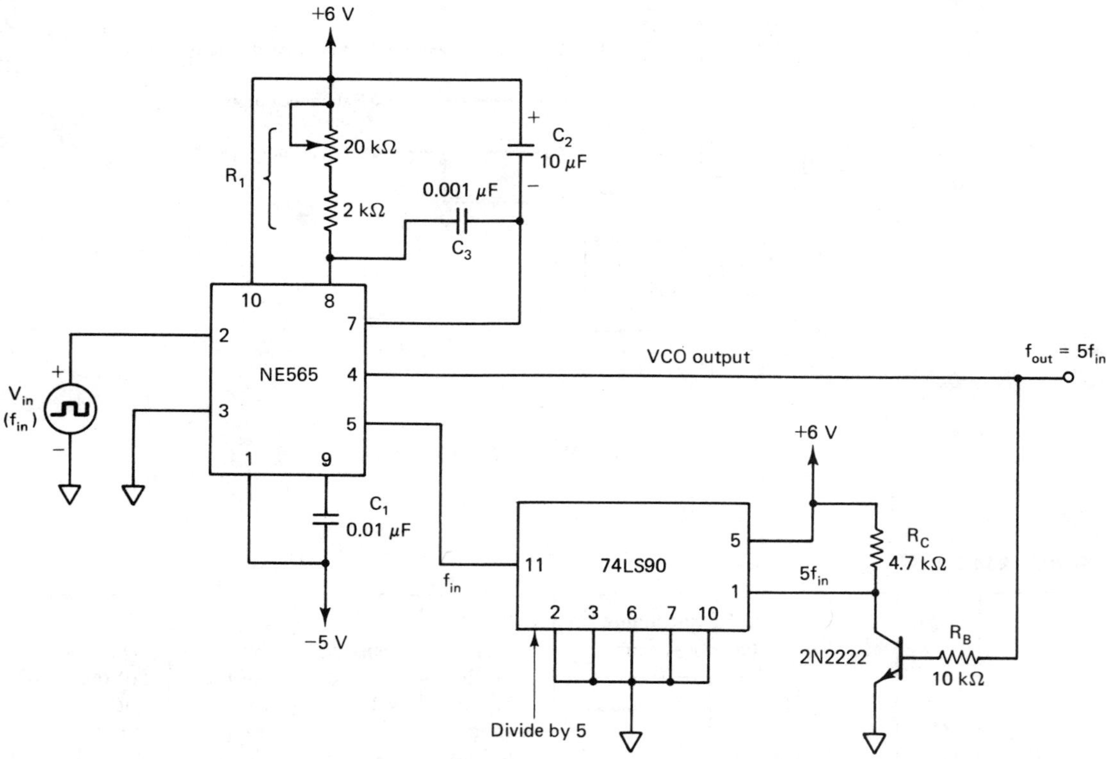

Figure E10-2.2 NE565 as a frequency multiplier.

TABLE E10-2.2

Input frequency, f_{IN} (Hz)	Output frequency, f_{OUT} (Hz)
500	
1000	
1500	

10. Set the generator for 1 V pp square wave at 500 Hz. Adjust the 20-kΩ potentiometer slowly until the loop is phase locked. Measure the output frequency and enter it in Table E10-2.2.

11. Change the input frequency to 1 kHz and again adjust the 20-kΩ potentiometer until the loop is phase-locked. Measure f_{OUT} and enter it in Table E10-2.2 Repeat the same procedure for an input frequency of 1.5 kHz.

Questions

1. Are the measured values of f_{OUT} within 10% of its calculated values? If not, give the reasons.
2. How does the value of R_1 affect the VCO free-running frequency f_{OUT}?
3. What effect does C_2 have on the capture range?
4. Is there any relationship between f_{OUT}, f_L, and f_C? Explain.
5. In the frequency multiplier of Figure E10-2.2, what would the minimum and maximum values of input frequency be? Explain.
6. What modifications are required in the frequency multiplier of Figure E10-2.2 if the input frequency is 10 kHz?

11

Selected IC System Projects

11-1 INTRODUCTION

The preceding chapters have emphasized the analysis and use of individual ICs. The next logical step would be to consider how some of these discrete ICs might be combined to form a system that serves some useful purpose. The aim of this chapter is to demonstrate some of the interesting and challenging projects devoted to constructing IC systems that serve a number of useful purposes.

To most people the word *system* means a very complex, cumbersome, and expensive combination of networks. On the contrary, a system may be simply a combination of a few ICs together with some discrete components, the effect of which is to perform a specific useful service. A system may even be built from a single IC and a few discrete components. The simpler system is increasingly more common today because of the increasing pace of refinement and advancement in IC technologies. Even though innumerable systems could be formed from combinations of the many ICs presented thus far, this chapter focuses on a limited number of systems only. These are simple enough to be discussed meaningfully at this level, and they are illustrative of IC systems designed for specific uses. In addition, this chapter introduces a function generator system using a specially designed IC that requires very few external components.

11-2 POWER SUPPLY: APPLICATION OF VOLTAGE REGULATORS

Since a power supply is a vital part of all electronic system, it will be discussed first. Most digital ICs, including microprocessors and memory ICs, operate on a ±5-V supply, while almost all linear ICs (op-amps and special-purpose ICs) require ±15- or ±12-V supplies. Therefore, the power supply presented in this section will have ±5 and ±15 V.

Figure 11-1 shows the block diagram of a typical power supply. The schematic diagram of the power supply that provides output voltages of ±5 V at 1.0 A and

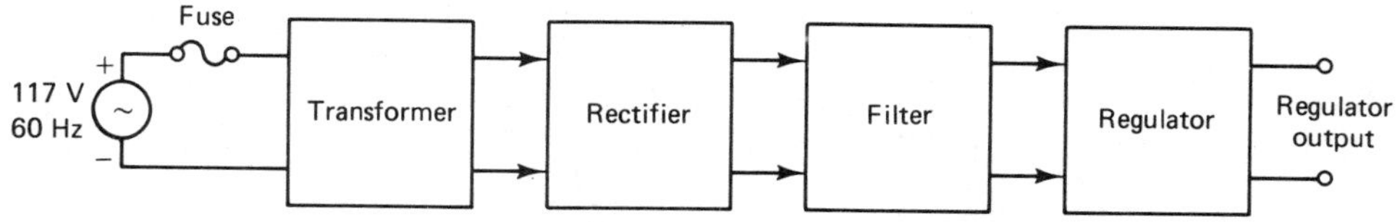

Figure 11-1 Block diagram of a power supply.

±15 V at 0.500 A is shown in Figure 11-2. In this figure two separate transformers are used because they are readily available; however, it is possible to custom design a single transformer with the same specifications to replace the two. The ±15 V supply voltages are obtained from a 26.8-V center-tapped (CT) transformer, and the ±5-V supply voltages are obtained from the 12.6-V CT transformer. The output of these secondaries is then applied to the bridge rectifiers, which convert the sinusoidal inputs into full-wave rectified outputs. The filter capacitors at the output of the bridge rectifiers are charged to the peak value of the rectified output voltage whenever the diodes are forward biased. Since the diodes are not forward biased during the *entire* positive and negative half-cycles of the input waveform, the voltage across the filter capacitors is a pulsating dc that is a combination of dc and a ripple voltage. From the pulsating dc voltage, a regulated dc voltage is extracted by a regulator IC.

Consider first how the ±15-V supply voltages are obtained in the circuit of Figure 11-2. As shown in Section 10-6, the 7815 is a +15-V regulator, the 7915 is a −15-V regulator, and both can deliver output current in excess of 1.0 A. They will hence perform satisfactorily in the circuit of Figure 11-2 by providing ±15 V at 0.500 A. However, since the drop-out voltage ($V_{in} - V_o$) is 2 V, the input voltage for the 7815 must be at least +17 V and that for the 7915 must be at least −17 V. This means that the rectified peak voltage must be greater than +17 V and −17 V, which in turn implies that the secondary voltage must be larger than 34 V peak or 24 V rms. The voltage across the center-tapped secondary in Figure 11-2 is 26.8 V rms, thus satisfying the minimum voltage requirement of 24 V rms. Besides that, the peak voltage between either of the secondary terminals and the center-tap (ground terminal) is $13.4(\sqrt{2}) = 18.95$ V peak, which is less than the maximum peak voltages of +35 V and −35 V for the 7815 and 7915, respectively.

Note that the voltages across the two halves of the center-tapped secondary are equal in amplitude but opposite in phase. During the positive half-cycle of the input voltage, diode D_1 conducts and capacitor C_1 charges toward a positive peak

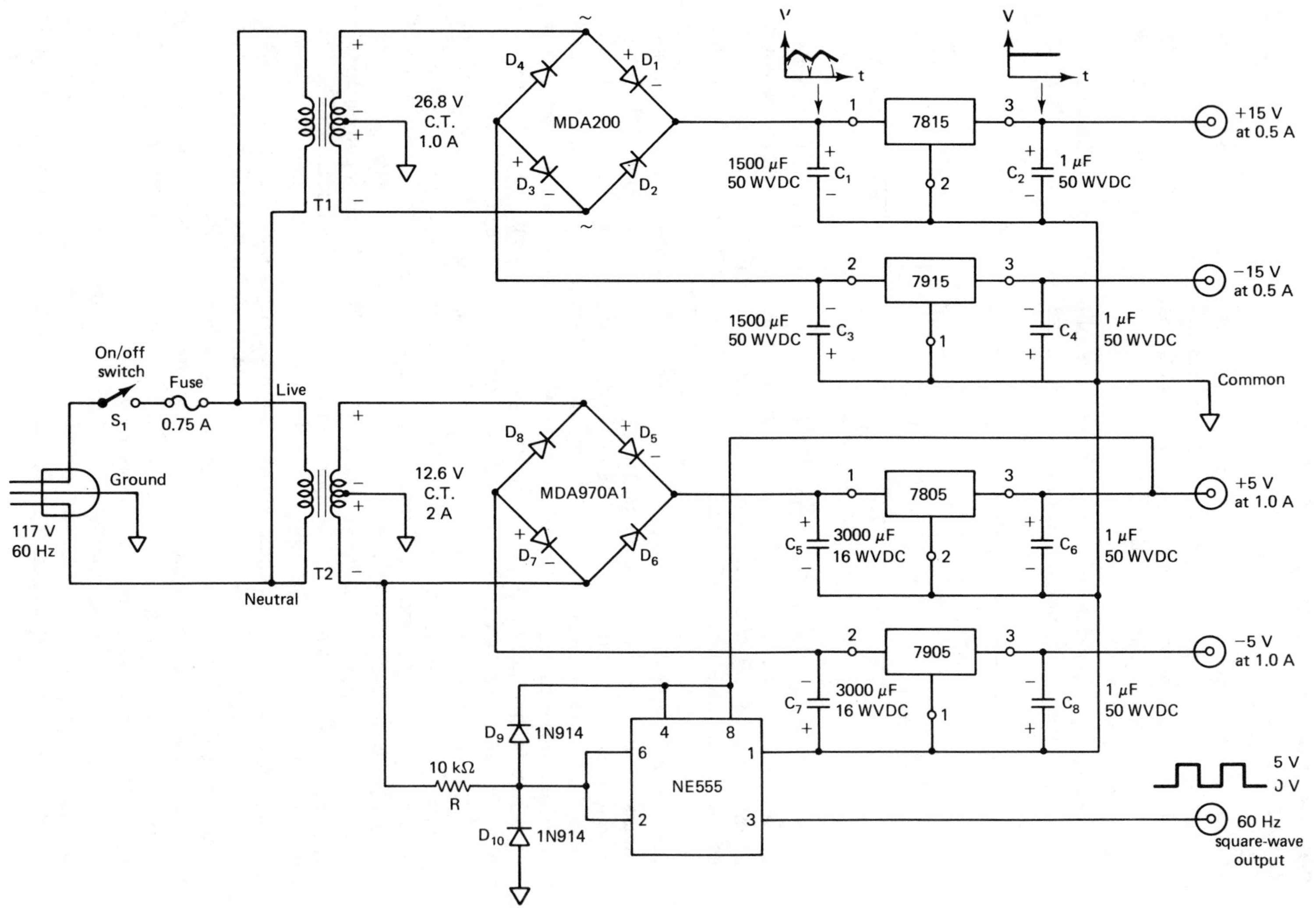

Figure 11-2 Power supply: ± 5 V at 1.0 A and ± 15 V at 0.5 A.

value $\cong$ 18.95 V. At the same time, diode D_3 is also conducting, hence capacitor C_3 charges toward a negative peak value $\cong -18.95$ V. This means that the voltage across nonconducting diodes D_2 and D_4 is 37.90 V peak, which implies that the peak-reverse-voltage (PRV) rating of the bridge rectifiers must be larger than 37.90 V peak or 26.8 V rms. The PRV rating of the bridge rectifier diodes, also known as a *working inverse voltage* (WIV), is specified on the data sheets.

The bridge rectifier, MDA200 (Motorola's rectifier) in Figure 11-2, has a PRV rating of 50 V, which is higher than needed. This bridge rectifier is, in fact, used here because it is readily available and more commonly used. On the other hand, during the negative half cycle of the input waveform, diodes D_2 and D_4 conduct and charge capacitors C_1 and C_3 toward the peak voltage of 18.95 V with indicated polarities. Note, however, that the diode pair that conducts during either the positive or negative half-cycle does not do so for the entire half-cycle. The diodes conduct only during the time when the anode is positive with respect to the cathode. In other words, when the diode is forward biased, the capacitor is charged by current pulses. Data sheets give the maximum average rectified current $I_{o\,\max}$ which the diode can safely handle. For the MDA200, $I_{o\,\max}$ is 2.0 A. In addition, when the power supply is first turned on, the initial charging of the capacitor causes a large transient current called the *surge current* to pass through the diodes. The surge current I_{FS} flows only briefly and is therefore much larger than the maximum average current $I_{o\,\max}$. The maximum surge-current I_{FSM} is normally included on the data sheets; it is 60 A for the MDA200.

Finally, the size of the filter capacitor depends on the secondary current rating of the transformer. As a rule of thumb, a 1500-μF capacitor should be used for each ampere of current. The working voltage rating (WVDC) of the capacitor, on the other hand, depends on the peak rectified output voltage and must be at least 20% higher than the peak value of the voltage it is expected to charge to. Capacitors C_1 and C_3 satisfy these requirements (see Figure 11-2). Capacitors C_2 and C_4 at the output of the 7815 and 7915 regulators, respectively, help to improve the transient response and should be in the range of 1 μF.

Next consider the $\pm$5-V supply. The circuit arrangement of the $\pm$5-V supply is identical to that of the $\pm$15-V supply except that here the specifications for the transformer T_2 secondary are different. Therefore, the operation and design considerations for the $\pm$5-V supply are the same as those presented above for the $\pm$15-V supply.

The voltage regulators in Figure 11-2 will require heat sinks. The power dissipated by the 15-V regulators is as follows:

$$\begin{aligned}\text{Power dissipated} &= (\text{dropout voltage})(\text{current}) \\ &= (18.95 - 15)(0.5) = 1.98 \text{ W}\end{aligned}$$

Similarly, the power dissipated by the 5-V regulators is

$$(8.91 - 5)(1.0) \cong 3.91 \text{ W}$$

Therefore, for the proper operation the regulators must be heat-sinked in order to keep their temperature down. If a regulator is a metal package (TO-3 type), the appropriate heat sink is mounted on the case of the package. However, if the regulator is an epoxy package, a silicon grease may be used on the back of the package and then the package can be bolted to the chassis of the power supply cabinet with insulating hardware.

Besides the ± 15-V and ± 5-V regulated supply voltages, there is often a need for a 60-Hz square-wave signal which is used as a time base in scanning the digital displays and as a trigger for sequential and timing circuits. If needed, a 1-Hz (1-s) signal for the real-time clock can be readily obtained from the 60-Hz signal by using a divide-by-60 network. Although not commonly done, a higher-frequency signal can also be obtained from the 60-Hz signal by using a multiplier. For these reasons, in Figure 11-2 a 60-Hz square-wave signal is produced by using two small-signal diodes and a 555 timer as the Schmitt trigger.

Parts List

Transformer T_1	Primary:	117 V, 60 Hz	Hobart P-300
	Secondary:	26.8 V CT, 1.0 A	
Transformer T_2	Primary:	117 V, 60 Hz	Hobart P-305
	Secondary:	12.6 V CT, 2.0 A	
Bridge rectifiers	MDA200	PRV = 50 V, $I_{o\,max} = 2.0$ A, $I_{FSM} = 60$ A	
	MDA970A1:	PRV = 50 V, $I_{o\,max} = 4.0$ A, $I_{FSM} = 100$ A	

NE555 timer
Two 1N914 signal diodes
10-kΩ resistor

Capacitors	3000 μF at 16 V (2):	Sprague TVA1175
	1500 μF at 50 V (2):	Sprague TVA1318
	1.0 μF at 50 V (4):	Sprague TVA1300
Regulators +15 V:	MC7815	
−15 V:	MC7915	
+ 5 V:	MC7805	
− 5 V:	MC7905	
Fuse	0.750 A slow blow	
Switch	On–off toggle type	

Silicon grease with insulating hardware or four heat sinks.

11-3 AUDIO FUNCTION GENERATOR: APPLICATION OF THE FUNCTION GENERATOR IC

A sine-wave signal is generally used in testing linear circuits such as amplifiers and filters; on the other hand, a square-wave input is often essential for testing digital circuits such as flip-flops, counters, and registers. In this section an audio function generator using a specially designed integrated circuit ICL8038 will be presented. This IC produces not only sine and square waves but also the triangular wave.

As a first step, let us briefly consider the internal structure. operation, and electrical characteristics of the ICL8038. Figure 11-3 shows the functional diagram

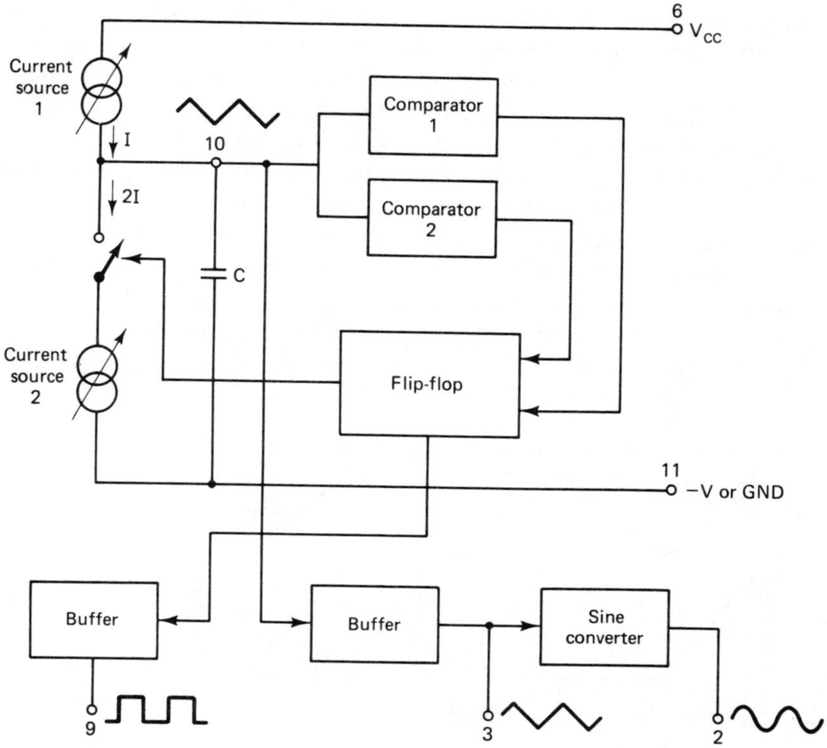

(a)

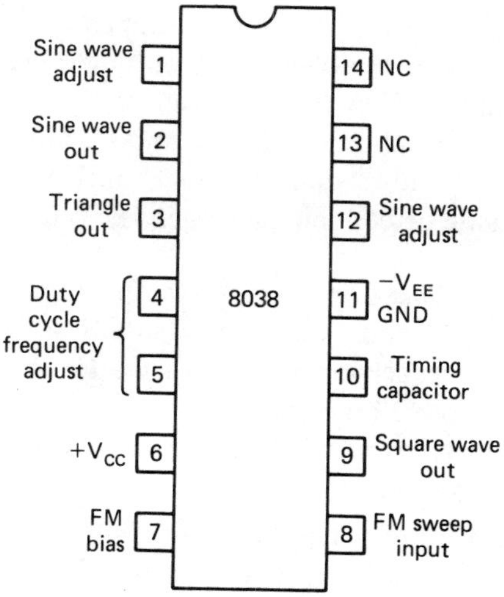

(b)

Figure 11-3 ICL8038 waveform generator. (a) Block diagram. (b) Connection diagram. (Courtesy of Intersil Inc.)

and connection diagram of the ICL8038. The 8038 function generator is a monolithic integrated circuit that utilizes advanced monolithic technology, such as thin-film resistors and Schottky-barrier diodes. As shown in Figure 11-3(a), it consists of two current sources, two comparators, two buffers, a flip-flop, and a sine converter.

The triangular wave is generated by alternately charging the external capacitor from one current source and then linearly discharging it with another. The triangular wave is then applied to comparators 1 and 2 and also simultaneously buffered to the sine converter. The comparators and the flip-flop together convert the triangular wave into a square wave, whereas the sine converter converts the triangular wave into a sine wave. The sine converter is composed of 16 transistors. These transistors function in a nonlinear manner by providing a decreasing shunt impedance as the triangular wave potential moves toward the positive and negative peaks. Figure 11-4 indicates the relationship between the sine, square, and triangular waveforms generated by the ICL8038.

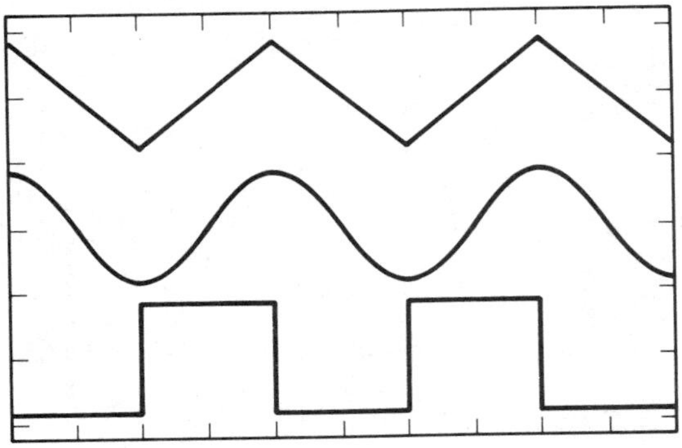

Figure 11-4 Output waveforms of the ICL8038. (Courtesy of Intersil Inc.)

As shown in the functional block diagram, both triangular and square-wave outputs are buffered so that the output impedance of each is 200 Ω. However, the sine-wave output has a relatively high output impedance of 1 kΩ. For this reason the sine-wave output is often fed into a separate noninverting amplifier which provides buffering, gain, and amplitude adjustment, as will be seen shortly.

Figure 11-3(b) shows that the 8083 function generator is a 14-pin DIP; it is available as a plastic or ceramic package. The pin functions are as follows:

Pins 1 and 12: Sine-wave adjust. By adjusting a 100-kΩ potentiometer connected between pins 12 and 11 (negative supply or ground), sine-wave distortion of less than 1% is achievable. However, to obtain a distortion close to 0.5%, two 100-kΩ potentiometers may be connected between pins 6 ($+V_{CC}$) and 11 ($-V_{EE}$ or ground) with the wiper of one potentiometer connected to pin 1 and that of the other to pin 12.

Pin 2: Sine-wave out. The sine-wave output is available at this pin, the amplitude of which is $0.22V_S$, where $\pm 5\text{ V} \leq V_S \leq \pm 15\text{ V}$.

Pin 3: Triangle out. The triangular output is taken out at this pin. Like the sine-wave, the output amplitude of the triangular wave is also a function of supply voltage V_S and is equal to $0.33V_S$, where $\pm 5\text{ V} \leq V_S \leq \pm 15\text{ V}$.

Pins 4 and 5: Duty cycle/frequency adjust. The *symmetry* (50% duty cycle for square wave) of all waveforms can be adjusted by connecting two external resistors (generally equal in value) between $+V_{CC}$ and pins 4 and 5 (see the connection diagram shown in Figure 11-5). These external resistors, together with the external capacitor connected to pin 10, determine the frequency of the output waveforms. The recommended value of external timing resistors is between 1 kΩ and 1 MΩ so that the resultant charging current is between 10 μA and 10 mA. The optimum performance is obtained for these charging currents.

Pin 6: $+V_{CC}$. Positive supply voltage is applied to this pin so that $10\text{ V} \leq V_{CC} \leq 30\text{ V}$ with respect to voltage at pin 11 ($-V_{EE}$ or ground).

Pin 7: FM bias. This pin is a junction of two resistors (10 kΩ and 40 kΩ) which form a voltage divider with the supply voltage(s). When this pin is shorted to pin 8, a high output frequency is obtained, as opposed to connecting pin 8 to the $+V_{CC}$ for the given values of external timing resistors and capacitor. This is because the output frequency of the waveform generator is a direct function of the dc voltage at pin 8 with respect to $+V_{CC}$. This procedure is generally used to test the ICL8038.

Pin 8: FM sweep input. For frequency sweeping or for larger FM deviations, the modulating (sweep) signal is applied between the $+V_{CC}$ and pin 8. In this way a very large sweep range is created. The sweep FM range is the ratio of maximum frequency to minimum frequency, which can be extended to 1000:1 for the 8038. However, for proper operation the sweep voltage should be within the range

$$\left(\frac{2}{3}V_{CC} + 2\text{V}\right) < V_{\text{sweep}} < V_{CC}$$

where V_{CC} is the total supply voltage. The sweep frequency of FM is typically 10 kHz for the 8038.

On the other hand, for small deviations (±10% typically) the FM signal can be applied directly to pin 8.

Pin 9: Square-wave output. The square-wave output is available at this pin. The square-wave output is in fact open collector, in that this pin can be connected through a load resistor to a different power supply than that used for the function generator itself. However, the maximum value of the power supply used must be less than 30 V. This means that to obtain a TTL-compatible square-wave output, a load resistor (10 kΩ typically) simply needs to be connected to the +5-V supply.

In addition, the duty cycle of the square wave can be varied from 2 to 98% by external components connected to pins 4, 5, and $+V_{CC}$.

Pin 10: Timing capacitor. An external capacitor connected to this pin, together with the timing resistors connected to pins 4 and 5, determines the frequency of output waveforms. Oscillations can be halted by connecting pin 10 to ground.

Pin 11: $-V_{EE}$*/ground.* If (±) supply voltages are used, a negative supply $-V_{EE}$ is connected to this pin. On the other hand, if a single supply is used, this pin is connected to ground. With (±) supply voltages, all output waveforms are symmetrical (bipolar) about the 0 V. Obviously, if a single supply is used, the output waveforms are unipolar; that is, their average voltage is equal to $+V_{CC}/2$.

Pins 13 and 14: NC. No connections.

The important features of the 8038 are these:

Simultaneous outputs: sine wave, square wave, and triangle wave.

Frequency range of operation: 0.001 Hz to 500 kHz.

Low distortion: 1%.

Low frequency drift with temperature: 50 ppm/°C maximum.

Easy to use: Requires very few external components.

Now let us see how the 8038 can be used as an audio function generator. Figure 11-5 shows a function generator that produces the sine wave, square wave, and triangular wave and has a frequency range of less than 20 Hz to above 100 kHz. However, around 100 kHz the quality of output waveforms starts deteriorating, in that the sine and triangular waves distort and the square wave becomes a pulse waveform with $< 50\%$ duty cycle.

Although the function generation of Figure 11-5 is not of a high laboratory standard and may not be used for calibration purposes or in critical design applications, it is nevertheless useful for beginning designers and hobbyists. Besides that, it is simple and low in cost.

In the circuit of Figure 11-5, switch SW_2 helps to select the desired output waveform. For the output waveform chosen, the frequency select switch SW_1 is then used to select an appropriate frequency range; a specific value of frequency is then obtained by adjusting potentiometer R_4. R_4 allows the potential at pin 8 to be varied from $+V_{CC}$ to $+V_{CC}/3$ with respect to ground, which in turn varies the frequency of the output waveform.

Potentiometer R_3 is used to vary the duty cycle of the square-wave output and hence may be used to adjust the duty cycle of the square wave to 50%, especially at higher frequencies. Besides that, potentiometers R_7 and R_9, whose wipers are connected to pins 1 and 12, respectively, are adjusted to minimize sine-wave distortion.

As mentioned earlier, the square-wave output is uncommitted and is available at pin 9. Therefore, in order to obtain a TTL-compatible square-wave output, pin 9 may be connected through a pull-up resistor to a +5-V supply if desired. However, in Figure 11-5, pin 9 is connected through resistor R_{10} and potentiometer R_{11} to +12 V. The output amplitude of the square wave can be adjusted to a desired value by varying R_{11}. Resistor R_{10} is used in series with potentiometer R_{11} so that the $+V_{CC}$ of 12 V is not directly applied to the (+) input of the op-amp (see Figure 11-5).

The triangular-wave output is available at pin 3, the amplitude of which is $(0.33)(\pm 12\text{ V}) = \pm 3.96\text{ V}$. However, potentiometer R_{12} is used to vary the amplitude of the triangular wave. Finally, the sine-wave output is available at pin 2. Although the amplitude of the sine wave is $(0.22)(\pm 12\text{ V}) = \pm 2.64\text{ V}$, it can be adjusted to a desired value with the help of potentiometer R_{13}.

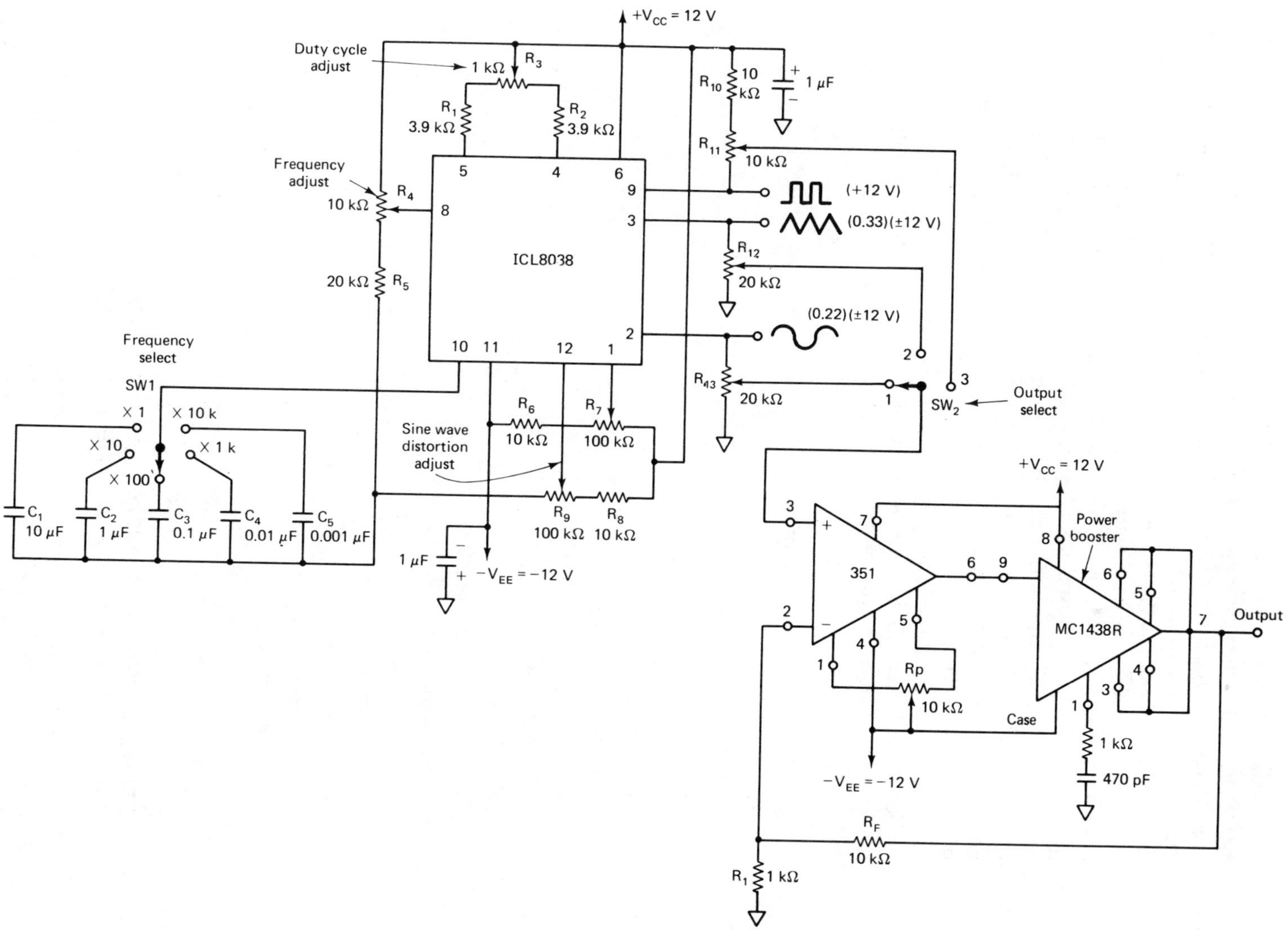

Figure 11-5 ICL8038 as an audio function generator.

Switch SW_2 helps to select one of the three output waveforms, which is then applied to the output stage composed of the op-amp and a power booster. The output stage serves two functions: (1) it provides a low output impedance and (2) it increases the output power drive capability of the 8038. Recall that without the output stage the output resistance of the triangular and square waves is 200 Ω, while that of the sine wave is 1 kΩ. Because the output stage is configured as a noninverting amplifier, however, the output resistance is reduced to a negligibly small value. In addition, the use of the power booster MC1438R (or BB3553) inside the op-amp's feedback loop improves the output drive capability of the 8038 without degrading op-amp characteristics. The op-amp used here is a LF351 with a slew rate of 13 V/μs and a unity gain bandwidth of 4 MHz. It is used as a noninverting amplifier with a gain of 11 (see Figure 11-5).

For smaller-amplitude output ac signals, it may be necessary to reduce the output offset voltage to zero initially. This is accomplished by adjusting the offset null circuitry, namely potentiometer R_p of the LF351 op-amp.

Parts List

ICL8038 function generator
LF351 op-amp
MC1438R power booster
Two 1-kΩ resistors
1-kΩ potentiometer
Two 3.9-kΩ resistors
Four 10-kΩ resistors
Three 10-kΩ potentiometers
20-kΩ resistor
Two 20-kΩ potentiometers
Two 100-kΩ potentiometers
470-pF capacitor
0.001-μF capacitor
0.01-μF capacitor
0.1-μF capacitor
Three 1.0-μF capacitors
10.0-μF capacitor
Five-position switch
Three-position switch

11-4 LED TEMPERATURE INDICATOR: APPLICATION OF THE V/F CONVERTER AND THE 555 TIMER

This project illustrates the use of a V/F converter in monitoring temperature in degrees Fahrenheit (°F). The block diagram of the temperature indicator is shown in Figure 11-6(a). The indicator is composed of a temperature sensor, amplifier, V/F converter, three-digit binary-coded-decimal (BCD) counter, time base, and LED display. In addition to the 9400 V/F converter, other ICs needed for this project include the LM334 temperature sensor, LF353 dual op-amp, NE555 timers, 74LS00 NAND gate, MC14553 three-digit BCD counter, MC14543 BCD-to-seven segment decoder/driver/latch, and three seven-segment (common anode or common cathode) LED displays with three PNP switching transistors.

Figure 11-6(b) shows the schematic diagram, which is designed to display temperatures from 0 to 100°F. Operation of the circuit is as follows. The output of the temperature sensor changes linearly as a function of temperature (10 mV/°K). This output is an input to the summing amplifier, which is used to calibrate the output of the temperature sensor for a desired temperature type (°K, °C, or °F) and an intended range. That is, in order to display the temperature in either °K, °C, or °F, potentiometer R_4 is adjusted accordingly so that a suitable voltage appears at the output of the summing amplifier. Since the output of the temperature sensor is directly proportional to temperature changes, R_4 needs to be adjusted at only one temperature. The output of the summing amplifier then drives the inverting amplifier. The purpose of the inverting amplifier is twofold: (1) to invert the input so that its output voltage is positive, which is necessary for the V/F converter, and (2) to provide a suitable gain, which depends on the voltage-to-frequency scaling used for the V/F converter.

The output of the inverting amplifier is the input to the V/F converter; therefore, the output frequency of the converter is directly proportional to the output voltage of the inverting amplifier. For example, as the temperature goes up the output voltage of the summing amplifier increases in the negative direction, whereas that of the inverting amplifier increases in the positive direction, which in turn causes the frequency of the V/F to increase in the positive direction.

The output frequency of the converter is then ANDed with the gating signal to produce the clock signal for the three-digit BCD counter. The BCD output of the counter then drives the three LED displays sequentially via the BCD-to-seven segment decoder/latch/driver stage, and the temperature is displayed on the LEDs, depending on the relationship between the frequency of the V/F converter and the gate signal. The gate, latch, and reset signals are generated by the time-base circuit, which consists of a free-running multivibrator and two one-shot multivibrators.

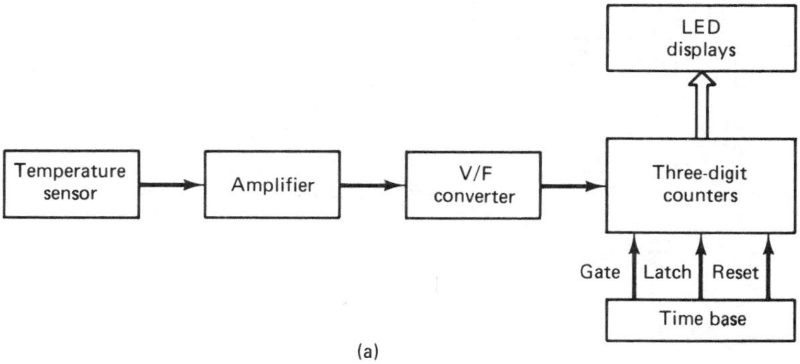

(a)

Figure 11-6 LED temperature indicator. (a) Block diagram.

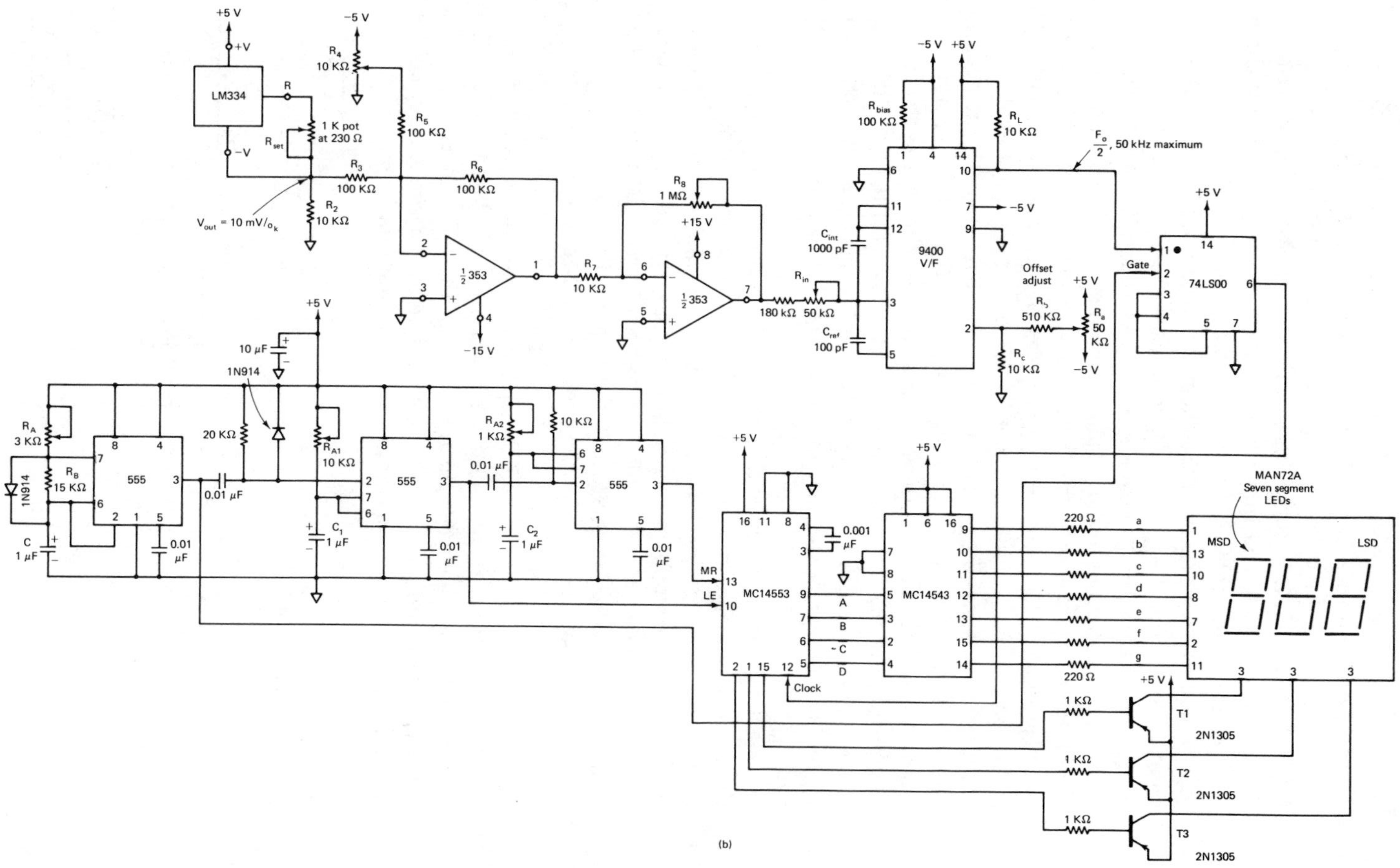

Figure 11-6 (b) Schematic diagram.

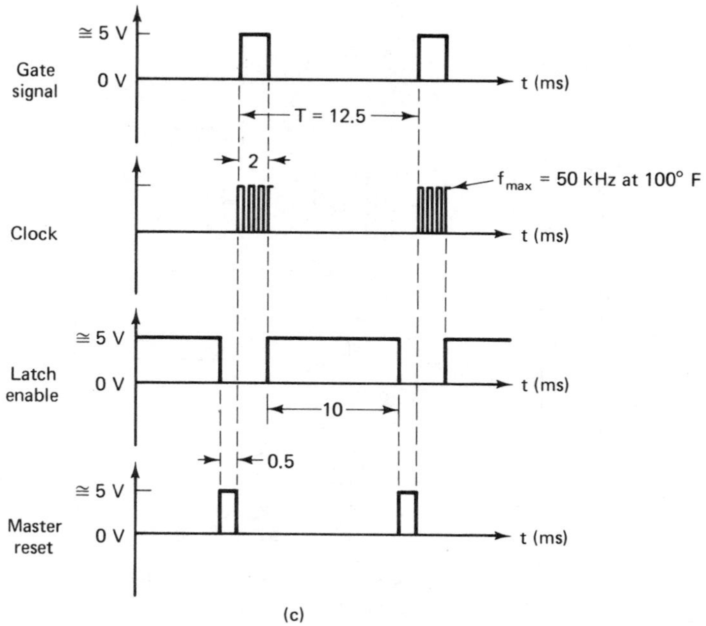

(c)

Figure 11-6 (c) Timing diagram.

Next let us examine the design considerations and procedures for each of the sections in the temperature indicator of Figure 11-6(b). The temperature sensor LM334 is a three-terminal adjustable current source whose current can be programmed from 1 μA to 10 mA with one external resistor R_{set}. The three terminals are labeled $+V$, R, and $-V$. The pin diagram and the connection diagram of the LM334 are shown separately in Figure 11-7.

The LM334 has a wide operating voltage range of 1 to 40 V. It can also withstand reverse voltage of up to 20 V (terminal $+V$ is negative with respect to $-V$). It is designed to operate over a temperature range of 0 to 70°C. For a wider temperature range, such as -55 to 150°C, Intersil's AD590 temperature sensor is recommended.

For the values indicated in Figure 11-6(b), the output of the LM334 changes 10 mV/°K. This means that at 0°F = 255.22°K, the output of the sensor will be 2552.2 mV, which must be scaled down to 0 V so that the temperature displayed will be in degrees Fahrenheit. This is accomplished by the use of the summing amplifier. Specifically, potentiometer R_4 of the summing amplifier is adjusted so that the output is 0 V. The same procedure is used to calibrate the output of the summing amplifier at any other value of °F. Table 11-1 shows the relationship between °K, °C, °F and the output of the temperature sensor and the summing amplifier at corresponding values of temperature. Because the output of the sensor is directly proportional to the temperature, the output of the summing amplifier needs to be calibrated at the temperature at which the circuit is initially started up (refer to Table 11-1).

Note that the output of the summing amplifier is a negative dc voltage since the net input voltage is always positive for temperatures > 0°F (see Table 11-1).

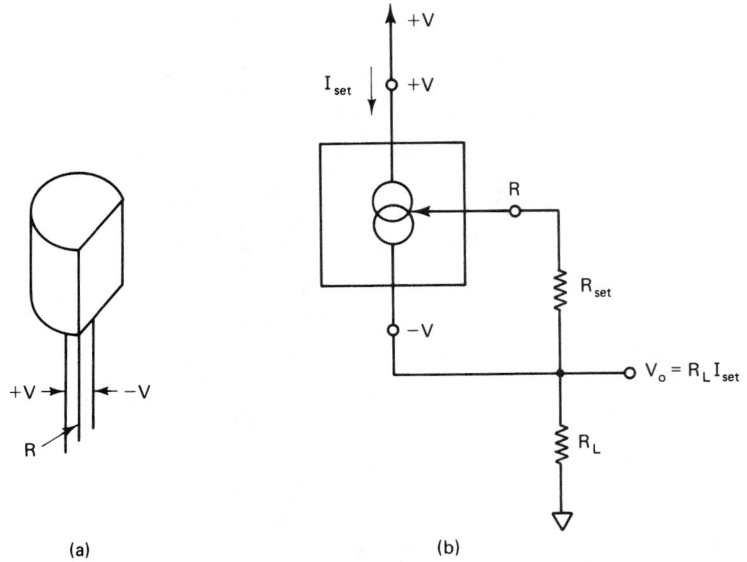

Figure 11-7 LM334 as a temperature sensor. (a) Pin diagram. (b) Connection diagram. (Courtesy of National Semiconductor.)

TABLE 11-1 RELATIONSHIP BETWEEN DIFFERENT TEMPERATURE UNITS AND OUTPUTS OF THE SENSOR AND SUMMING AMPLIFIER

Degrees Kelvin (°K)	Degrees Celsius (°C)	Degrees Fahrenheit (°F)	Output of the temperature sensor (mV)	Output of the summing amplifier (mV) to be adjusted to
1. 255.22	−17.78	0	2552.2	0
2. 273	0	32	2730	−177.8
3. 298	25	77	2980	−427.8
4. 310.78	37.78	100	3107.8	−555.6

However, the 9400 V/F requires a positive input voltage. The summing amplifier must therefore be followed by an inverting amplifier. The gain of the inverting amplifier, however, depends on the voltage-to-frequency scaling of the converter. The V/F converter of Figure 11-6(b) is calibrated for a maximum frequency of 50 kHz, which represents a temperature of 100°F when the input voltage is 10 V maximum. For 9400 V/F calibration procedure refer to Section 9-10.1(b). Since the output of the summing amplifier is −555.6 mV at 100°F, the gain of the inverting amplifier must be equal to

$$\frac{10\text{ V}}{555.6\text{ mV}} \cong 18$$

The output frequency of the V/F converter is then ANDed with the output frequency (called the gate signal) of the 555 free running multivibrator to produce the clock signal for the three-digit BCD counter. Since the maximum 50 KHz output frequency of the converter represents 100°F, the three-digit BCD counter must be clocked 100 times in order to display 100°F. To accomplish this, the pulse-width of the free running multivibrator must be 100/50 K = 2 ms so that 100 pulses will be produced in 2 ms. At the end of 2 ms, the count of the counter is latched and displayed on the LED display. After the count is displayed as a temperature on the LEDs (50 KHz V/F output frequency represents 100°F), the BCD counter is reset and the cycle repeats. In other words, the counter continuously cycles through three states: count, latch, and reset. Therefore, the free running multivibrator (gate signal) must provide for the time period required to count, latch, and reset the BCD counter. The latch enable and master reset pulses for the BCD counter MC14553 are produced by using two 555 one-shot multivibrators. Figure 11-6(c) indicates the timing diagram for the BCD counter MC14553, where the time period of the free running multivibrator is approximately 12.5 ms with a pulse width of 2 ms. The pulse width of the latch enable pulse is approximately 10 ms, and the master reset pulse width is approximately 0.5 ms. To accomplish 2 ms, 10 ms, and 0.5 ms pulse width, adjust potentiometers R_A, R_{A1}, and R_{A2}, respectively. (Figure 11-6b).

The three-digit BCD counter used in Figure 11-6(b) is the MC14553. The block diagram and pin diagram of the MC14553 is shown in Figure 11-8. As shown in the block diagram, the MC14553 consists of three negative-edge-triggered BCD counters with a quad latch at the output of each counter, which enables the storage of any given count. The outputs of the latches are time multiplexed so that one BCD digit at a time is produced. The operation of the multiplexer output selector is controlled by the on-chip oscillator, whose frequency depends on the external capacitor.

The pin functions are as follows:

Pins 2, 1, and 15: $\overline{DS1}$, $\overline{DS2}$, *and* $\overline{DS3}$. These are the active-low digit-select outputs which sequentially control the three LED/LCD displays.

Pins 4 and 3: C1A and C1B. An external capacitor C_1 between these pins controls the frequency of the on-chip oscillator, which in turn drives the multiplexer output selector. If an external clock is to be used instead of the capacitor, the clock must be applied to pin 4.

Pins 9, 7, 6, and 5: Q_0, Q_1, Q_2, *and* Q_3. The BCD outputs are available at these pins and are active-high.

Pin 8: V_{SS}. This pin is connected to ground potential. The (+) supply voltage is also measured with respect to this pin.

Pin 10: LE. This is a latch-enable input pin. When the latch-enable input goes high, the information in the BCD counters is stored in the quad latches and is retained as long as the latch is high.

Pin 11: Dis. When an external input pulse applied to this pin goes high, the input clock is disabled, which in turn freezes the counters. However, the last count in the counters is retained. When not used, this pin should be connected to ground.

Pin 12: Clock. The input trigger source for the counters is connected to this

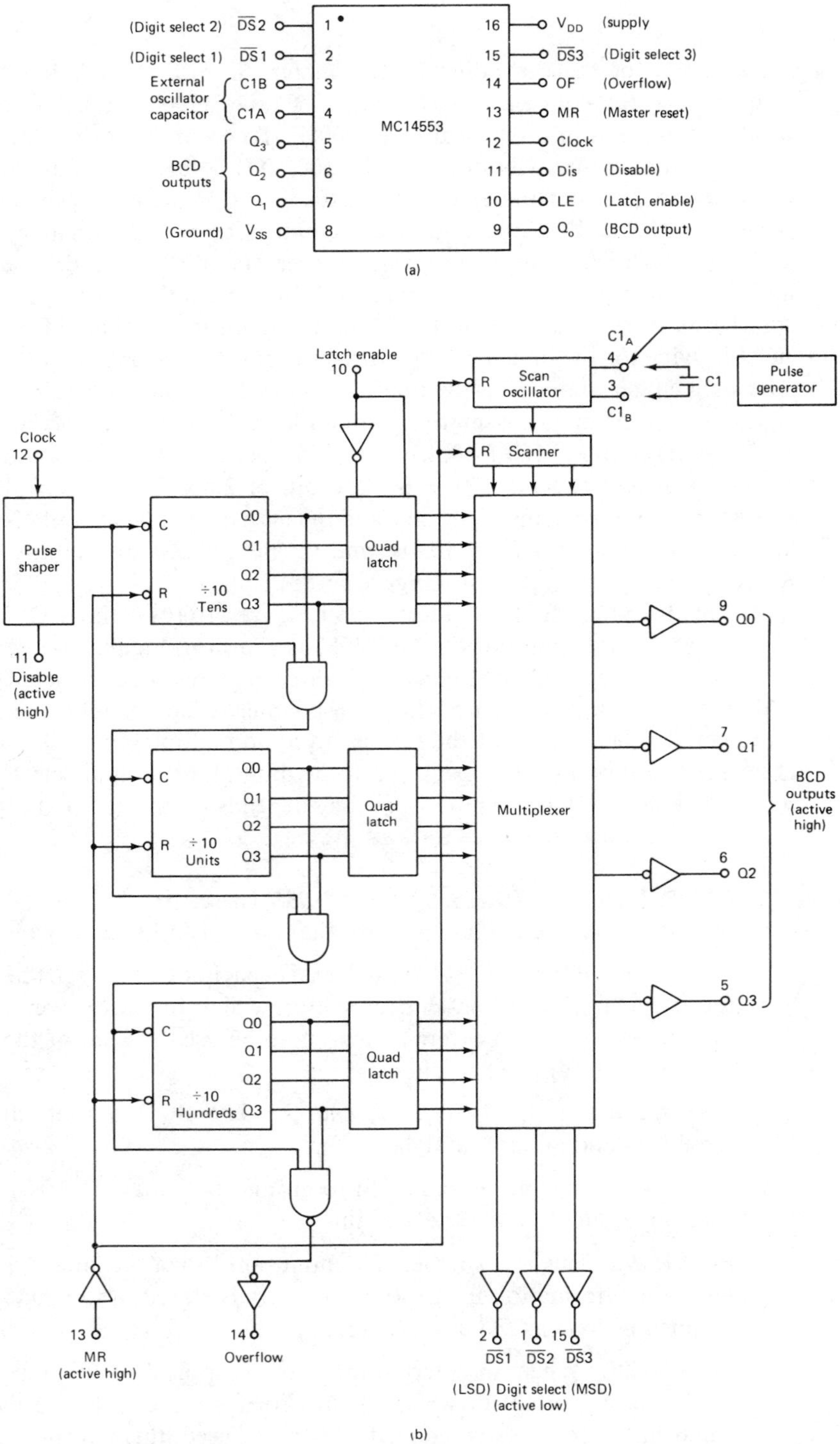

Figure 11-8 MC14553 three-digit BCD counter. (a) Pin diagram. (b) Block diagram. (Courtesy of Motorola Semiconductor.)

pin. The counters are advanced on each high-to-low (logic 1 to 0) transition of the input trigger.

Pin 13: MR. When an input pulse applied to this pin goes high, the three BCD counters and the multiplexer scanning circuit are initialized (reset). In addition, the scan oscillator is inhibited, and digit-select outputs $\overline{DS1}$, $\overline{DS2}$, and $\overline{DS3}$ are disabled. For that reason, this pin is called *master reset* (MR).

Pin 14: OF. This pin indicates the overflow condition of the counters. The voltage at pin 14 goes high for every 1000 counts and hence can be used to cascade multiple MC14553s.

Pin 16: V_{DD}. A (+) supply voltage up to +18 V dc can be applied to this pin with respect to pin 8 (V_{SS}).

As shown in Figure 11-6(b), the master reset (MR) and latch enable (LE) pulses for the MC14553 are produced by using two 555 one-shot multivibrators. In addition, the clock signal for the MC14553 is produced by ANDing the output of the V/F converter with the gating signal, which is obtained by using the 555 free-running multivibrator. (For more information on the 555 timer, refer to Section 10-3.) The digit select outputs $\overline{DS1}$, $\overline{DS2}$, and $\overline{DS3}$ sequentially drive the 2N1305 PNP transistors T_1, T_2, and T_3, which in turn control the three LED displays. The BCD outputs of the MC14553 are connected to the BCD inputs of the MC14543, which is a BCD-to-seven segment latch/decoder/driver [see Figure 11-6(b)]. The seven-segment outputs of the MC14543 then drive the seven segments of the LED selected by the digit-select of the MC14553.

The MC14543 is designed to provide three functions: a 4-bit storage latch, an 8421 BCD-to-seven segment decoder, and a driver. The device is capable of driving LCD and LED displays. Figure 11-9(a) shows the pin diagram, with pin functions as follows:

Pin 1: LD. Since this pin is used to enable and disable the latches of the device, it is called *latch disable* (LD). When this pin is high (connected to V_{DD}), the BCD at the input is latched into the device. On the other hand, when this pin is low (connected to V_{SS}), the latches are disabled.

Pins 5, 3, 2, and 4: A, B, C, and D. These pins are the input pins to which the BCD inputs are applied.

Pin 6: Ph. This pin is called a *phase invert* (Ph) pin because it is used to invert the logic levels of the seven-segment output combinations. That is, when this pin is logic 0 (connected to V_{SS}), the MC14543 can drive common cathode LED readouts. On the other hand, for common anode LED readouts, this pin must be logic 1 (connected to V_{DD}).

Pin 7: BI. When this pin is logic 1, the display is blanked; hence it is called *blanking input* (BI). However, if this pin is logic 0, the display reads the BCD number corresponding to the code at pins 2, 3, 4, and 5.

Pin 8: V_{SS}. This is a ground pin, and (+) supply V_{DD} is measured with respect to it.

Pins 9, 10, 11, 12, 13, 15, and 14: a, b, c, d, e, f, g. These pins represent the

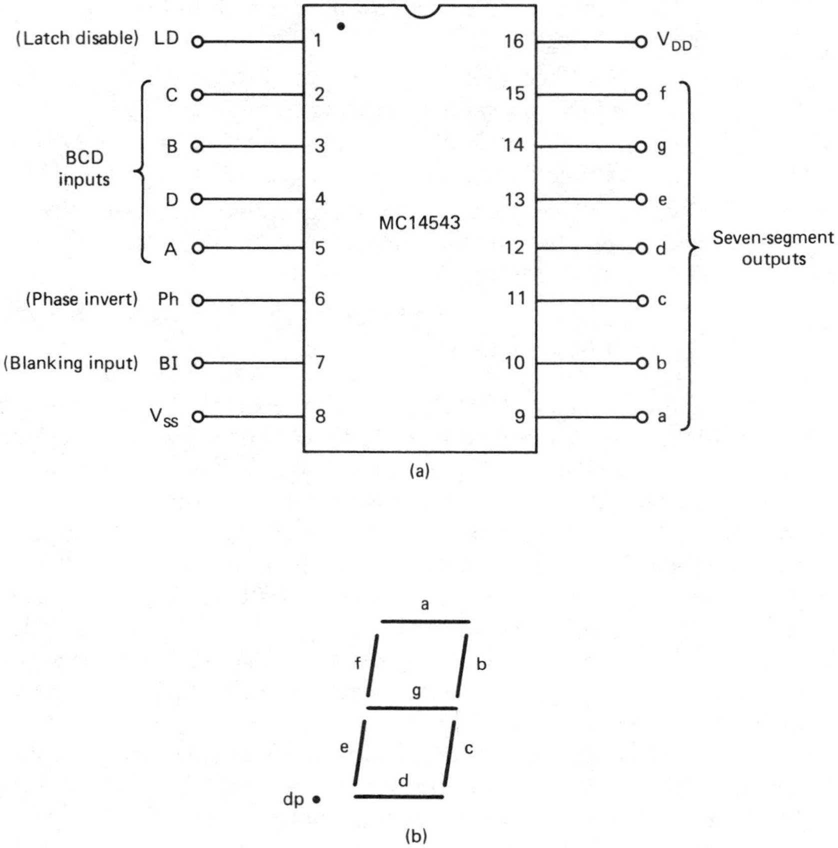

Figure 11-9 (a) MC14543 BCD-to-seven segment latch/decoder/driver pin diagram. (Courtesy of Motorola Semiconductor.) (b) LED segments.

seven-segment outputs corresponding to the BCD code at the inputs A, B, C, and D [see the LED segment assignment shown in Figure 11-9(b)].

Pin 16: V_{DD}. A (+) voltage of 3.0 to 18 V dc can be applied to this pin with respect to pin 8 (V_{SS}).

As shown in Figure 11-6(b), the Ph pin 6 of the MC14543 is connected to V_{DD} (logic 1) because the LED displays are the common anode type. In order to limit the current through each of the LED segments, a separate resistor is used in series with each segment. Note, however, that only seven resistors are required for all the segments of the LEDs. This is possible because the digit-select outputs of the MC14553 function sequentially. In addition, the BCD outputs are also multiplexed, one BCD digit at a time.

Finally, since the accuracy of the temperature displayed depends mainly on the frequency stability of the free-running multivibrator, all resistors must be of 1% or better tolerance, and capacitors must be either Mylar or tantalum types.

Also, remember that the temperature indicator must be calibrated at the temperature at which it is initially turned on.

Parts List

Teledyne 9400 V/F converter
LM334 temperature sensor
LF353 dual op-amp
Three NE/SE 555 timers
74LS00 NAND gate
MC14553 three-digit BCD counter
MC14543 BCD-to-seven segment decoder/driver/latch
Three seven-segment common anode LEDs: MAN72A or equivalent
Three 2N1305 switching transistors
Two 1N914 signal diodes
Seven 220 Ω resistors
Three 1-kΩ resistors
Two 1-kΩ potentiometers
3-kΩ potentiometer
Five 10-kΩ resistors
Two 10-kΩ potentiometers
15-kΩ resistor
20-kΩ resistor
Two 50-kΩ potentiometers
Four 100-kΩ resistors
180-kΩ resistor
510-kΩ resistor
1-MΩ potentiometer
100-pF capacitor
Two 0.001-μF capacitors
Five 0.01-μF capacitors
Three 1-μF capacitors
10-μF capacitor

All resistors with $\pm 1\%$ tolerance and all capacitors either Mylar or tantalum.

11-5 DIGITAL DC MOTOR SPEED CONTROL: APPLICATION OF THE D/A AND F/V CONVERTERS

Various techniques can be used to control the speed of a dc motor, such as using the phase-locked-loop principles, digital inputs, or analog inputs. If desired, the speed of the motor may also be monitored with LED or LCD displays. This project illustrates the use of digital inputs to control the speed of a dc motor. To process the digital inputs, a D/A converter will be used, while a combination of a speed sensor and F/V converter will be used to sense and convert the speed into an appropriate voltage.

Figure 11-10 shows the block diagram of a digitally controlled dc motor. The output of the D/A converter is proportional to the binary equivalent of its digital inputs. The differential amplifier compares the D/A converter output with the output voltage of the F/V converter. The resulting difference voltage is an input to the power amplifier/driver stage. The output of the power amplifier/driver then drives the dc motor. The speed sensor converts the motor's speed into a pulse waveform, which is in turn converted into a proportional voltage by the F/V converter. Since the output of the F/V converter is processed using a negative feedback formed with the differential amplifier, the motor is kept at a constant speed corresponding to the setting of the digital inputs. In fact, the key to the operation of the circuit is that the differential amplifier maintains a specific difference between two input voltages so that motor speed is constant at the selected digital input setting.

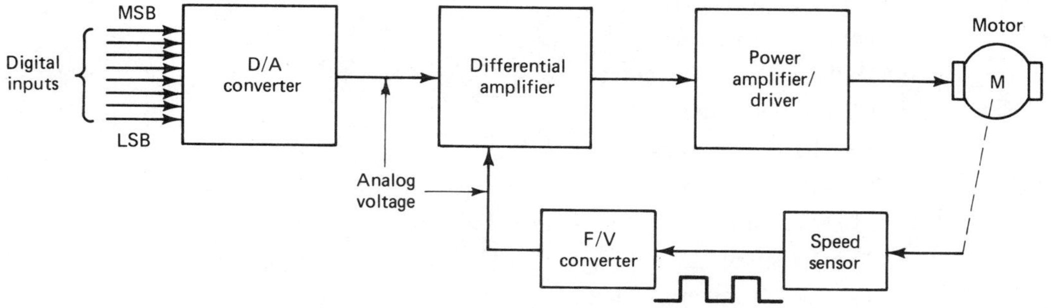

Figure 11-10 Block diagram of a digitally controlled dc motor.

Since the output of the D/A converter is directly proportional to the binary equivalent of its digital inputs, the output voltage of the D/A converter will be maximum positive when all the inputs are logic 1. This means that when all inputs are logic "1," the motor will run at a maximum speed.

Now suppose that the motor is initially running at a certain speed, and digital inputs have just been set to lower the speed. This action will reduce the output voltage of the D/A converter, which in turn reduces the difference between the two input voltages of the differential amplifier, resulting in a reduced drive for the motor. Therefore, the speed of the motor will be lowered until the output of the F/V converter is such that a specific input difference voltage for the differential amplifier, which is required to keep the motor running at a constant speed, is reached. The difference voltage necessary to maintain the constant motor speed is a function of the physical dimensions and electrical characteristics of the motor. These include torque, speed, inertia, current and voltage ratings, and gain of the differential amplifier. Thus the constant difference voltage and, in turn, a constant motor speed are maintained through the use of negative feedback.

The digital inputs may be calibrated in terms of revolutions per minute

(RPM). In addition, the output of the speed sensor may be applied to the frequency meter/RPM indicator in order to monitor the motor's speed.

Figure 11-11 shows the schematic diagram of a digitally controlled dc motor. As shown in the figure, the following ICs are used: SN74LS241 octal tri-state buffer, MC1408 8-bit D/A converter, MC1403 2.5-V voltage reference, LF353 dual op-amp, 9400 F/V converter, and the Hall effect transducer. The desired digital inputs are selected by the use of a switch assembly. The eight LEDs indicate the state of the digital inputs applied to the D/A converter. When a switch is open, the corresponding LED lights up. The 74LS241 octal tri-state buffer is used here because it isolates the LEDs from switches and also provides a current drive for the LEDs.

In Figure 11-11 the only device that is not discussed so far is the *Hall effect transducer.* A number of materials exhibit the Hall effect in that when a current-carrying semiconductor strip (usually silicon) is placed in the transverse (perpendicular) magnetic field, the combination produces an EMF between the opposite edges of the strip. This EMF is proportional to the product of the current and field strength. On the other hand, when the magnetic field is zero, or of specific polarity, an EMF between the opposite edges of the strip is also zero. Thus the Hall effect transducer is a magnetically activated electronic switch that can be used for sensing a magnetic field.

Figure 11-12 shows the equivalent circuit, operating arrangement, and an electrical switching characteristic with hysteresis of Texas Instrument's TL170 bipolar Hall effect switch. The TL170 is a three-terminal plastic package which consists of a silicon sensor, signal conditioning and hysteresis function, and an open-collector output stage integrated onto a monolithic chip [see Figure 11-12(a)]. The output of the device is compatible with bipolar or MOS logic circuits. Figure 11-12(b) shows the practical setup for the *on state.* The sensor is "on" (output voltage ≤ 0.04 V) when the magnetic field (B_{ON}) associated with the permanent-magnet north pole is perpendicular to the surface of the sensor and below a certain level, called the *operate point* or the *threshold.* On the other hand, the sensor is "off" when the magnetic field (B_{OFF}) emitted from the south pole of a permanent magnet is perpendicular to the surface of the sensor and above a certain level, called the *release point.* The TL170 has a typical operate point of ≤ -350 gauss and a release point of ≥ 350 gauss with a magnetic switching hysteresis ($B_{ON} - B_{OFF}$) of 200 gauss typically. Obviously, the negative and positive magnetic fields are defined as those fields that are emitted from the north and south poles, respectively, of a permanent magnet. The magnetic switching hysteresis curve of the TL170 is shown in Figure 11-12(c). The sensor is designed so that its output stage can withstand up to 20 V in the *off state* and can sink up to 16 mA in the *on state.*

To operate, the TL170 sensor is positioned so that the plain surface of the sensor faces the permanent magnet. In addition, to obtain two samples per revolution and hence help to control the motor speed more accurately, four permanent magnets are used in Figure 11-11. These magnets are glued to the 4-in.-diameter disk with alternately south and north poles up, as shown in Figure 11-12(d). The disk is then mounted on the motor's shaft. When the motor is running, the TL170 is turned on due to the magnetic field strength of the north pole and turned off due to the magnetic

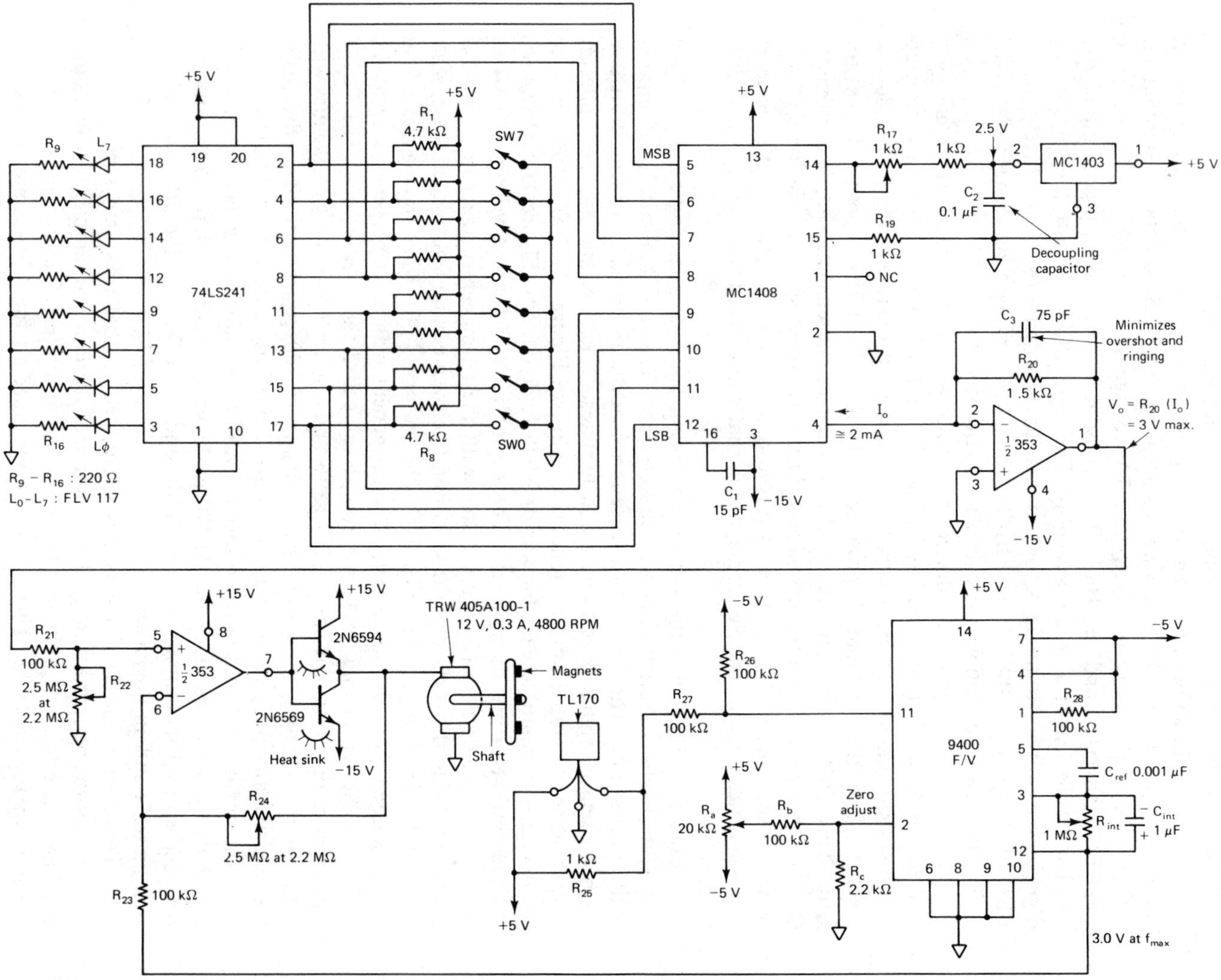

Figure 11-11 Digital dc motor speed control.

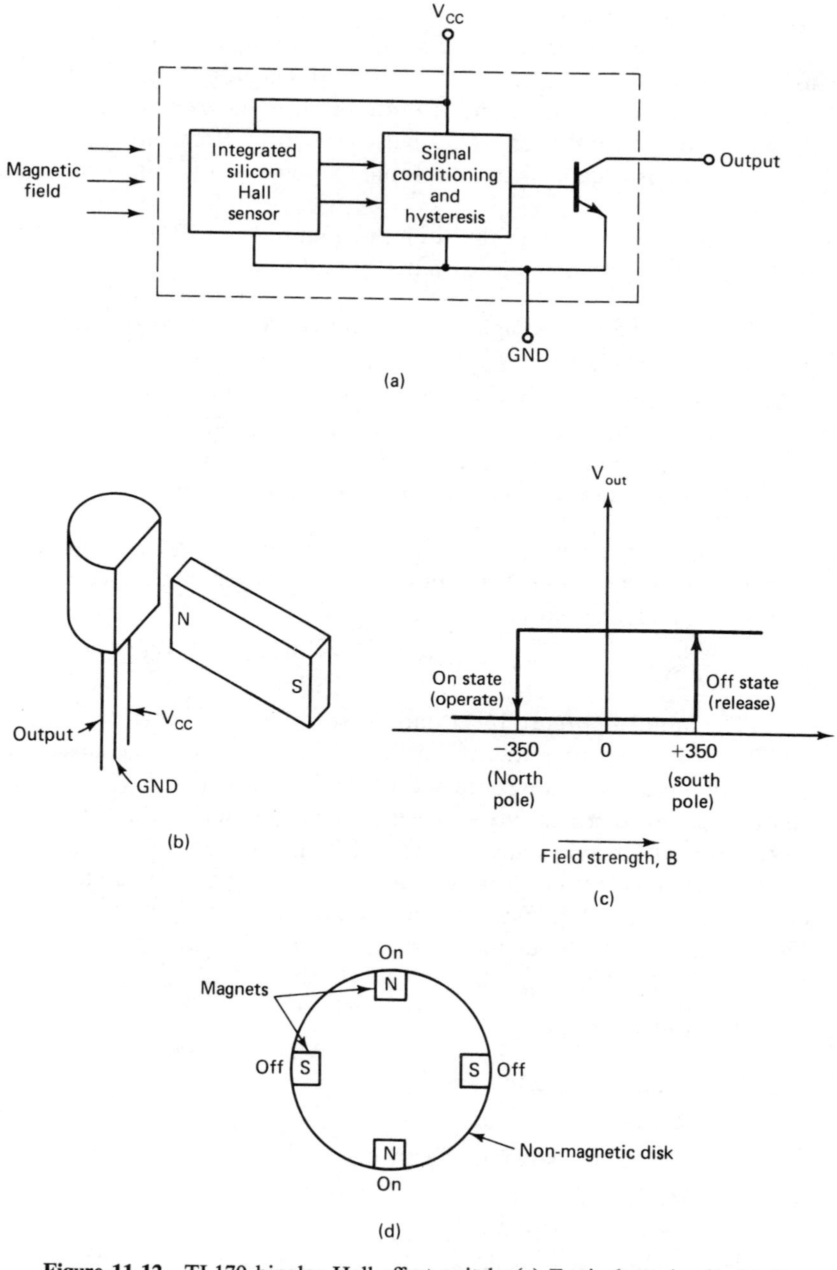

Figure 11-12 TL170 bipolar Hall effect switch. (a) Equivalent circuit. (b) On-state arrangement. (c) Electrical switching characteristic. (d) Practical setup. (Courtesy of Texas Instruments.)

field strength of the south pole. Therefore, because of four permanent magnets the sensor will generate two cycles per revolution. The distance between the magnets and the sensor, however, depends on the field strength of the magnets. For the TL170 used in Figure 11-11, a magnetic field strength magnitude of $\geq$ 350 gauss is necessary. When the motor is running, the distance between the disk and sensor can be adjusted so that the output of the sensor is a pulse waveform. Remember that the output amplitude of the sensor depends on the supply voltage and is independent of the RPM of the motor.

Now let us reconsider the circuit of Figure 11-11. The D/A and F/V converters in this figure should be adjusted initially as follows:

1. With all the inputs high (logic 1), adjust R_{17} so that the output voltage of the I-to-V converter is 3.0 V.
2. Disconnect pin 11 of the F/V converter from the junction of R_{27} and R_{26}. First, use the zero-adjust circuit connected to pin 2 to reduce the output voltage to zero. Second, apply 160 Hz 5 V pp symmetrical square wave to pin 11 and adjust R_{int} until the output voltage is equal to 3.0 V.

Once the adjustments are performed in this order, they should not have to be repeated. Note that the F/V is calibrated for maximum expected speed of the motor. In Figure 11-11, the motor speed is

$$4800 \text{ RPM} = \frac{(4800)(2)}{60} = 160 \text{ Hz}$$

The motor in Figure 11-11 initially starts running when the input binary code is $(00000110)_2$. Thereafter, the motor speed increases with the digital input until the motor attains a maximum speed at $(00111111)_2$. After $(00111111)_2$ however, the motor speed does not increase further even though the digital input is increased. In other words, we get 6-bit resolution instead of 8-bit. To obtain 8-bit resolution, an appropriate DAC with better resolution, a motor having favorable electromechanical specifications, and a differential amplifier with proper gain must be selected. The principles illustrated in the digital dc motor speed control of Figure 11-11 are used in the cruise control of automobiles.

Parts List

SN74LS241: tri-state buffer
MC1408: 8-bit D/A converter
MC1403: 2.5-V voltage reference
LF353: dual op-amp
Teledyne 9400: F/V converter
TRW 405A100-1: dc motor
2N6569 and 2N6594: complementary power transistors
TL170: bipolar Hall effect switch
FLV 117: light-emitting diodes (8)
Miniature 8-slide or rocker DIP switch

Heat sinks for power transistors
Four small magnets with field strength magnitude $\geq$ 350 gauss
Nonmagnetic disk 4 in. in diameter
Eight 220-Ω resistors
Three 1-kΩ resistors
1.5-kΩ resistor
2.2-kΩ resistor
Eight 4.7-kΩ resistors
Six 100-kΩ resistors
1-kΩ potentiometer
20-kΩ potentiometer
1-MΩ potentiometer
Two 2.5-MΩ potentiometers
15-pF capacitor
75-pF capacitor
0.001-μF capacitor
0.1-μF capacitor
1.0-μF capacitor

11-6 APPLIANCE TIMER: APPLICATION OF THE 555 TIMERS

The previous project showed how the Hall effect transducer works as a magnetic sensor. This project introduces another important type of transducer, an optical coupler. An optical coupler together with a square-wave generator and counters can be used to construct a timer for appliances. That is, an appliance can be turned on at a desired time for a specific time internal. Figure 11-13 shows a block diagram of such a timer.

As shown in Figure 11-13, the output of the square-wave generator is used as a clock (trigger signal) for the divide-by-N counters. A divide-by-N counter is a digital IC that produces a single output pulse for every N input pulses, where N is an integer. The integer N is commonly called the *modulus* of the counter. There are

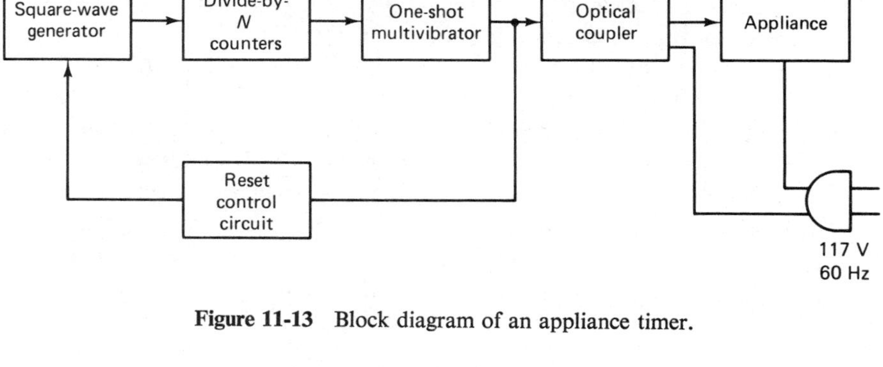

Figure 11-13 Block diagram of an appliance timer.

basically two types of divide-by-N counters: fixed and programmable. For example, the 7490 is the fixed modulus type and can be used as a divide-by-5 or divide-by-10 counter. On the other hand, the 74192 is a programmable counter which can be programmed to divide from 1 to 15. Often, in order to increase the resultant output count, the fixed-modulus type counters are connected in series (cascaded). In this arrangement the output of the first counter is the input to the second, the output of the second counter is the input to the third, and so on. Using the cascading technique, time delays of hours, days, or even weeks can be achieved.

In Figure 11-13, the output of the divide-by-N counters is then applied to a one-shot (monostable) multivibrator, the output of which in turn drives the optical coupler. The output pulse width of the one-shot controls the time an optical coupler is on. The output of the monostable multivibrator is also applied to the reset control circuit, which halts the square-wave generator and also resets the circuit for restart operation.

The resistor–capacitor combination used for the monostable determines its output pulse width. When the output pulse goes high (logic 0 to 1), the optical coupler is triggered and the appliance turns on. On the other hand, when the output pulse width of the monostable goes low (logic 1 to 0), the optical coupler is disabled and the appliance turns off. At the same time, the negative transition of the pulse also triggers the reset control circuit, which in turn shuts off the square-wave generator. Thus the circuit is ready for the next cycle.

Before considering the schematic diagram of the appliance timer, examine the operation of Motorola's MOC3011, an optically isolated triac driver. Figure 11-14 shows the internal construction and the pin diagram of the MOC3011. The

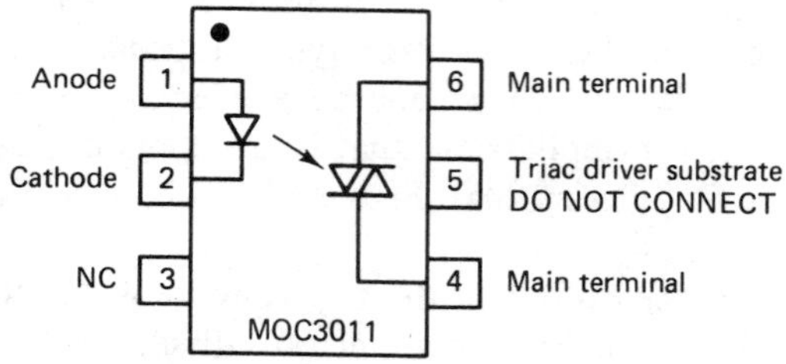

Figure 11-14 MOC3011 internal structure and pin diagram. (Courtesy of Motorola Inc.)

3011 consists of an infrared emitting diode, optically coupled to a silicon bilateral switch, and is available as a 6-pin DIP. The bilateral switch, commonly known as a *diac*, is a member of the thyristor family. Other members of the thyristor family are the silicon-controlled rectifier (SCR) and the triac. A peculiarity of the thyristor is that it consists of four layers—P, N, P, and N—and is primarily designed for high-voltage and high-current switching applications. The diac consists of high-voltage/current diodes connected in opposite directions. On the other hand, the major difference between the SCR and triac is that the SCR conducts in only one direction, while the triac conducts in both the directions. In fact, the triac is two SCRs connected in opposite directions. Both the SCR and triac are three-terminal devices (anode, cathode, and gate), whereas a diac is a two-terminal device (anode and cathode).

In the internal structure of the 3011, even though a symbolic notation for the diac is shown between output pins 4 and 6, this component actually works as a tirac since its operation is controlled by the infrared emitting diode. The maximum electrical ratings of an infrared emitting diode are as follows:

1. Reverse voltage (V_R) = 3.0 V.
2. Continous forward current (I_F) = 50 mA at forward voltage (V_F) of 1.5 V.

On the other hand, the output driver maximum electrical ratings are:

1. On-state output terminal voltage (V_{DRM}) = 250 V.
2. On-state rms current (I_T) = 100 mA.
3. Peak nonrepetitive surge current (I_{TSM}) = 1.2 A.
4. Isolation surge voltage (an internal device dielectric breakdown rating) (V_{ISO}) = 7500 V peak.

In addition, the total device power dissipation is 330 mW, and the LED trigger current—the current required to latch output—is 5 mA typically.

Now let us see how the 3011 is used in the appliance timer. Figure 11-15 shows the schematic diagram of the timer, which is designed to switch an appliance on 23 min after the initial startup time. The appliance remains on for approximately 12 min. The 556, a dual 555 timer, is used to form the astable and monostable multivibrators. The time between the pulses as well as the output pulse width of the timers is limited by the size and cost of the timing components. In the circuit of Figure 11-15, the time period of the pulse output waveform of the astable multivibrator is $T = 0.69(R_A + 2R_B)C \cong 11.5$ min, even though a maximum time period of approximately 20 min is possible. The output pulse waveform of the astable multivibrator works as a clock for the SN74177 4-bit binary counters/latches.

The SN74177, a 14-pin DIP, consists of four dc-coupled master–slave flip-flops, which accept frequencies of 0 to 35 MHz at the clock 1 input and 0 to 17.5 MHz at the clock 2 input (see Figure 11-15). It is fully programmable and triggers on the negative-going edge of the clock pulse. A, B, C, and D are the data inputs and Q_A, Q_B, Q_C, and Q_D are the outputs. When the count/load (pin 1) is low, the outputs directly follow the inputs. However, when the count/load is high (logic 1) and the clock inputs are inactive, the outputs remain unchanged. The *clear* terminal (pin 13), when taken low (logic 0), sets all outputs low regardless of the states of the clocks. In Figure 11-15, the SN74177 is used as a 4-bit ripple-through counter in which output Q_A must be externally connected to the clock 2 input. The output of the astable multivibrator is applied to the clock 1 input. In addition, the *count/load* is connected to +5 V and *clear* is taken low each time the timer is used. In fact, this arrangement enables the 74177 to perform simultaneous divisions by 2, 4, 8, and 16 at the Q_A, Q_B, Q_C, and Q_D outputs, respectively. That is, the outputs at Q_A, Q_B, Q_C, and Q_D will have time periods of 23, 46, 92, and 184 min. However, only the Q_A output is used in Figure 11-15. To achieve longer time periods, output Q_D of the 74177 can be used to clock the next SN74177, and so on. In addition, by using all four outputs of the 74177, three more appliances can be controlled, although at different times.

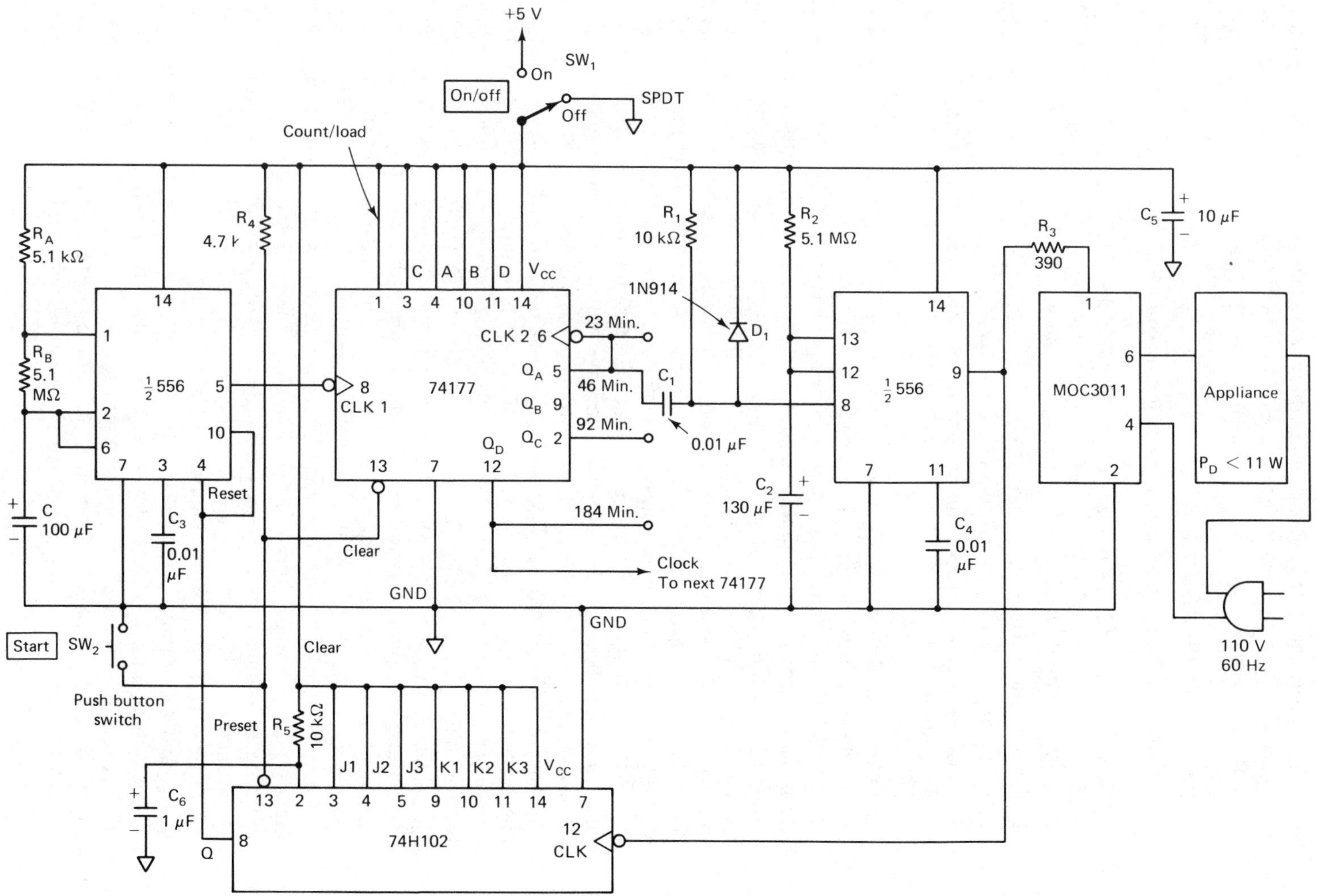

Figure 11-15 Appliance timer.

The Q_A output of the 74177 with a time period of 23.0 min is applied to the monostable multivibrator, which has a pulse width of approximately 12 min. Note that the waveshaping network composed of R_1, C_1, and D_1 is used between trigger pin 8 and the V_{CC} pin 14. This arrangement assures proper operation of the monostable multivibrator, since the negative pulse width of the input trigger will be shorter than that of the output pulse width of the monostable multivibrator. Thus in Figure 11-15, 23 min after the timer is initially set the appliance will come on and will remain on for approximately 12 min. At the end of 12 min, the output pulse width of the monostable goes low (logic 1 to 0), which disables the 3011 and the appliance turns off. At the same time, the negative transition in the output pulse of the monostable multivibrator also triggers the reset control circuit and shuts the astable multivibrator off.

The reset control circuit is composed of a single J-K negative-edge-triggered flip-flop with *preset* and *clear*. The Q output of the flip-flop is connected to reset pin 4 of the astable multivibrator and is normally high (logic 1). However, when the output pulse of the monostable goes low, it switches the Q output low (logic 0), which in turn shorts out the reset terminal. This halts the astable operation and prevents further clocking of the 74177 counters. The end result is that the appliance goes through one cycle (off–on–off) and then stops. Initially, to start the timer first apply the power to the timer by placing switch SW_1 in the *ON* position, Second, depress the START switch SW_2. With SW_2 depressed, the J-K flip-flop is preset ($Q = 1$) and the reset terminal of the astable multivibrator is pulled high (logic 1). The switch SW_2 also clears the Q_A, Q_B, Q_C, and Q_D outputs of the 74177 counters. Now the timer is on and the appliance will come on in approximately 23 min and will remain on for approximately 12 min. At the end of 12 min the appliance turns off and the Q output of the J-K flip-flop switches low. This action shuts off the astable multivibrator and hence the timer. To restart the timer, simply depress the START switch SW_1. When the timer is not in use, switch SW_1 should be in the off position. The timing diagram for the circuit in Figure 11-15 is shown in Figure 11-16. In the appliance timer of Figure 11-15 since the maximum *on* state rms current applied by the MOC3011 is 100 mA, the power rating of an appliance whose operation is to be controlled must be ≤ 11 W $= (110\text{ V})(100\text{ mA})$. However, for appliances with higher power ratings (>11 W), a triac with appropriate electrical specifications (V and I) may be used between the MOC3011 and an appliance. For more information on using the triacs with the MOC3011, refer to the MOC3011 data sheets. Finally, to verify the operation of the appliance timer, a 5-W light bulb may be used as the appliance.

The timer concepts discussed above can be used to control the operation of a furnace or central air conditioning unit so that it comes on at a desired time. For example, a timer may be used to shut off the furnace in the house after 10 P.M. and turn it back on at 5 A.M. This arrangement will help to save energy and money, especially in winter.

The concepts of appliance timers can also be used in switching house lights on and off at regular intervals, especially when the owner is on vacation. This provision should discourage burglars. In fact, when cyclic operation of the timer is desired, the reset control circuit shown in Figure 11-15 is not needed.

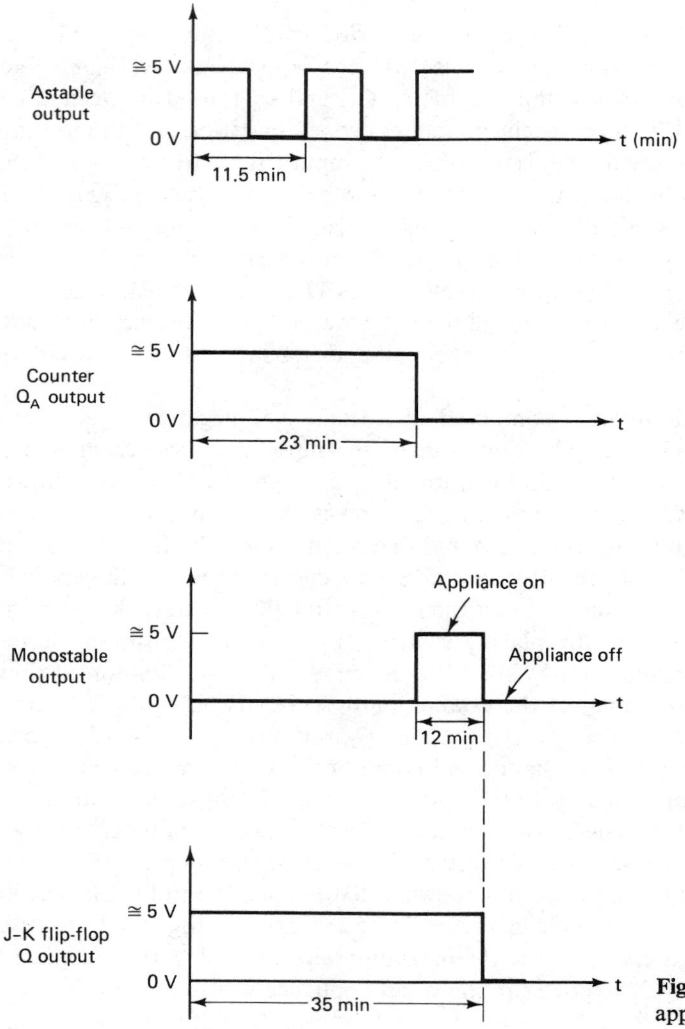

Figure 11-16 Timing diagram for the appliance timer.

Parts List

NE556 timer
SN74177 4-bit binary counters/latches
MOC3011 optically isolated triac driver
SN74H102 single J-K flip-flop or 7476 dual J-K
1N914 signal diode
390-Ω resistor
4.7-kΩ resistor
5.1-kΩ resistor
Two 10-kΩ resistors
Two 5.1-MΩ resistors

Three 0.01-μF capacitors
1.0-μF capacitor
10-μF capacitor
100-μF capacitor
130-μF capacitor
SPDT switch
Pushbutton switch (normally open)

11-7 SIREN/ALARM: APPLICATION OF THE LM380 POWER AMPLIFIER

Devices such as burglar alarms and sirens, whose basic purpose is to monitor certain conditions, make enjoyable projects because of the variety of sounds they can generate. Figure 11-17 shows a simple siren/alarm circuit using a dual op-amp MC1458, audio amplifier LM380, and a 1-W speaker. The dual op-amp is used as a signal generator which produces square, pulse, and triangular or sawtooth waveforms as discussed in Section 8-17, while the operation of the audio power amplifier LM380 is explained in Section 10-5.

The operation of the circuit is as follows. The A_1 and A_2 op-amps make up a waveform generator in which the output of A_1 is a square wave or pulse waveform and that of A_2 is either a triangular or sawtooth waveform. The potentiometer R_2 controls the frequency as well as the type of output waveform of op-amps A_1 and A_2. The switch SW_1 then connects the output of A_1 or A_2 to the audio power amplifier LM380, which in turn drives the speaker.

Although not used in the circuit of Figure 11-17, a potentiometer may be connected between (+) and (−) inputs of the power amplifier to control its voltage gain, which in turn controls the sound volume. The sound level produced depends on the position of switch SW_1, the wiper setting of potentiometer R_2, and the value of capacitor C_F. Therefore, sounds of varying intensities can be obtained by adjusting SW_1, R_2, and C_F.

Recall that pulse and sawtooth waveforms can be generated by using the 555 timer, as shown in Section 10-3.4. Similarly, the VCO 556 persented in Section 8-18 also generates a square wave and a triangular wave. These circuits may therefore be used as the signal generator in the circuit of Figure 11-17, instead of an op-amp signal generator.

For higher output power that is sound intensities, audio power amplifiers may be used in the bridge form shown in Figure 10-35. In fact, this configuration will also require a higher wattage speaker.

Parts List

MC1458 dual op-amp
LM380 audio power amplifier
Three-position switch

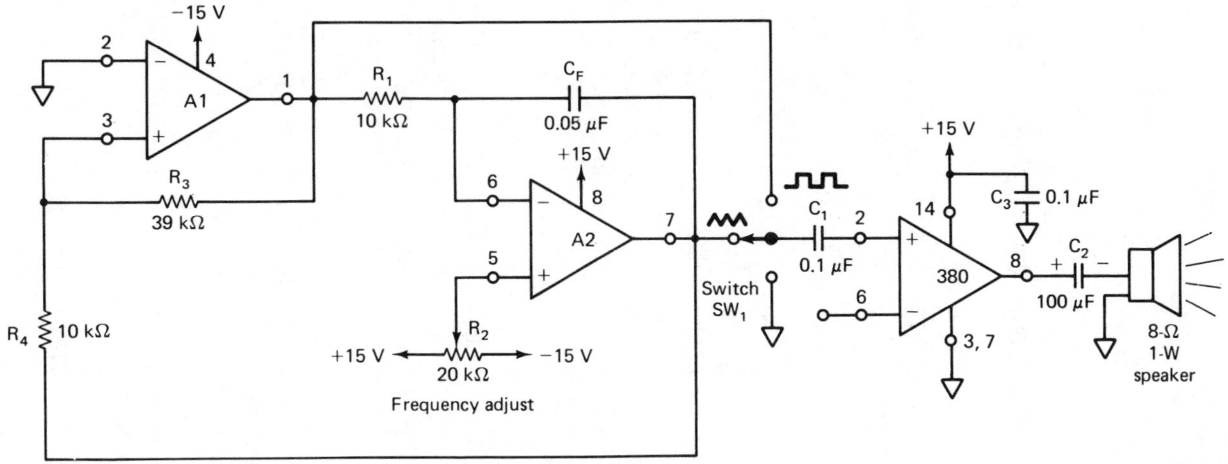

Figure 11-17 Siren/alarm schematic diagram. (A_1 and A_2 1458/353 dual op-amp.)

1-W speaker
Two 10-kΩ resistors
39-kΩ resistor
20-kΩ potentiometer
0.05-μF capacitor
Two 0.1-μF capacitors
1000-μF capacitor

11-8 SUGGESTED IC PROJECTS

The IC systems presented above are a fair sample of practical applications of ICs discussed in the previous chapters. These IC projects demonstrate that a practical and useful system does not have to be complex or expensive. In addition, the projects illustrated here should give you enough experience and confidence so that you might try your hand at a project of your own.

Although numerous practical IC systems can be formed with appropriate combinations of linear and/or digital ICs, block diagrams for suggested IC projects that should prove to be simple, interesting, practical, yet challenging to you are shown in Figure 11-18. These are:

(a) Rpm speed indicator
(b) Frequency meter
(c) Motor speed control using phase-locked-loop principles
(d) Digital light meter
(e) Digital voltemeter
(f) Sixteen-LED sequencer or roulette wheel
(g) Sound generator
(h) Amplifier with digitally controlled gain
(i) Logic probe using window comparator

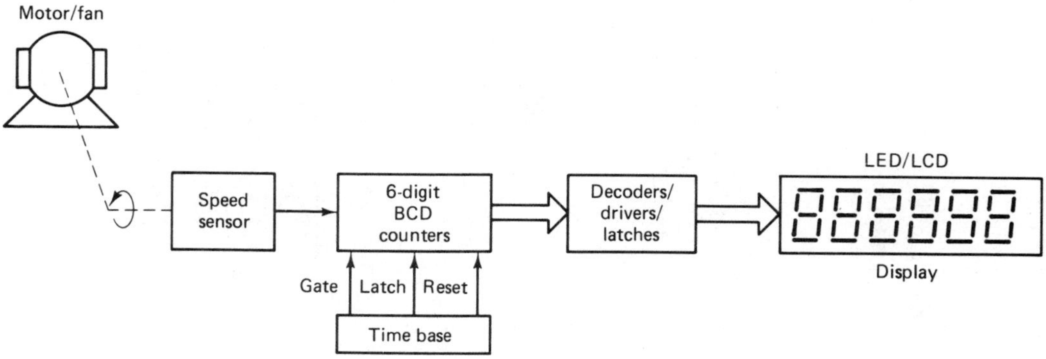

Figure 11-18 Block diagrams for IC projects. (a) RPM speed indicator.

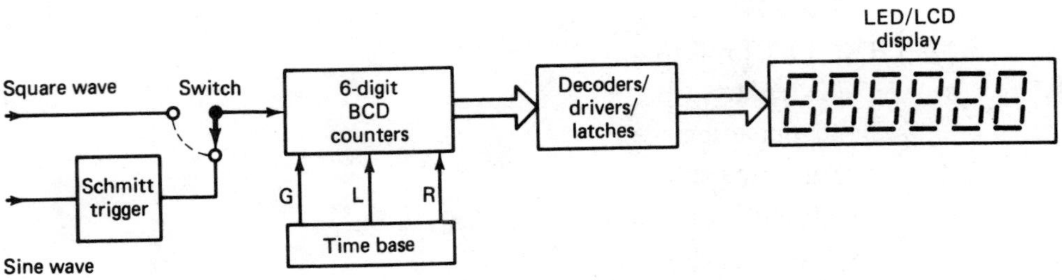

(b) Frequency meter.

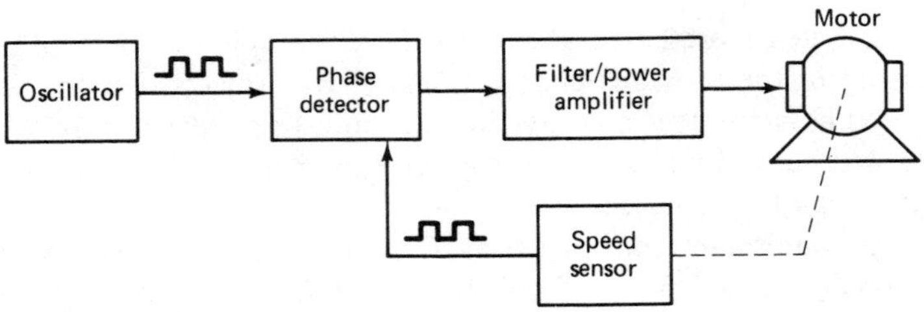

(c) Motor speed control using phase-locked loop principles.

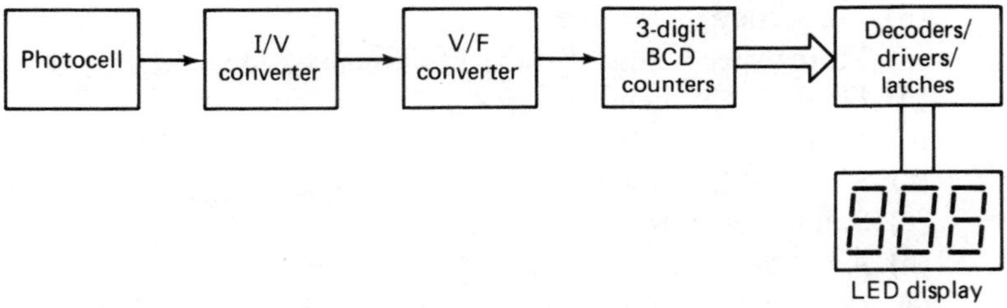

(d) Digital light meter.

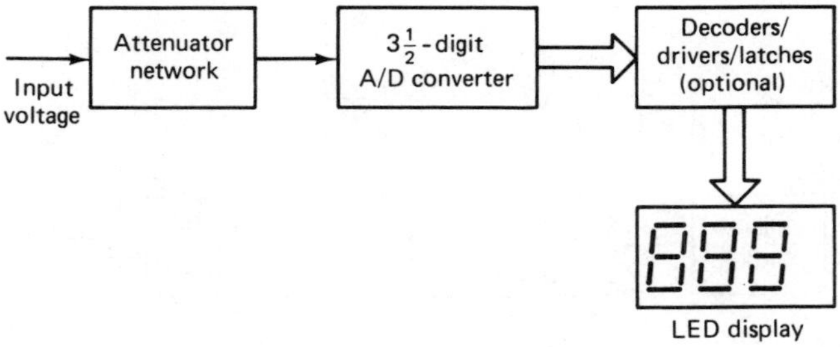

(e) Digital voltmeter.

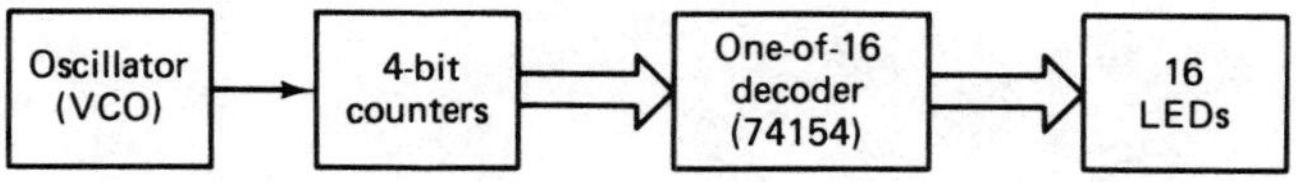

(f) Sixteen LED sequencer or roulette wheel.

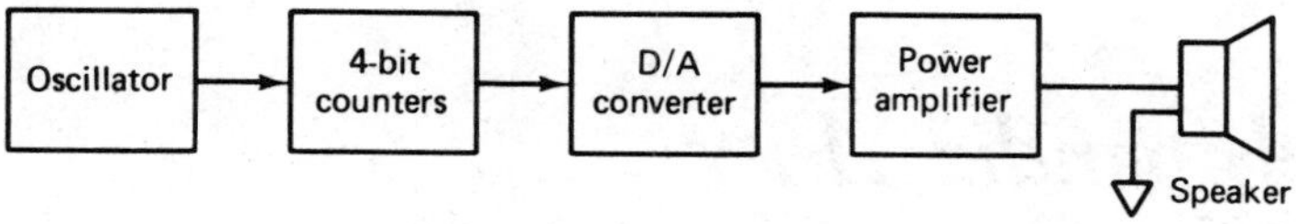

(g) Sound generator.

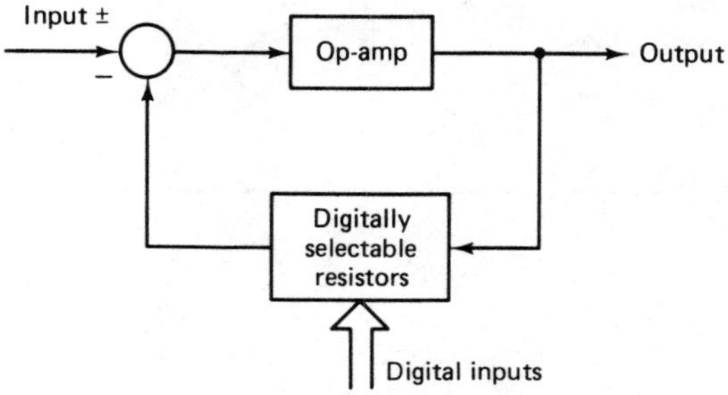

(h) Amplifier with digitally controlled gain.

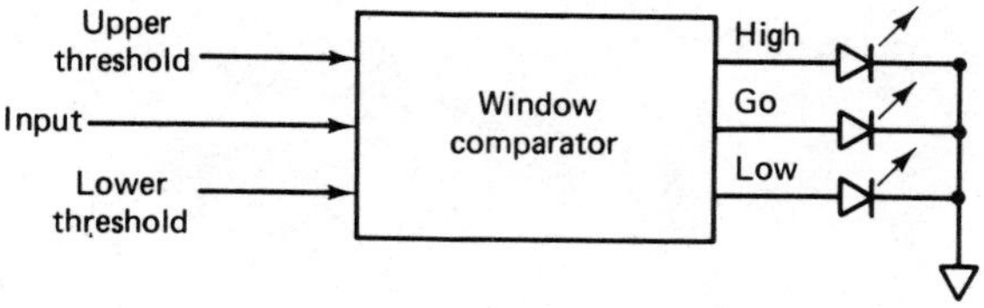

(i) Logic probe using window comparator.

Appendix A

RESISTANCE CHART (10 Ω to 22 MΩ)[a]

Resistance (Ω)	Resistance (kΩ)	Resistance (kΩ)	Resistance (MΩ)
* 10	* 1.0	*100	* 1.0
11	1.1	110	1.1
* 12	* 1.2	*120	* 1.2
13	1.3	130	1.3
* 15	* 1.5	*150	* 1.5
16	1.6	160	1.6
* 18	* 1.8	*180	* 1.8
20	2.0	200	2.0
* 22	* 2.2	*220	* 2.2
24	2.4	240	2.4
* 27	* 2.7	*270	* 2.7
30	3.0	300	3.0
* 33	* 3.3	*330	* 3.3
36	3.6	360	3.6
* 39	* 3.9	*390	* 3.9
43	4.3	430	4.3
* 47	* 4.7	*470	* 4.7
51	5.1	510	5.1
* 56	* 5.6	*560	* 5.6
62	6.2	620	6.2
* 68	* 6.8	*680	* 6.8
75	7.5	750	7.5
* 82	* 8.2	*820	* 8.2
91	9.1	910	9.1
*100	*10.0		*10
110	11		11
*120	*12		*12
130	13		13
*150	*15		*15
160	16		16
*180	*18		*18
200	20		20
*220	*22		*22
240	24		
*270	*27		
300	30		
*330	*33		
360	36		
*390	*39		
430	43		
*470	*47		
510	51		
*560	*56		
620	62		
*680	*68		
750	75		
*820	*82		
910	91		

[a]Tolerance of ±10% available only in values with *; ±5% available for all.

Appendix B

CAPACITANCE CHART (1 pF to 1000 μF)[a]

Capacitance (pF)	Capacitance (pF)	Capacitance (μF)	Capacitance (μF)	Capacitance (μF)	Capacitance (μF)
1	47	0.001	0.025	0.82	300
1.2	50	0.0012	0.027	1.0	400
1.5	56	0.0015	0.03	2	500
1.8	68	0.0018	0.033	3	600
2.2	75	0.002	0.035	4	750
2.7	82	0.0022	0.039	5	800
3.0	100	0.0025	0.04	6	900
3.3	120	0.0027	0.047	8	1000
3.9	150	0.003	0.05	10	
4.7	180	0.0033	0.056	12	
5.0	200	0.0039	0.06	15	
5.6	220	0.004	0.068	16	
6.8	270	0.0047	0.075	20	
8.2	300	0.005	0.082	25	
10	330	0.0056	0.1	30	
12	390	0.006	0.12	40	
15	430	0.0068	0.15	50	
18	470	0.0075	0.18	60	
20	500	0.008	0.22	75	
22	510	0.0082	0.25	80	
25	560	0.01	0.27	85	
27	620	0.012	0.33	100	
30	680	0.015	0.39	150	
33	750	0.018	0.47	160	
39	820	0.02	0.56	200	
	910	0.022	0.68	250	

[a]The standard WVDC ratings for the capacitors are: 1 V, 3 V, 6 V, 10 V, 12 V, 16 V, 20 V, 25 V, 50 V, 60 V, 75 V, 90 V, 100 V, 150 V, 175 V, 200 V, 250 V, 300 V, 350 V, 400 V, 450 V, 475 V, 500 V, 525 V, 550 V, 600 V, 1000 V.

Appendix C
Answers to Selected Problems

Chapter 1

2. (a) $I_{E1} = I_{E2} = 1.19$ mA
   $V_{B1} = V_{B2} = 0$ V (assuming $B_1 + B_2$ are connected to ground)
   $V_{E1} = V_{E2} = -0.715$ V
   $V_{C1} = V_{C2} = 6.79$ V
   (b) $A_d = 64.26$
   (c) $R_{i1} = R_{i2} = 4.2$ kΩ
4. (a) $I_{E1} = I_{E2} = 1.19$ mA
   $V_{B1} = V_{B2} = 0$ V (assuming B_1 and B_2 are connected to ground)
   $V_{E1} = V_{E2} = -0.715$ V
   $V_{C1} = V_{C2} = 3.34$ V
   (b) $A_d = 266.56$
   (c) $R_{i1} = R_{i2} = 4.2$ kΩ
6. (a) $I_{E1} = I_{E2} = 1.19$ mA
   $V_{B1} = V_{B2} = 0$ V (assuming B_1 and B_2 are connected to ground)
   $V_{E1} = V_{E2} = -0.715$ V
   $V_{C1} = V_{C2} = 3.34$ V
   (b) $A_d = 133.28$
   (c) $R_{i1} = R_{i2} = 4.2$ kΩ
8. (a) $v_o = 385.5$ mV pp
   (b) $v_{o\,max} = 12.85$ V pp
10. (a) $v_o = 40.2$ mV PP
   (b) $v_{o\,max} = 12.53$ V pp
12. $A_o = 15.6$
   $R_i = 1$ MΩ
14. Q_1, Q_2, Q_3, Q_4: CA3086
   $R_2 = 4.6$ kΩ
   $R_C = 1$ kΩ
16. $I_{E3} = 0.988$ mA
18. (a) $I_{CQ1} = I_{CQ2} = 1.15$ mA
   $V_{CEQ1} = V_{CEQ2} = 5.31$ V
   (b) $A_d = 216.2$
   (c) $R_i = 4.35$ kΩ
   (d) $v_{o\,max} = 20$ V
20. (a) $I_{E1} = I_{E2} = 0.988$ mA
   $I_{E3} = 1.95$ mA
   $I_{E4} = I_{E6} = 0.975$ mA
   $I_{E5} = I_{E7} = 9.75\ \mu$A
   $V_{B1} = V_{B2} = 0$ V (assuming B_1 and B_2 are connected to ground)
   $V_{E1} = V_{E2} = -0.715$ V
   $V_{C1} = V_{C2} = V_{B5} = V_{B7} = 7.33$ V
   $V_{E4} = V_{E6} = V_{C3} = 5.9$ V
   $V_{C4} = V_{C5} = 10$ V
   $V_{C6} = V_{C7} = 8.25$ V
   $V_{B3} = -5$ V, and $V_{E3} = -5.715$ V
   (b) $A_{dT} = 1865.2$
   (c) $v_{o\,max} = 3.5$ V pp
22. (a) $I_{E3} = I_{E5} = 2.11$ mA
   $I_{E1} = I_{E2} = I_{E6} = I_{E7} = 1.055$ mA

$V_{B1} = V_{B2} = 0$ V with B_1 and B_2 connected to ground.
$V_{E1} = V_{E2} = V_{C3} = -0.715$ V
$V_{C1} = V_{C2} = V_{B6} = V_{B7} = 4.04$ V
$V_{E6} = V_{E7} = V_{C5} = 3.325$ V
$V_{C6} = V_{C7} = 5.52$ V
$V_{E3} = V_{E5} = -9$ V
(b) $A_{dT} = 6{,}932.2$
(c) $v_{o\,max} = 4.39$ V pp

24. (a) $I_{E1} = I_{E2} = 0.368$ mA
$I_{E3} = I_{E4} = 3.8$ mA
$I_{E5} = 0.736$ mA
$V_{B1} = V_{B2} = V_{CM} = 3.5$ V
$V_{E1} = V_{E2} = V_{C5} = 2.785$ V
$V_{C1} = V_{C2} = V_{B3} = V_{B4} = 8.32$ V
$V_{B5} = 2.26$ V
$V_{E5} = 1.545$ V
$V_{C3} = V_{C4} = 12$ V
$V_{E3} = V_{E4} = 7.61$ V
(b) $A_{dT} = 143.63$
(c) $v_{o\,max} \cong 7.36$ V pp (unbalanced output riding on 7.61 V dc.)

26. D_1, D_2, Q_5, Q_6: CA3086 transistor array
$R_2 = 5.6$ kΩ
$R_E = 680$ kΩ
$R = 4.18$ kΩ (5 kΩ pot)

28. Q_6, Q_7, Q_8: CA3086 transistor array
$R_2 = 5.6$ kΩ
$R = 3.51$ kΩ (5 kΩ pot)

30. (a) $I_{E1} = I_{E2} = 1.02$ mA
$V_{B1} = 2.87$ V, $V_{B2} = 7.01$ V
$V_{E1} = 2.24$ V
$V_{E2} = V_{C1} = 6.38$ V
$V_{C2} = 10.16$ V
(b) $A_v = -73.44$
(c) $A_i = 104.86$
(d) $R_i = 2.57$ kΩ

Chapter 2

2. (a) $A_{dT} = 4{,}812.8$ (exact)
$A_{dT} = 5{,}013.4$ (ideal)
(b) $R_i = 2.79$ MΩ
(c) $v_{o\,max} = 16.96$ V pp

Chapter 3

2. $A_{cm} = 1$
4. $V_o \cong -14$ V
6. $v_o = 3$ V

Chapter 4

2. (a) $A_F = 10.99$
(b) $A_F = 11$
(c) Results in a and b are the same because $AB \gg 1$.

4. $A_F = 10.99$
$R_{iF} = (1.2)(10^{12})\ \Omega$
$R_{oF} = (1.8)(10^{-3})\ \Omega$
$f_F = 54.55$ KHz
$V_{ooT} = +0.358$ mV or -0.358 mV

6. $v_o = 1.1$ V pp

8. (a) $A_F = -9.99$
(b) $A_F = -10$
(c) Results in a and b are the same because $AB \gg 1$.

10. $A_F = -4.7$
$R_{iF} \cong 1$ kΩ
$R_{oF} = 0.705$ mΩ
$f_F = 127.66$ kHz
$V_{ooT} = +152.8\ \mu$V or $-152.8\ \mu$V

12. (a) $v_o = -4.7$ V pp

14. (a) $A_D = -39$
(b) $R_{iF} = 4$ kΩ

16. (a) $A_D = -44$
(b) $A_D = -66$

18. (a) $A_D = 14$
(b) $R_{iF_1} = R_{iF_2} = 342.86$ GΩ
(c) $R_{oF} = 5.25$ mΩ
(d) $f_F = 71.43$ kHz

Chapter 5

2. $V_{oo} = +240$ mV or -240 mV
4. $v_o \cong 1$ V pp
6. $V_{oI_{to}} = 7$ mV
8. $V_{ooT} = 71$ mV
$V_{ooT} = 68$ mV

10. (a) $E_v = \pm 30.6$ mV
$V_o = 101$ mV $\pm$ 30.6 mV
(b) $E_v = \pm 30.6$ mV
$V_o = 1.01$ V $\pm$ 30.6 mV

12. (a) $E_v = \pm 51.44$ mV
$v_o = 1.66$ V peak $\pm$ 51.44 mV

14. (a) $\Delta V_{oo} = \pm 60.6$ mV
(b) $v_o = -1.01$ V peak $\pm$ 60.6 mV
16. $\Delta V_{oo} = 7.61\ \mu$V r.m.s.
18. $\Delta V_{oo} = 12.514$ mV
20. (a) $v_o = 470$ mV sinewave at 1 kHz
(b) $V_{no} = 4.7\ \mu$V at 60 Hz
22. $V_{cm} = 10.64$ V r.m.s.

Chapter 6

2. $BW \cong 80$ kHz
4. $BW \cong 5$ kHz
6. $$A_{oL} = \frac{10^4}{\left(1 + j\frac{f}{0.7\text{ M}}\right)\left(1 + j\frac{f}{3\text{ M}}\right)\left(1 + j\frac{f}{18\text{ M}}\right)}$$
8. $UBW = 72$ MHz
10. (a) $-97.46°$
(b) $-260.92°$
12. $C_1 = 10$ pF, $R_1 = 0\ \Omega$, $C_2 = 3$ pF
14. $f_{\max} = 79.5$ MHz
16. $v_{in\ \max} = 0.8$ V pp
18. $SR = 36.01$ V/μS
20. $v_{o\ \max} = 3.02$ V

Chapter 7

2. $V_o = 5.7$ V
4. (a) $BW = 99.84$ kHz
(b) $v_{o\ \max} = 15$ V
(c) $v_o' = 7.5$ V dc $-$ 2 V peak sinewave
$v_o = -2$ V peak sinewave
6. (a) $V_o = -1.14$ V
(b) Scaling amplifier or weighted amplifier
8. $R_F = 2.4$ kΩ
10. (a) V_{ab} varies from -1 V to 4.09 V
(b) V_o varies from -2.61 V to 10.68 V
14. $v_{in} = 11.34$ mV
16. $v_{in} \cong 0.52$ V r.m.s.
18. (a) $I_o = 3.33$ mA
(b) $V_o = 10.1$ V
20. $V_o = 316.4$ mV
22. $v_o = 3.98$ coswt $-$ 3.98
24. $R_1 = 7.9$ kΩ (10 kΩ pot)
$C_1 = 0.01\ \mu$F
$C_F = 500$ pF
$R_F = 159$ kΩ (200 kΩ pot)
$R_{OM} = 8.2$ kΩ

Chapter 8

2. $R = 5.3$ kΩ (10 kΩ pot)
$C = 0.01\ \mu$F
Use a voltage follower.
4. $R = 13.25$ kΩ (use 20 kΩ pot)
$C = 0.01\ \mu$F
$R_1 = 4.7$ kΩ and $R_F = 2.7$ kΩ
8. For $f_H = 2$ kHz: $R' = 7.95$ kΩ (use 10 kΩ pot)
$C' = 0.01\ \mu$F
For $f_L = 400$ Hz: $R \cong 39$ kΩ and $C = 0.01\ \mu$F
For $A_{FT} = 4$: $R_1 = R_1' = R_F = R_F' = 10$ kΩ
10. For $f_L = 2$ kHz: $R = 7.95$ kΩ (use 10 kΩ pot), $C = 0.01\ \mu$F
For $f_H = 400$ Hz: $R' \cong 39$ kΩ, and $C' = 0.01\ \mu$F
For $A_{FT} = 2$: $R_1 = R_F = R_1' = R_F' = R_2 = R_3 = R_4 = 10$ kΩ
12. $\phi = -126.83°$
To obtain positive phase shift interchange resistor R and capacitor C.
14. $R_s = 7.95$ kΩ (use 10 kΩ pot)
$C_s = 0.01\ \mu$F
$R_1 = 10$ kΩ
$R_F = 20$ kΩ
16. $f_o = 1$ kHz
18. (a) $V_o = 4.94$ V pp
(b) $f_o = 1.18$ kHz
20. f_o changes from 9.8 kHz to 2.18 kHz if R_1 is varied from 4 kΩ to 18 kΩ.

Chapter 9

2. (a) $V_{ut} = 30.8$ mV
$V_{1t} = -30.8$ mV
(b) $V_{hy} = 61.6$ mV
4. The output voltage swings between -12 V and $+6.9$ V
6. $V_o = 700$ mV
Change R_{int} from 1 MΩ to 100 kΩ if F_{in} is changed from 10 kHz to 100 kHz.
8. (a) Size of each step = 0.84375 V
(b) $V_o = -12.66$ V
(c) The advantage of $R/2R$ DAC is that it requires only two sets of resistance values.

10. The output voltage is a negative going half wave rectified waveform having a peak value of −300 mV.
12. $V_c \cong 1$ V

Chapter 10

2. $R_2 = 6.32$ kΩ (use 10 kΩ pot)
   $R_3 = 124.2$ kΩ (use 200 kΩ pot)
   $R_4 = R_5 = 5.03$ kΩ (use 5.1 kΩ)
   $R_6 = 1$ kΩ
   $R_7 = 9$ kΩ (use 10 kΩ pot)
   $R_8 = 900$ Ω (use 820 Ω)
4. $R_2 = 100$ kΩ
   $R_3 = 1.46$ kΩ (use 1.5 kΩ)
   $R_4 = R_5 = 25.15$ kΩ (use 50 kΩ pots)
6. $R_A = 909.09$ kΩ (use 1 MΩ pot)
   $C = 0.1$ μF, and NE555 timer
8. $R_A = 3.33$ kΩ (use 3.3 kΩ)
10. $R_A = 3.6$ kΩ (use 5 kΩ pot)
   $R_B = 1.8$ kΩ
   $C = 0.02$ μF, and NE555 timer
12. 1N914 small signal diode
   2N404 PNP transistor
   NE555 timer
   $R = 516.2$ Ω (use 1 kΩ pot)
   $C = 1$ μF
   $C_1 = 0.01$ μF, and $C_3 = 10$ μF
14. $f_{OUT} = 53.19$ kHz
   $f_L = \pm 21.28$ kHz
   $f_C = \pm 306.7$ Hz
16. $R_1 = 12$ kΩ
18. $R = 34.3$ Ω, and $V_{in} \geq 8.2$ V
20. $n = 3$

Bibliography

Burr-Brown General Catalog. Tucson, Ariz.: Burr-Brown, 1979.

Datal-Intersil Data Acquisition Component Handbook. Mansfield, Mass.: Datel-Intersil, Inc., 1980.

Fairchild Linear Op-Amp Data Book. Mountain View, Calif.: Fairchild Camera and Instrument Corporation, 1979.

Intersil Data Book. Cupertino, Calif.: Intersil Inc., 1979.

Motorola Linear Integrated Circuit. Motorola CMOS Data. Phoenix, Ariz.: Motorola Semiconductor Products, 1979.

National Linear Data Book. Santa Clara, Calif.: National Semiconductor Corporation, 1978.

RCA Linear Integrated Circuits. Somerville, N.J.: RCA Solid State, 1978.

Signetics Analog Data Manual. Sunnyvale, Calif.: Signetics Corporation, 1979.

Teledyne Data Conversion Design Manual. Mountain View, Calif.: Teledyne Semiconductor, 1980.

The TTL Data Book for Design Engineers. Texas Instruments, 1976.

BERLIN, H. M., *The Design of Active Filters with Experiments*. Derby, Conn.: E and L Instruments, Inc., 1977.

BYLERLY, J. E., and M. VANDER KOOI, LM380 Power Audio Amplifier. *National Semiconductor Linear Applications*, Vol. 1, 1976.

KAHN, M., *The Versatile Op-Amp*. New York: Holt, Rinehart and Winston, Inc., 1970.

MALVINO, A. P., *Electronic Principles*. New York: McGraw-Hill Book Company, 1979.

Phase-Locked Loops. Benton Harbor, Mich.: Heathkit Continuing Education, 1980.

PRENSKY, S. D., *Manual of Linear Integrated Circuits: Operational Amplifiers and Analog ICs*. Reston, Va.: Reston Publishing Company, Inc., 1974.

RUTKOWSKI, G. B., *Handbook of Integrated Circuit Operational Amplifier*. Englewood Cliffs, N.J.: Prentice-Hall, Inc., 1975.

Index

A

B

C

D

E

F

G

H

O

P

W

Z